LIVING DINOSAURS

Living Dinosaurs

The Evolutionary History of Modern Birds

EDITED BY

Gareth Dyke and Gary Kaiser

A John Wiley & Sons, Ltd., Publication

This edition first published 2011, © 2011 by John Wiley & Sons Ltd

Registered office: John Wiley & Sons Ltd, The Atrium, Southern Gate, Chichester, West Sussex, PO19 8SQ, UK

Editorial offices: 9600 Garsington Road, Oxford, OX4 2DQ, UK
The Atrium, Southern Gate, Chichester, West Sussex, PO19 8SQ, UK
111 River Street, Hoboken, NJ 07030-5774, USA

For details of our global editorial offices, for customer services and for information about how to apply for permission to reuse the copyright material in this book please see our website at www.wiley.com/wiley-blackwell

Library of Congress Cataloguing-in-Publication Data

Living dinosaurs : the evolutionary history of modern birds / edited by Gareth Dyke and Gary Kaiser.
    p. cm.
  Includes index.
  ISBN 978-0-470-65666-2 (cloth)
  1. Birds–Evolution. 2. Paleobiology. I. Dyke, Gareth. II. Kaiser, Gary
  QL677.3.L58 2011
  598.13′8–dc22

                                                                                    2010043277

ISBN: 978-0-4706-5666-2 (hbk)

A catalogue record for this book is available from the British Library.

This book is published in the following electronic formats: eBook 9781119990451;
Wiley Online Library 9781119990475; ePub 9781119990468

Set in 9/11.5pt,TrumpMediaeval by Thomson Digital, Noida, India.
Printed and bound in Malaysia by Vivar Printing Sdn Bhd

1   2011

Contents

Foreword

Beyond Bird-Like Dinosaurs: The Emerging Evolutionary History of Modern Birds

As every school-child now knows, birds are related to some dinosaurs. To specialists, this is a very old idea that was born in the 19th century, lost in the first half of the 20th century, and then rediscovered again in the second half by John Ostrom. For those of us concerned with the evolutionary history of birds, reiterating the link between birds and dinosaurs is passé. We are, of course, interested in the history leading up to modern birds (Neornithes), but we must remember the latter is just one node on a very large Tree of Life (TOL). But it is a very important node. First, it is seemingly the most important higher-level node of the TOL to the general public, at least as measured in terms of information retrieval. Thus, using a Google search, "birds" returned 128 million hits, "mammals" a paltry 14.1 million, and "dinosaurs" was even lower at 13.5 million. "Fishes," "reptiles," and "insects" are all more "popular" than dinosaurs, but lag far behind birds.

Given this high profile of birds in modern society, it is disquieting that we don't know more about their evolutionary history, and it is worthwhile recounting why this might be so and how the study of avian evolution has been weighed-down by traditions that still influence us to this day. When we entered the modern era of avian phylogenetics, in the decades following Willi Hennig's contributions to cladistic analysis in the 1960s, our understanding of avian relationships was based on the comparative anatomical studies and classifications of the giants of late 19th century ornithology, especially Max Fürbringer and Hans Gadow, but others as well. Soon, however, the study of avian relationships essentially fell off the map for nearly 70 years. As Erwin Stresemann – himself a giant in ornithology – noted in a 1959 paper in *The Auk*, "all the avian systems...are similar...they are all based on Fürbringer and Gadow. My system of 1934 does not differ in essence from those [of] Wetmore (1951) and Mayr and Amadon (1951)...." This quote manifests a 70-year span of intellectual stagnation regarding avian relationships. There are two main reasons for the general absence of "tree thinking" in ornithology during this period, as well as, perhaps, in biology as a whole.

First, phylogeny was considered an unknown and an unknowable. To many, only fossils were capable of revealing past history, yet fossils were rare – an argument made by Darwin himself. Phylogeny was thus unknowable: again, as Stresemann remarked in 1959, "the construction of phylogenetic trees has opened the door to a wave of uninhibited speculation. Everybody may form his own opinion...because, as far as birds are concerned, there is virtually no paleontological documentation...Only lucky discoveries can help us...." Relationships were in the eye of the beholder; there was no objective method. Never mind that numerous workers 70 years prior to this were building trees, and making inferences about relationships, based on comparative morphology of recent birds!

The second, and perhaps main, reason for the long history of phylogenetic neglect within ornithology was as much sociological as scientific; namely, phylogeny was eclipsed by a redefinition of systematics as "population thinking." This view developed primarily in Germany and then in the United States as systematists such as Theodosius Dobzhansky and Ernst Mayr saw the relevance of the new field of population genetics for the study of speciation. Thus, Mayr declared in his influential 1942 book *Systematics and the Origin of Species* that "The population...has become the basic taxonomic unit." Mayr went further, and with typically assertive language left no doubt where he stood on phylogenetics: "The study of phylogenetic trees... comprise a field which was the happy hunting ground of the speculative-minded taxonomist of bygone days. The development of the 'new systematics' has opened up a field which is far more accessible to accurate research and which is more apt to produce tangible and immediate results." It is probably no accident that his major professor, Stresemann, echoed virtually the same opinions in 1959, but harsher: "Science ends where comparative morphology, comparative physiology, comparative ethology have failed us after nearly 200 years of effort. The rest is silence."

The silence was short-lived. Along came Hennig (at least in English) seven years later. But surprisingly, the first cladistic tree for birds was published by Wilhelm Meise in 1963 in the *Proceedings of the 13th International Ornithological Congress* in which he proposed a synapomorphy scheme for the ratite birds based on behavioral characters. This timing was perhaps no coincidence, as Meise's office was next door to Hennig's.

Over the past several decades the higher-level relationships of birds have proved very difficult to resolve using a host of different data and analytical techniques. It now appears that the problem is complex because of the history itself. The origins of many major groups are old, in the Early Tertiary or Cretaceous, although just how old is a subject of great debate. What does seem clear is that toward the base of the tree both morphological and molecular internodal distances are quite short, and the most parsimonious interpretation is that this is real – that neornithines experienced a relatively rapid radiation early in their history. Reconstructing this history is thus analytically difficult and is made more difficult by the fact that there are many

long-branched lineages. We systematists have also contributed to the problem with sampling that has often assumed monophyly of ingroups, close relationship of outgroups, and inadequate coverage of taxa and characters.

"But the times they are a-changin'." Large-scale phylogenetics is now becoming the norm, and over the past decade many studies have included taxon sampling numbering in the hundreds and characters in the thousands. These are beginning to provide some resolution to higher-level relationships and a lot more clarity at lower taxonomic levels.

This is the milieu into which this book, *Living Dinosaurs: The Evolutionary History of Modern Birds*, is thrust. It bookmarks how far we have come and how far we might expect to go over the next 10 years. *Living Dinosaurs* highlights the growth in our phylogenetic understanding of birds and their fossil record, as well as the breadth of advances in the comparative biology of birds, themselves a consequence of this growth. Importantly, it captures how knowledge about the evolutionary history of birds has benefited from phylogenetic approaches that integrate fossil and recent taxa as well as combined data sets of morphological, molecular, and other characters.

Living Dinosaurs articulates a transition to the future, one that is hard to predict but easy to underestimate in its potential. One thing is certain: the next five years will be transformative relative to the preceding five. One reason is that the rate of fossil discoveries and phylogenetic knowledge about birds is nonlinear, and for the latter, at least, it is probably close to exponential. The number of papers on avian relationships, worldwide, is astonishing and a not-unreasonable prediction is that nearly all biological species of birds will be on some tree within the next five years. The current trend is moving toward the phylospecies level, and we will probably see most avian phylospecies on a tree within a decade or so. Having species-level trees will accelerate evolutionary analysis throughout ornithology.

Perhaps the biggest reason we will see a transformation in avian phylogenetics, and avian evolution in general, is our entrance into a new era of cheaper genomics, as various chapters in *Living Dinosaurs* take note of. Ongoing international collaborations, along with the precipitous fall in the cost of whole-genome sequencing, suggest that we will see 50–100 whole genomes of birds, if not more, within five years. Indeed we will likely have a quarter of that number before this book has its first birthday. One should have a measured view of the ensuing hyperboly as it will take a new generation of avian systematists trained in genomic analysis to make sense of all these new data. But this problem should be transient and ornithologists should expect that some of our recalcitrant phylogenetic problems will be solved. It is also worth remembering that we have been through these "revolutions" before and each one leaves a plethora of unsolved problems, and crushed expectations, in their wake.

DNA sequences, by themselves, are just Gs, As, Ts, and Cs. The editors and authors of this book are quite right in calling for better integration of morphology and other aspects of the phenotype, as well as fossils, into the evolutionary analysis of birds. There are fundamental questions about the evolution of the phenotype that have barely been touched. The rise of molecular sequence data has garnered most of the excitement and research attention, yet at the same time molecular research has also contributed to the growth of morphological-based systematics and paleontology because evolution is primarily about phenotypic patterns seen in whole organisms. This integrative approach is pushing research efforts to become diversely collaborative (the Assembling the Tree of Life – AToL – initiative is a good example). The need for larger and more diverse data sets also necessitates expanded research groups.

What will be the big questions in avian evolutionary research in the coming years? Readers can look to chapters in this book for some of the answers. Clearly, the central question is resolving the Tree of Life at the finest taxonomic level for all known taxa, living and fossil. Minimally that number is probably around 25,000 to 30,000, if

we take subspecies-level names as a proxy for phylotaxa. For the living avifauna the problem is not as daunting at it might first seem, given the pace of activity of the world's avian systematists. A tree this large is a reachable goal and, as noted, will be largely realized in the near term. Obviously placing the known fossil taxa on this tree is a much more complex problem, as discussed in this volume, but given support for more comparative morphological research, the prognosis for major advancement is good. There are two other potential impediments, what might be called the Problem of Uncertain Knowledge, and the Problem of Investigator Tenacity.

With respect to the first impediment, more taxa and more characters do not guarantee that targeted research questions will be solved. More taxa, especially at fine taxonomic level, virtually guarantee uncertainty for having well-resolved relationships (multiple gene-tree versus species-tree conflicts are a clear example). And just what does it mean to have addressed "uncertainty" – to think that we have resolved relationships? Perhaps the greatest challenge for many investigators will be sociological: just how much effort do they want to expend to "know" a node (i.e., resolve well)? For some questions the answer will be a lot; for others, not so much.

Chapters in this book identify another major question: understanding the timetree of avian evolution. This too is a very difficult problem, not only with respect to the data needed (phylogenetic relationships, accurate geochronological calibrations) but also to the complexities of analysis. Some investigators think these challenges are so daunting that building a timetree is essentially fruitless. However, because having a timetree for all taxonomic levels is crucially important for addressing many evolutionary problems, we can expect to see significant work in the future to mitigate these difficulties.

Perhaps the take-home message of this book is that for those of us interested in avian evolution the future is bright indeed. It is exciting and expansive in its possibilities. The more we know about relationships, about the temporal pattern of avian evolution provided by a growing fossil record, and about the structure and function of the avian genome and phenome, the more questions we generate. Good questions precede good answers, and within evolutionary biology in general, the study of birds will continue to provide good questions as well as good answers. Young investigators can start with this book to get their feet wet.

Joel Cracraft
American Museum of Natural History
New York, USA
January, 2011

List of Contributors

HERCULANO ALVARENGA *Museu de História Natural de Taubaté, Taubaté, Brazil*

TATSURO ANDO *Ashoro Museum of Paleontology, Hokkaido, Japan*

F. KEITH BARKER *Department of Ecology, Evolution and Behavior; and, Bell Museum of Natural History, University of Minnesota, St. Paul, USA*

ALYSSA BELL *University of Southern California, Los Angeles, California, USA*

ROBERT BERNER *Yale University, New Haven, USA*

SARA BERTELLI *Museum für Naturkunde, Berlin, Germany*

ESTELLE BOURDON *American Museum of Natural History, New York, USA*

JOSEPH W. BROWN *University of Michigan Museum of Zoology, Ann Arbor, USA*

LUIS M. CHIAPPE *Los Angeles County Museum, Los Angeles, USA*

KENNETH P. DIAL *University of Montana, Missoula, USA*

GARETH DYKE *University College Dublin, Ireland*

SCOTT V. EDWARDS *Harvard University, Cambridge, USA*

EOIN GARDINER *University of Bristol, Bristol, UK*

BRANDON E. JACKSON *University of Montana, Missoula, USA*

GARY KAISER *Royal British Columbia Museum, Victoria, Canada*

DANIEL I. KSEPKA *North Carolina State University, Raleigh, USA*

BENT E. K. LINDOW *Natural History Museum of Denmark, Copenhagen, Denmark*

BRADLEY C. LIVEZEY *Carnegie Museum, Pittsburgh, USA*

PETER J. MAKOVICKY *Field Museum, Chicago, USA*

ANGELA MILNER *The Natural History Museum, London, UK*

JINGMAI O'CONNOR *Institute of Vertebrate Paleontology and Paleoanthropology, Beijing, China*

CHRIS L. ORGAN *Harvard University, Cambridge, USA*

GAVIN H. THOMAS *Imperial College London, Ascot, and University of Bristol, UK*

BRET W. TOBALSKE *University of Montana, Missoula, USA*

MARCEL VAN TUINEN *University of North Carolina at Wilmington, Wilmington, USA*

PETER WARD *University of Washington, Seattle, USA*

STIG WALSH *National Museums Scotland, Edinburgh, Scotland, UK*

DOUGLAS R. WARRICK *Oregon State University, Corvallis, USA*

LINDSAY E. ZANNO *Field Museum, Chicago, USA*

Preface

The scope of this book ranges widely, from bio-molecular aspects of avian biology to details of the anatomy of dinosaurs. However, it is not just a simple compilation of current material. Its purpose is to help bridge a gap that has developed between those who study birds as fossils and those who study the living animals. The size of that gap has much to do with two controversies related to the evolution of birds that remained unresolved for much of the 20th century. One involved the origin of birds, from dinosaurs or not, and the other involved the inability of ornithologists to reach a general agreement on the relationships among the living groups of birds. The first was resolved at the very end of the 20th century, in favor of an origin of birds from ancestors among theropod dinosaurs, and although the second remains unresolved, it is being pursued by a host of optimistic scientists using novel techniques. These include new methods of interpreting morphological characters (cladistics) and sophisticated analyses of genetic material. Each chapter of this book puts current understanding in context with directions that research may take over the next few years.

Most of the organisms discussed are true birds in the sense that they include the most recent common ancestor of *Archaeopteryx* and all of its descendants among the Neornithes (Gauthier & de Quieroz, 2001). The remainder are those dinosaurs considered to be the closest relatives of that ancestor. The emphasis of the discussions is on evolutionary aspects of function and ecology rather than the technical description of individual fossils, many of which are discussed in *Mesozoic Birds* (Chiappe & Witmer, 2002) and other recent works.

Shortly after Charles Darwin offered natural selection as the driving force behind the evolutionary process, the origin of birds and the path of their subsequent development seemed remarkably clear. The 140 million year-old *Archaeopteryx* bridged the gap between birds and their reptilian ancestors among the dinosaurs (Huxley, 1868), and within a few years toothed birds, such as *Ichthyornis* and *Hesperornis* (Marsh, 1888), seemed to confirm that birds enjoyed a simple and straightforward evolutionary story from primitive, reptilian forms to modern varieties. It was one of the great triumphs of early evolutionary theory – the living birds that we are all so familiar with are descended from 'reptile-like' forms including the toothed and long-tailed *Archaeopteryx*. Nothing in biology is ever that simple.

In 1926, the Dane, Gerhaard Heilmann, shattered the complacency of Victorian biologists with carefully constructed arguments that birds could not be related to dinosaurs. Among other things, dinosaurs lacked the distinctively avian furcula or wishbone. His arguments held sway through much of the 20th century but could not stand up to sophisticated advances in interpretive methodology and the discovery of dinosaurs with wishbones and other features shared with birds (Gauthier & Gall, 2001; Chiappe & Witmer, 2002). The intensity of this controversy was one of the factors that led to the isolation of students who studied fossil birds from those who studied the living animals. Gradually most ornithologists became so focused on experimental ecology and energetics that they no longer had the background to appreciate the subtle anatomical differences between *Archaeopteryx* and nonavian dinosaurs, such as *Velociraptor* and *Troodon* (Makovicky & Zanno: Chapter 1, O'Connor et al.: Chapter 3).

In spite of resolution of the fundamental issue – that birds are a living lineage of dinosaurs – the immediate source of the living groups of birds (Neornithes) remains a controversial subject among both ornithologists and paleontologists (Lindow: Chapter 14). The absence of undoubtedly modern birds, except for Anseriformes, in the Mesozoic fossil record contradicts many of the biomolecular studies that show very early divergence dates (Brown & van Tuinen: Chapter 12). Many hope for greater success from analyses of the actual genome (Organ & Edwards: Chapter 13). Biomolecular techniques appear to be well founded in theory, but at a practical level they have yet to produce a generally accepted phylogeny for the living groups of birds. They have produced a variety of results from the much-maligned "tapestry" of Sibley & Ahlquist (1990) (Houde, 1987; Sarich *et al.*, 1989; Lanyon, 1992; Harshman, 1994) to the more technologically sophisticated analyses of Fain & Houde (2004) or Hackett *et al.* (2008). They all vary greatly from morphology-based hypotheses such as that of Livezey & Zusi (2007). Without an agreed-upon phylogeny, the evolutionary story of the whole group is moot (Livezey: Chapter 5).

In comparison to the fundamental problem of avian phylogeny, recent puzzles posed by the discovery of fossils of avian and nonavian animals bearing feathers, seem almost trivial. The presence of feathers no longer defines birds; feathers are now known from a variety of flightless animals and seem most likely to have arisen as insulation (O'Connor *et al.*: Chapter 3). *Anchiornis* (Xu *et al.*, 2008; Hu *et al.*, 2009), *Confuciusornis* (Hou *et al.* 1995), and the entire lineage of Cretaceous Enantiornithes (Walker, 1981) can tell us much about the advantages of being somewhat bird-like even if the actual relationship of those fossil forms to modern lineages is tenuous, and the implications for the eventual success of the Neornithes are slight.

Flight appears to have been a fundamental factor in the early success of birds but its origins are not likely to be revealed by the fossil record. For many years, two competing theories dominated discussion of the origin of flight. Either birds learned to fly by jumping from trees or by flapping their wings while running across the ground. Recently, the tendency of flightless nestlings to flap their wings while running uphill has been developed into the theory of "Wing-assisted Inclined Running" (Tobalske *et al.*: Chapter 10). On the other hand, recently discovered fossils imply that a variety of small, feathered animals could be found in the canopies of Mesozoic trees. Not all had effective wings but wings are not the only organs required for flight. An endocast of the brain of *Archaeopteryx* suggests that it could already have been capable of processing the large amounts of data accumulated by a flying animal. Subsequent changes in the shape of the brain can be tracked into modern times (Walsh & Milner: Chapter 11).

The first section of the book reviews the early ancestry of birds and the conditions under which they, and their nearest relatives, diversified in the Cretaceous. It is intended to provide ornithologists with an overview of the fossil record and describes the history of some highly specialized groups such as extinct giants and seabirds from the Tertiary (Alvarenga *et al.*: Chapter 7, Bourdon: Chapter 8). Other middle chapters focus on adaptations contributing to the success of living forms such as penguins (Ksepka & Ando: Chapter 6) or other seabirds (Kaiser: Chapter 15), and are intended to provide paleontologists with a basic introduction to groups whose specialized lifestyles are reflected in their skeletal anatomy. It is the adaptations of these and other living species that are being tested by the alienation of habitats for human use and the effects of global warming (Thomas: Chapter 16).

We hope that by bringing together such a wide range of areas in one volume, the reader will find an entrée into less familiar topics and appreciate the potential value of linkages among seemingly unrelated approaches to the study of avian evolution.

Gareth Dyke
Gary Kaiser
May 2010

REFERENCES

Chiappe LM, Witmer LM. 2002. *Mesozoic Birds: Above the Heads of Dinosaurs*. Berkley: University of California Press.

Fain MG, Houde P. 2004. Parallel radiations in the primary clades of birds. *Evolution* **58**: 2558–2573.

Gauthier J, de Quieroz K. 2001. Feathered dinosaurs, flying dinosaurs, crown dinosaurs, and the name "*Aves.*" In *New Perspectives on the Origin and Early Evolution of Birds: Proceedings of an International Symposium in Honor of John H. Ostrom*, J. Gauthier, J. and L. F. Gall (eds). New Haven: Yale University Press; 7–41.

Gauthier J, Gall LF (eds). 2001. *New Perspectives on the Origin and Early Evolution of Birds: Proceedings of an International Symposium in Honor of John H. Ostrom*. New Haven: Yale University Press.

Hackett SJ, Kimball RT, Reddy S, Bowie RCK, Braun EL, Braun MJ, Chojnowski JL, Cox WA, Kin-Lan Han, Harshman J, Huddleston CJ, Marks BD, Miglia KJ, Moore WS, Sheldon FH, Steadman DW, Witt CJ, Yuri T. 2008. A phylogenetic study of birds reveals their evolutionary history. *Science* **320**: 1763–1768 (and supporting material).

Harshman J. 1994. Reweaving the tapestry: what can we learn from Sibley and Ahlquist (1990)? *Auk* **111**: 377–388.

Heilmann G. 1926. *The Origin of Birds*. London: Witherby; 210 pp.

Hou L, Zhou Z, Gu Y, Zhang H. 1995, *Confuciusornis sanctus*, a new Late Jurassic sauriurine bird from China. *Chinese Science Bulletin* **40**(18): 1545–1551.

Houde P. 1987. Critical-evaluation of DNA hybridization studies in avian systematics. *Auk* **104**: 17–32.

Hu D, Hou L, Zhang L, Xu X. 2009. A pre-*Archaeopteryx* toodontid theropod from China with long feathers on the metatarsus. *Nature* doi: 10.1038/nature08322.

Huxley TH. 1868. On the animals which are most nearly intermediate between birds and reptiles. *Annals and Magazine of Natural History* **4**(2): 66–75.

Lanyon SM. 1992. Review of Sibley and Ahlquist 1990. *Condor* **94**: 304–307.

Livezey BC, Zusi RL. 2007. Higher-order phylogenetics of modern Aves based on comparative anatomy. *Netherlands Journal of Zoology* **51**: 179–205.

Marsh OC. 1880. *Odontornithes: A Monograph on the Extinct Toothed Birds of North America. Report of the Geological Exploration of the 40th Parallel*, Vol. 7. Washington, DC: Government Printing Office.

Sarich VM, Schmid CW, Marks J. 1989. DNA hybridization as a guide to phylogenies: a critical analysis. *Cladistics* **5**: 3–32.

Sibley CG, Ahlquist JE. 1990. *Phylogeny and Classification of Birds: A Study in Molecular Evolution*. New Haven, CT: Yale University Press.

Walker CA. 1981. New subclass of birds from Cretaceous of South America. *Nature* **292**(2): 51–53.

Xu X., Zhao Q, Norell MA, Sullivan C, Hone D, Erickson G, Wang X-L, Han F-L, Guo Y. 2008. A new feathered maniraptoran dinosaur fossil that fills a morphological gap in avian evolution. *Chinese Science Bulletin* **54**: 430–435.

Part 1 Introduction: the Deep Evolutionary History of Modern Birds

Introduction: Changing the Questions in Avian Paleontology

GARY KAISER[1] AND GARETH DYKE[2]

[1]Royal British Colombia Museum, Victoria, Canada
[2]University College Dublin, Dublin, Ireland

In our world today there is no living animal that looks like a bird but is actually something else. If it has feathered wings, a wishbone, and walks on its toes, then it is definitely a bird. In the Mesozoic world life was much more diverse. Among the giant dinosaurs, there were dozens of small, feathered animals that had wishbones and walked on their toes. Some of these were so small that the feathers on their limbs could lift them into the air. Somewhere among those ancient feathered animals, there was just one that was the earliest ancestor of modern birds. The rest, and all of their varied descendants, were "also rans" that disappeared in the events that carried away the giant dinosaurs and many other lineages at the end of the Cretaceous Period. Equally mysterious, and even more poorly known, are the birds, or groups of birds, that survived that great extinction to become the ancestors of today's 10,000 or so species.

The precise nature of the earliest birds has been one of the most contentious topics in biology for almost 150 years. At first the discovery of *Archaeopteryx* seemed to endorse Charles Darwin's newly published theory (Huxley, 1868). A reptile-like animal with feathers looked like the perfect link between two major lineages. However, concerns about apparent differences between birds and dinosaurs arose in the early 20th century (Heilman, 1926) and generated a heated debate that lasted until the beginning of the 21st century.

The issue was not with *Archaeopteryx* as the earliest known bird but the suite of differences that distinguish birds from dinosaurs. Until recently, birds seemed unlikely as relatives of the iconic giant dinosaurs of the Mesozoic; indeed as discussed in later chapters the nature of their clavicles and furculae remained contentious until the end of the 20th century (Chure & Madsen, 1996; Makovicky & Currie, 1998). In addition, Richard Owen, the noted Victorian authority on all things anatomical, declared that the two groups could not be related. Dinosaurs had simplified their hands by losing the two outer digits (IV and V), whereas birds lost the innermost and the outermost digits (I and V) (see Makovicky & Zanno, Chapter 1, this volume).

We can blame shortcomings in the fossil record for the duration of this controversy. Dinosaurs were thought to be giants because the bones of giants are more likely to be fossilized than the skeletons of small animals. Fragile structures like bird skeletons are easily destroyed and scattered long before they can be preserved. Fortunately the fossil record has been greatly enhanced and reinterpreted since 1980. John Ostrom (1976) led the way by describing the specialized bird-like wrist in the theropod *Deinonychus* and within a few years, thousands of new specimens were discovered in unexplored areas of China and other remote parts of the world (see O'Connor *et al.*, Chapter 3, this

Living Dinosaurs: The Evolutionary History of Modern Birds, First Edition. Edited by Gareth Dyke and Gary Kaiser.
© 2011 John Wiley & Sons, Ltd. Published 2011 by John Wiley & Sons, Ltd.

volume). This new material included fossils of dozens of small, feathered animals in both avian and nonavian lineages.

During that same period at the end of the 20th century, biomolecular techniques were brought to bear on the genetic history of living birds. Analyses of nuclear and mitochondrial DNA allowed the construction of family trees that established some new connections between groups and demolished long-held beliefs in other relationships. At the time pioneering biomolecular analyses revealed that the ostriches and their relatives (Paleognathae) represent a very ancient lineage branch from the living birds (Sibley & Ahlquist, 1990; see Figure 0.1). Similarly the highly aquatic ducks were shown to be close relatives of the entirely terrestrial chickens and both belong to the Galloanserae, a group that split from the base of the main lineage shortly after the divergence

of the paleognaths (Sibley & Ahlquist, 1990; Hackett et al., 2008).

Recently, genomic analyses have revealed the genetic history of the chicken, providing insight into the ancient history of birds and hinting at events in deep time when the avian lineage diverged from that of mammals (International Chicken Genome Sequencing Consortium, 2004). To no one's surprise, chickens completely lack genes for making tooth enamel giving scientific credibility to the adage, "scarce as hens' teeth." Recently, genetic studies of Hox gene activity in the embryo have contributed to discussions of the vexed question of the bird's lost digits (Wagner, 2005; Vargas et al., 2008; Young et al., 2009).

In spite of amazing advances, biomolecular studies still have one great shortcoming. They cannot be applied directly to the interpretation of organisms known only as fossils. However, they

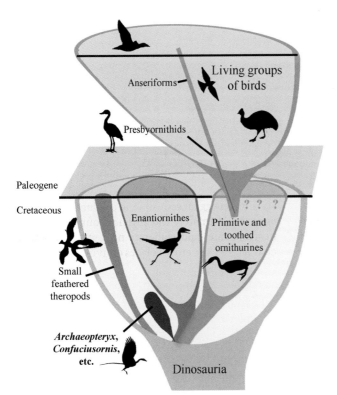

Figure 0.1 Cartoon to illustrate the basics of our current consensus regarding the pattern of the evolution of birds, relative to the Cretaceous-Paleogene (K-Pg) boundary. [This figure appears in color as Plate 0.1]

might lead us to the identity of extant lineages whose antecedents crossed the Cretaceous–Paleogene (K–Pg) boundary and gave rise to modern birds (Figure 0.1). The fossil record of terrestrial animals in the Paleocene is particularly weak and biomolecular insights may be our best chance of understanding post-Mesozoic survival.

In spite of the new approaches to cladistic analyses of morphological characters and a vast array of newly discovered fossils, *Archaeopteryx* remains secure in its position as the earliest bird. It is no longer the oldest known feathered animal. (Xu & Zhang, 2005) and *Anchiornis* from China (Hu *et al.*, 2009) have been placed in the Mid-Jurassic, some 10 million years (myr) earlier than *Archaeopteryx* (see Makovicky and Zanno, Chapter 1, this volume). Both fossils reveal feathered animals as small as *Archaeopteryx* but their feathers are symmetrical and not aerodynamically shaped. Such feathers cannot generate the thrust needed for forward flight and the aerial performance of these animals may have been limited to parachuting to the ground. Perhaps the plumage in these animals was more important as insulation.

These early feathered animals suggest that control of the third dimension was a theme with a long history in dinosaur evolution. The parasaggital stance of early forms raised them above their squamosal contemporaries, greatly increasing their field of view and allowing effective attacks from above. *Pedopenna*, *Anchiornis*, and *Archaeopteryx* represent a later development of small, feathered animals that may have kept their plumage clean by living in the forest canopy, such as it was in the Jurassic. The feathers offered a low overall density that reduced the risks of living far above the ground and allowed these animals to parachute, glide, or even fly down onto unsuspecting prey.

Archaeopteryx preserves the earliest evidence for feathered wings with asymmetric feathers capable of generating some useful thrust in flapping flight, but anatomically it is more similar to smaller-bodied dinosaurs, such as *Velociraptor*, *Deinonychus* or *Troodon*, than to any modern bird. It had none of the specialized skeletal struc-

tures found in later birds that anchor large flight muscles and its aerial capabilities may have been extremely limited.

The birds that followed *Archaeopteryx*, in the Early Cretaceous, were also very dinosaur-like (Figure 0.1). Taxa such as *Jeholornis* and *Rahonavis* retained independent fingers in their wings, long bony tails, and, in some cases, sharp little teeth. Surprisingly, other fossils of Early Cretaceous birds, such as those of *Gansus* and *Apsaravis* already show the fused hand bones (carpometacarpus) and footbones (tibiotarsus) that are typical of modern birds. *Gansus* also had a significant keel on its breastbone, which suggests that it was also capable of sustained flapping flight. The vertebrae at the end of its vertebral column had fused to become a short pygostyle similar the one that carries the tail fan in modern birds. The pygostyle and sophisticated wrist joints imply that some Cretaceous birds were able to use their tail fans and wing tips for aerial maneuvers and control.

The skeleton of *Gansus* differed from those of its contemporaries in another important way that has been passed down to its descendants. The tips of its pubic bones met but were not fused. By the Mid-Cretaceous (100 million years ago (Ma)), the pubic bones of ornithurine birds, such as *Ichthyornis*, had lost even that level of connection. It and all more recent birds have no bony structures that cross the lower abdomen. The eggs and the gut are held in place solely by flexible connective tissue and there are no skeletal constraints on egg size. The absence of skeletal constraints on egg-size allows the production of large eggs and ultimately the altricial young, intensive nestling care, and elaborate nest structures that characterize the crown clades of living birds (Dyke & Kaiser, 2010).

Although primitive versions of modern birds appear in the Early Cretaceous, 120 Ma (Figure 0.1), their fossils are rather rare and we know little about their global distribution. The rich fossil beds of the Jehol Formation in China suggest that early representatives of modern birds successfully competed for resources with a variety of other feathered animals. However, the primitive pygostylian *Confuciusornis* is its most abundant avian fossil, even though it soon disappeared without known descendants. Somewhat later in the Cretaceous, birds must have competed for resources with a variety of small, feathered, arboreal theropods, best represented by *Microraptor gui*. The exceptional diversity of the Jehol fauna may not have been a widespread phenomenon but there, and elsewhere, early ornithine birds competed for air space with other major lineages that achieved global distribution in the Cretaceous.

The membrane-winged pterosaurs were one of the most diverse groups of flying animals in the Cretaceous. They came in a wide variety of sizes and had already achieved global diversity long before the time of *Archaeopteryx*. Pterosaurs may have been able to exclude birds from some specific habitats but, in spite of their aerial habits, we have no evidence of interactions with early birds. They may have co-existed, just as bats and birds have little interaction today.

More importantly the Neornithes had to compete with a highly successful group of look-alikes called 'opposite birds' or Enantiornithes (Figure 0.1). Enantiornithines were not recognized until 1981, when Cyril Walker (1981) noticed subtle variations in the features in bird fossils from several different parts of the world. The Enantiornithines were capable of sophisticated flight based on a triosseal pulley system in the shoulder, but the structure of their individual bones was somewhat different. Enantiornithines had a boss on the coracoid that articulated with a facet on the scapula. In ornithurine birds, a boss on the scapula articulates with a facet on the coracoid.

For many years the Mesozoic fossil record for ornithurine birds was so sparse that it was easy to believe that they did not begin to achieve their modern diversity until after the great extinction at the end of the Cretaceous. Not only did potential competitors among the small, feathered dinosaurs and the aerobatic pterosaurs die out 65 Ma but Enantiornithes disappeared as well. In many places fossils of enantiornithine birds are abundant enough to suggest that they prevented or delayed the diversification

of ornithurine birds into some of the Cretaceous habitats. The enantiornithines appear to have been particularly abundant in terrestrial habitats, particularly in forests, where modern ornithurine birds are currently so successful. However, both fossils and tracks show that Mesozoic ornithurine were able to exploit aquatic and transitional shoreline habitats in which enantiornithine fossils were rare or absent. The very early ornithurine *Gansus* was fully aquatic (You *et al.*, 2006), while abundant remains of the later graculavids present evidence for a wide variety of shorebird-like species in the Mid-Cretaceous.

It has proven exceptionally difficult to find fossils of Mesozoic birds that can be related to extant lineages. Not only are ornithurine fossils scarce but most consist solely of isolated elements or broken fragments. The handful of Mesozoic species, such as *Ichthyornis* and *Hesperornis*, that are represented by nearly complete skeletons have become quite famous but appear to have no modern relatives. Unfortunately, the discovery of many well-preserved and nearly complete specimens of early birds in China has done little to clarify the situation. Most of them represent archaic, long-tailed birds with the rudimentary wings and unfused feet of other dinosaur lineages. A great many of them had teeth, a feature that clearly distinguishes them from any extant lineage. None of them have a uncontroversially "modern" anatomy and none can be linked to living forms.

For a while it seemed as though the Cretaceous avifauna would remain dominated almost entirely by creatures only remotely related to modern birds, but the story began to change early this century. First with the discovery of a 100 Ma wing in Mongolia's Gobi Desert (Kurochkin *et al.*, 2002) and then with the re-description of a fossil from Cretaceous deposits in Antarctica (Clarke *et al.*, 2005). Both these fossil birds have turned out to be the oldest known representatives of the living order Anseriformes (Figure 0.1). The detached wing was given the name *Teviornis* and is a member of a group of long-legged duck-like birds called presbyornithids. The more complete Antarctic fossil was named *Vegavis*. It is well enough preserved to be included in evolutionary studies that rely on anatomical characteristics taken from living birds. Both *Teviornis* and *Vegavis* appear to be early diverging members of the anseriform family tree.

Other named fossils from the Cretaceous that might just turn out to be representatives of living lineages – *Palintropus*, *Lonchodytes*, *Tytthostonyx* – remain contentious, either in the interpretation of their identifying characteristics or in the precise age of the matrix from which they were extracted. The determination of their exact relationship to modern forms must await further analysis or the discovery of new material. If these candidates are eventually demonstrated to be representatives of extent lineages, it will be as members of basal groups. None have characteristics attributable to any of the crown clades of birds.

The two major problems facing paleontologists over the next few years will be the same as those for the past century: "What did the ancestor of modern birds look like?" and "Where did the living groups birds come from?" However, the potential answers to those questions have already changed because our knowledge of Mesozoic birds has expanded greatly. Even in the late 20th century we were looking for a more reptilian *Ichthyornis* to answer the first question and a less-toothy version to answer the second. Now the answer to the first question needs fossils from the Late Jurassic or the very early Cretaceous to tell us what the earliest modern birds looked like, an ancestor of *Gansus* that had not achieved sustained flight. Unfortunately, to find such a fossil one must be extremely lucky and appropriate Jurassic strata are exceedingly rare globally. It may be easier to answer the second question. All we need are examples of a few recognizably modern groups from some of the richest fossil beds in the world.

REFERENCES

Chure DJ, Madsen JH. 1996. On the presence of furculae in some nonmaniraptoran theropods. *Journal of Vertebrate Paleontology* **16**: 573–577.

Clarke JA, Tambussi C, Noriega J, Erickson G, Ketchum R. 2005. Definitive fossil evidence for the extant avian radiation in the Cretaceous. *Nature* **433**: 305–309.

Dyke GJ, Kaiser GW. 2010. Cracking a developmental constraint: egg size and bird evolution. *Records of the Australian Museum* **62**(1): 207–216.

Hackett SJ, Kimball RT, Reddy S, Bowie RCK, Braun EL, Braun MJ, Chojnowski JL, Cox WA, Han K-L, Harshman J, Huddleston CJ, Marks BD, Miglia KJ, Moore WS, Sheldon FH, Steadman DW, Witt CC, Yuri T. 2008. A phylogenomic study of birds reveals their evolutionary history. *Science* **320**: 1763–1768.

Heilman G. 1926. The Origin of Birds. London: Witherby; 210 pp.

Hu DY, Hou LH, Zhang LJ, Xu X. 2009. A pre-*Archaeopteryx* troodontid theropod from China with long feathers on the metatarsus. *Nature* **461**: 640–643.

Huxley TH. 1868. On the animals which are most nearly intermediate between birds and reptiles. *Geological Magazine* **5**: 357–365.

International Chicken Genome Sequencing, Consortium., 2004. Sequence and comparative analysis of the chicken genome provide unique perspectives on vertebrate evolution. *Nature* **432**: 695–716.

Kurochkin EN, Dyke GJ, Karhu A. 2002. A new presbyornithid bird (Aves: Anseriformes) from the Late Cretaceous of Southern Mongolia. *American Museum Novitates* **3386**: 1–11.

Makovicky PJ, Currie PJ. 1998. The presence of a furcula in tyrannosaurid theropods, and its phylogenetic and functional implications. *Journal of Vertebrate Paleontology* **18**: 143–149.

Ostrom JH. 1976. *Archeopteryx* and the origin of birds. *Biological Journal of the Linnean Society* **8**: 91–182.

Sibley CG, Ahlquist JE. 1990. *Phylogeny and Classification of Birds: A Study in Molecular Evolution*. New Haven: Yale University Press.

Vargas AO, Kohlsdorf T, Fallon JF, VandenBrooks J, & Wagner GP. 2008. The evolution of Hoxd11 expression in the bird wing: insights from *Alligator mississippiensis*. *PLoS One* **3**: e3325.

Wagner GP. 2005. The developmental evolution of avian digit homology: an update. *Theory in Biosciences* **124**: 165–183.

Walker CA. 1981. New subclass of birds from the Cretaceous of South America. *Nature* **292**: 51–53.

Xu X, Zhang FC. 2005. A new maniraptoran dinosaur from China with long feathers on the metatarsus. *Naturwissenschaften* **92**: 173–177.

You HL, Lamanna MC, Harris JD, Chiappe LM, O'Connor JK, Ji SA, Lü JC, Yuan C-X, Li DQ, Zhang X, Lacovara KJ, Dodson P, Ji Q. 2006. A nearly modern amphibious bird from the Early Cretaceous of north-western China. *Science* **312**: 1640–1643.

Young RL, Caputo V, Giovannotti M, Kohlsdorf T, Varga AO, May GE, Wagner GP. 2009. Evolution of digit identity in the three-toed Italian skink *Chalcides chalcides*: a new case of digit identity frame shift. *Evolution and Development* **11**(6): 647–658.

1 Theropod Diversity and the Refinement of Avian Characteristics

PETER J. MAKOVICKY AND LINDSAY E. ZANNO

Field Museum, Chicago, USA

Bird origins have been debated ever since Darwin published his *"Origin of Species,"* and was subsequently challenged on the topic by Sir Richard Owen, who pointed out the lack of transitional fossil forms in the evolution of the highly derived avian body plan. Indeed, Owen likely carefully selected birds to make his point due to their many unique traits and physiological features such as flight, feathers, bipedality, and a remarkable respiratory system in which the lungs are connected to and ventilated by a complex system of air sacs that pneumatize the skeleton. Within two years of this debate, the discovery of the first specimens of *Archaeopteryx* provided conclusive evidence of avian evolution in the fossil record and became the focal point for research and deliberation on the topic for more than a century. While the fossils of *Archaeopteryx* provided incontrovertible evidence for a reptilian origin for birds, opinions varied as to which group of reptiles birds may have originated from.

Following the discovery of the small nonavian theropod *Compsognathus* in the same limestone deposits as *Archaeopteryx*, Huxley (1868) presciently proposed an evolutionary relationships between birds and nonavian theropods based on shared traits such as three principal, weight-bearing toes in the foot (confirmed by foot-prints), a tall ascending process of the astragalus, and hollow bones. Other contemporary evolutionary biologists such as Cope favored an evolutionary relationship between birds and ornithopod dinosaurs such as hadrosaurs, based again on a three-toed foot and a retroverted pubic shaft. While a variety of ancestors or sister taxa were proposed for birds, a broad consensus that they were related to dinosaurs prevailed until the publication of the English edition of Heilmann's (1926) *"Origin of Birds."* Heilmann's book presented a detailed study of neontological, embryological, and fossil evidence, all of which pointed to a theropod ancestry for birds. Nevertheless, based on the prevailing assumption of the time that dinosaurs ancestors had lost their clavicles, their reappearance in birds would violate Dollo's (1893) law of irreversibility. Heilmann therefore concluded that the similarities between birds and theropods were due to convergence, and that birds were derived from more basal archosaurs that still retain clavicles.

Due to the thoroughness of his book, Heilmann's (1926) hypothesis held sway for the next four decades until the discovery and description of the mid-sized dromaeosaurid theropod *Deinonychus* by Ostrom (1969). Ostrom's detailed study of the skeletal anatomy of *Deinonychus* led him to recognize derived characters shared between it and the basal bird *Archaeopteryx* (Ostrom, 1976), and to the discovery of a misidentified specimen of *Archaeopteryx* (Ostrom, 1970).

Living Dinosaurs: The Evolutionary History of Modern Birds, First Edition. Edited by Gareth Dyke and Gary Kaiser.
© 2011 John Wiley & Sons, Ltd. Published 2011 by John Wiley & Sons, Ltd.

Among the new traits that Ostrom mustered to revive a theropod ancestry for birds are the presence of a half-moon-shaped wrist bone that allows the hand to adduct against the forearm as in the wing-folding mechanism of living birds, and a three-fingered hand with characteristic proportions between the three metacarpals and phalanges. Following Ostrom's work, progressively more evidence has been amassed to support this hypothesis of avian origins, as a plethora of new fossil discoveries continue to blur the morphological distinction between birds and their closest theropod relatives.

Over the past three decades, widespread adoption of cladistic methodology for establishing and testing proposed evolutionary relationships has provided the conceptual framework for deciphering the origin and evolutionary history of birds. Gauthier (1986) was the first to apply an explicit cladistic parsimony analysis of theropod relationships to the question of bird origins. In doing so, he provided the first rigorous test of the hypothesis of theropod ancestry and set the stage for evaluating the evolutionary history of avian anatomy, physiology, and behavior in a quantitative framework. Subsequent decades have seen a remarkable surge in the discovery of new theropods, including fossil stem birds, with each discovery spawning novel systematic analyses (for reviews see Weishampel et al., 2004; Norell & Xu, 2005; Benton, 2008) and further supporting the hypothesis that birds are theropod dinosaurs.

To date, profound advances have been made in teasing out the evolution of the avian body plan as well as correlated physiological features and life history parameters of modern birds. Likewise, our knowledge of these traits in modern birds and their anatomical correlates has been used to yield insight into the biology of nonavian theropod dinosaurs and to infer whether particular avian traits originated before or after the origin of the avian lineage itself. Here we provide a general overview of the stepwise acquisition of the avian body-plan throughout the theropod family tree and discuss how the physiology of modern birds is being used to reconstruct aspects of dinosaur biology.

A ROADMAP TO THE DINOSAURIAN HERITAGE OF BIRDS

Despite their radically different body plans, birds inherited a mosaic of anatomical traits from various stages of vertebrate history. A host of discoveries over the past five decades provide a detailed road map to how the highly specialized avian anatomy was assembled over the evolutionary lineage leading to birds and demonstrates that many of the traits that are considered uniquely avian among extant amniotes actually arose before the origin of birds themselves. While most of our understanding of where birds fit into the tree of life comes from study of hard tissues, dramatic new discoveries in the past decade and a half are providing unprecedented insights on the evolution of soft tissue systems (Schweitzer et al., 1999; Xu et al., 2001) aspects of physiology (Varricchio et al., 2002; Varricchio & Jackson, 2004; Organ et al., 2007), and even behavior (Norell et al., 1995; Varricchio et al., 1997; Xu & Norell, 2005).

That birds are archosaurs and the closest living relatives to crocodilians has long been inferred from shared derived traits such as a four-chambered heart and thecodont dentition. Virtually all birds possess an external mandibular fenestra, a synapomorphy of Archosauria (Benton, 1990) and fossil avians also possess an antorbital fenestra (Figure 1.1), also considered a hallmark of this clade, although in extant birds the latter opening is lost or merged with the external nares through expansion of the premaxillae and concomitant reduction of other preorbital bones. Phylogenetic analyses have identified numerous derived traits shared by birds and various hierarchically arranged subsets of archosaur diversity (Figure 1.1). A comprehensive review detailing all of these characters is beyond the scope of this chapter, so here we concentrate on a select subset of traits and evolutionary branching points that are most critical to understanding the evolution of the unique avian body plan and its derived physiological and functional correlates.

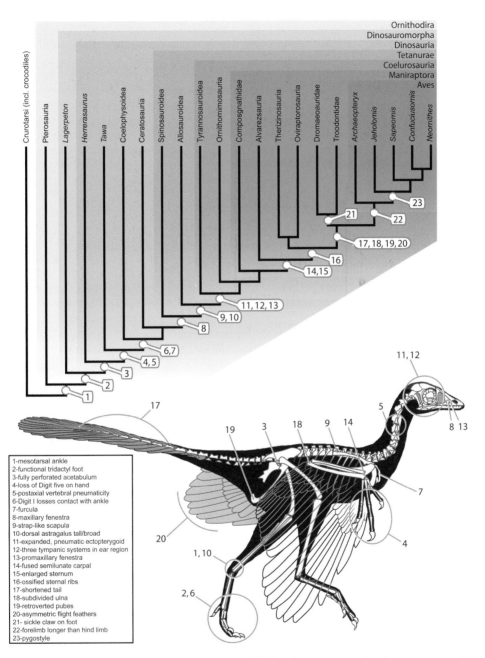

Fig. 1.1 Guide to avian evolutionary history based on a simplified evolutionary tree of Archosauria. Key traits in avian evolution are mapped onto the cladogram with their corresponding position indicated on the skeleton of *Archaeopteryx*. Note how traits are distributed throughout the skeleton revealing the mosaic assembly of avian anatomy throughout ornithodiran evolution. See text for more detailed discussion of individual traits. *Archaeopteryx* reconstruction by M. Donnelly.

The deepest division among archosaurs is between the lineages leading to the extant clades Crocodylia and Aves. Each of these branches also subtends numerous fossil clades, which were dominant faunal elements during the Mesozoic. The avian total group (= extant plus extinct diversity after the split from the lineage leading to Crocodylia) is known as Ornithodira and is characterized principally by the possession of a mesotarsal ankle joint (Figure 1.1), in which the articulation between the foot and crus occurs between the proximal and distal tarsals and approximates a roller joint, restricting foot motions to fore–aft swinging without a rotational component. Ornithodirans are also characterized by having a clearly defined femoral head that is distinctly offset from the femoral shaft. Besides birds, Ornithodira is comprised of extinct dinosaurian subclades, pterosaurs, and suite of lesser-known Triassic forms. Many of these taxa, especially those within the dinosauromorph clade, share the derived trait of being bipedal, thus freeing the forelimbs to evolve new functions including flight. Although many large, herbivorous dinosaurs later re-evolved a quadrupedal stance in concert with the evolution of large body mass, all derive from bipedal ancestors.

Dinosauromorphs, including birds, are united in their possession of three principal weight-bearing bones of the hind foot with the first and especially fifth toes at least partially reduced (Figure 1.1). Recent discoveries of dinosauriforms and basal dinosaurs from Argentina and New Mexico, USA provide new insights on the progressive nature of the reduction and loss of weight-bearing function in the first and fifth toes of the ornithodiran foot (Nesbitt et al., 2009a).

Dinosauria, which comprises the bulk of known ornithodiran diversity, is characterized by a fully perforated acetabulum, in which the head of the femur fits into a medially open socket formed by the three hipbones. The head of the femur is angled almost perpendicular to the shaft allowing the hindlimbs to be brought under the body for a fully upright, parasagittal gait. Triassic dinosauromorphs such as Lagerpeton with partially open

acetabula (Sereno & Arcucci, 1993) reveal that the acquisition of a fully open acetabulum occurred in a gradual fashion spanning several branching points at the root of the dinosaurian evolutionary tree (Figure 1.1). Counterintuitively, birds fall within the Saurischia, or "lizard-hipped" branch of dinosaur diversity, rather than the Ornithischia, the "bird-hipped" branch. Saurischians are united by their possession of hyposphen–hypantrum accessory articulations between the neural arches of the trunk vertebrae, which are also marked by lateral excavations on their neural arches presumably housing diverticula of the respiratory system. Hypantrum–hyposphen articulations were lost in a later stage of bird evolution, but some fossil birds such as Patagopteryx reveal that this trait was present ancestrally.

Birds form part of the theropod radiation within Saurischia. Progressive reduction of the hand toward the tridactyl condition in birds is encountered near the base of the theropod radiation. When present in basal theropods, the fourth digit is clawless and the fifth digit, known in Eoraptor, is reduced to a metacarpal splint devoid of knuckles (Figure 1.2). Coelophysoids and more derived theropods (Figure 1.1) display a reduction in the number of carpals to five or less, and the number of digits to four or less, although a fifth metacarpal has been tentatively identified in Dilophosaurus (Xu et al., 2009) indicating that this process may have occurred in parallel in a number of theropod lineages. Other shared derived traits that reflect the deep theropod origins of birds include a dorsally ascending process of the astragalus, a fifth pedal digit reduced to only the metatarsal, and the wishbone. As briefly mentioned above, the presence of clavicles (whether fused or not) has been a topic of debate in avian and theropod systematics for the better part of a century (Makovicky & Currie, 1998). Because many of the earliest dinosaur discoveries were of advanced ornithopods and sauropodomorphs, taxa that have lost their clavicles, these elements were generally held to be absent in all dinosaurs, leading to the subsequent misinterpretation of theropod furculae as either interclavicles (Makovicky & Currie, 1998) or fused gastralia

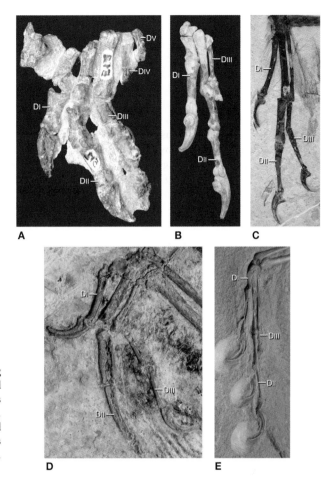

Fig. 1.2 Comparison of theropod mani showing progressive reduction and loss of digits IV and V and changes in the proportions of manus elements. *Eoraptor* (A), *Guanlong* (B), *Sinornithosaurus* (C), *Archaeopteryx* (D), and *Confuciusornis* (E). Abbreviations: DI–V, digits I–V. All specimens shown at the same scale. [This figure appears in color as Plate 1.2.]

(Chure & Madsen, 1996). Unfused clavicles are now known to occur in prosauropods (Yates & Vasconcelos, 2005) and some ceratopsians (Brown & Schlaikjer, 1940), and fused clavicles have been documented in an evergrowing list of theropod taxa, indicating that this trait is likely a synapomorphy of almost the entire clade (Smith *et al.*, 2007; Nesbitt *et al.*, 2009b). Almost all theropods with the exception of *Herrerasaurus* also exhibit some degree of pneumatization on the sides of the postaxial cervical vertebral centra, with basal forms such as *Tawa* and coelophysoids exhibiting fossae (Nesbitt *et al.*, 2009a), while more derived taxa have invasive foramina that pierce the vertebral centra invading and excavat-

ing their interiors (Britt, 1997). These pneumatic features correlate with one of the avian respiratory air-sac systems (see below). Coelophysoids and more derived theropods are characterized by a first toe in which the metatarsal is reduced and no longer contacts the ankle (Figure 1.1) – in most birds this toe is rotated on to the plantar face of the foot and allows for perching in arboreal forms.

Theropods exclusive of the basal coelophysoid radiation that spanned the Late Triassic–Early Jurassic, can be largely divided into the two major lineages, Ceratosauria and Tetanurae. Birds belong to the latter lineage, which also includes such well-known denizens as

Allosaurus, Tyrannosaurus, and *Velociraptor.* Both clades radiated throughout the Jurassic and Cretaceous giving rise to small- and large-bodied forms. Ceratosaurs are characterized by a progressive reduction of forelimb traits such as muscle attachment areas and overall robustness of the forelimbs, accompanied by a reduction of their relative size in larger taxa.

Large bodied ceratosaurs such as *Carnotaurus, Majungasaurus,* and *Rajasaurus* are grouped together in the clade Abelisauridae, and are characterized by very short, deep skulls adorned with surficial sculpturing and even horn-like structures in some species (Bonaparte *et al.,* 1990; Sampson *et al.,* 1998; Sereno & Brusatte, 2008), as well as heavy reduction of forelimb elements including a completely unossified wrist and loss of phalanges on most fingers (Chiappe *et al.,* 1998). These taxa have an almost exclusively Gondwanan distribution during the Cretaceous, and have been the subject of intense biogeographic debate (Sampson *et al.,* 2001; Sereno *et al.,* 2004; Sampson & Krause, 2007).

The Tetanurae have a greater known Mesozoic diversity than do the ceratosaurs, and include a wide range of both large and small taxa. Tetanuran theropods are characterized by a large suite of synapomorphies, including presence of a maxillary fenestra (Figure 1.1) rostral to the archosaurian antorbital fenestra and a fully horizontally directed femoral head. All birds exhibit this last trait, although the maxillary fenestra is only evident in *Archaeopteryx,* having been incorporated into the naris in all living and most fossil avian species. Most tetanuran taxa are members of three principal groups, the Spinosauroidea, Allosauroidea, and Coelurosauria, although membership and exact relationships among these three subclades remain debated (Sereno *et al.,* 1996; Rauhut, 2003; Smith *et al.,* 2007, 2008). Allosauroids and spinosauroids are predominantly large-bodied carnivores with body masses ranging from a few hundred kilograms to as much as seven tons in derived end-members of these clades, such as the spinosaurid *Spinosaurus* and the carcharodontosaurid allosauroid *Giganotosaurus.* Allosauroids are recovered as the sister clade to the

Coelurosauria in most cladistic analyses (Sereno *et al.,* 1996; Smith *et al.,* 2007, 2008), through their shared possession of traits such as perforated maxillary fenestra, pneumatic openings on the axial centrum, and a reduction of the ischiadic apron to form an open obturator notch rather than an enclosed fenestra. Both spinosauroids and allosauroids were globally distributed during the Late Jurassic and Early Cretaceous, but wane in diversity during the latest Cretaceous, with the coelurosaurian tyrannosaurids filling the dominant carnivore niche at least on northern landmasses.

Throughout most of the 20th century, carnivorous dinosaurs were taxonomically divided according to body size with small to medium-sized taxa grouped in Coelurosauria, and large species lumped within Carnosauria. This taxonomic scheme is artificial (Holtz, 1994) and recent work demonstrates that large body sizes evolved multiple times within Theropoda. Coelurosauria has recently been redefined (Gauthier, 1986; Sereno, 1999a) as the clade encompassing birds and all theropods closer to birds than to *Allosaurus,* and encompasses the smallest known theropods (hummingbirds) as well as some of the largest (*T. rex*), although all lineages appear to derive from small to medium-sized ancestors. Coelurosaurs are united by having a third opening, the promaxillary fenestra, within the antorbital fossa, although this is only recognizable in *Archaeopteryx* among birds as is the case with the maxillary fenestra (Figure 1.1). Other traits uniting this grouping include the presence of three tympanic pneumatic systems emanating from the middle ear (more basal taxa only possess one or two of these), an expanded and pneumatic ectopterygoid (lost in neornithine birds), and an expanded astragalar ascending process that is twice as tall as it is wide and covers almost the full width of the ankle. The manus is fully tridactyl in most Coelurosauria (Figures 1.1 and 1.2C–E), although a few basal taxa such as the basal tyrannosauroid *Guanlong* (Xu *et al.,* 2006) retain a splint-like vestige of the fourth metacarpal (Fig. 1B). The large predatory tyrannosauroids, the beaked and herbivorous

(Kobayashi *et al.*, 1999, Zanno & Makovicky, 2010) ornithomimosaurs, and small compsognathids are generally considered to be basal lineages within the coelurosaur radiation, whereas birds and their closest taxa form a more exclusive clade within Coelurosauria known as Maniraptora. In the past decade, discoveries of numerous coelurosaur taxa sporting protofeathers, fully formed flight feathers (Ji *et al.*, 2001), and even specialized feathers convergent on display structures in oscines (Zhang *et al.*, 2008) have been made in Jurassic and Cretaceous lake-bed deposits of northeastern China. Together, these discoveries indicate that possession of plumage covering most of the body with the exception of the feet and snout likely characterizes at minimum the coelurosaurian node. The presence of hollow, fibrous integumentary structures along the dorsal midline of two ornithischian taxa and on the body of pterosaurs indicates the presence of such structures alone may be much wider among ornithodirans, though their exact homology remains unclear. Doubts have been cast regarding the presence of feather homologues in various theropods (Lingham-Soliar *et al.*, 2007; Feduccia *et al.*, 2005), and these have instead been interpreted as collagen fibers derived from decomposition of the skin in a specimen of the compsognathid *Sinosauropteryx*. These claims are based only on very selective comparisons between the soft tissues of feathered nonavian theropods and experimentally manipulated integument on extant reptiles and extant and extinct marine amniotes. More appropriate comparisons to either birds or reptiles from the same shale deposits that yield the feathered theropods, representing equivalent preservational conditions, were not conducted by Lingham-Soliar and colleagues, and indeed dismissed with the tautological argument that animals with feathers are by definition birds and thus need not be considered when testing for feather preservation (Fedduccia *et al.*, 2005). Two recent studies (Zhang *et al.*, 2010; Li *et al.*, 2010) demonstrate that the preservation of nonavian coelurosaur integumentary structures matches those of unquestionable stem birds from the same rock units and that they preserve melanosomes imbedded within the keratinous matrix of the feathers themselves, thus providing not only evidence on the homology of these integumentary structures, but also on the color and appearance of these animals in life. Conversely, preservation of a body outline composed of frayed and decomposing dermal layers has never been reported in any of the hundreds of choristodere specimens collected from these shale beds, casting doubt on the conclusion that such decomposition patterns should be observed in a dinosaur as posited by Lingham-Soliar *et al.* (2007).

Maniraptorans are characterized by distinguishing traits including a half-moon shaped wrist bone that is thought to represent a fusion of the first and second distal carpal (Figure 1.1, trait 11). A pulley-like proximal surface on this element allows the hand to be flexed sideways toward the forearm, and is responsible for the wing-folding mechanism in birds. As with many other anatomical traits relevant to understanding the origins and relationships of birds, the refinement of this particular synapomorphy accrued over a range of branches in the phylogeny, and incipient versions of this structure have been recognized in more basal tetanurans such as *Allosaurus* (Sereno, 1999b). A number of novel evolutionary features in the thoracic skeleton, which are known to play a role in avian respiration (O'Connor & Claessens, 2005), further diagnose some maniraptorans. These include presence of enlarged sternal plates with extensive medial contact and distinct facets for ossified sternal ribs (Barsbold, 1983; Norell & Makovicky, 1999) (Figure 1.1, trait 16) and uncinate processes spanning the thoracic ribs (Clark *et al.*, 1999).

Maniraptoran fossils that preserve integumentary structures reveal an increased complexity in both feather types and morphology over the simple filamentous structures observed in basal coelurosaurs (Xu *et al.*, 2001; 2010). Xu and colleagues (2001) demonstrated a correlation between the order of appearance of progressively more complex feather types in maniraptoran evolution with their order of development in avian ontogeny, and it is clear that almost all basic feather types known in birds had evolved earlier in theropod evolution.

Several aberrant clades of theropods are included within the Maniraptora. These include the herbivorous Therizinosauria, whose theropod affinities were strongly debated due to their unusual anatomy, but which are now known to possess unquestionable theropod hallmarks such as pneumatic vertebrae, furculae, feathers, and a semilunate carpal that fuses in at least adult specimens of some taxa (Kirkland *et al.*, 2005). Another group with unusual anatomy and debated affinities are the Alvarezsauridae. Derived, small-bodied members of this group discovered in Late Cretaceous sediments of the Gobi Desert exhibit a remarkable mosaic of characters including loss of a postorbital bar, a double-headed quadrate, a keeled sternum, short but massive arms with an enlarged pollex but reduction of the other fingers, a splint-like fibula, and diminutive, supernumerary teeth (Perle *et al.*, 1993; Chiappe *et al.*, 1998). A number of these traits, such as the reduced postorbital, streptostylic quadrate, keeled sternum, and reduced fibula, are also encountered in birds more derived than *Confuciusornis*, leading to initial hypotheses that these fossil taxa represent flightless birds more derived than *Archaeopteryx*. Subsequent discoveries of more basal alvarezaurids in Argentina led to the recognition that many of the avian-like characteristics of derived alvarezaurids evolved convergently in birds, and most recent studies agree that they represent basal maniraptorans rather than members of the avian lineage (Norell *et al.*, 2001; Novas & Pol, 2002; Senter, 2007; Zanno *et al.*, 2009).

Oviraptorosaurs represent another anatomically bizarre lineage, some members of which exhibit remarkable convergence on avian anatomy in parts of their skeleton. When analyzed in a limited context, such traits have also prompted hypotheses that this clade represents secondarily flightless birds (Maryanska *et al.*, 2002), a conclusion that is not supported in more rigorous and comprehensive phylogenetic studies incorporating a greater array of both taxa and characters. Such studies overwhelmingly posit oviraptorosaurs (often, but not always, in combination with therizinosaurs) as sister to the clade Paraves that encompasses birds and their sister taxon, the sickle-clawed Deinonychosauria. Birds and deinonychosaurs are united by numerous apomorphic features such as possessing retroverted pubes (Figure 1.1, trait 19), an expanded and flexed coracoid that repositions the humeral articulation closer to the vertebral column and imbues the scapulocoracoid with an L-shaped profile, a proximal ulna articulation subdivided into two distinct facets (Figure 1.1, trait 18), and a shortened tail with 25 or less vertebrae of which the anterior ones are short and box-like and distal ones are elongate and cylindrical (Figure 1.1, trait 17). Most paravians, including all birds, are also known to possess primary and secondary feathers with asymmetrically developed vanes on either side of the rachis (Figure 1.1, trait 20), a feature considered to be an adaptation for aerodynamic function, but the recently described *Anchiornis* exhibits symmetrical vane distribution on its primaries (Xu *et al.* 2009) as in the basal oviraptorosaur *Caudipteryx*, complicating our understanding of how many times this trait evolved.

Deinonychosauria comprises two distinct clades, the Troodontidae and Dromaeosauridae, which are united by the presence of a sickle-shaped claw on the second digit of the foot (Figures 1.1 and 1.2, trait 21), and a triangular lateral exposure of the splenial along the edge of the lower jaw. A close relationship between dromaeosaurs and birds was initially recognized by Ostrom (1976) following his discovery and description of the first relatively complete dromaeosaurid *Deinonychus* (Ostrom, 1969), but some debate persisted regarding the affinities of Troodontidae, derived members of which share characters with other coelurosaur clades and also lack some paravian synapomorphies such as a retroverted pubis. Discovery of a number of basal troodontids from the Early Cretaceous Yixian and Jiufotang Formation of China reveals that these traits are homoplastic in derived troodontids, and that Deinonychosauria is a natural grouping (Xu *et al.*, 2002). Many of the deinonychosaurs recently discovered in China and elsewhere are also significant because they represent the smallest nonavian dinosaurs yet discovered (Xu *et al.*, 2000) and are comparable in body size to basal avian taxa such as *Archaeopteryx* and *Jeho-lornis* (Turner *et al.*, 2007; Xu *et al.*, 2009). Some of the small deinonychosaurs from these rock units, such as *Microraptor* and *Sinornithosaurus*, possess vaned feathers on the hindlimb as well as the forelimb (Xu *et al.*, 2001; Xu & Zhang, 2005; Ji *et al.*, 2001), along with a frond-like arrangement of the rectrices in a pattern like that of *Archaeopteryx* (Figure 1.3A). This four-winged body plan may represent a transitional step in the evolution of powered flight (Longrich, 2006; Hu *et al.*, 2009), though its optimization on the evolutionary tree is complicated by the extreme similarity between basal members of the three principle paravian lineages and hence some phylogenetic lability between them. The earliest instance of this unique body plan is represented by *Pedopenna* (Xu & Zhang, 2005; Figure 1.3B) and *Anchiornis* (Hu *et al.*, 2009), which are Middle Jurassic in age and thus older than *Archaeopteryx*. More derived deinonychosaurian taxa evolved larger body sizes culminating in the 30 ft long *Utahraptor*.

The fossil record of Deinonychosauria was until recently largely restricted to Cretaceous deposits of the northern continents, but a slew of recent discoveries of dromaeosaurids from

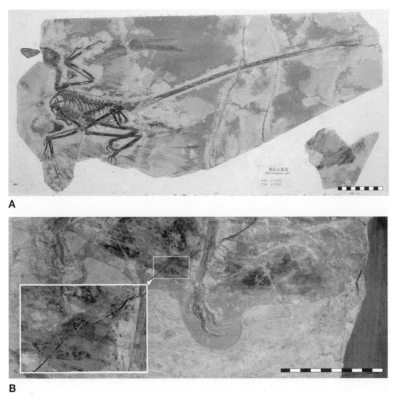

Fig. 1.3 (A) Skeleton of the dromaeosaurid *Microraptor gui* from the Yixian Formation of Liaoning, China, exhibiting vaned, asymmetric feathers on both fore- and hindlimbs. (B) Detail of hindlimb primary feathers of *Pedopenna* from the Middle Jurassic of Inner Mongolia, China. *Pedopenna* is the earliest paravian fossil to exhibit vaned feathers and a four-winged body plan. The inset shows a close-up of the aligned and parallel barbs on each vane that indicate the presence of interlocking barbules, as well as the rachis. Note the large sickle claw characteristic of deinonychosaurians (= dromaeosaurs and troodontids) on digit II of the foot. Scale bars equal 5 cm. (Photographs: P. Makovicky.) [This figure appears in color as Plate 1.3.]

Argentina (Novas & Puerta, 1997; Makovicky et al., 2005; Novas & Pol, 2005) are evidence for a Gondwanan radiation of these animals. The discovery of the near-complete holotype of the Gondwanan dromaeosaurid *Buitreraptor* (Makovicky et al., 2005) provided evidence to unite all of these different taxa into a single basal lineage, the Unenlagiinae, whose split from the better-known Laurasian dromaeosaurids may correlate with the break up of Pangaea. The discovery of Gondwanan dromaeosaurids also prompted a reinterpretation of the purported basal bird *Rahonavis* as member of the Unenlagiinae, demonstrating that the skeletons of basal deinonychosaurs and the earliest birds are almost indistinguishable. *Rahonavis* is characterized by hyperelongate forelimbs suggesting that such flight-related proportions may have arisen more than once in paravian evolution, with the main occurrence being characteristic of the avian lineage (Figure 1.1, trait 22).

Apart from *Archaeopteryx* and a few other species such as *Jeholornis*, most Cretaceous avian fossils exhibit rapid evolution of the avian body plan. *Sapeornis* is the most primitive bird to possess a foreshortened tail with the distalmost segments fused into a pygostyle (Figure 1.1, trait 23). Without a long tail to counterbalance the body as in typical nonavian theropods, the last common ancestor of *Confuciusornis* and more derived birds evolved a posture where the knee is permanently angled to bring the center of mass above the foot and offset the loss of a long counterbalanced tail. An ossified kneecap, which is unknown in non-avian dinosaurs, is present in *Confuciusornis* and more derived birds and serves to stabilize the bent knee. *Sapeornis* and *Confuciusornis* are also the basalmost avian taxa to exhibit a fused sternum with an incipient sternal keel for anchoring enlarged flight musculature, marking another key step in the assembly of the modern avian body plan. Both retain primitive theropod traits, however, such as a functional grasping tridactyl hand (Figure 1.2E), and *Sapeornis* and most Cretaceous birds retain dentition. Though some theropods convergently lost their teeth, the avian bill appears to have arisen at or very close to the origin of the avian crown group.

NEW INFERENCES ON SOFT TISSUE, PHYSIOLOGY, AND BEHAVIOR

The recent surge in dinosaur discoveries and research has not only yielded a better understanding of the skeletal evolution of theropods and a more nuanced understanding of the stepwise assembly of the unique avian body plan, but also provided insights into the evolution of avian physiology, reproductive biology, and even aspects of their related behaviors. Through integrative research incorporating fossil and neontological data, advances in our understanding of modern birds are being applied back in time to generate hypotheses regarding aspects of dinosaurian biology lost in the fossil record using phylogenetic history as a guide (Witmer, 1995). Here we review some recent advances in our understanding of dinosaur biology based on some of the most remarkable and informative theropod fossil discoveries made to date and new methodological approaches to the study of fossilized remains.

Metabolism and respiration

A long-standing debate regarding dinosaurian metabolic regimes has persisted for over 30 years (Chinsamy-Turan & Hillenius, 2004; Padian & Horner, 2004), since the recognition that birds are derived theropods prompted speculations that they inherited their homeothermic physiology from dinosaurian ancestors. Debates on physiological inferences made on evidence such as histological traits, the possible presence or absence of turbinals, choanal position, and basic physiological calculations have been inconclusive and marred by attempts to draw wide-ranging conclusions through oversimplified interpretations of relatively limited (and often inaccurate) data. The presence of a plumage of filamentous or downy feather homologues covering the body in a variety of coelurosaurs, including taxa that presage evolution of vaned feathers with aerodynamic functions, suggests that feathers evolved in response to selective pressures other than adaptation to aerodynamic locomotion (Norell

& Xu, 2005; Li *et al.*, 2010). Given the insulating properties of feathers and the small size and correspondingly high surface area to volume ratio of most of the nonavian theropods discovered with plumage, many authors have concluded that feather evolution may in part have been driven by a need for insulation, which in turn implies an ability to generate metabolic energy. A recent study of the histology of dinosaurs has demonstrated that theropods tend to have smaller osteocyte lacunae in their bones, indicating smaller cell sizes (Organ *et al.*, 2007). Living birds have relatively smaller cells and markedly lighter cell nuclei with far less redundant DNA compared to other amniotes. Small cell size facilitates increased basic metabolic rates due to the higher surface to volume ratio of the cells, and correlates with nuclear mass, so it is thought that birds underwent active selection for smaller nucleus size. Organ and colleagues' (2007) results robustly suggest that this selective process began much earlier in theropod history.

The high avian basal metabolic rate is in part sustained by a unique respiratory system, in which the incompressible lungs are ventilated by a complex system of interconnected air sacs. Phylogenetic continuity has been established between the pneumatic openings in the vertebral columns of theropod dinosaurs and those of birds (Britt *et al.*, 1998; O'Connor & Claessens, 2005), which are formed through ontogeny as the air sacs invade adjacent bones. Five main air sac systems are connected to the lungs either directly or through their connections to one another in birds. Of these five, the cervical, clavicular, and abdominal air sac systems invade and pneumatize vertebral, girdle, and even limb bones in birds. Skeletal pneumatic features such as openings into bones and honeycombed interior architecture correlated with these systems have been recognized in theropods, with vertebral pneumaticity related to the cervical air sacs being virtually ubiquitous in theropods (Britt *et al.*, 1998). Hard tissue correlates of the other two systems are less common, but widespread enough throughout theropod diversity to suggest that at least the last

common ancestor of ceratosaurs and tetanurans possessed abdominal air sacs (O'Connor & Claessens, 2005; Sereno *et al.*, 2008), and that most tetanurans potentially had a clavicular air sac (Makovicky *et al.*, 2005; Sereno *et al.*, 2008). It should be noted that air sacs do not always invade skeletal elements in living birds and the degree of pneumaticity is observed to correlate with life history parameters such as body size and ecological habits (O'Connor, 2009), so absence of pneumatic traces in bones of extinct theropods cannot be taken as evidence for absence of air sacs, especially if such taxa are bracketed phylogenetically by taxa with positive evidence for air sacs.

Reproductive biology

Recent discoveries of nesting or gravid maniraptoran dinosaurs from Mongolia (Norell *et al.*, 1995; Figure 1.4) and elsewhere (Varricchio *et al.*, 1997; Currie & Chen, 2001; Sato *et al.*, 2005; Grellet-Tinner & Makovicky, 2006) have yielded crucial insights into the evolution of avian reproductive biology. Examination of the histology of such specimens (Erickson *et al.*, 2007) has demonstrated that many of them are not fully grown. Assuming their association with nests is demonstrative of a parental relationship, it suggests that nonavian theropods and perhaps even the earliest birds reached reproductive maturity before attaining somatic maturity (= cessation of growth). This pattern was reinforced by a paleohistologic study of four dinosaur taxa, in which reproductive maturity was established from the presence of bony tissues interpreted as medullary bone, which in some living birds serves as a calcium reserve for generating eggshell in gravid females (Lee & Werning, 2007). The concordance between these studies indicates that nonavian theropods, including paravians, retained the primitive reptilian pattern of reproducing before attainment of somatic maturity as opposed to the modern avian reproductive cycle in which somatic maturity is decoupled from and precedes reproductive maturity (Erickson *et al.*, 2007; Lee & Werning, 2007). Given that a number of primitive avian taxa including *Archeopteryx* (Erickson *et al.*, 2009), *Confuciusornis*

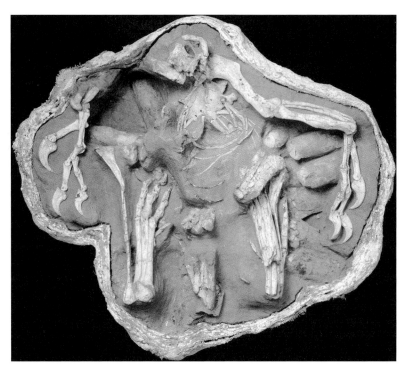

Fig. 1.4 Partial skeleton of an oviraptorosaur in brooding posture on a nest of its eggs. Egg identity has been independently confirmed through embryonic remains. Specimens such as this reveal that these dinosaurs laid eggs in pairs over protracted periods (diachronous laying), and brooded them with direct contact indicative of synchronous hatching. Such associations of eggs and sexually mature individuals are now known from multiple nonavian maniraptoran taxa. (Photograph M. Ellison/AMNH) [This figure appears in color as Plate 1.4.]

(Chiappe *et al.*, 2008), and *Patagopteryx* (Chinsamy *et al.*, 1994) reveal cyclical growth patterns and multiple age classes like nonavian dinosaurs, but unlike living birds which grow to maturity very rapidly, the decoupling between growth rate and reproductive maturity likely occurred later in avian evolution.

Living birds exhibit a relatively complex set of reproductive adaptations and behaviors relative to other reptiles (Varricchio *et al.*, 1997; Varricchio & Jackson, 2004). Although dinosaur eggs and nests have been known for well over a century, and correctly recognized since 1923, remarkable discoveries of dinosaur embryos or adults associated with nests or eggs represent intermediate stages in the evolution of the uniquely avian mode of reproduction. While most nonavian dinosaurs exhibit clutches comparable to those of crocodilians in terms of egg numbers and individual egg volumes, at least some maniraptoran taxa have significantly larger egg volumes (Varricchio & Jackson, 2004) indicating a shift toward the derived avian condition in this important parameter. Although clutch sizes for well preserved nests of the nonavian maniraptorans *Troodon* and *Citipati* scale according to the same equations as in living birds, individual eggs are about half the volume of an extant avian egg for an animal scaled to corresponding size. The eggs of these taxa are arranged in pairs within the nest, demonstrating that nonavian theropods still retained two functional oviducts, rather than the single oviduct of extant avians (Varricchio *et al.*, 1997; Clark *et al.*, 1999; Sato *et al.*, 2005). The dimensions of the pubic canal in basal birds such as *Archaeopteryx* and *Confuciusornis*, which retain a fused contact between the pubes distally, compares more favorably with those of nonavian theropods of similar size, rather than to the unfused and expanded pelvic canals of more derived birds. This suggests that individual egg volume was smaller in these ancestral birds than in modern taxa, and that they may have retained two functional oviducts.

The relatively large volume of nonavian maniraptoran clutches compared to body size (Varricchio & Jackson, 2004) (Figure 1.4) precludes that all eggs were retained within the female and then deposited during a single laying event. Rather these animals must have laid eggs over a protracted period of time as living birds do, a conclusion supported by analysis of the orientation of egg pairs within individual nests. Coupled with evidence for a brooding posture in individuals atop nests (Figure 1.4) (Norell *et al.*, 1995; Varricchio *et al.*, 1997), and for synchronous stages of embryonic development within one clutch (Varricchio *et al.*, 2002), this provides compelling evidence that extinct maniraptorans exhibited synchronous hatching like their living relatives, but unlike more basal egg-laying reptiles.

Taken together, these findings demonstrate that birds inherited some components of their complex reproductive biology such as nest care/brooding, diachronous laying, and synchronous hatching from nonavian ancestors, whereas other components such as loss of one oviduct and concomitant increase in egg volume plus decoupling of somatic and sexual maturity occurred within the avian lineage itself. Much as with any of the other biological systems discussed here, avian reproduction is a mosaic of inherited traits many of which pre-date bird origins combined with others that post-date this event.

Brain evolution

Among amniotes, birds are characterized by large relative brain sizes (measured as an encephalization quotient (EQ) that takes the allometry of brain to body-mass scaling into account (Jerison, 1973)), with particularly enlarged optical lobes and cerebellum thought to be adaptations for neural control of flight. Despite popular misconceptions regarding the brain size of nonavian dinosaurs, theropods show a progressive increase in EQ throughout their evolutionary history and reconstructed EQ values from brain endocasts of various extinct maniraptorans approach those of the basal lineages of living birds (Dominguez Alonso *et al.*, 2004). Detailed three-dimensional examination of the brain of *Archaeopteryx* using computed X-ray tomography (Dominguez Alonso *et al.*, 2004) demonstrates that the brain of the basalmost avian taxon already possesses a bird-like architecture with a pronounced pontine flexure that displaces the hindbrain below the mid brain, and enlargements of features thought to be adaptations for enhanced neurosensory control of active flight in modern birds, such as enlarged optic lobes, a proportionately well developed cerebellum, and an enlarged inner ear with expanded semicircular canals set in a modern avian configuration. Some combination, though not all, of these traits have also been observed in nonavian coelurosaurs such as troodontids (Norell *et al.*, 2009), various oviraptorosaurs (Balanoff *et al.*, 2009), and ornithomimosaurs (Balanoff *et al.*, 2009). Parallel trends in relative brain size evolution in birds and other maniraptoran lineages such as oviraptorosaurs have been noted, but there is little doubt that an elevated EQ and expansion of certain parts of the brain in birds was inherited from a more distant maniraptoran ancestor. Thus, with regard to the evolution of the unique avian brain, phylogeny again demonstrates how highly derived avian traits were acquired in stepwise fashion throughout theropod evolutionary history.

ARE BIRD ORIGINS STILL CONTROVERSIAL?

While widely accepted by biologists and paleontologists, the theropod ancestry of birds is not without its critics. The hypothesis has been challenged by a vocal, if small, opposition who have pointed to a number of perceived inconsistencies in the theropod ancestry of birds. In general, their challenges fall into several categories, which include disagreements over homology of various traits and structures, the seeming 'temporal paradox' in which most nonavian maniraptorans post-date *Archaeopteryx*, and inconsistency between inferred theropod paleobiology and preferred scenarios of how some aspect of avian evolution (often involving hypothetical intermediate forms) must have progressed. Indeed, most of these challenges rely on a combination of all three categories of arguments. A number of traits shared by birds and various subsets of dinosaurs and theropods, such as a broad ascending process of the astragalus (Martin, 1991), the presence of a furcula (Feduccia &, Martin 1998), homology of the wrist and digit elements of the forelimb, and even the presence of thecodont dentition in various theropods, have been challenged (Martin & Stewart, 1999). Many of these assertions, such as whether a furcula is present in nonavian theropods or whether theropod teeth are truly thecodont are simply based on inaccurate observations such as that the interdental plates of some theropod taxa including *Archaeopteryx* (Elzanowski &

Wellnhofer, 1996) represent separate ossifications rather than being part of the dentary (Figure 1.5). Others depend on indefensible assumptions that structures be completely identical to qualify as homologues, such as Feduccia & Martin's (1998) claim that variations in interclavicular angle of the furcula between some birds and nonavian theropods represent evidence of separate evolutionary origins of these structures. Many such misconceptions have been disproven by the wealth of evidence amassed against them, but continue to be cited indefinitely by those favoring a nontheropod origin of birds.

Other challenges relating to the homology of structures such as digit and wrist identity (Burke & Fedduccia, 1997) and homology of the ascending process of the astragalus, conflate primary homology statements based on comparisons in fossils with embryological observations on a limited set of avian model taxa. For example, Martin & Stewart's (1985; see also Martin *et al.*, 1980) claim that the ascending process of the nonavian theropod astragalus is fundamentally different from the large spur of bone that emanates dorsally from the avian astragalus, because the former is termed a 'process of the astragalus' and the latter

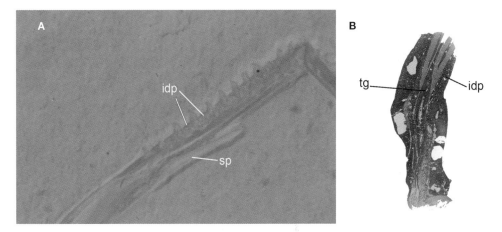

Fig. 1.5 (A) Lower jaws of the Munich specimen of *Archaeopteryx* revealing the presence of interdental plates. (B) Cross-section of the dentary of *Allosaurus* revealing continuous histological ultrastructure between the bone below the alveoli and the interdental plates and demonstrating that the latter are not separate ossifications. Abbreviations: idp, interdental plates; sp, splenial; tg, germinating tooth. Specimens not to scale. (Photographs P. Makovicky.) [This figure appears in color as Plate 1.5.]

derives from a distinct center of ossification during early embryology and is dubbed the 'pretibial bone', is based more on the polemics of how these structures are named rather than on relevant observations of their topological relationships. Since the embryology of nonavian theropods is unknown, it remains undeniable that for the life stages that can be compared across both living and fossil archosaurs, this tall, flat spur of bone that rises from the ankle along the front the tibia is present in virtually all birds and tetanurans, and is either significantly smaller or absent in more distantly related taxa (*contra* James & Pourtless, 2009).

Without question, the inconsistency between identifying avian digits by comparison to archosaur fossils versus identifying them through embryological studies of modern neornithines has been the greatest point of contention between the two opposing camps. In short, the dilemma is rooted in the fact that the tridactyl hand of *Archaeopteryx* exhibits a phalangeal formula and interelemental proportions that identify its digits as representing the first three fingers of the primitive pentadactyl amniote hand, whereas embryological studies (Burke & Feduccia, 1997; Feduccia & Nowicki, 2002; Larsson & Wagner, 2002) identify these digits as arising from the limb bud condensations that develop into digits II–IV in nonavian amniotes.

Interpreting the results of these two different methods for establishing digit identity at face value, detractors of the theropod origin of birds (Burke & Feduccia, 1997) conclude that the avian and nonavian theropod hands cannot be homologous despite the dozens of primary homologies in the shape, proportion, and number of wrist, hand, and finger bones, which they dismiss as convergence. Such a conclusion implicitly assumes a one-to-one correspondence between condensations in the developing limb bud and ossified adult structures, however, something that is untestable in fossils for which corresponding life stages are not preserved. The fundamental assumption of one-to-one correspondence has been challenged by experimental data that demonstrates considerable latent lability in the expression of chondrified digits from various primary condensations (Dahn

& Fallon, 2000; Wagner, 2005), and by the recent discovery of the basal ceratosaur *Limusaurus* (Xu *et al.*, 2009) which exhibits a reduced splint-like digit I and unusual phalangeal formula, demonstrating that theropod hand evolution is not as stereotypical as was once believed. With this assumption in doubt, the 'digital mismatch' can no longer be invoked to disqualify the numerous derived similarities of element shapes and proportions in hands of basal birds such as *Archaeopteryx* and other nonavian coelurosaurs. Moreover, these experimental results support novel models of how the embryology of the theropod hand evolved, most notably the Frame Shift Hypothesis (Wagner & Gauthier, 1999), which proposes a serial shift in digit identity between the embryonic primordia and chondrified digits over the course of development. Predictions of this hypothesis with respect to *Hox* gene expression patterns have been recently confirmed with chicken digit II exhibiting the digit I *Hox*-gene expression pattern of pentadactyl taxa such as mouse and alligator (Vargas *et al.*, 2008). Furthermore, criticisms that a wholesale frame shift affecting all digits in a limb is not documented in any other amniote taxon have recently been muted by the confirmation of a parallel case of a frame shift in the Italian three-toed skink *Chalcides* (Wagner, 2005; Young *et al.*, 2009).

Another mainstay of the opposition to the theropod ancestry of birds has been to point to a supposed 'temporal paradox' (Feduccia, 1996), namely the later occurrence in the fossil record of the coelurosaurian and maniraptoran sister clades to birds when compared to *Archaeopteryx*. While the argument as a whole is based on the mistaken assumption that taxa such as *Velociraptor* represent avian ancestors rather than sister taxa, and should therefore occur earlier in the fossil record, it has also been rendered moot by the recent discovery of several paravian taxa that pre-date *Archaeopteryx* (Xu & Zhang, 2005; Zhang *et al.*, 2008; Xu *et al.*, 2009).

A third persistent trend in the polemics surrounding avian origins has been the construction of scenarios circumscribing how a complex function such as avian flight or avian respiration evolved, followed by application of these scenarios as a "test" of the fossil record. The size discrepancy between basal birds and much larger paravians such as *Velociraptor* and *Deinonychus* was long cited as evidence that flight (and implicitly birds themselves) could not have evolved from such large and earthbound animals (Feduccia, 1996), but discoveries of small maniraptoran taxa of comparable size to *Archaeopteryx* (Turner *et al.*, 2007) and with possible arboreal traits (Xu *et al.*, 2001) have erased this argument. Similar evidentiary concerns apply to other scenarios based on the incorrect projection of parameters of the anatomy and physiology of living birds onto distant fossil ancestors that have been summoned against a theropod ancestry of birds. For example, Ruben and colleagues (Ruben *et al.*, 1997) attempted to argue that theropods had a crocodylian hepatic piston pump style of breathing based on interpretation of discolorations inside the body cavity of two exceptionally preserved compsognathid specimens as defining the limits of a large liver subdividing the thoracic cavity. However, in both cases the limits of these discolorations have been demonstrated to be preservational artifacts (Currie & Chen, 2001).

With the notable exception of a recent paper by James & Pourtless (2009), none of these challenges to the bird-theropod hypothesis have been set within a modern phylogenetic context and all have relied on selectively picking certain traits, specimens, and observations while ignoring others to construct narrative, scenario-laden attacks on the theropod ancestry of birds. They contribute little to the overall understanding of how the derived avian body plan evolved and have generally offered few alternatives for avian ancestry, usually positing some vague, paraphyletic assemblage of small bodied Triassic reptiles as possible avian sister groups (Feduccia, 1996), or arguing for a close relationship between birds and crocodylomorphs based on select dental and cranial traits that have a homoplastic distribution.

James & Pourtless (2009) recently presented a detailed phylogenetic analysis to challenge the premise of whether such analyses unequivocally support a theropod ancestry for birds. While this analysis certainly represents a step forward in the debate, their effort is deeply flawed on a number of counts. For one, they based their analysis on a now outdated dataset developed to examine generic-level interrelationships of coelurosaurs (Clark *et al.*, 2002), and thus focused on traits uniting various maniraptoran genera to one another, rather than on the traits more broadly nesting birds within Theropoda. Citing many of the older challenges to synapomorphies that support the birds-as-theropods, they eliminate a number of relevant characters of the wrist, hand, and ankle while reinterpreting others to favor character interpretations put forth by proponents of a nontheropod ancestry of birds, sometimes in illogical fashion. For example James & Pourtless (2009) go to great length to defend Martin and collegues' (1998) hypothesis that the hypocleideum of enanationithines has a distinct embryological identity from that of crown birds and proceed to redefine the relevant character definition, yet they never provide any insight on how such developmental distinctions are to be made on fully ossified structures in fossil specimens. To this decimated dataset they add a broad, but skewed, sample of more basal theropods, crocodilians, and the enigmatic and poorly preserved Triassic fossil *Longisquama,* without a correspondingly sufficient increase in character sampling to accurately test the relationships of the diversity of added taxa. Critically, they omit any basal crurotarsan taxa or other Triassic archosauriforms necessary to properly evaluate the phylogenetic affinities of either crocodylomorphs or the enigmatic Triassic fossil *Longisquama,* despite the fact that taxon sampling has been recognized as a key parameter for achieving accuracy in phylogenetic analysis (Poe, 1998; Graybeal, 1998). Remarkably, despite such manipulations, inadequate taxon and character sampling, and mistakes in data scoring (e.g. James & Pourtless' (2009) statement that interdental plates are absent in *Archeopteryx;* Figure 1.5A), the primary signal of the original data set examining the position of birds within maniraptoran dinosaurs remains largely intact, attesting to its robustness.

To date, no credible alternative to the theropod ancestry of birds enjoys much support from the fossil record. Although, new fossil discoveries offer the potential to challenge existing hypotheses of relationships, when it comes to bird origins, such discoveries have only served to further strengthen the theropod origin hypothesis through novel synapomorphies (e.g. wishbones, wrist anatomy, feathers), reduction of gaps in the fossil record, and bridging gaps in parameters such as body size. In contrast, opposing views have not been able to muster any new fossil taxa in support of alternative hypotheses (vague as these have been; Prum, 2002) in the past 25 years, relying instead on reinterpretations of a handful of fossils that are either so poorly preserved (e.g. *Longisquama*) or whose identity is so contested (e.g. *Protoavis*) that consensus on their anatomy and affinities is lacking.

CONCLUSIONS

Birds represent the most speciose and widespread clade of amniotes and are characterized by unique locomotory and physiological adaptations, which have long fascinated humans and been the focus of intense evolutionary and ecological research. To fully understand these aspects of avian biology we need to comprehend the origin of birds and their traits in a historical context. While avian origins remained unresolved for most of the first century that followed Darwin's publication of "*On the Origin of Species*", a string of discoveries of small- to medium-sized theropod dinosaurs since the 1960s has identified an inordinate number of derived characters shared with birds.

Today there is little debate over the theropod ancestry of birds. The accelerated rate of discovery and description of well-preserved Mesozoic theropods over the past decade and a half has not only strengthened this hypothesis, but has also immeasurably improved our knowledge of how avian anatomy and biology evolved.

We now know that some avian hallmarks such as the wishbone and skeletal pneumaticity have much deeper origins near the base of the theropod radiation, while others, such as feathers, evolved closer to birds but still characterize a more inclusive group of theropods. These discoveries also allow inferences on which derived physiological and behavioral traits of modern birds evolved before the avian lineage itself, and which ones came later. Paleohistological data suggest that theropods had elevated basic metabolic rates over those of living ectotherms, an inference corroborated by the presence of feather homologues in this clade, and by the growth rates approaching those of metatherian mammals and basal avian lineages (Erickson *et al.*, 2001). Nevertheless, extremely high modern avian growth rates and decoupling between somatic and reproductive maturity evolved much later within the avian lineage itself. In similar fashion, birds inherited proportionately large brain sizes from their coelurosaurian ancestors, but the evolutionary trend toward increased brain sizes continued within the avian lineage such that modern birds generally have larger encephalization quotient (EQ) values than nonavian theropods.

The continuing pace of discovery as well as technological advances in paleomolecular biology, CT scanning, and biogeochemistry hold great promise for future research surrounding the origin of birds. While our understanding of this important branching point in the tree of life has made a quantum leap in recent years, much still remains to be discovered about the earliest chapters in the evolution of birds and their biology.

ACKNOWLEDGMENTS

We thank the editors G. Dyke and G. Kaiser for inviting this chapter and for their patience in receiving it. Colleagues across the globe, including Mark Norell, Carl Mehling, Xu Xing, Ricardo Martinez, Oscar Alcober, Angela Milner, Alejandro Kramarz, Oliver Rauhut, Fernando Novas, Rodolfo Coria, Diego Pol, and Oliver Hampe provided access to specimens and insights that helped shape this manuscript. M. Donnelly executed the reconstruction in Figure 1.1. Research that contributed to this chapter was supported by the U.S. National Science Foundation grant EAR 0228607.

REFERENCES

Alonso PD, Milner AC, Ketcham RA, Cookson MJ, Rowe TB. 2004. The avian nature of the brain and inner ear of *Archaeopteryx*. *Nature* **430**: 666–669.

Balanoff AM, Xu X, Kobayashi Y, Matsufune Y, Norell MA. 2009. Cranial osteology of the theropod dinosaur *Incisivosaurus gauthieri* (Theropoda: Oviraptoro-sauria). *American Museum Novitates* **3651**: 1–35.

Barsbold R. 1983. Carnivorous dinosaurs from the Cretaceous of Mongolia. *Joint Soviet–Mongolian Paleontological Expedition Transactions* **19**: 5–120. (In Russian)

Benton MJ. 1990. Phylogeny of the major tetrapod groups - morphological data and divergence dates. *Journal of Molecular Evolution* **30**: 409–424.

Benton MJ. 2008. How to find a dinosaur, the role of synonymy in biodiversity studies. *Paleobiology* **34**: 516–533.

Bonaparte JF, Novas FE, Coria RA. 1990. *Carnotaurus satrei* Bonaparte, the horned, lightly built carnosaur from the Middle Cretaceous of Patagonia. *Natural History Museum of Los Angeles County, Contributions in Science* **416**: 1–41.

Britt B. 1997. Postcranial pneumaticity. In *The Encyclopedia of Dinosaurs*, Currie PJ, Padian K (eds). San Diego: Academic Press; 590–598.

Britt BB, Makovicky PJ, Gauthier J, Bonde N. 1998. Postcranial pneumatization in *Archaeopteryx*. *Nature* **395**: 374–376.

Brown B, Schlaikjer EM. 1940. The structure and relationships of *Protoceratops*. *Annals of the New York Academy of Sciences* **40**: 133–266.

Burke AC, Feduccia A. 1997. Developmental patterns and the identification of homologies in the avian hand. *Science* **278**: 666–668.

Chiappe LM, Norell MA, Clark JM. 1998. The skull of a relative of the stem-group bird *Mononykus*. *Nature* **392**: 275–278.

Chinsamy A, Chiappe LM, Dodson P. 1994. Growth rings in Mesozoic birds. *Nature* **368**: 196–197.

Chinsamy-Turan A, Hillenius WJ. 2004. Physiology of non-avian dinosaurs. In *The Dinosauria*, Weishampel DB, Dodson P, Osmolska H (eds). Berkeley: University of California; 643–659.

Chure DJ, Madsen JH. 1996. On the presence of furculae in some non-maniraptoran theropods. *Journal of Vertebrate Paleontology* **16**: 573–577.

Clark JM, Norell MA, Chiappe LM. 1999. An oviraptorid skeleton from the Late Cretaceous of Ukhaa Tolgod, Mongolia, preserved in an avianl-like brooding position over an oviraptorid nest. *American Museum Novitates* **3265**: 1–36.

Clark JM, Norell M, Makovicky P. 2002. Cladistic approaches to the relationships of birds to other theropods. In *Mesozoic Birds: Above the Heads of Dinosaurs*, Chiappe LM, Witmer LD (eds). Berkeley: University of California Press; 31–61.

Currie PJ, Chen PJ. 2001. Anatomy of *Sinosauropteryx prima* from Liaoning, northeastern China. *Canadian Journal of Earth Sciences* **38**: 1705–1727.

Dahn RD, Fallon JF. 2000. Digital identity is regulated by interdigital bmp signaling. *Developmental Dynamics* **219**: 8.

Dollo LAMJ. 1893. Les lois d'Evolution. *Bulletin de la Société Belge de Géologie, de Páleontologieet d'Hydrologie, Mémoire* **7**: 164–166.

Domínguez Alonso P, Milner AC, Ketcham RA, Cookson MJ, Rowe TB. 2004. The avian nature of the brain and inner ear of Archaeopteryx. *Nature* **430**, 666–669. doi:10.1038/nature02706

Elzanowski A, Wellnhofer P. 1996. Cranial morphology of *Archaeopteryx*: Evidence from the seventh skeleton. *Journal of Vertebrate Paleontology* **16**: 81–94.

Erickson GM, Rogers KC, Yerby SA. 2001. Dinosaurian growth patterns and rapid avian growth rates. *Nature* **412**: 429–433.

Erickson GM, Curry-Rogers K, Varricchio DJ, Norell MA, Xu X. 2007. Growth patterns in brooding dinosaurs reveals the timing of sexual maturity in non-avian dinosaurs and genesis of the avian condition. *Biology Letters* **3**: 558–561.

Erickson GM, Rauhut OWM, Zhou Z, Turner AH, Inouye BD, Hu D, Norell MA. 2009. Was dinosaurian physiology inherited by birds? Reconciling slow growth in *Archaeopteryx*. *PLoS ONE* **4**(10): e7390. doi:10.1371/journal.pone.0007390

Feduccia A. 1996. *The Origin and Evolution of Birds*. New Haven: Yale University Press.

Feduccia A, Martin LD. 1998. Theropod-bird link reconsidered. *Nature* **391**: 754.

Feduccia A, Nowicki J. 2002. The hand of birds revealed by early ostrich embryos. *Naturwissenschaften* **89**: 391–393.

Feduccia A, Lingham-Soliar T, Hinchcliffe JR. 2005. Do feathered dinosaurs exist? Testing the hypothesis on neontological and paleontological evidence. *Journal of Morphology* **266**: 125–166.

Gauthier JA. 1986. Saurischian monophyly and the origin of birds. In *The Origin of Birds and Evolution of Flight*, Padian K (ed.). San Francisco: California Academy of Sciences Memoir; 1–55.

Grellet-Tinner G, Makovicky P. 2006. A possible egg of the dromaeosaur *Deinonychus antirrhopus*: phylogenetic and biological implications: Canadian. *Journal of Earth Sciences* **43**: 705–719.

Heilman G. 1926. *The Origin of Birds*. London: Witherby; 210 pp.

Holtz TR. 1994. The phylogenetic position of the Tyrannosauridae – implications for theropod systematics. *Journal of Paleontology* **68**: 1100–1117.

Hu DY, Hou LH, Zhang LJ, Xu X. 2009. A pre-*Archaeopteryx* troodontid theropod from China with long feathers on the metatarsus. *Nature* **461**: 640–643.

Huxley TH. 1868. On the animals which are most nearly intermediate between birds and reptiles. *Geological Magazine* **5**: 357–365.

James FC, Pourtless IV, JA. 2009. Cladistics and the origin of birds: a review and two new analyses. *Ornithological Monographs* **66**: 1–78.

Jerison HJ. 1973. *Evolution of the Brain and Intelligence*. New York, NY: Academic Press.

Ji Q, Norell MA, Gao KQ, Ji SA, Ren D. 2001. The distribution of integumentary structures in a feathered dinosaur. *Nature* **410**: 1084–1088.

Kirkland JI, Zanno LE, Sampson SD, Clark JM, DeBlieux DD. 2005. A primitive therizinosauroid dinosaur from the Early Cretaceous of Utah. *Nature* **435**: 84–87.

Kobayashi Y, Lu JC, Dong ZM, Barsbold R, Azuma Y, Tomida Y. 1999. Palaeobiology – herbivorous diet in an ornithomimid dinosaur: *Nature* **402**: 480–481.

Larsson HCE, Wagner GP. 2002. Pentadactyl ground state of the avian wing. *Journal of Experimental Zoology* **294**: 146–151.

Lee AH, Werning S. 2008. Sexual maturity in growing dinosaurs does not fit reptilian model. *Proceedings of the National Academy of Sciences* **105**: 582–587.

Li Q, Gao K-Q, Vinther J, Shawkey MD, Clarke JA, D'Alba L, Meng Q, Briggs EEG, Prum RO. 2010. Plumage color patterns of an extinct dinosaur. *Science* **327**: 1369–1372.

Lingham-Soliar T, Feduccia A, Wang XL. 2007. A new Chinese specimen indicates that 'protofeathers' in the Early Cretaceous theropod dinosaur *Sinosauropteryx* are degraded collagen fibres. *Proceedings of the Royal Society of London Series B (Biological Sciences)* **274**: 1823–1829.

Longrich N. 2006. Structure and function of hindlimb feathers in *Archaeopteryx lithographica*. *Paleobiology* **32**: 417–431.

Makovicky PJ, Currie PJ. 1998. The presence of a furcula in tyrannosaurid theropods, its phylogenetic and functional implications. *Journal of Vertebrate Paleontology* **18**: 143–149.

Makovicky PJ, Apesteguía S, Agnolin FL. 2005. The earliest dromaeosaurid theropod from South America. *Nature* **437**: 1007–1011.

Martin LD. 1991. Mesozoic birds and the origin of birds. In *Origins of the Higher Groups of Tetrapods, Controversy and Consensus*. Schultze H-P, Trueb L (eds). Ithaca: Cornell University Press; 485–540.

Martin LD, Stewart JD. 1985. Homologies in the avian tarsus. *Nature* **315**: 159.

Martin LD, Stewart JD. 1999. Implantation and replacement of bird teeth. *Smithsonian Contributions to Paleobiology* **89**: 295–300.

Martin LD, Stewart JD, Whetstone KN. 1980. The origin of birds: structure of the tarsus and teeth. *Auk* **97**: 86–93.

Martin LD, Zhou Z, Hou L, Feduccia A. 1998. *Confuciusornis sanctus* compared to *Archaeopteryx lithographica*. *Naturwissenschaften* **85**: 286–289.

Maryanska T, Osmolska H, Wolsan M. 2002. Avialan status for Oviraptorosauria. *Acta Palaeontologica Polonica* **47**: 97–116.

Nesbitt SJ, Smith ND, Irmis RB, Turner AH, Downs AM, Norell A. 2009a. A complete skeleton of a Late Triassic saurischian and the early evolution of dinosaurs. *Science* **326**: 1530–1533.

Nesbitt SJ, Turner AH, Spaulding M, Conrad JL, Norell MA. 2009b. The theropod furcula. *Journal of Morphology* **270**: 856–879.

Norell MA, Makovicky PJ. 1999. Important features of the dromaeosaurid skeleton II: Information from newly collected specimens of *Velociraptor mongoliensis*. *American Museum Novitates* **3282**: 1–45.

Norell MA, Xu X. 2005. Feathered dinosaurs. *Annual Review of Earth and Planetary Sciences* **33**: 277–299.

Norell MA, Clark JM, Chiappe LM, Dashzeveg D. 1995. A nesting dinosaur. *Nature* **378**: 774–776.

Norell MA, Makovicky PJ, Clark JM. 2001. Relationships among Maniraptora: problems and prospects. In *New Perspectives on the Origin and Evolution of Birds: Proceedings of the International Symposium in Honor of John H. Ostrom*, Gauthier JA, Gall LF (eds). New Haven: Yale University Press; 49–68.

Norell MA, Makovicky PJ, Bever GS, Balanoff AM, Clark JM, Barsbold R, Rowe T. 2009. A review of the Mongolian Cretaceous dinosaur *Saurornithoides* (Troodontidae: Theropoda). *American Museum Novitates* **3654**: 1–63.

Novas FE, Pol D. 2002. Alvarezsaurid relationships reconsidered. In *Mesozoic Birds: Above the Heads of Dinosaurs*, Chiappe LM, Witmer LM (eds). Berkeley: University of California Press; 121–124.

Novas FE, Pol D. 2005. New evidence on deinonychosaurian dinosaurs from the Late Cretaceous of Patagonia. *Nature* **433**: 858–861.

Novas FE, Puerta PF. 1997. New evidence concerning avian origins from the Late Cretaceous of Patagonia. *Nature* **387**: 390–392.

O'Connor PM. 2009. Evolution of archosaurian body plans: skeletal adaptations of an air sac-based breathing apparatus in birds and other archosaurs. *Journal of Experimental Zoology, Part A – Ecological Genetics and Physiology* **311A**: 629–646.

O'Connor PM, Claessens LP. A. M. 2005. Basic avian pulmonary design and flow-through ventilation in non-avian theropod dinosaurs. *Nature* **436**: 253–256.

Organ CL, Shedlock AM, Meade A, Pagel M, Edwards SV. 2007. Origin of avian genome size and structure in non-avian dinosaurs. *Nature* **446**: 180–184.

Ostrom JH. 1969. Osteology of *Deinonychus antirrhopus*, an unusual theropod from the Lower Cretaceous of Montana. *Peabody Museum of Natural History, Yale University Bulletin* **30**: 1–163.

Ostrom JH. 1970. *Archaeopteryx* – notice of a new specimen. *Science* **170**: 537.

Ostrom JH. 1976. *Archeopteryx* and the origin of birds. *Biological Journal of the Linnean Society* **8**: 91–182.

Padian K, Horner JH. 2004. Dinosaur physiology. In *The Dinosauria*, Weishampel DB, Dodson P, Osmolska H (eds). Berkeley: University of California Press; 660–671.

Poe S. 1998. Sensitivity of phylogeny estimation to taxonomic sampling. *Systematic Biology* **47**: 18–31.

Perle A, Norell MA, Chiappe LM, Clark JM. 1993. Flightless Bird from the Cretaceous of Mongolia. *Nature* **362**: 623–626.

Prum RO. 2002. Why ornithologists should care about the theropod origin of birds: *Auk* **119**: 1–17.

Rauhut OWM. 2003. The interrelationships and evolution of basal theropod dinosaurs. *Special Papers in Palaeontology* **69**: 3–213.

Ruben JA, Jones TD, Geist NR, Hillenius WJ. 1997. Lung structure and ventilation in theropod dinosaurs and early birds. *Science* **278**: 1267–1270.

Sampson SD, Krause DW. 2007. *Majungasaurus crenatissimus* (Theropoda: Abelisauridae) from the Late Cretaceous of Madagascar – Preface. *Society of Vertebrate Paleontology Memoir* **27**: XIII–XIV.

Sampson SD, Witmer LM, Forster CA, Krause DW, O'Connor PM, Dodson P, Ravoavy F. 1998. Predatory dinosaur remains from Madagascar: Implications for the Cretaceous biogeography of Gondwana. *Science* **280**: 1048–1051.

Sampson SD, Carrano MT, Forster CA. 2001. A bizarre predatory dinosaur from the Late Cretaceous of Madagascar. *Nature* **409**: 504–506.

Sato T, Cheng YN, Wu XC, Zelenitsky DK, Hsiao YF. 2005. A pair of shelled eggs inside a female dinosaur. *Science* **308**: 375–375.

Schweitzer MH, Watt JA, Avci R, Knapp L, Chiappe LM, Norell MA, Marshall M. 1999. Beta-keratin specific immunological reactivity in feather-like structures of the Cretaceous alvarezsaurid, *Shuvuuia deserti*. *Journal of Experimental Zoology* **285**: 146–157.

Senter P. 2007. A new look at the phylogeny of Coelurosauria (Dinosauria: Theropoda). *Journal of Systematic Palaeontology* **5**: 429–463.

Sereno PC. 1999a. A rationale for dinosaurian taxonomy. *Journal of Vertebrate Paleontology* **19**: 788–790.

Sereno PC. 1999b. The evolution of dinosaurs. *Science* **284**: 2137–2147.

Sereno PC, Arcucci AR. 1993. Dinosaurian presursors from the Middle Triassic of Argentina: *Lagerpeton chanarensis*. *Journal of Vertebrate Paleontology* **13**: 385–399.

Sereno PC, Brusatte SL. 2008. Basal abelisaurid and carcharodontosaurid theropods from the lower Cretaceous Elrhaz Formation of Niger. *Acta Palaeontologica Polonica* **53**: 15–46.

Sereno PC, Dutheil DB, Iarochene M, Larsson HCE, Lyon GH, Magwene PM, Sidor CA, Varricchio DJ, Wilson JA. 1996. Predatory dinosaurs from the Sahara and Late Cretaceous faunal differentiation. *Science* **272**: 986–991.

Sereno PC, Wilson JA, Conrad JL. 2004. New dinosaurs link southern landmasses in the Mid-Cretaceous. *Proceedings of the Royal Society of London, Series B – Biological Sciences* **271**: 1325–1330.

Sereno PC, Martinez RN, Wilson JA, Varricchio DJ, Alcober OA, Larsson HCE. 2008. Evidence for avian intrathoracic air sacs in a new predatory dinosaur from Argentina. *PLoS ONE* **3** (9): e3303. doi:10.1371/journal.pone.0003303

Smith ND, Makovicky PJ, Hammer WR, Currie PJ. 2007. Osteology of *Cryolophosaurus ellioti* (Dinosauria:

Theropoda) from the Early Jurassic of Antarctica and implications for early theropod evolution. *Zoological Journal of the Linnean Society* **151**: 377–421.

Smith ND, Makovicky PJ, Agnolin FL, Ezcurra MD, Pais DF, Salisbury SW. 2008. A *Megaraptor*-like theropod (Dinosauria: Tetanurae) in Australia: support for faunal exchange across eastern and western Gondwana in the Mid-Cretaceous. *Proceedings of the Royal Society of London, Series B – Biological Sciences* **275**: 2085–2093.

Turner AH, Pol D, Clarke JA, Erickson GM, Norell MA. 2007. A basal dromaeosaurid and size evolution preceding avian flight. *Science* **317**: 1378–1381.

Vargas AO, Kohlsdorf T, Fallon JF, VandenBrooks J, Wagner GP. 2008. The evolution of HoxD-11 expression in the bird wing: insights from *Alligator mississippiensis*. *PLoS ONE* **3**(10): e3325. doi:10.1371/journal.pone.0003325

Varricchio DJ, Jackson FD. 2004. Two eggs sunny-side up: reproductive physiology in the dinosaur *Troodon formosus*. In *Feathered Dragons: Studies on the Transition from Dinosaurs to Birds*, Currie PJ, Koppelhus EB, Shugar MA, Wright JA (eds). Bloomington: Indiana University Press; 215–233.

Varricchio DJ, Jackson F, Borkowski JJ, Horner JR. 1997. Nest and egg clutches of the dinosaur *Troodon formosus* and the evolution of avian reproductive traits. *Nature* **385**: 247–250.

Varricchio DJ, Horner JR, Jackson FD. 2002. Embryos and eggs for the Cretaceous theropod dinosaur *Troodon formosus*. *Journal of Vertebrate Paleontology* **22**: 564–576.

Wagner GP. 2005. The developmental evolution of avian digit homology: an update. *Theory In Biosciences* **124**: 165–183.

Wagner GP, Gauthier JA. 1999. 1, 2, 3 = 2, 3, 4: A solution to the problem of the homology of the avian hand. *Proceedings of the National Academy of Sciences, USA* **96**: 5111–5116.

Weishampel DB, Dodson P, Osmolska H. (eds). 2004. *The Dinosauria*. Berkeley: University of California Press.

Witmer LM. 1995. The extant phylogenetic bracket and the importance of reconstructing soft tissues in fossils. In *Functional Morphology in Vertebrate Paleontology*, Thomason JJ (ed.). Cambridge: Cambridge University Press; 19–33.

Xu X, Norell MA. 2004. A new troodontid dinosaur from China with avian-like sleeping posture. *Nature* **431**: 838–841.

Xu X, Zhang FC. 2005. A new maniraptoran dinosaur from China with long feathers on the metatarsus. *Naturwissenschaften* **92**: 173–177.

Xu X, Zhou ZH, Wang XL. 2000. The smallest known non-avian theropod dinosaur. *Nature* **408**: 705–708.

Xu X, Zhou HH, Prum RO. 2001. Branched integumental structures in *Sinornithosaurus* and the origin of feathers. *Nature* **410**: 200–204.

Xu X, Norell MA, Wang XL, Makovicky PJ, Wu XC. 2002. A basal troodontid from the Early Cretaceous of China. *Nature* **415**: 780–784.

Xu X, Norell MA, Kuang XW, Wang XL, Zhao Q, Jia CK. 2004. Basal tyrannosauroids from China and evidence for protofeathers in tyrannosauroids. *Nature* **431**: 680–684.

Xu X, Clark JM, Forster CA, Norell MA, Erickson GM, Eberth DA, Jia CK, Zhao Q. 2006. A basal tyrannosauroid dinosaur from the Late Jurassic of China. *Nature* **439**: 715–718.

Xing X, Qi Z, Norell MA, Sullivan C, Hone DWE, Erickson GM, Wang X-L, Han F-L and Yu G. 2009. A new feathered maniraptoran dinosaur fossil that fills a morphological gap in avian evolution. *Chinese Science Bulletin* **54**: 430–435.

Xu X, Clark JM, Mo JY, Choiniere J, Forster CA, Erickson GM, Hone DWE, Sullivan C, Eberth DA, Nesbitt S, Zhao Q, Hernandez R, Jia CK, Han FL, Guo Y. 2009. A Jurassic ceratosaur from China helps clarify avian digital homologies. *Nature* **459**: 940–944.

Yates AM, Vasconcelos CC. 2005. Furcula-like clavicles in the prosauropod dinosaur *Massospondylus*. *Journal of Vertebrate Paleontology* **25**: 466–468.

Young RL, Caputo V, Giovannotti M, Kohlsdorf T, Vargas AO, May GE, Wagner GP. 2009. Evolution of digit identity in the three-toed Italian skink *Chalcides chalcides*: a new case of digit identity frame shift. *Evolution and Development* **11**: 647–658.

Zanno LE, Gillette DD, Albright LB, Titus AL. 2009. A new North American therizinosaurid and the role of herbivory in 'predatory' dinosaur evolution. *Proceedings of the Royal Society of London, Series B – Biological Sciences* **276**: 3505–3511.

Zhang FC, Zhou ZH, Xu X, Wang XL, Sullivan C. 2008. A bizarre Jurassic maniraptoran from China with elongate ribbon-like feathers. *Nature* **455**: 1105–1108.

Zhang F, Kearns S, Orr P, Benton M, Zhou Z, Johnson D, Xu X, Wang X. 2010. Fossilized melanososmes and the colour of Cretaceous dinosaurus and birds. *Nature* **463**: 1075–1078.

2   Why Were There Dinosaurs? Why Are There Birds?

PETER WARD[1] AND ROBERT BERNER[2]

[1]University of Washington, Seattle, USA
[2]Yale University, New Haven, USA

Astrobiologists commonly use the term "Earth-like Planet" in their search of the heavens for another potentially habitable planet nearby in the Milky Way Galaxy. But just what does "Earth-like" mean? This question is far more complex than it seems. If we define Earth-like as having oceans, continents, and atmosphere in modern quantities, it turns out that Earth-like qualities are quite new. Over time the continents have repeatedly changed position and size, and the oceans have risen and fallen in their basins based on the effects of gravity and tectonic changes within the ocean basins themselves. But these latter two aspects of the Earth show far less change than does the atmosphere. Our current atmosphere is but a slice of a forever changing entity, and greatly different from the atmosphere at most times in Earth's history. Since the composition of the atmosphere greatly affects planetary temperatures, it is probably safe to say that Earth life is well adapted to the current atmosphere. Yet should atmospheric conditions found at the end of the Permian and end of the Triassic Period suddenly exist in our current world, the result would be calamitous. Those two not-so-ancient versions of an Earth-like atmosphere very nearly wiped out our furry ancestors some 250 million years ago, and then again some 200 million years ago. Yet while it is well-known that the mass extinctions, of which the Permian and Triassic events are but two of many (albeit two of the five most catastrophic), there is far less discussion about the effects that "optimal" as well as dangerous levels of various atmospheric components may have had on Earth life.

In this chapter we will briefly summarize how one of the most important of atmospheric gases (and for all those organisms with aerobic respiration, the *most* important), oxygen, may have stimulated formation of the most efficient respiratory system ever evolved, and in so doing paved the way for the existence of birds.

OXYGEN THROUGH TIME

The first step in uncovering past atmospheric gas compositions comes from understanding the carbon cycle, which comes with a long-term component, the movement of carbon into, and out of rocks, and a shorter term cycle, the movement of carbon in and out of air, water, and organisms. It is the long-term cycle that most importantly affected the levels of oxygen and carbon dioxide over the past 500 million years. The most important processes affecting levels of oxygen have been the rates of organic carbon burial and weathering.

Burning organic carbon-rich minerals or compounds liberates greenhouse gases into the atmosphere, while burying these organic rich

Living Dinosaurs: The Evolutionary History of Modern Birds, First Edition. Edited by Gareth Dyke and Gary Kaiser.
© 2011 John Wiley & Sons, Ltd. Published 2011 by John Wiley & Sons, Ltd.

compounds has quite another consequence: it causes oxygen levels in the atmosphere to rise. While the process of photosynthesis is producing some oxygen, much of this free oxygen is then destroyed when it combines with the organic compounds and "oxidizes" them, by producing a new compound that has oxygen bound to it. Thus, in times when greater or lesser amounts of reduced carbon compounds are buried, atmospheric oxygen either goes up – or down. Just how much it goes up or down is just now being realized, and we theorize that the consequences of these rises and fall to the history of life have been immense and remain vastly underappreciated as a driver of evolutionary change and the resultant body plans on Earth at any given time.

Measuring the amount of oxygen now present in ancient sediments is easily accomplished. But measuring how much oxygen there was in the atmosphere when that sediment was laid down (or when some organism synthesized bone, leaf, or shell) is impossible. There is no magic analysis where some rock or fossil goes into a machine, and out rolls a viable number representing the amount of oxygen in the air of some chosen time long ago. Instead, the best that we can do is estimate the amount of oxygen in the atmosphere from models taking into account many factors, including the net accumulation (or weathering) of organic material. Given an overall increase in burial rates of organic (or reduced) carbon over time, such as coal, atmospheric oxygen levels would have risen. The inputs into these models include the aforementioned rates of burial and weathering of many kinds of rocks, as well as the movement within a second great elemental cycling system on Earth, the sulfur cycle. These and other values, either measured or estimated, have allowed development of various analytical computer models with the goal of arriving at credible estimates of O_2 levels at known times. The oldest of these models, called the "GEOCARB" program (Berner, 2004) estimates carbon dioxide, while a derivative of this model called GEOCARBSULF yields both oxygen and carbon dioxide estimates. Of interest here, the GEOCARBSULF model calculates O_2 through time. A diagram showing the various inputs and outputs is shown in Figure 2.1.

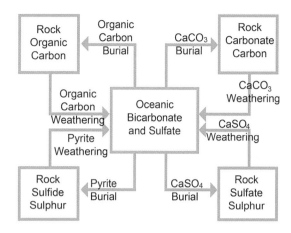

Fig. 2.1 Cartoon to show long-term global cycles of carbon and sulfur over time.

The results of GEOCARBSULF themselves evolved through time as ever better estimates and measurements of the many inputs into the model have been discovered. The latest estimate prior to our writing this chapter was produced in mid-2009, and is shown in Figure 2.2. The most striking aspect of this curve is the large "spike' in oxygen levels about 300 million years ago (Ma), and the nadirs both before and after this oxygen

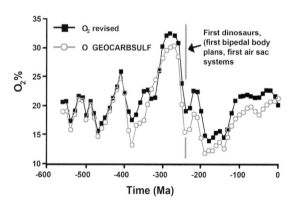

Fig. 2.2 Oxygen through time – two different estimates from GEOCARBSULF model outputs (Berner, 2009). The arrow approximates the time of the first dinosaur appearances: 230 million years ago in Madagascar, and 227 million years ago in South America.

maxima. It is these three points on the graph that may have had the greatest effect on the history of life. All three are relevant to the discussion of why dinosaurs with their particular body plans evolved.

The first of the great nadirs in oxygen levels occurred about 360 Ma, a time when the Devonian mass extinction, one of the five most devastating of all mass extinctions, was playing out. This event may even have been caused by the combination of low oxygen and high global temperatures in the Givetian Stage of the Devonian Period, based on new research by graduate student Kelly Hillbun of the University of Washington. Its relevance to the bird story comes from the effects this downturn may have had on land vertebrate life. We have previously shown (Ward *et al.*, 2006) that the "conquest" of land by early vertebrates and arthropods was stymied by falling oxygen levels in the Devonian Period. Perhaps originally instigated by what appears to have been the 400 Ma spike in O_2 during the Silurian and Early Devonian Periods to levels as much as 26% (compared to the modern day level of 21%), the falling oxygen level beginning in the Middle Devonian was relatively rapid and catastrophic, bottoming out at levels of as little as 13%, or half the levels of only some few millions of years before this bottom. The effect on life was catastrophic (>50% extinction of species) and resulted in what came to be known as "Romer's Gap", a period during the oxygen low when few fossils of either arthropods or land vertebrates are known. Re-diversification only came about tens of millions of years later, in the Carboniferous Period, and once again this diversification appears to have ridden on the wings of rising oxygen. This is the second of the most noticeable aspects of the GEOCARBSULF output.

The oxygen high of the Carboniferous had ramifications for terrestrial vertebrates that continue into the present day. It looks as if most clades of amphibians and stem reptiles evolved either immediately before or during the pronounced oxygen high. While data on this are still few, the work of Vandenbrooks (2006) demonstrated that alligators show optimal development when oxygen in rearing chambers was raised to 27%. Both higher and lower O_2 levels slowed development. But this time of high oxygen did not last. By the last half of the Permian, oxygen levels began to plummet, with their second major minimum occurring in the Triassic, or perhaps twice in the Triassic, with a slight Mid-Triassic rise in between. In any event, the Late Triassic was a time with low oxygen compared to the modern day. It was also the time when two of the most successful of all clades, the dinosaurs and true mammals, first appear in the record. They appear at the third of the great inflections in the Berner oxygen curve: the late Triassic. The currently oldest known dinosaur fossils come from Madagascar strata dated to 230 Ma. The previously oldest known, and still very old for dinosaurs, was from Argentina, where 227 Ma dinosaur fossils have been found.

RESPIRATION AND BIPEDALISM

The fossil record shows that the earliest true dinosaurs were bipedal, and came from more primitive bipedal thecodonts slightly earlier in the Triassic. Thecodonts (diapsids) were the ancestors of the lineage giving rise to the crocodiles as well, and may have been either warm blooded or heading in that way. We see bipedalism as a recurring body plan in this group, and there were even bipedal crocodiles early on. Why bipedalism, and how could it have been an adaptation to low oxygen? Our view is that the initial dinosaur body plan, of a gracile body, long tail and bipedalism evolved as a response to the problem that many four legged lizards encounter – it is more difficult to breath while running because of the position of the four legs, and the swinging side to side motion of the thorax during locomotion in quadrupeds compresses the lung on one side. This is known as Carrier's constraint, and surely it affected extinct as well as extant species with the splayed leg body plan. In our view, bipedalism evolved as a response to the low oxygen in the Middle–Late Triassic. With a bipedal stance the first dinosaurs overcame the respiratory limitations imposed by Carrier's constraint. The Triassic oxygen low thus triggered the origin of dinosaurs through formation of this new body plan.

Modern day mammals show a distinct rhythm by synchronizing breath–taking with limb movement. Horses, jackrabbits, and cheetahs (among many other mammals) take one breath per stride. Their limbs are located directly beneath the mass of the body, and to allow this the backbone in these quadruped mammals has been enormously stiffened compared to the backbones of the sprawling reptiles. The mammalian backbone slightly bows downward and then straightens out with running, and this slight up and down bowing is coordinated with air inspiration and exhalation. But this system did not appear until true mammals appeared, in the Triassic. Even the most advanced cynodonts of the Triassic were not yet fully upright, and thus would have suffered somewhat when trying to run and breath. Obviously, all birds, not just the obligate ground runners, suffer from Carrier'a constraint.

By running on two legs instead of four, the lungs and rib cage are not affected. Breathing can be disassociated from locomotion – the bipeds can take as many breaths as they need to in a high-speed chase. At a time of low oxygen, but high predation, any slight advantage either in chasing down prey, or running from predators – even in the amount of time looking for food, and how food is looked for, would have surely increased survival.

Had bipedalism and the respiratory advantage given by this new body been the only possible adaptation to low oxygen, it would be hard to make the case that the dinosaur body plan was related to oxygen levels. But the dinosaurs evolved a second, and even more advantageous adaptation to low ambient oxygen – what is now called the avian air sac system, first invented by the avian precursors, the dinosaurs.

Air sacs and pneumatized bones

Birds, like reptiles, have septate lungs that are small and somewhat rigid. Thus bird lungs do not greatly expand and contract as ours do on each breath. But the rib cage is very much involved in respiration, and especially those ribs closest to

the pelvic region are very mobile in their connection to the bottom of the sternum, and this mobility is quite important in allowing respiration. But these are not the biggest differences. Very much *unlike* extant reptiles and mammals, these lungs have appendages added to them known as air sacs, and the resultant system of respiration is highly efficient. When a bird inspires air, it goes first into the series of air sacs. It then passes into the lung tissue proper, but in so doing the air passes but one way over the lung, since it is not coming down a trachea but from the attached air sacs. Exhaled air then passes out of the lungs. The one way flow of air across the lung membranes allows a counter current system to be set up – the air passes in one direction, and blood in the blood vessels within the lungs passes in the opposite direction. This countercurrent exchange allows for more efficient oxygen extraction and carbon dioxide venting than are possible in dead-end lungs.

In Figure 2.3 the various air sacs are shown with their communication to the lungs. It is clear that the volume of air sacs far exceeds the volume of the lungs themselves. The air sacs are not involved in removing oxygen; they are an adaptation that allows the countercurrent system to work.

There is no question that the greater efficiency of this system compared to all other lungs in vertebrates is related to the two-cycle, counter current system produced by the air-sac–lung anatomy in birds as evidenced by the well known, higher tolerance to hypoxia of birds compared to mammals and reptiles: birds cannot only exist but can undertake the costly exercise of flying to at least 33,000 feet (bar headed goose, observed above the Himalaya Mountains by high flying airplanes), where there are oxygen levels that are fatal to mammals. It has been estimated that a bird is 33% more efficient in extracting oxygen from air than a mammal, at sea level. But at higher altitude this differential increases: a bird at 5000 feet in altitude may be 200% more efficient at extracting oxygen than a mammal. This gives the birds a huge advantage over mammals at living at altitude. And if such a system were present deep in the past, when oxygen even at sea level was lower than we find today at 5000 feet, surely such a

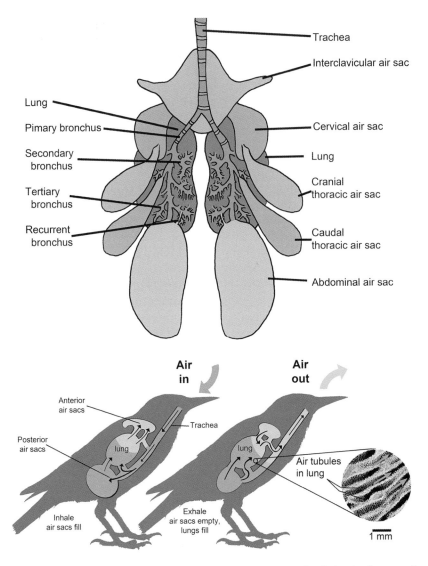

Fig. 2.3 The air sac system in the lungs of birds (above) and respiratory cycles (below). The cranial, cervical, and vertebral air sac positions are based on new data, while the caudal air sacs are partially encased in bones of the bird, thus creating "pneumatized" bones. On the first inspiration, air is taken into the air sacs. It then moves to the lungs on the second inspiration.

design would have been advantageous – perhaps enormously so – to the group that had it in competing or preying on groups that did not.

But when did this system first appear, and in how many groups? We know that birds evolved from small bipedal dinosaurs that were of the same lineage as the earliest dinosaurs – a group called Saurischians (Chapter 1). The first bird skeletons come from the Jurassic. But the air sacs attached to bird lungs are soft tissue, and would fossilize only under the most unusual circumstances of preservation. Thus we do not have direct evidence for when the air sac system came about. But we do have indirect evidence, enough to have stimulated the "air-sac in dinosaurs group" to posit that *all* saurischian dinosaurs had the same air sac system as do modern birds. The evidence for an air sac system in ancient organisms can be deduced entirely from well-preserved skeletons. The evidence is the presence of:

1 pneumatic bones, especially in the vertebral column;
2 shortening of the trunk of the body;
3 shortening of the first dorsal ribs;
4 elongation and increased mobility of posterior ribs – this mobility is enabled by the presence of ribs with double heads at their ends;
5 uncinate processes on several of the ribs (these are small, hook-shaped bones attached to the trunk ribs);
6 a hinge joint making up the attachment of the ribs with the sternum.

THE EVOLUTION OF THE AIR SAC SYSTEM

The complex air-sac–lung system found in birds had to have evolved from a reptilian, sac-like lung, and all evidence indicates that this happened in the middle Triassic. The most primitive theropods from this time (the first dinosaurs) do not show bone pneumatization, but their ribs became double headed, showing that the rib cage itself was capable of a great ventilation capacity. Perhaps as a consequence of going bipedal, these dinosaurs may have switched from the more primitive abdominal pump system to the first air sac system – one with only the abdominal air sac found in modern day birds. Soon after, descendents of these first dinosaurs, forms such as the well known, Late Triassic *Coelophysis*, show the evidence of the bone pneumatization, consistent with the proposal that more air sacs had evolved, this time those in the neck region. With the Jurassic forms such as *Allosaurus*, the air sac system may have been essentially complete (but still much different from the bird system, modified as it has been for flying, for even the modern day flightless birds came from flyers in

the deep past), with large thoracic and abdominal air sacs.

By the time that *Archaeopteryx* had evolved in the middle part of the Jurassic, there may have been a great diversity of respiratory types among the dinosaurs, some with pneumatized bones, some without. There also may have been a great deal of convergent evolution going on. For instance, the extensive pneumatization in the large sauropods studied with such care by Wedel may have arisen somewhat independently from the saurischian carnivores (Wedel, 2003)

Perhaps the greatest contribution to the subject of air sacs both modern and ancient was the seminal paper by O'Connor & Claessens (2005). They pointed out that the entire community has misunderstood which air sacs penetrate which bones in birds. By injecting liquid rubber into the air sacs of modern birds, they showed that the sacral, or tail ward parts of the body are most important in producing the characteristic "bird breathing" pattern. They also showed a great variability in modern bird pneumaticity, with diving birds showing different patterns from small flyers, and different again among ground dwellers and birds of large size. With this new and improved understanding of the size, morphology, and positions of pneumatized bones and their attendant air sacs, O'Connor & Claessens (2005) then showed evidence of nearly identical holes in homologous bones of found in saurischian dinosaurs.

Why would this the air sac system have evolved? First, metabolism. If we accept that at least some dinosaurs were warm-blooded, then it is immediately apparent that they would have needed of highly efficient oxygenation mechanisms; and this would be even more important in a low oxygen world. All modern (cold blooded) reptiles have to warm up at the start of the day, and thus there is little early morning activity other than behavioral movement to acquire heat from the external environment. If the first bipedal dinosaurs – all predators – did not have to do this, they would have been able to forage freely on the slower ectotherms in the morning or nighttime hours. But what is the price paid for this? At rest, all endotherms use as much as 15 times the amount of oxygen as do ectotherms (there is a 5–15 times range based on experimental observation). In our oxygen-rich world this is not a problem for the warm-blooded animals. There is so much oxygen available that there is no penalty. But in the oxygen-poor Late Triassic and into the Jurassic, such was surely not the case. And the energy and oxygen necessary for endothermy would not have been necessary if the dinosaurs moved toward large size. With larger body size the ratio of surface area (from which heat is lost) to body volume becomes increasingly favorable.

DINOSAUR SURVIVORSHIP ACROSS THE TRIASSIC–JURASSIC BOUNDARY

The end Triassic mass extinction was one of the so called "Big Five": one of the five most catastrophic of all mass extinctions. It took a toll on all kinds of life, and vertebrates were not spared. As with any mass extinction, clues to cause can be found by comparing "winners" (survivors) and losers (those clades undergoing complete extinction). In this case the biggest winners, in that they were the only clade to actually increase in numbers across the Triassic–Jurassic (T–J) boundary, were the saurischian dinosaurs. This can be tracked with synoptic data, but this finding is supported by a local finding. Olson *et al.* (2002) noted that the size, abundance, and diversity of dinosaur footprints increased across the T–J boundary in the Newark Basin, the site where footprint records are perhaps best preserved for this time interval. These authors, who (then) favored large body impact as the largest single cause of the T–J mass extinction, have more recently abandoned the impact hypothesis for this event. Whatever the cause, it must have somehow favored survivability of those clades that were pneumatized.

The synoptic data at hand for this event are shown in Figure 2.4, which looks at families of vertebrates and classifies them by hypothesized lung type. The two aspects of this graph that are most notable are that the saurischian dinosaur body plan somehow conferred higher survivability

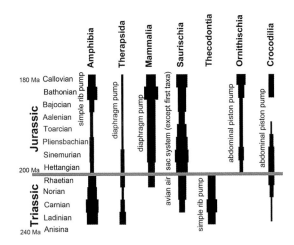

Fig. 2.4 Survivorship across the Triassic–Jurassic boundary mass extinction. The groups are put into larger clades, but the data making up the spindles are number of genera taken at the stage level. (Diagram from Ward (2006), with permission.)

than the others, and that even after the extinction was long over, in fact over most of the Jurassic Period, vertebrate diversity including dinosaurs remained low. This is contrary to our usual view of the Jurassic as a time of highly diverse dinosaur genera. This problem can be further assessed by looking at the number of dinosaur genera with respect to atmospheric oxygen in the Jurassic.

OXYGEN AND DINOSAUR DIVERSITY

The oxygen curve shown on the page before (Figure 2.2) indicates that oxygen did not begin to significantly rise till the second half of the Jurassic Period. With this in mind it is instructive to compare a detail of the oxygen curve (from the Late Triassic through Cretaceous Period) with dinosaur diversity. For the latter we have relied on the work of Fastovsky *et al.* (2004) plotted against a GEOCARBSULF oxygen estimate for the time over which dinosaurs (at least the non-avian dinosaurs) existed.

The Fastovsky *et al.* (2004) data (Figure 2.5) showed dinosaur genera staying roughly constant from the time of the first dinosaurs in the late

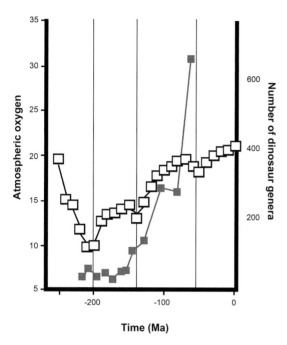

Fig. 2.5 Atmospheric oxygen percentage plotted against number of dinosaur genera. This figure supports the hypothesis that higher oxygen supported a higher diversity of dinosaurs. Part of the reason for this may be due to the fact that rising oxygen levels opened up more habitable areas at altitude, a prediction from Huey & Ward (2005). (Diagram from Ward (2006), with permission.)

Middle Triassic and Late Triassic through most of the Jurassic. It is not until the latter part of the Late Jurassic that dinosaur numbers started to rise significantly, and this trend then continued inexorably to the end of the Cretaceous, with the only (and slight) pause in this rise coming in the early part of the Late Cretaceous, and this slight drop may be due to the very small number of localities known of this age yielding dinosaurs. By the end of the Cretaceous (in the Campanian Stage) there are hundreds of times more dinosaurs than during the Triassic to Upper Jurassic. So what was the cause of this great increase?

The relationship shown in Figure 2.5 suggests that oxygen levels have either played a major role

in dictating dinosaur diversity, directly (by physiological effects on the animals themselves) or indirectly (by somehow affecting success of dinosaurs through food supply or available habitat area). Through the Late Triassic and first half of the Jurassic, dinosaur numbers were both stable and low. While originating in the latter part of the Triassic, they stayed relatively few in number until a moderate rise at the end of the period – a rise that seemed to coincide with the end-Triassic mass extinction itself. Gradually, if the oxygen results from GEOCARBSULF are even approximately correct, oxygen rose in the Jurassic, hitting 15% or more in the latter part of the period. It is then that the numbers of dinosaurs really begin to increase. It is also at this time that the sizes of dinosaurs increased as well, culminating in the largest dinosaurs ever evolved appearing from latest Jurassic into the first half of the Cretaceous. Oxygen levels steadily climbed through the Cretaceous, and so too did dinosaur numbers, with a great rise in dinosaur numbers found in the Late Cretaceous, the true dinosaur heyday. It must be noted that of the two great stocks of dinosaurs, the saurischians and ornithischians, it was the latter that made up the greatest number of new forms in the Cretaceous. Because these species were without pneumatization, as well as being mainly herbivores (and thus less active than saurischian carnivores, all with pneumatization) it appears that oxygen had to rise to certain thresholds for ornithischians to get first a toe-hold, and then dominance among herbivorous animals of the latter parts of the Mesozoic.

There were surely other reasons for this Cretaceous rise. For instance, in mid-Cretaceous times the appearance of angiosperms caused a floral revolution, and by the end of the Cretaceous Period the flowering plants had largely displaced the conifers that had been the Jurassic dominants. The rise of angiosperms created more plants, and sparked an insect diversification. More resources were available in all ecosystems, and this may have been a trigger for diversity as well. But even with these caveats, the correlation between numbers of dinosaurs and oxygen levels is clear.

CONCLUSIONS

The fact that the air sac system survived the Triassic mass extinction, which killed off many of the more common lineages of Triassic animals, such as the phytosaurs, most cynodonts, and many primitive reptilian groups, is, to us, persuasive. The filter that was the Triassic mass extinction was one where method of respiration seemingly mattered. One of us (Ward, 2006) showed that of the various lung types present at the beginning of this extinction, the clades with air sac respiratory system not only survived unscathed, but actually increased in numbers just before and just after the mass extinction itself. The most parsimonious explanation as to why there were dinosaurs, and why there are birds, is because of the low atmospheric oxygen during the Triassic Period.

REFERENCES

Berner RA. 2004. *The Phanerozoic Carbon Cycle: CO$_2$ and O$_2$*. Oxford: Oxford University Press.

Berner RA. 2009. Phanerozoic atmospheric oxygen: new results using the GEOCARBSULF model. *American Journal of Science* **309**: 603–606.

Fastovsky D, Huang Y, Hsu J, Martin-McNaughton J, Sheehan PM, Weishampel DB. 2004. Shape of dinosaur richness. *Geology* **32**: 877–880.

Huey RB, Ward PD. 2005. Climbing a Triassic Mount Everest: into thinner air. *Journal of the American Medical Association, Research Letters* **294**(14): 1761–1762.

O'Connor PM, Claessens LPAM. 2005. Basic avian pulmonary design and flow-through ventilation in non-avian theropod dinosaurs. *Nature* **436**: 253–256.

Olson PE, Kent DV, Sues H-D, Koeberl C, Huber H, Montanari A, Rainforth EC, Fowell SJ, Szajna MJ, Hartline BW. 2002. Ascent of dinosaurs linked to an iridium anomaly at the Triassic–Jurassic boundary. *Science* **5571**: 1305–1307.

Ward, P. 2006. *Out of Thin Air*. Washington, DC: National Academy Press.

Ward P, Labandeira CC, Lauren M, Berner R. 2006. Confirmation of Romer's Gap as a low oxygen interval constraining the timing of initial arthropod and vertebrate terrestrialization. *Proceedings of the National Academy of Sciences* **103**(45): 16818–16822.

Wedel, M. 2003. Vertebral pneumaticity, air sacs, and the physiology of sauropod dinosaurs. *Paleobiology* **29**: 243–255.

3 Pre-modern Birds: Avian Divergences in the Mesozoic

JINGMAI O'CONNOR,[1,2] LUIS M. CHIAPPE,[2]
AND ALYSSA BELL[3]

[1]Institute of Vertebrate Paleontology and Paleoanthropology, Beijing, China
[2]Los Angeles County Museum, Los Angeles, USA
[3]University of Southern California, Los Angeles, California, USA

Birds are the most diverse group of living land vertebrates on the planet, yet the origin of the clade is one of the most heated and longest debates in scientific history (Witmer, 2001, 2002). Recent years have witnessed incredibly rapid growth in the fossil record of Cretaceous birds and theropod dinosaurs and today, the idea that birds are part of the diverse evolutionary radiation of maniraptoran theropods is nearly universally accepted and supported by similarities in skeletal morphology, egg structure, behavioral patterns, integument, bone histology, and genome architecture (Figures 3.1–3.5; Gauthier, 1986; Chiappe, 2001, 2007, 2009; Holtz, 2001; Norell *et al.*, 2001; Padian *et al.*, 2001; Clark *et al.*, 2002; Erickson, 2005; Schweitzer *et al.*, 2005, 2007; Xu, 2006; Organ *et al.*, 2007; Long & Schouten, 2008). This wealth of multidisciplinary evidence supports the hypothesis that extant birds are living maniraptorans and suggests that many "avian" morphologies and behaviors evolved early in their evolutionary history. The clues necessary to better understand the unique attributes of modern birds clearly lay buried with the fossils of their extinct Mesozoic relatives.

Historically, our knowledge of the Mesozoic avifauna has been greatly limited to the Late Jurassic *Archaeopteryx* (von Meyer, 1861) and the Late Cretaceous "Odontornithiformes" of North America (*Hesperornis* and *Ichthyornis*; Marsh, 1880). The anatomy of these Late Cretaceous birds testified to an enormous gap in the early history of the group when compared to the morphology of the older and much more primitive *Archaeopteryx*, making it difficult for early scientists to understand the early evolution of and within the clade. In the late 20th century, new avian fossils began to be uncovered around the world (Elzanowski, 1977; Walker, 1981; Martin, 1983; Hou & Liu, 1984; Sanz *et al.*, 1988; Chiappe & Calvo, 1991; Sereno & Rao, 1992; Chiappe *et al.*, 1999); in China alone, hundreds of specimens of Early Cretaceous birds have been uncovered during the past 20 years and discoveries continue at an unprecedented rate (Zhou & Zhang, 2006; Chiappe, 2007). These discoveries have revealed numerous new clades and filled much of the anatomical and temporal gaps that previously existed, making the study of early birds one of the most dynamic fields in vertebrate paleontology today.

At the same time, spectacular discoveries of nonavian theropods with avian features from China and Mongolia have documented a range of morphologies and behaviors previously thought to be unique to birds (Norell *et al.*, 1995; Chiappe

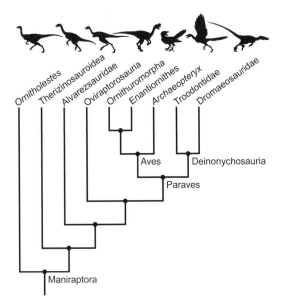

Fig. 3.1 A cladogram depicting the hypothetical relationships of maniraptoran dinosaurs including Aves.

et al., 1998; Xu & Norell, 2006; Turner *et al.*, 2007; Xu *et al.*, 2009b). To date fossils in support of the phylogenetic placement of birds within the maniraptoran theropods are largely of Cretaceous age, although important evidence from Jurassic maniraptorans contemporaneous or even older to *Archaeopteryx* has begun to accumulate (Xu *et al.*, 2001, 2009a; Xu & Zhang, 2005; Zhang *et al.*, 2008a; Choiniere *et al.*, 2010). Altogether, this evidence highlights a great deal of evolutionary experimentation within the theropod clade

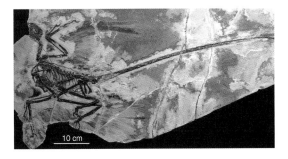

Fig. 3.2 The holotype of the Jiufotang Formation dromaeosaur *Microraptor gui* IVPP 13352. (Photograph by Luis Chiappe.)

around the divergence of birds, with a number of different lineages evolving a wide range of features previously viewed as strictly avian.

THE THEROPOD HYPOTHESIS OF THE ORIGIN OF BIRDS

Prior to the inundation of new fossil material spanning across the evolutionary transition between nonavian theropods and birds, there existed a number of competing hypotheses suggesting non-dinosaurian origins for Aves. These alternative hypotheses envisioned the avian ancestor as either a crocodylomorph (Walker, 1972; Martin & Stewart, 1999; Martin, 2004; Kurochkin, 2006), a basal archosauriform (Welman, 1995), or as other types of archosaur-related animals of Triassic age (Feduccia, 1996). These hypotheses have been largely abandoned in the absence of support for such an origin, yet there still exists opponents to the theropod hypothesis (while accepting a dinosaurian origin for birds) and those who place Aves outside of the dinosaurian clade entirely (Feduccia, 1996; Martin, 1997, 2004; Kurochkin, 2006; James & Pourtless, 2009). Despite the persistence of such hypotheses, cladistic analyses and numerous lines of evidence support a dinosaurian ancestry for birds specifically within the maniraptoran theropod clade (Figure 3.1; Gauthier, 1986; Sereno, 1999; Clarke *et al.*, 2002; Turner *et al.*, 2007).

The skeletal morphologies of birds and nonavian theropods share many important features, some of which were at one time considered defining characteristics of birds (e.g. furcula, retroverted pubis; de Beer, 1954). The similarity between nonavian theropods and birds has blurred with increasing discoveries of a wide range of birdlike theropods. The distribution of "avian" characters such as a beak (Clark *et al.*, 2001), furcula (Nesbitt *et al.*, 2009), sternal plates (Norell & Makovicky, 1997; Burnham *et al.*, 2000), uncinate processes (Codd *et al.*, 2007), retroverted pubis (Chiappe *et al.*, 1998; Norell & Makovicky, 1999), distally noncontacting pubes (Chiappe

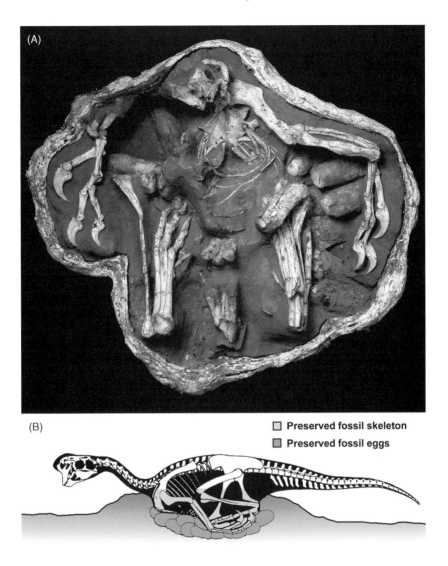

Fig. 3.3 Nesting oviraptorosaur, *Citipati osmolska* IGM 100/979, from the Late Cretaceous Mongolian Djadokhta Formation. (Mark Norell, AMNH, with permission.)

et al., 2002a), parallel pubis and ischium (Chiappe *et al.*, 1998), and a pygostyle (Barsbold *et al.*, 2000) are so spread out through Maniraptora that for most of them it is difficult to determine a clear pattern of character origination. This suggests a deeper origin for many of these characters, while their occurrence within apparently unrelated nonavian theropod groups and their absence in primitive birds suggests a highly homoplastic evolutionary history.

Feathers were, until the last few decades, known exclusively within Aves (de Beer, 1954), yet now integumentary structures hypothesized as homologous to feathers are known throughout Dinosauria (Mayr *et al.*, 2002; Xu *et al.*, 2004; Ji *et al.*, 2007; Zheng *et al.*, 2009) and pennaceous

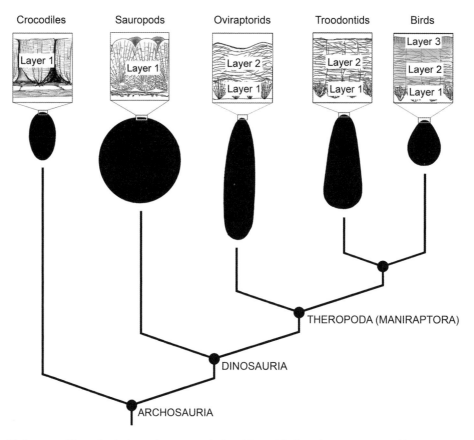

Fig. 3.4 Cladogram of hypothetical maniraptoran relationships with the relative position of known avian eggshell morphologies. (Modified from Grellet-Tinner, 2004)

feathers of modern morphology have a wide distribution among maniraptorans (Figure 3.2) (Ji *et al.*, 1998, 2001; Norell *et al.*, 2002; Xu *et al.*, 2009b; Hu *et al.*, 2009). Evolutionary stages in feather development are documented in the fossil record with "proto-feathers," unbranched filamentous structures, present among a wide range of nonavian theropods (e.g. *Dilong paradoxus*, *Sinosauropteryx prima*, *Beipiaosaurus inexpectus*), and branched integumentary structures similar to down feathers and/or fully modern pennaceous feathers occuring in a number of small, bird-like maniraptorans (e.g. *Sinornithosaurus millenii*, *Protoarchaeopteryx robusta*, *Caudipteryx zoui*, *Microraptor gui*, *Anchiornis*

huxleyi) (Xu *et al.*, 2001, 2003; Prum & Brush, 2002; Chuong *et al.*, 2003; Prum, 2005; Zhang *et al.*, 2006; Hu *et al.*, 2009). More recently, the homology between even the simplest of these structures and modern feathers has been compellingly documented by microstructural studies that have identified melanosomes (pigment-bearing organelles) identical to those of modern feathers, and have been able to determine the basic patterns of plumage coloration of some dinosaurs (Li *et al.*, 2010; Zhang *et al.*, 2010).

The discoveries of *in ovo* embryos and close associations between eggs and adults have allowed for eggs and eggshell microstructure to be studied in a phylogenetic context (Figure 3.3). Within

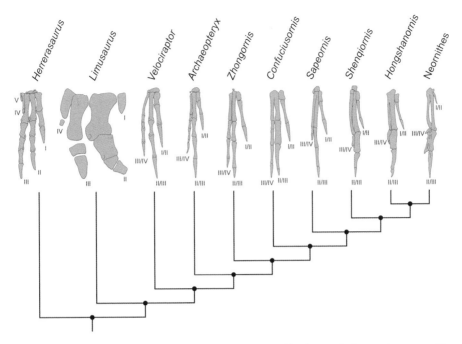

Fig. 3.5 The hand morphology of basal theropods, maniraptorans, and birds in a phylogenetic context illustrating the controversy of digit identity in birds.

Dinosauria, only the maniraptorans possess eggs with modern avian characteristics (e.g. egg asymmetrical with an air sac, shell composed of multiple layers with prismatic condition) (Figure 3.4) (Varricchio *et al.*, 1997; Grellet-Tinner & Chiappe, 2004; Grellet-Tinner *et al.*, 2006; Varricchio & Jackson, 2004a,b). The development of increasing parental care within the phylogenetic tree of theropods as it approaches Aves can be documented through the discovery of nesting adults such as the oviraptorid *Citipati osmolska* (Figure 3.3) (Clark *et al.*, 1999). What can be inferred from the fossil record on nonavian maniraptoran behavior suggests both brooding and parental care, as in living birds (Grellet-Tinner & Chiappe, 2004), as well as precociality (Erickson *et al.*, 2001) and paternal care (Varricchio *et al.*, 2008), which are inferred to be the ancestral conditions for modern birds (McKitrick, 1992).

Histological studies of bone tissue have revealed that avian growth rates had their origins within Dinosauria (Erickson *et al.*, 2001, 2007).

There is a general trend that can be observed within Mesozoic avians, from primitive growth in *Archaeopteryx* with multiple visible lines of arrested growth (LAGs), as present in many other dinosaurs, to the development of the modern strategy of rapidly achieving terminal size in shorter periods of time higher in the clade (Chinsamy *et al.*, 1995; Chinsamy & Elzanowski, 2001; Erickson *et al.*, 2009). Like *Archaeopteryx*, the more derived Cretaceous enantiornithines grew slowly, with adult specimens preserving multiple lines of growth (Chinsamy *et al.*, 1995). The basal ornithuromorphs *Patagopteryx* and *Hollanda*, taxa more closely related to modern birds than either *Archaeopteryx* or the enantiornithines, show a single LAG, suggesting more rapid growth (Chinsamy *et al.*, 1994, 1995; Bell *et al.*, 2010) compared to birds outside the ornithuromorph clade. The modern bone microstructure of the ornithurine *Hesperornis*, a derived ornithuromorph closely related to modern birds, indicates that growth patterns comparable to those of living birds may have been achieved prior to the origin of the crown clade (Chinsamy et al., 1995).

The developmental plasticity that characterizes the osteogenesis of modern birds (Starck & Chinsamy, 2002) may also have developed even earlier within avian evolutionary history; de Ricqles et al. (2003) suggested that the Early Cretaceous Confuciusornis had uninterrupted growth with rates comparable to modern birds. This claim, however, is not supported by a recent morphometric study, which suggests a slower and possibly discontinuous growth pattern for Confuciusornis (Chiappe et al., 2008). The bone histology of this abundant but geographically restricted Mesozoic fossil may need to be re-evaluated in light of the significant difference in size between specimens (Chiappe et al., 2010).

Most of the arguments made by opponents of the theropod origin of birds are weak (Welman, 1995; Feduccia, 1996; Kurochkin, 2006) and there is very little cladistic support for an alternative hypothesis (Witmer, 1991; James & Pourtless, 2009). Until recently, the most compelling criticism of the theropod ancestry of birds was based on the differential interpretation of the ossified digits of the hand in modern birds and nonavian theropod dinosaurs (Figure 3.5). During the development of the manus in extant bird embryos, five points of condensation appear and only the middle three develop, presumably II–III–IV of the I–II–III–IV–V buds. This has led embryologists to infer a II–III–IV manual formula for living birds (alular, major, and minor, respectively; Burke & Feduccia, 2007). However, the three digits present in nonavian maniraptorans are interpreted as corresponding to digits I–II–III of the ancestral pentadactyl hand (I–V) (Shubin, 1994), based on interpretations of the early theropod fossil record (Sereno & Novas, 1992; Sereno, 1993; Padian & Chiappe, 1998). Basal theropods such as Herrerasaurus and Eoraptor retain five digits but display reduction in the outer two digits IV and V, leading to the interpretation that more advanced theropods retain digits I–II–III (Figure 3.5; Sereno 1993). This discrepancy in the perceived homologies of the hands of nonavian theropods and birds had led some researchers to exclude Aves from Theropoda (Feduccia & Nowicki, 2002; Kurochkin, 2006; Burke & Feduccia, 2007), while others have attempted to resolve this apparent inconsistency with hypotheses regarding gene regulatory mechanisms, in which the identity of the digit is transferred so that a given digit arises from a different condensation (Wagner & Gauthier, 1999). While this frame shift is possible, and known to occur in a number of vertebrates (Shapiro, 2002), new evidence also suggests the possibility that the inferred pattern of reduction within theropods may have been misinterpreted, as the Triassic fossil record is admittedly extremely fragmentary (Galis et al., 2005). This is supported by Morse's Law (which states that in most tetrapod lineages digits V and I become reduced prior to other digits), for which theropods are currently considered one of a few exceptions (Shubin, 1994), and the recent discovery of a ceratosaur theropod with digital reduction in digits I and IV and no digit V (Xu et al., 2009a). Despite the systematic placement of this taxon outside Tetanurae (the theropod clade that includes maniraptorans and birds), the new ceratosaur theropod Limusaurus inextricablis shares derived features with tetanurans suggesting a close relationship, and the reinterpretation of the tetanuran manus as II–III–IV (Figure 3.5; Xu et al., 2009a). The disparity in the manual morphology between birds and nonavian dinosaurs has long been pitched by opponents as a major flaw in the theropod hypothesis (Feduccia, 1996, 2001; Kurochkin, 2006). While this debate cannot be considered closed based on the reduction of digit I in one taxon, it is important to consider the highly fragmentary nature of the Triassic and Early Jurassic theropod fossil record. As demonstrated by Limurasaurus, new discoveries have the potential to radically change our interpretation of early theropod evolution. For this reason the digits of the avian hand are here referred to as the alular (I or II), major (II or III) and minor (III or IV).

Even though it is now generally accepted that birds are theropod dinosaurs nested within the clade Maniraptora (Turner et al., 2007; Chapter 1), the sister taxon to Aves within this clade is debated and differs between cladistic analyses

(Holtz, 1994; Gauthier, 1986; Sereno, 1997; Forster *et al.*, 1998a; Norell *et al.*, 2001; Huang *et al.*, 2002; Makovicky *et al.*, 2005; Novas & Pol, 2005; Gölich & Chiappe, 2006; Turner *et al.*, 2007). Most cladistic analyses place one of two groups as the closest relative to birds: Dromaeosauridae or Troodontidae, and these two clades are often considered to form a more inclusive clade, Deinonychosauria (Forster *et al.*, 1998a; Sereno, 1999; Benton, 2004). Each of these clades possesses a different combination of avian characters distributed amongst the included taxa, suggesting that these groups are more closely related to birds than other theropods. Recent discoveries have identified the bizarre maniraptorans *Epidexipteryx* (Zhang *et al.*, 2008a) and *Anchiornis* (Xu *et al.*, 2009b) as apparently even closer to birds than the deinonychosaurs, forming a clade with Aves termed Avialae (Xu *et al.*, 2009b). However, these claims need to be examined in greater detail. The search for the definitive closest relative of birds continues but the overwhelming evidence in favor of a maniraptoran origin and its wide acceptance within the scientific community allows us to follow this hypothesis for the remainder of this chapter.

THE MESOZOIC AVIARY

Long-tailed birds

Known exclusively from the 150 million year old (Ma) Solnhofen limestones of central Bavaria (Germany), *Archaeopteryx lithographica* constitutes the oldest and most primitive definitive bird in the fossil record. The taxon is known solely from ten, largely two-dimensional, skeletal specimens (Wellnhofer, 2008); *Archaeopteryx*, at the center of debates about avian origins, has a remarkably complex taxonomic history, with nearly every specimen identified as a distinct species at some point in time (Elzanowski, 2002; Wellnhofer, 2008). As the oldest known bird, the diversity of this taxon is of great interest. Currently, some regard all specimens as belonging to a growth series within a single taxon (*Archaeopteryx lithographica*; Houck *et al.*, 1990; Senter

& Robins, 2003; Chiappe, 2007; Erickson *et al.*, 2009), while others argue that the size range of the specimens (from jay-sized to the size of a small gull) and specific morphologies indicate the presence of more than one species (Elzanowski, 2002; Wellnhofer, 2008).

The anatomy of *Archaeopteryx* illustrates the most primitive condition known in Aves, with many similarities to the dromaeosaurids, troodontids, and other nonavian maniraptorans regarded as close relatives. The toothed skull of *Archaeopteryx* retains primitive bones, such as the postorbital and remains highly unfused, while at the same time it shows an increase in orbit and brain size, features associated with birds (Alonso *et al.*, 2004) but also that may have predated their origin (Hu *et al.*, 2009). The postcranial skeleton lacks almost all of the avian modifications associated with flight: the long trunk, with no synsacrum, lacks the rigidity of extant birds; the scapula and coracoid remained fused; the sternum is unossified; and its long bony tail is composed of 21–23 elongate free vertebrae. The forelimb of *Archaeopteryx* is extremely similar to those of its immediate nonavian predecessors. The wing of *Archaeopteryx* lacks an alula and retains an ancestral design: the humerus is longer than the ulna–radius, the wrist is characterized by a differentiated (unfused) semilunar carpal, and the elongate manus bears three unreduced digits with large, recurved claws. Despite the primitive osteology of the forelimb and gross size, the wing is remarkably similar to those of living birds in shape and number of flight feathers (11–12 primaries and 12–14 secondaries; Elzanowski, 2002; Wellnhofer, 2008). The hindlimb is elongate, as in the typically cursorial maniraptorans, a true tibiotarsus and tarsometatarsus are absent, and the hallux is only weakly reversed. The pelvis is unfused, the pubis is not fully retroverted and distally it bears an elongate symphysis. The long tail retains the elongated prezygapophyses (anterior intervertebral articulations) and T-shaped chevrons present in the maniraptoran theropods inferred to be closely related to Aves (Elzanowski, 2002; Wellnhofer, 2008). While undoubtedly preserving evidence of crural feathers (i.e. plumage covering portions of the tibiotarsus), claims that the hindlimbs of *Archaeopteryx* carried a set of aerodynamically significant, long and asymmetrically vaned feathers (Christiansen & Bonde, 2004; Longrich, 2006) cannot be confirmed at the present, however, *Archaeopteryx* clearly did not possess the large tarsometatarsal feathers observed in the Early Cretaceous dromaeosaurid, *Microraptor gui* (Xu et al., 2003).

The ecology of *Archaeopteryx* has been the center of great controversy as it pertains indirectly to the evolution of flight; there are two main camps of theory, those who believe *Archaeopteryx* was predominantly arboreal and those who believe it was largely terrestrial. In order to determine mode-of-life, first it must be determined if *Archaeopteryx* could fly or only glide. Numerous and diverse hypotheses have been proposed (Ostrom, 1974; Martin, 1983, 1991; Ruben, 1991; Feduccia, 1996, 1999; Padian & Chiappe, 1998; Burgers & Chiappe, 1999; Rayner, 2001; Elzanowski, 2002; Wellnhofer, 2008), however these and other ecological and functional questions remain largely conjectural. Given the functional and aerodynamic considerations, we support the interpretation of *Archaeopteryx* as a bird that spent a good amount of time foraging on the ground (Ostrom, 1974; Hopson, 2001), was able to take off from the ground (Burgers & Chiappe, 1999) and capable of powered flight (Rayner, 2001). However, an arboreal lifestye for *Archaeopteryx* has taken center stage in light of the recent discoveries of small "aboreal" non-avian theropods such as *Epidendrosaurus* and *Microraptor* (Xu et al., 2000; Zhang et al., 2002), which have been interpreted as better adapted for an arboreal existence than other non-avian theropods.

Until recently, *Archaeopteryx* was the only known bird with a long bony tail, however, discoveries from China and Madagascar have documented a diversity of similarly long-tailed birds, albeit in much younger deposits. The 75 Ma *Rahonavis ostromi* from Madagascar (Figure 3.6; Forster et al., 1998a, b), like *Archaeopteryx*, is characterized by having a long bony tail formed of elongated vertebrae, primitive proportions in the pelvis, and incomplete fusion of some compound bones (e.g. tarsometatarsus, tibiotarsus). The avian status of *Rahonavis* is controversial. Makovicky et al. (2005) argued in favor of a nonavian relationship,

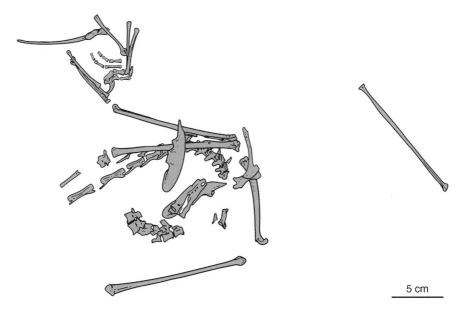

5 cm

Fig. 3.6   Quarry map of the holotype specimen of the Late Cretaceous long-tailed bird *Rahonavis ostromi*.

placing it within a clade of Gondwanan dromaeosaurids. This view was supported by Turner *et al.* (2007) although using essentially the same data. If *Rahonavis* were to be a dromaeosaurid, it would be one in which the wings would be proportionally much longer than those of any recognized non-avian theropod. Here *Rahonavis* is considered a bird and a swift predator (Forster *et al.*, 1998a), with a wingspan comparable to that of a red-tailed hawk. The most striking anatomical feature of *Rahonavis* is the enlarged, sickle-shaped claw of its second toe, a specialization presumably used for slashing prey as in dromaeosaurid and troodontid theropods.

Despite the presence of this and other basal features, *Rahonavis* appears to be more derived than *Archaeopteryx* (Holtz, 1998; Chiappe, 2002a; Zhou & Zhang, 2002a). Although no feathers were found in association with the skeleton of the single known specimen, quill knobs preserved on the forelimb indicate at least 10 flight feathers attached to its ulna, which is within the range of living birds. The structure of the shoulder girdle – in particular, the mobile glenoid for the articulation of the head of the humerus – more closely approaches that of extant birds than the rigidly fused glenoids of *Archaeopteryx* and some other early birds (e.g. *Confuciusornis* and allies). The reconstructed position of the scapula with respect to the rib cage indicates that *Rahonavis* was capable of flapping its wings with greater amplitude than *Archaeopteryx* or any nonavian theropod (Forster *et al.*, 1998a). All these features are consistent with well-developed aerodynamic capabilities, although it is possible that like *Archaeopteryx*, such a large, primitive bird may have required a take-off run to become airborne (Burgers & Chiappe, 1999).

All other examples of long-tailed birds have been found in the celebrated Early Cretaceous Jehol deposits of northeastern China (Zhang *et al.*, 2003), where they coexisted with other more derived taxa. The turkey-sized *Jeholornis prima* (Figure 3.7A; Zhou & Zhang, 2002a) from the Jiufotang Formation (120 Ma) has edentulous upper jaws while its dentary teeth are small and rostrally restricted (Zhou & Zhang, 2003a). The skeletal architecture of *Jeholornis* differs significantly from that of *Archaeopteryx* (Zhou & Zhang, 2003a). As in *Rahonavis*, the shoulder girdle articulation of *Jeholornis* is mobile. Unlike *Archaeopteryx*, the coracoid of *Jeholornis* is elongate, the scapula is curved and tapered distally, both features characteristic of the more advanced ornithuromorphs, and the more dorsal extent of the glenoid (facet for forelimb articulation) would have allowed greater amplitude during wing beats. The forelimb is proportionately longer than the hindlimb and the hand is shorter, though still bearing powerful claws. Despite these morphological advances, *Jeholornis* shares primitive features with *Archaeopteryx* and *Rahonavis*, such as a vertically oriented pubis, incompletely formed compound bones (e.g. tibiotarsus, tarsometatarsus), a splint-like fifth metatarsal, and a very short hallux (Zhou & Zhang, 2002a, 2003a). *Jeholornis* also exhibits a condition even more primitive than that of *Archaeopteryx*; the tail is longer than that of any other bird and carried a fan-shaped tuft of terminal feathers reminiscent of the tail in feathered dromaeosaurids (Xu *et al.*, 2003; Norell & Xu, 2005) and troodontids (Ji *et al.*, 2005). Two more recent specimens have shown that the tail of *Jeholornis* contains 27 elongate vertebrae (several more than *Archaeopteryx* and *contra* the initial estimation of 22 caudals; Zhou & Zhang, 2002a) and chevrons that approach the morphology in dromaeosaurids (Zhou & Zhang, 2003a).

The Jehol Group has produced several other long-tailed birds very similar to *Jeholornis*, such as *Dalianraptor cuhe* (Gao & Liu, 2005) and *Shenzhouraptor sinensis* (Ji *et al.*, 2002a), also from the Jiufotang Formation, and the older *Jixiangornis orientalis* (Figure 3.7B; Ji *et al.*, 2002b) from the Yixian Formation. Only preliminary descriptions of these fossils are available and it has been suggested that *Shenzhouraptor* and *Jixiangornis* may be junior synonyms of *Jeholornis* (Zhou & Zhang, 2006), but such taxonomic assessments will remain inconclusive until detailed studies of all these birds have been completed. *Dalianraptor* can be distinguished from *Jeholornis* on the basis of a smaller forelimb/hindlimb ratio, much longer alular digit, and different phalangeal

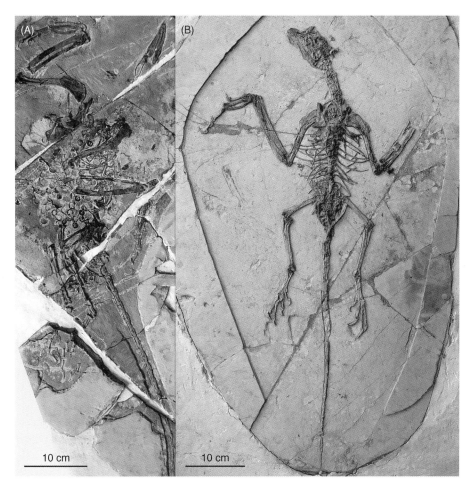

Fig. 3.7 Jiufotang Formation long-tailed birds: (A) holotype of *Jeholornis prima* IVPP V13274; (B) holotype of *Jixiangornis orientalis* CDPC-02-04-001. (Ji Shu-an, CAGS, with permission.)

proportions of manual digit III. This poorly pre-served taxon has a very short forelimb relative to its body and new material and information may prove this taxon not to be a bird. More fragmentary fossils have been interpreted as belonging to other long-tailed avian species; these include the Early Cretaceous *Jinfengopteryx elegans* (Ji *et al.*, 2005), an incomplete skeletal tail (Lü & Hou, 2005), and the Late Cretaceous *Yandangornis longicaudus* (Cai & Zhao, 1999). The former displays a number of features (small and closely spaced teeth, short forelimbs) that question its avian status and sug-gest it may be a troodontid maniraptoran, a con-clusion supported by recent cladistic analysis (Turner *et al.*, 2007). The other two specimens are too fragmentary to diagnose and the validity of their avian taxonomic status will depend on the discovery of additional material.

Most interpretations place these Chinese long-tailed birds in an arboreal habitat (Zhou & Zhang, 2004). However, such a conclusion is not sup-ported by the structure of the feet, which do not suggest perching specializations. In these birds, not only is the hallux very short but the penulti-mate phalanx of each toe is shorter than the remaining phalanges, a condition indicative of cursorial habits in modern birds (*contra* Zhou & Zhang, 2002a). The holotype of *Jeholornis prima*, preserved with a large number of seeds in the visceral cavity, provides the only direct evidence for the diet of long-tailed birds, however, the complete range of trophic preferences cannot be determined on the basis of a single specimen (Zhou & Zhang, 2002a).

Cretaceous long-tailed birds revealed an unforeseen quantity of evolutionary experimentation and homoplasy, bearing features typical of nonavian maniraptorans but absent in *Archaeopteryx*, or morphologies more primitive than those exhibited by the latter (despite the closer phylogenetic proximity to modern avians). Although more primitive than the long-tailed birds from China, the Late Cretaceous *Rahonavis* lived many millions of years later. This stratigraphic pattern may suggest a hidden diversity of Late Cretaceous long-tailed birds remaining to be discovered. Alternatively, *Rahonavis* may be an example of a Malagasy relic – a Late Cretaceous equivalent of today's lemurs – that survived in geographic isolation after the extinction of all other long-tailed birds (Forster *et al.*, 1998a) or, as argued by Makovicky *et al.* (2005), a member of a lineage of Late Cretaceous Gondwanan dromaeosaurids. Only an enlarged sample of Late Cretaceous avifaunas, particularly those from Gondwana, and additional comparative studies of these long-tailed birds will tell.

The origin of short-tailed birds

Despite the fact that the abbreviation of the long skeletal tail and the evolution of a pygostyle is one of the most apparent avian transitions and closely related to the fine-tuning of avian flight, the evolutionary transition between long-tailed birds and the first birds with a pygostyle is very poorly documented (Chiappe, 2007). Until the recent discovery of *Zhongornis haoae* from the Early Cretaceous Yixian Formation (125 Ma) of northeastern China (Figure 3.8; Gao *et al.*, 2008), there were no intermediate taxa known between the basalmost birds with a pygostyle and the long-tailed birds. The small *Zhongornis*

is characterized by having edentulous jaws, a primitive shoulder and forelimb bearing a unique phalangeal formula (alular, major and minor digits carry 2, 3, and 3 phalanges, respectively), a foot with a reversed hallux and weakly curved claws, and only 13–14 differentiated tail vertebrae that do not form a terminal pygostyle. *Zhongornis* is the only known bird to possess a short tail with a reduced number of vertebrae, yet lacks the pygostyle present in other short-tailed birds. The available anatomical information from this fossil suggests that phylogenetically, *Zhongornis* is the closest known relative to all pygostylians – birds whose bony tail ends in a pygostyle (Chiappe, 2002a; Gao *et al.*, 2008). The anatomy of *Zhongornis* and its intermediate placement between primitive long-tailed birds and the basalmost pygostylians suggest that in at least one lineage of early bird, the reduction in the long tail occurred first through the loss in number of caudals rather than a reduction in size of the tail while retaining a large number of vertebrae. Whether this pattern is unique to *Zhongornis* or ancestral for Pygostylia will have to be determined as more evidence of the immediate relatives of these birds comes to light.

Basal pygostylians: Confuciusornithidae, Sapeornithidae, and *Zhongjianornis*

The beaked confuciusornithids, including the abundant *Confuciusornis sanctus* (Chiappe *et al.*, 1999), appear to be the most basal pygostylians and the oldest known birds with a horny beak (Figure 3.9) (Gao *et al.*, 2008). Hundreds, if not thousands, of specimens of *Confuciusornis sanctus* have been collected in the past decade from the same Early Cretaceous deposits in northeastern China that yielded *Zhongornis* and the Chinese long-tailed birds (Chiappe *et al.*, 1999, 2008). While particularly abundant in the Yixian Formation, *Confuciusornis sanctus* is also recorded in the slightly younger Jiufotang Formation (Dalsatt *et al.*, 2006). Several species of *Confuciusornis* have been named (Chiappe, 2007; Zhang *et al.*, 2008b), yet only *C. sanctus* and *C. dui* have sufficient diagnostic support to be considered valid

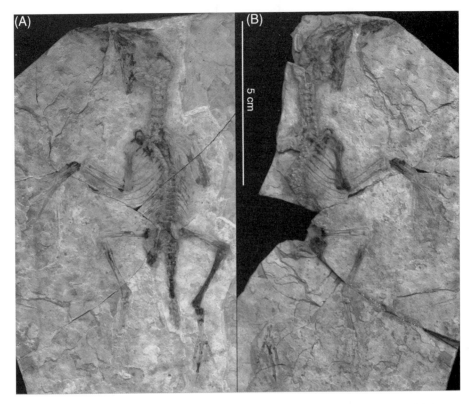

Fig. 3.8 Holotype of the Yixian Formation nonpygostlian *Zhongornis haoae* with abbreviated tail: (A) DNHM D2455; (B) DNHM D2456. (Photograph by Luis Chiappe.)

(Chiappe *et al.*, 2008). Another confuciusornithid, *Changchengornis hengdaoziensis,* is known from a single specimen from the Yixian Formation (Chiappe *et al.*, 1999; Ji *et al.*, 1999). This confuciusornithid is much smaller than *Confuciusornis* and it differs from the latter in a number of skeletal details including a shorter and more curved beak – this, however, may be a preservational artifact pending the discovery of additional specimens. The older *Eoconfuciusornis zhengi*, also known from a single specimen but from the oldest layers of the Jehol Group, the Dabeigou Formation (131 Ma), is possibly the most primitive confuciusornithid known (Zhang *et al.*, 2008b). It lacks a number of specializations present in both *Confuciusornis* and *Changchengornis*, such as the large fenestra piercing the deltopectoral crest of the humerus, and slightly more primitive morphologies, such as a short coracoid, and the absence of lateral depressions on thoracic vertebrae (Zhang *et al.*, 2008b). The holotype of *Eoconfuciusornis zhengi*, however, is clearly a juvenile and thus, the extent to which some of the features of this taxon, particularly the absence of fused compound bones, are not ontogenetic will have to be determined through additional discoveries.

Confuciusornithids are very primitive birds in many respects. Their skull is remarkable in exhibiting complete diapsid temporal fenestrae (upper and lower openings in the rear of the skull that provided space for the origin of jaw muscles; Chiappe *et al.*, 1999), which excludes the potential for cranial kinesis. The shoulder bones of these birds are fused into a rigid scapulocoracoid, a condition more primitive than that of the long-tailed *Rahonavis* and *Jeholornis*. The

Fig. 3.9 The most common pygostylian from the Early Cretaceous Yixian Formation, *Confuciusornis sanctus* IVPP 11372. (Photograph by Luis Chiappe.)

forelimb is proportionally much shorter than in the latter, approaching the length of the hindlimb as in *Archaeopteryx*. The forelimb bones also retain primitive proportions – the hand is the longest segment and the ulna–radius is shorter than the humerus. The robust wishbone has the boomerang appearance of *Archaeopteryx* and the sternum is essentially flat, lacking the prominent ventral keel that is seen in more advanced birds (although some specimens do have a faint ridge that could have anchored a deeper cartilaginous carina). As in all basal avians, confuciusornithids

had a full set of gastralia, although with fewer rows than preserved in *Archaeopteryx*. The hindlimbs are robust and the reversed hallux is one-half to two-thirds the length of the second toe. While retaining many primitive features, confuciusornithids show dramatic departures from earlier phases of avian evolution in the expansion of the sternum, development of several compound bones, and perhaps most notably, the pygostyle at the end of the tail. Like other basal birds, confuciusornithids possess essentially modern plumage, although the skeletal morphology of their clawed wing does not differ considerably from their theropod predecessors. Confuciusornithids evolved a propatagium, the lift-generating skin fold joining the shoulder and wrist, and remarkably long flight feathers that gave their wing the appearance of the long and narrow wing of living terns (Chiappe *et al.*, 1999). Estimates of important aerodynamic parameters such as wing loading also hint at refinements in the flight capabilities of confuciusornithids. Sanz *et al.* (2000) calculated the wing loading (weight/wing surface) of *Confuciusornis* to be lower than that of *Archaeopteryx*, thus suggesting that basal pygostylians were more agile and energetically more efficient than the more primitive long-tailed birds. These estimates, however, are based only on estimated mass and femoral length and comparisons with modern birds, and likely reflect the relatively longer hindlimb of *Archaeopteryx* compared to confuciusornithids. Unfortunately, preservation in most specimens prevents accurate assessment of wing area or femoral diameter and thus wing loading cannot be measured directly.

A pair of long streamer-like feathers are variably preserved extending from the tail region in specimens of *Confuciusornis sanctus* (Figure 3.9; Chiappe *et al.*, 1999, 2008; Zhang *et al.*, 2008b). This intraspecific difference has been hailed as evidence of sexual dimorphism (Feduccia, 1996; Hou *et al.*, 1996; Zhou & Zhang, 2004), in which specimens with streamer-like feathers are interpreted as males that died during lekking (male reunions for competitive display). This hypothesis is not supported by a recent morphometric study in which no statistical correlation between size distribution and the presence or absence of streamer-like feathers was found, indicating that, if these feathers are sexual characteristics, they are not correlated with a size difference between genders as would be expected in a sexually dimorphic species (Chiappe *et al.*, 2008, 2010).

Sapeornithidae is another group of very primitive pygostylians, also from the Jehol Biota (Zhou & Zhang, 2002b; Chiappe, 2007; Yuan, 2008). *Sapeornis chaoyangensis* from the Jiufotang Formation is the best-known species of this group (Figure 3.10) (Zhou & Zhang 2002b, 2003b) and,

with a wingspan comparable to that of a turkey vulture, is the largest Late Jurassic–Early Cretaceous bird known. The skull is relatively short with conical and robust teeth restricted to the tip of the upper jaw. The temporal region of the skull is primitive, with at least a complete supratemporal fenestra. The articulation of the scapula and the coracoid was mobile but the latter bone remained primitively short and axe-shaped. While more than a dozen articulated specimens have been discovered, none of them preserves an ossified sternum, suggesting that the sternum may have been cartilaginous and that the large flight muscles operating the wings could have extended their point of origin to the expanded distal coracoids. As in confuciusornithids and other very primitive birds, the furcula of *Sapeornis* is very robust and shaped like a boomerang, but unlike other basal birds, it bears a hypocleidium. Its elongate forelimb is much longer than that of *Archaeopteryx*, *Jeholornis*, or the confuciusornithids, reaching approximately 1.5 times the length of the hindlimb. The humerus is shorter than the ulna–radius and as in most confuciusornithids, is pierced by a large proximal foramen of uncertain function. The hand of *Sapeornis* is approximately the same length as the humerus but the presence of a clawless, reduced third digit illustrates a more advanced stage in the reduction of the hand than that of *Zhongornis*. As in *Archaeopteryx* and other long-tailed birds, the pelvic bones of *Sapeornis* remained unfused although its pubic symphysis is reduced. In the tail, 6–7 free caudals precede a small pygostyle reminiscent of the plow-shaped pygostyle of ornithuromorphs, rather than the robust and long pygostyle of confuciusornithids. Other named sapeornithids include *Sapeornis angustis* (Provini *et al.*, 2009), *Dydactylornis jii* (Yuan, 2008), and Shenhiornis primita (Hu *et al.*, 2010), all from the Jiufotang Formation. Differences between these species and *Sapeornis chaoyangensis* are largely limited to variance in the ratios of bones (Yuan, 2008; Provini *et al.*, 2009). *Sapeornis angustis* is approximately 30% smaller than *Sapeornis chaoyangensis*, and thus, the extent to which these quantitative differences are not related to allometry within a

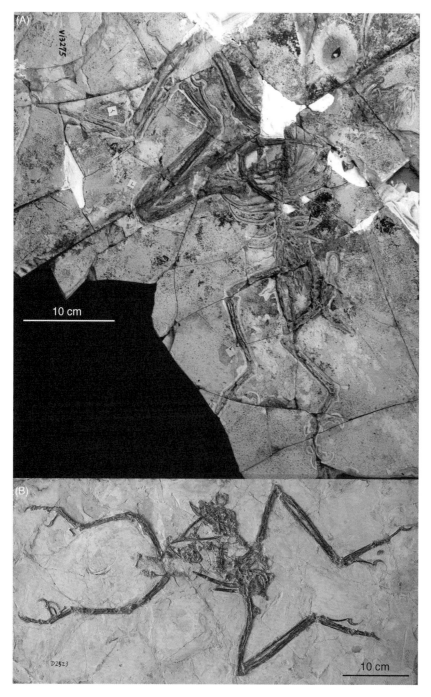

Fig. 3.10 Referred specimens of the Jiufotang Formation basal pygostylian *Sapeornis chaoyangensis*: (A) IVPP V13275; (B) DNHM D2523. (Photograph by Luis Chiappe.)

specific growth series needs further examination. *Dydactylornis jii* exhibits drastic differences with respect to *Sapeornis chaoyangensis* in the proportion of the first phalanx of manual digit I (it may be indistinguishable from metacarpal I, and thus give a "longer" appearance) and its apparent complete absence of manual digit III. However, as with *Dydactylornis*, more detailed studies are needed prior to dismissing taphonomic factors for this morphological discrepancy. Shenhiornis primita reports no major morphological differences other than a tail composed of ten free caudals; unfortunately the tail in the only known specimen of this taxon is poorly preserved and damaged.

More recently, another basal pygostylian was described from the Jiufotang Formation although no pygostyle is preserved; based on a single poorly preserved specimen, *Zhongjianornis yangi* is similar in size to a large *Confuciusornis* and also reportedly lacks dentition (Zhou *et al.*, 2010). Although poor preservation prevents detailed comparison, overall, this taxon appears very similar to confuciusornithids although it differs in the presence of a pointed beak and a proportionately longer forelimb and reduced manus, and poor preservation prevents detailed comparison. *Zhongjianornis*, like *Confuciusornis*, more advanced birds, and some non-avian theropods, preserves uncinate processes on its ribs. As in other basal birds, the furcula is robust and boomerang like. The humerus bears a large deltopectoral crest, similar to confuciusornithids, although a fenestra is reportedly absent, as in *Eoconfuciusornis*. The wing is elongate and the proportions are fairly advanced; the ulna–radius exceeds the humerus in length. The manus is reduced compared to other basal birds, with only two thin, clawed digits. The pelvic girdle is similar in some ways to more derived ornithuromorphs, with a fairly distally located, hooked dorsal process present on the tapering ischium in some ways to more derived ornithuromorphs while also reminiscent of more basal birds, with robust, rod-like pubes.

Very little is known about the lifestyles of basal pygostylians. The proportions of the pedal phalanges of confuciusornithids suggest a generalist function, equally capable of spending time on the ground or in trees (Hopson, 2001). Based on overall wing morphology and the skeletal advancements relative to *Archaeopteryx*, confuciusornithids are interpreted as capable of taking off from the ground and thus not needing to climb a tree to become airborne (Chiappe, 2007). The shape of the wing of sapeornithids and *Zhongjianornis* remains unknown – all known skeletons lack any evidence of plumage.

The presence of large clusters of stomach stones or gastroliths, like those common among extant herbivorous birds, inside the visceral cavity of several specimens of *Sapeornis* (Zhou & Zhang, 2003b; Yuan, 2008) has consistently being used to support a diet of plants for the sapeornithids (Zhou & Zhang, 2003b). However, the correlation between gastroliths and herbivory is not supported by the presence of the former in theropods that eat fish (as in the basal ornithuromorph *Yanornis martini* and the spinosaurid *Baryonyx walkeri*; Charig & Milner, 1986; Zhou *et al.*, 2004). Based on the morphology of the robust and massive beaks of confuciusornithids, envisioned as ideal for cracking seeds or other hard plant material, these birds have also been interpreted as herbivorous (Hou *et al.*, 1999; Zhou & Zhang, 2003b), although no specimen of Confuciusornis has ever been discovered with gastroliths (Dalsatt *et al.*, 2006). The discovery of a specimen of *Confuciusornis sanctus* with fish remains preserved at the base of its neck – allegedly contained in its crop – was used to indicate that confuciusornithids may have eaten a much wider range of items (Zhou *et al.*, 2004), but whether the fragmentary fish remains found in association with this specimen are truly indicative of its diet (as opposed to taphonomic events) needs to be further documented.

Enantiornithes: the first large radiation

Before modern birds (Neornithes) radiated in the Late Cretaceous and Early Tertiary, Mesozoic avifaunas were dominated by the enantiornithines (Figures 3.11–3.13), a group evolutionarily intermediate between the basalmost pygostylians, *Sapeornis* and *Confuciusornis*, and Ornithuromorpha, the

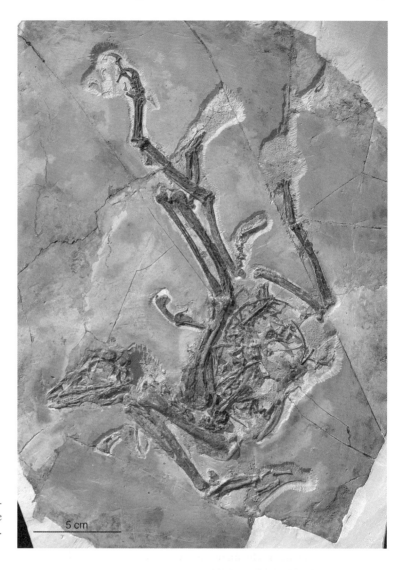

Fig. 3.11 Holotype of the Jiufotang Formation enantiornithine *Pengornis houi* IVPP V15336. (Photograph by Luis Chiappe.)

group of birds that includes living taxa (Walker, 1981; Chiappe & Walker, 2002). Enantiornithes are the prominent clade of pre-modern birds, with over 60 taxa named and specimens known from every continent with the exception of Antarctica (Chiappe & Walker, 2002). The enantiornithines spanned over most of the Cretaceous Period, the oldest known fossils dating back to about 130 Ma from Early Cretaceous deposits in Hebei Province, northeastern China (Figure 3.12; Wang *et al.*, 2010; Zhang & Zhou, 2000). Their fossil record ends with

that of other nonavian dinosaurs, the last occurrence being in the North American Hell Creek Formation, dated between 67 and 65.5 Ma (Brett-Surman & Paul, 1985).

Throughout this period, enantiornithines were widely successful, representing the first large-scale avian radiation. Their diversity encapsulated a wide size range, from very small taxa, the size of a sparrow, to others the size of a turkey vulture (Chiappe & Walker, 2002). Fossils have been collected from a variety

Fig. 3.12 Holotype of Qiaotou Formation enantiornithine *Shenqiornis mengi* DNHM D2951. (Photograph by Luis Chiappe.)

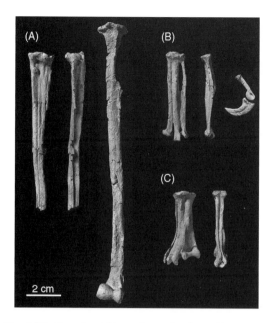

Fig. 3.13 Enantiornithines from the Late Cretaceous Lecho Formation in Argentina: (A) *Lectavis bretincola*; (B) *Soroavisaurus australis*; (C) *Yungavolucris brevipedalis*.

of depositional environments indicating these birds had diversified into a wide spectrum of ecologies and occupied a vast geographic range (including polar regions – Close *et al.*, 2009). Specimens have been found in continental river and dune sandstones, coastal and inland lake shales, and nearshore marine limestones (Chiappe *et al.*, 2001; Chiappe *et al.*, 2002b; Harris *et al.*, 2006; Zhou & Zhang, 2006). Enantiornithines are known to have evolved a wide range of wing proportions suggesting a diversity of flight styles, although the known diversity is less than that of modern birds (Dyke & Nudds, 2009). One Late Cretaceous taxon, *Elsornis keni* from the Gobi Desert (Chiappe *et al.*, 2006) is thought to have been nearly flightless. Though most enantiornithines are considered to be capable fliers, members of the Late Cretaceous genus *Martinavis* are theorized to have been superb fliers that migrated across large oceans, as fossils have been collected in North and South America and Europe (Walker *et al.*, 2007). Alter-

natively, given that this taxon is only known from humeri, additional material may separate these specimens phylogenetically, thus rejecting the previously mentioned hypothesis.

Enantiornithines encompass a remarkable range of morphologies, which in part reflects their long temporal range and ecological specialization. However, even within a single locality, enantiornithines show remarkable diversity (Figure 3.13; Chiappe, 1993). Within the Jehol Group, there exist taxa that appear to have primitive diapsid skulls (Figure 3.12; Wang *et al.*, 2010), while in other specimens the postorbital is reduced and the skull may have been capable of cranial kinesis (Sanz *et al.*, 1997; Zhou *et al.*, 2008). Likewise, within a single Jehol clade of "longirostrine" enantiornithines, the Longipterygidae, taxa have a high amount of morphological disparity. *Rapaxavis pani* and *Longipteryx chaoyangensis* are envisioned by cladistic analyses to be closely related (Chiappe *et al.*, 2007; O'Connor *et al.*, 2009), however, *Longipteryx* possesses large teeth, and a primitive manus that is longer than the humerus and bears two large claws, while *Rapaxavis* has reduced teeth, and a reduced manus, shorter than the humerus, with claws absent.

Though a majority of enantiornithine fossils are found in the Early Cretaceous, a broad trend towards increasingly more advanced morphologies in the Late Cretaceous is observed (Chiappe & Walker, 2002). The later evolution of enantiornithines has been previously characterized by a distinct increment in size, a trend that was interpreted as a result of the aerodynamic power achieved by this group (Chiappe & Walker, 2002). The discovery of large primitive Early Cretaceous enantiornithines (Zhou *et al.*, 2008) reduces the disparity in size between Early and Late Cretaceous enantiornithines, however, the largest taxa are still found in the Late Cretaceous. The braincase of the Late Cretaceous *Neuquenornis volans* is similar to that of modern birds in terms of fusion and proportions of the occipital condyle and foramen magnum, compared to Early Cretaceous species which show avian morphologies, but are primitive in that the bones do not fuse together and the foramen magnum is proportionately smaller relative to the occipital condyle (Zusi, 1993; Chiappe & Calvo, 1994; Wang et al., 2010). Compound bones found in modern birds such as the fused mandibular symphysis, the carpometacarpus (fused distal carpals and metacarpals), and tibiotarsus (fusion of the tibia with the astragalus and calcaneum) are absent among some Early Cretaceous enantiornithines, but present in Late Cretaceous species (i.e. *Gobipteryx minuta, Enantiornis leali*). The absence of fusion in many Early Cretaceous specimens may be ontogenetic, given that enantiornithine ontogeny is still poorly known.

Several advanced features indicate that the enantiornithines were more closely related to modern birds than other basal pygostylians, and more importantly, may have been able to fly in a comparable manner. Enantiornithines all have individualized scapulae and coracoids, with the scapula articulating lower on the coracoid, creating more leverage for the powerful flight muscles (Feduccia, 1999). Compared to the boomerang-like furculae of basal birds, enantiornithines, like modern birds, possess narrow and slender furculae, which may have enhanced the bone's capacity to assist the flight stroke by acting as a spring (Jenkins et al., 1988). The enantiornithine furcula is angular, typically forming a V- or Y-shape, compared to the U-shaped furcula of more advanced birds, and is typically dorsolaterally excavated, possibly representing a pneumatic specialization (Chiappe et al., 2006; Close et al., 2009). Enantiornithines may have evolved a triosseal foramen between the scapula, coracoid, and furcula but with the possible exception of *Protopteryx fengningensis*, this canal was formed without the participation of the procoracoid process or a similar medial projection of the coracoid (Zhang & Zhou, 2000). Unfortunately, no three-dimensional, articulated fossil detailing the spatial relationships of the coracoid, scapula, and furcula has been found. Modern birds have saddle-shaped (heterocoelous) articulations between cervical vertebrae, and this is known to have developed at least in the anterior portion of the neck in several enantiornithine taxa (Sanz et al., 2001; Zhang et al., 2004; Zhou et al.,

2008). This innovation may have enhanced the functional performance of the latter. Most enantiornithines possess a broad sternum, with a small ventral keel usually limited to the distal half of the bone (Sanz & Buscalioni, 1992; Chiappe & Walker, 2002; Li et al., 2006; Chiappe et al., 2006), though the Late Cretaceous *Neuquenornis* possesses a large keel of modern aspect (Chiappe & Calvo, 1994). The wing of most enantiornithines displays modern proportions: the humerus is shorter than the ulna and longer than the hand (Dyke & Nudds, 2009). The enantiornithine humerus is distinct; compared to the globe-like proximal head of modern birds, the head of the enantiornithine humerus is saddle-shaped, concave at the mid-point, with proximal surfaces that are cranially convex and caudally concave respectively. While the manus typically bears claws on the alular and major digits (claws are absent in some species), the hand is more compact when compared to that of more basal birds, with the carpometacarpus typically fused proximally, the alular digit short, and the minor digit extremely reduced (manual phalangeal formula typically 2–3–2–x–x).

Enantiornithines also evolved aerodynamic specializations beyond the skeletal level, having developed integumentary specializations for increased aerodynamic function in the wing and tail. The wing has primaries and secondaries comparable to modern taxa in size and number, and an alula is preserved in several specimens (Sanz et al., 1996; Zhang & Zhou, 2000; Zhou et al., 2005). While most known taxa lack elongate tail feathers, a diversity of feather morphologies are known within the clade. Several taxa (e. g. *Protopteryx, Dapingfangornis, Paraprotopteryx*) developed streamer-like feathers that may have played a sexual role (Zhang & Zhou, 2000; Li et al., 2006; Zheng et al., 2007), and at least one taxon evolved a fan-like tail morphology capable of generating substantial lift (Gatesy & Dial, 1996; O'Connor et al., 2009). The advanced features of the flight apparatus – the triosseal foramen, sternal keel, modern proportions of the wing, compact hand, and alula – suggest that these birds were capable of well-controlled and active flapping flight, however, future studies will be necessary to clarify how the anatomical differences between enantiornithines and modern birds affected the flight performance of this extinct clade.

The enantiornithine hindlimb has a wide range of morphologies reflecting the adaptation for a diversity of habitats within the group, as evidenced by the disparity among the tarsometatarsi of the El Brete enantiornithines (Figure 3.13; Chiappe, 1993). However, hindlimb proportions appear to be conservative relative to the wing, occupying a much more limited area of the modern avian morphospace (Dyke & Nudds, 2009). The femur in enantiornithines is excavated proximolaterally by a posterior trochanter (Chiappe, 1996; Chiappe & Walker, 2002), the tibiotarsus lacks the prominent cnemial crests present in modern birds with only a single small crest present in some taxa (Chiappe & Walker, 2002), and the tarsometatarsus is typically incompletely fused with a tubercle for the *m. tibialis cranialis* that is located medial and distal to its position in more advanced birds. Although definitive evidence for perching is lacking among more basal birds, these capabilities were clearly present among the earliest enantiornithines as evidenced by their pedal morphology (Sereno & Rao, 1992; Chiappe & Calvo, 1994; Chiappe, 1995; Martin, 1995; Sanz *et al.*, 1995; Zhou, 1995; Morschhauser *et al.*, 2009).

The exceptional preservation of the Early Cretaceous avifaunas in northeastern China reveals diverse enantiornithine trophic communities. The Jehol biota preserves taxa with a range of tooth morphologies and dental patterns, hinting at the diversity of dietary niches utilized by the clade. Despite tooth reduction in Chinese long-tailed birds and all basal pygostylians, enantiornithines typically possessed teeth similar to those of *Archaeopteryx*: found throughout both jaws, small, slightly recurved and unserrated. However, new discoveries continue to expand the known range of morphologies and patterns; *Shenqiornis mengi* possessed large bulbous teeth that may have been adapted for eating hard food items such as insects and other arthropods (Figure 3.12; Wang *et al.*, 2010), while another species,

Pengornis houi, had numerous low crowned teeth, possibly indicating a herbivorous diet of hard food (Figure 3.11; Zhou *et al.*, 2008). While several species retained large teeth, other enantiornithines reduced their teeth, some in terms of size and others in distribution. Enantiornithines even convergently evolved an edentulous rostrum (beak) with modern birds, confuciusornithids and *Zhongjianornis*, known from the toothless *Gobipteryx minuta* from Late Cretaceous deposits of the Gobi Desert (Elzanowski, 1974; Chiappe *et al.*, 2001). The only direct evidence to enantiornithine dietary preference is the preservation of the remains of freshwater arthropods interpreted as stomach contents found in association with the holotype of *Eoalulavis hoyasi* from the Early Cretaceous of Spain (Sanz *et al.*, 1996), and amber corpsules preserved with the holotype of *Enantiophoenix electrophyla* (Dalla Vecchia & Chiappe, 2002; Cau & Arduini, 2008). Since no cranial morphology is preserved in either specimen, correlations between dietary preferences and dental morphology cannot be inferred.

Developmentally, enantiornithines are interpreted as precocial based on the presence of asymmetrical vaned feathers on hatchlings, juveniles and even one embryo (Zhou & Zhang, 2004), as well as the high degree of ossification preserved in known embryos (Elzanowski, 1981). This is supported by studies that suggest altriciality and the bi-parental care necessary to sustain it evolved within crown group Aves (Cracraft, 1988; Sibley & Ahlquist, 1990; McKitrick, 1992; Hackett *et al.*, 2008). However, given correlations between egg size and developmental strategy in modern birds (larger eggs for precocial chicks), as in other basal birds, this developmental strategy may have been constrained relative to modern birds by the distally contacting pubes of known enantiornithines (Starck & Ricklefs, 1998; Dyke & Kaiser, 2010). Histology shows that the development of some of these birds was punctuated by periods of interrupted growth, however, perinatal specimens (late stage embryo or early hatchling) show fibrolamellar bone indicative of rapid growth, suggesting that enantiornithines grew rapidly until fledging

(Chinsamy & Elzanowski, 2001). When rapid growth ceased is unclear; although histological analysis suggests that rapid growth ceased when terminal size was nearly achieved (Cambra-Moo et al., 2006), the large number of very young fledged individuals suggests at least in some lineages rapid growth may have ceased very early in development (Chinsamy & Elzanowski, 2001; Chiappe et al., 2007).

Nearly every known Cretaceous avifauna had an enantiornithine component and even where they coexisted with other birds, both more primitive and more advanced, enantiornithines often represent the dominant group in terms of diversity and numbers. Their success during the Cretaceous makes their subsequent extinction alongside other nonavian dinosaurs perplexing. Their flight or respiratory inefficiencies relative to modern birds may have contributed to their demise, or their unique developmental strategy may have made the clade susceptible to extinction (Chinsamy & Elzanowski, 2001; Chiappe & Walker, 2002). Currently, there is little concrete evidence from which to draw conclusions, and their extinction continues to be one of the great puzzles of avian paleontology.

Ornithuromorpha: the rise of modern birds

Birds belonging to Ornithuromorpha, the clade that includes modern birds, are known in the fossil record as far back as the Early Cretaceous. Their abundant Cretaceous fossil record spans over nearly the same duration as that of their sister group, the enantiornithines. Ranging greatly in size and appearance, the Cretaceous ornithuromorphs achieved global distribution and occupied a wide range of ecological niches, inferred from their diverse morphological specializations. From their onset, these birds exhibit a number of characters that highlight their evolutionary proximity to modern birds (Neornithes). For example, in the skull they have lost the postorbital bone (Chiappe, 2002b), in the shoulder the coracoid possesses procoracoid and sternolateral (lateral) processes (Clarke et al., 2006), the scapula is curved and distally tapered, the

furcula is U-shaped (Zhou & Zhang, 2005), and the sternum typically has a well-developed keel (reduced among flightless taxa) that projects cranially from the bone's rostral margin. As in modern birds, the head of the humerus of basal ornithuromorphs is globose, the synsacrum has a greater number of incorporated vertebrae than more basal birds, the pygostyle is small and plow-shaped as in modern taxa, and the tarsometatarsus is often fully fused, with a primitive hypotarsus (Clarke et al., 2006). In addition, the growth patterns of basal ornithuromorphs, as they can be inferred from their bone microstructure, approach those of their living descendants with only a single line of arrested growth compared to multiple lines in more primitive birds (Chinsamy et al., 1995; Bell et al., 2010). However, basal ornithuromorphs also retain primitive features such as teeth (Jianchangornis), manual claws (Hongshanornis), pubes forming a distal symphysis (Yanornis), and a pelvis that lacks an ilioschiadic fenestra (the ischium and ilium do not contact distally) (Zhou & Zhang, 2001, 2005; Clarke et al., 2006; Zhou et al., 2009).

In the past few years, our knowledge of basal Ornithuromorpha has been greatly improved by important discoveries from the Early Cretaceous of China (Figures 3.14–3.17) (Zhou & Zhang, 2001, 2005, 2006; Clarke et al., 2006; You et al., 2006; O'Connor et al., 2010). These fossils have significantly augmented the previously limited global diversity of known basal ornithuromorphs, which included primarily incompletely known taxa such as Ambiortus dementjevi from the Early Cretaceous of Mongolia (Kurochkin, 1982, 1985), and Vorona berivotrensis from the Late Cretaceous of Madagascar (Forster et al., 1996), and the flightless, hen-sized Patagopteryx deferrariisi (Chiappe, 1995, 2002b) from the Late Cretaceous of Argentina. The latter is known from several incomplete specimens and is the best-represented basal ornithuromorph from the Mesozoic of the Southern Hemisphere. The recently discovered species of basal ornithuromorphs from the Early Cretaceous of China are represented by complete and well-preserved skeletons. These include the relatively small and possibly toothless Archaeorhynchus

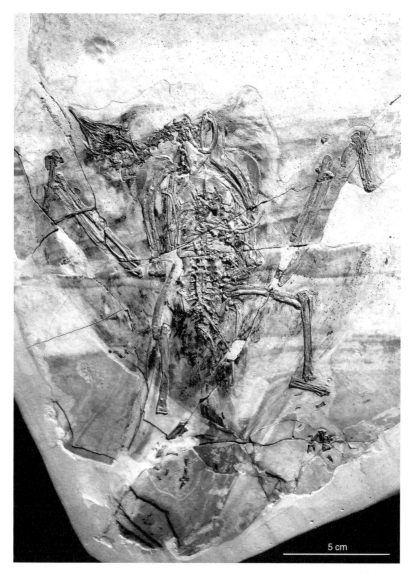

Fig. 3.14 Holotype of the Yixian Formation ornithuromorph *Archaeorhynchus spathula* IVPP V14287. (Photograph by Luis Chiappe.)

spathula (Figure 3.14; Zhou & Zhang 2006) and hongshanornithids (Zhou & Zhang, 2005; O'Connor *et al.*, 2010), and the much larger and toothed *Yixianornis grabaui* (Figure 3.15; Zhou & Zhang, 2001; Clarke *et al.*, 2006) and *Yanornis martini* (Figure 3.16; Zhou & Zhang, 2001; Zhou *et al.*, 2004). Many of these Chinese fossils preserve good portions of their plumage – fully modern flight, downy, and contour feathers are all represented. Although preserved tail feathers are rare, all known specimens preserve a large fan-shaped tail (Figure 3.15; Zhou & Zhang, 2005; Clarke *et al.*, 2006; O'Connor *et al.*, 2010), which would have generated substantial lift;

Fig. 3.15 Holotype of the ornithuromorph *Yixianornis grabaui* IVPP V12631, from the Jiufotang Formation. (Photograph by Luis Chiappe.)

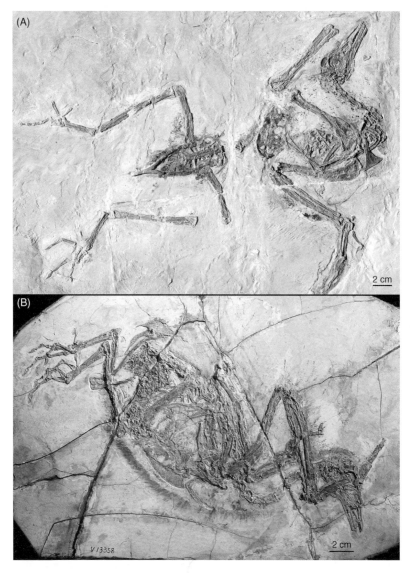

Fig. 3.16 (A) holotype specimen of the Jiufotang ornithuromorph *Yanornis martini* IVPP V12558; (B) referred specimen with gastroliths IVPP V13358a. (Photograph by Luis Chiappe.)

given the modern, plow-shaped appearance of the pygostyle in these more advanced birds, it is possible that a bulb rectricium was present (Clarke *et al.*, 2006). The exceptional preservation of some specimens of *Yanornis* has also allowed the identification of fish bones in the stomach of one specimen and numerous gastroliths (often asso-ciated with a more herbivorous diet) in others, suggesting that some early birds may have been capable of altering their diet based upon seasonal availability (Figure 3.16; Zhou & Zhang, 2001; Zhou *et al.*, 2004). *Archaeorhynchus* is possibly one of the most primitive known members of the group although the holotype is a subadult

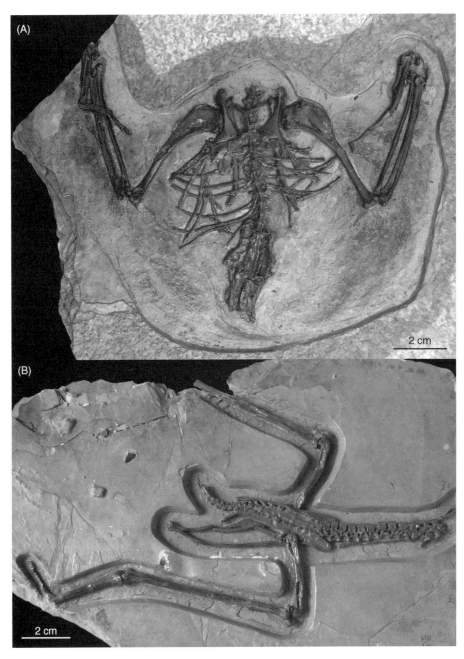

Fig. 3.17 Referred specimens of the Xiagou Formation ornithuromorph *Gansus yumenensis*: (A) CAGS 04-CM-004; (B) CAGS 04-CM-015. (Photograph by Hailu You, CAGS, with permission.)

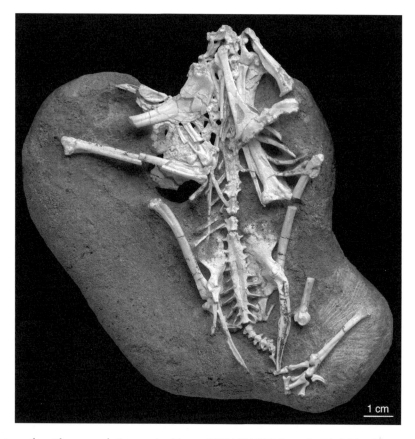

Fig. 3.18 Holotype of ornithuromorph *Apsaravis ukhaana* IGM 100/1017, from the Djadokhta Formation of Mongolia. (Mark Norell, AMNH, with permission.)

specimen (Figure 3.14; Zhou & Zhang, 2006; Zhou *et al.*, 2010), and the interrelationships between basal ornithuromorphs are currently not well defined.

Another important new basal ornithuromorph is the flighted *Apsaravis ukhaana* (Figure 3.18) (Norell & Clarke, 2001; Clarke & Norell, 2002), known from an exquisitely well-preserved specimen from the Late Cretaceous of Mongolia. The skeleton of *Apsaravis* shows clear similarity to that of modern birds – the dentaries are fused to one another, the thoracic vertebral series is reduced, and the pelvis is very broad. However, perhaps the most notable feature of *Apsaravis* is the development of a pronounced extensor process on the metacarpal I, which is involved in the

automatic extension of the hand by the propatagial ligaments (Fisher, 1957; Vazquez, 1992, 1994). The development of the extensor process in *Apsaravis* indicates that this bird was able to extend its wing automatically during the downstroke (Clarke & Norell, 2002).

Discoveries of more advanced ornithuromorphs include the Chinese Early Cretaceous *Gansus yumenensis*, known from dozens of well-preserved specimens from the Xiagou Formation, Gansu Province in China (Figure 3.17; You *et al.*, 2006) and the long-legged *Hollanda luceria* (Bell *et al.*, 2010) from the Late Cretaceous of Mongolia. *Hollanda* is very incomplete, known only from a hindlimb, the proportions of which are very close to those of a modern roadrunner, suggesting a cursorial lifestyle (Bell et al., 2010). *Gansus*, from the Early Cretaceous of China, is one of the most advanced non-ornithurine birds known, as well as one of the best represented (You et al., 2006). In addition to a number of advanced features relative to basal ornithuromorphs, *Gansus* shows several marked similarities to modern-day aquatic birds, such as a large and elongate proximocranially projecting cnemial crest on the tibiotarsus and the proximal position of the trochlea of metatarsal II. The pygostyle is greatly reduced suggesting that these birds, like modern grebes (Podicipediformes) and other modern water birds, had an extremely abbreviated tail. These taxa appear to be immediate outgroups of a clade that includes the famous *Ichthyornis* and *Hesperornis* (Marsh, 1880; Figure 3.19), Ornithurae.

The phylogenetic analysis presented as part of this chapter places *Ichthyornis* as a more remote outgroup of Neornithes than *Hesperornis*, an unusual tree topology that differs from most previous phylogenies in which *Ichthyornis* and Neornithes shared a most immediate common ancestor (Chiappe, 1996; Clarke et al., 2006; You et al., 2006; Zhou et al., 2008). Known since the 19th century (Marsh, 1880), *Ichthyornis* is abundant in the Late Cretaceous marine sediments of the Western Interior of North America (Clarke, 2004). Although toothed, *Ichthyornis* is remarkably modern in its anatomy. This fully flighted bird has been compared to a large, modern tern in terms of ecology (Clarke, 2004). Derived ornithuromorphs form a clade Ornithurae (Chiappe, 2002a), which contains the common ancestor of hesperornithiforms (including *Hesperornis*) and

5 cm

Fig. 3.19   Holotype of the Maastrichtian duck *Vegavis iaai* MLP 93-I03-1 from Antarctica. (Mark Norell, AMNH, with permission.)

Neornithes. The Hesperornithiformes comprise a unique group of highly derived foot-propelled diving birds. First recognized by O. C. Marsh over 100 years ago (Marsh, 1872a, b), today this group includes over 25 species classified within three major groups (i.e. enaliornithids, baptornithids, and hesperornithids). This makes hesperornithiforms one of the most diverse lineages of Mesozoic birds known, with a geographic distribution throughout the Northern Hemisphere and possibly extending to the Southern Hemisphere (Lambrecht, 1929). Members of the group occupy a wide range of body sizes; the smallest hesperornithiform, *Enaliornis*, was the size of a grebe, while the largest, *Hesperornis* and *Asiahesperornis*, had body lengths 1.5 times that of the emperor penguin. With the exception of the Early Cretaceous *Enaliornis*, all these taxa are known from the Late Cretaceous. Fossils of hesperornithiformes are known exclusively from aquatic or marine deposits and possess a suite of unusual skeletal adaptations which indicate specialization for a swimming lifestyle, such as a highly streamlined body, powerful hind limbs oriented behind the body rather than underneath it, and a reduced, nonfunctional wing. The majority of hesperornithiform fossils are known from deposits of the Western Interior Seaway of North America; these include taxa such as *Hesperornis regalis*, *Parahesperornis alexi*, and *Baptornis advenus*, which are known from nearly complete specimens. Hesperornithiform fossils found elsewhere, such as *Asiahesperornis* of Kazakhstan (Dyke *et al.*, 2006), *Judinornis* of Mongolia (Nessov & Borkin, 1983), and *Enaliornis* of England (Galton & Martin, 2002) are much more fragmentary.

The Mesozoic record of Neornithes, the crown-group clade of birds, is thus far limited to the Late Cretaceous. A number of fossils previously considered to be neornithine such as the Early Cretaceous *Neogaeornis wetzeli* (Lambrecht, 1929) of Chile and the Late Cretaceous *Teviornis gobiensis* (Kurochkin *et al.*, 2002) of Mongolia are largely incomplete and their alleged relationship with neornithines needs to be treated with caution. Less equivocal specimens have been collected, such as the purported loon, *Lonchodytes* from the Lance Creek Formation (Brokorb, 1963), however, the best record of Cretaceous neornithines is the latest Cretaceous *Vegavis iaai* from Antarctica (Figure 3.20; Clarke *et al.*, 2005). The holotype specimen is much more complete, preserving many advanced features absent in basal ornithuromorphs such as a fully developed hypotarsus, the morphology of which allows for the taxonomic assignment of the specimen within Anseriformes. This fossil indicates that by the close of the Cretaceous the Galliformes–Anseriformes split had already occurred and the radiation of modern birds was fully underway (Clarke *et al.*, 2005).

PHYLOGENETIC HYPOTHESES OF THE EARLY EVOLUTION OF BIRDS

Phylogenetic hypotheses regarding the interrelationships of Mesozoic avians appear often alongside descriptions of new taxa, however, these analyses tend to be limited, with heavy taxonomic sampling only within the clade of interest. Since the last major large-scale attempt to resolve Mesozoic bird relationships (Chiappe, 2002a), entire new lineages have been discovered (i.e. the Jeholornithidae and Sapeornithidae), many new species have been described, and a great deal of new information has accumulated on a number of critical taxa. The present analysis is the most comprehensive phylogenetic study to date – significantly increasing taxonomic sampling and including new and modified characters. Using an expanded version of the Gao *et al.* (2008) character list (which is modified from Chiappe (2002a) to incorporate new characters as well as characters from Clarke *et al.*, 2006), we have scored 54 taxa for 245 characters in order to place as much of the current taxonomic diversity within a single phylogenetic hypothesis (see Appendices 3A and 3B). The analysis was rooted using the Dromaeosauridae. All lineages of Mesozoic birds were sampled including six long-tailed birds (*Archaeopteryx lithographica*, *Rahonavis ostromi*, *Jeholornis prima*, *Jixiangornis orientalis*, *Shenzhouraptor sinensis*, and *Dalianraptor cuhe*), the short-tailed nonpygostylian *Zhongornis haoae*, two

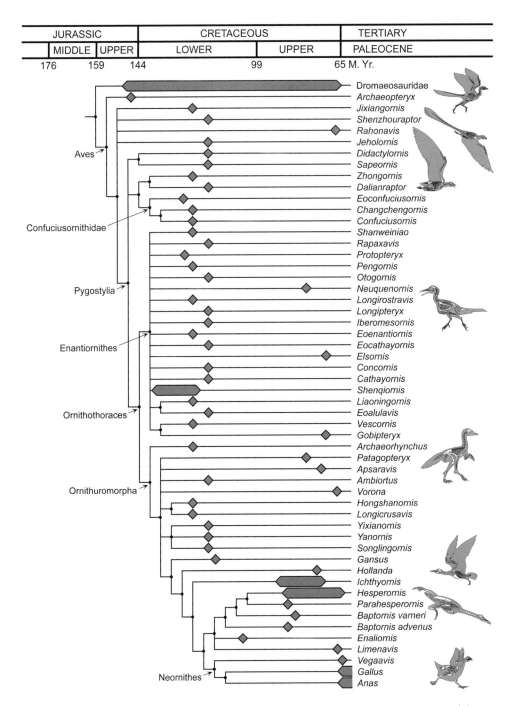

JURASSIC		CRETACEOUS		TERTIARY
MIDDLE	UPPER	LOWER	UPPER	PALEOCENE
176	159 144		99	65 M. Yr.

Dromaeosauridae
Archaeopteryx
Jixiangornis
Shenzhouraptor
Rahonavis
Jeholornis
Didactylornis
Sapeornis
Zhongornis
Dalianraptor
Eoconfuciusornis
Changchengornis
Confuciusornis
Shanweiniao
Rapaxavis
Protopteryx
Pengornis
Otogornis
Neuquenornis
Longirostravis
Longipteryx
Iberomesornis
Eoenantiornis
Eocathayornis
Elsornis
Concornis
Cathayornis
Shenqiornis
Liaoningornis
Eoalulavis
Vescornis
Gobipteryx
Archaeorhynchus
Patagopteryx
Apsaravis
Ambiortus
Vorona
Hongshanornis
Longicrusavis
Yixianornis
Yanornis
Songlingornis
Gansus
Hollanda
Ichthyornis
Hesperornis
Parahesperornis
Baptornis varneri
Baptornis advenus
Enaliornis
Limenavis
Vegaavis
Gallus
Anas

Aves
Confuciusornithidae
Pygostylia
Enantiornithes
Ornithothoraces
Ornithuromorpha
Neornithes

Fig. 3.20 Cladogram of Mesozoic bird relationships; the cladogram represents the strict consensus of the 1242 most parsimonious trees (length = 788 steps) derived from a cladistic analysis of 245 characters for 54 taxonomic units.

sapeornithids (*Sapeornis chaoyangensis* and *Dydactylornis jii*) and three confuciusornithids (*Eoconfuciusornis zhengi*, *Confuciusornis sanctus* and *Changchengornis hengdaoziensis*). Eighteen enantiornithines (approximately half the known diversity) were sampled (*Cathayornis yandica*, *Concornis lacustris*, *Elsornis keni*, *Eoalulavis hoyasi*, *Eocathayornis walkeri*, *Eoenantiornis buhleri*, *Gobipteryx minuta*, *Hebeiornis fengningensis*, *Iberomesornis romerali*, *Longirostravis hani*, *Longipteryx chaoyangensis*, *Neuquenornis volans*, *Otogornis genghisi*, *Pengornis houi*, *Protopteryx fengningensis*, *Rapaxavis pani*, *Shanweiniao cooperorum*, and *Shenqiornis mengi*). Twenty-three fossil ornithuromorph taxa were included (*Ambiortus dementjevi*, *Apsaravis ukhaana*, *Archaeorhynchus spathula*, *Baptornis varneri*, *Baptornis advenus*, *Enaliornis baretti*, *Gansus yummenensis*, *Hesperornis regalis*, *Hollanda luceria*, *Hongshanornis longicresta*, *Ichthyornis dispar*, *Liaoningornis longidigitrus*, *Limenavis patagonicus*, *Longicrusavis houi*, *Parahesperornis alexi*, *Patagopteryx deferrariisi*, *Songlingornis linghensis*, *Vegavis iaai*, *Vorona berivotrensis*, *Yanornis martini*, and *Yixianornis grabaui*), with *Anas platyrhynchos* and *Gallus gallus* scored for living Neornithes. The data were analyzed using the program TNT (Goloboff *et al.*, 2008); a heuristic parsimony search was conducted based on 1000 replications of tree bisection reconnection (TBR) retaining the single shortest tree from each replication followed by an additional round of TBR. The analysis produced 1242 trees of 788 steps. The differences in trees are found primarily within the enantiornithine clade, with these taxa almost entirely forming a polytomy in the strict consensus as in previous attempts to understand this clade (Chiappe & Walker, 2002; Chiappe *et al.*, 2006; Clarke *et al.*, 2006; Zhou *et al.*, 2008; O'Connor *et al.*, 2009).

The strict consensus tree confirms the basal placement of *Archaeopteryx* relative to other known birds and resolves *Rahonavis*, *Jixiangornis*, *Jeholornis* and *Shenzhouraptor* in a polytomy with Pygostylia. Sapeornithidae, Confuciusornithidae, and Ornithothoraces form a polytomy. *Dalianraptor* and *Zhongornis* fall with in this

polytomy, forming a dichotomy with Confuciusornithidae. *Eoconfuciusornis* is resolved as a basal confuciusornithid.

Ornithothoraces consists of a dichotomy between Enantiornithes and Ornithuromorpha. Within Enantiornithes there is little resolution; *Liaoningornis* is resolved as an enantiornithine (formerly interpreted as an ornithuromorph; Zhou & Zhang, 2006) forming a relationship with *Eoalulavis*. *Gobipteryx* and *Hebeiornis* form another clade. *Archaeorhynchus* is resolved as the most basal ornithuromorph. *Patagopteryx*, *Apsaravis*, *Ambiortus*, *Vorona*, Hongshanornithidae (*Hongshanornis* and *Longicrusavis*), and Songlingornithidae (*Songlingornis*, *Yanornis*, and *Yixianornis*) form a polytomy with more derived ornithuromorphs: *Gansus*, *Ichthyornis*, and *Hollanda* form successive outgroups to Ornithurae, a dichotomy between Neornithes and Hesperornithiformes. *Limenavis* and *Enaliornis* form a polytomy with other hesperornithiforms. *Parahesperornis* and *Hesperornis* form the derived clade, with *Baptornis varneri* and *Baptornis advenus* forming successive outgroups. *Vegavis* is resolved outside the clade formed by *Anas* and *Gallus*.

There are several interesting hypotheses suggested by these results. *Rahonavis* is typically resolved as more primitive than the Early Cretaceous long-tailed birds, but is here resolved in a polytomy with these taxa. This polytomy does support previous inferences that *Shenzhouraptor* and *Jixiangornis* may be junior synonyms of *Jeholornis*, however, the fact that *Rahonavis* also falls in this polytomy suggests a lack of character sampling for the most basal, long-tailed birds. In depth studies into the taxonomic validity of *Shenzhouraptor* and *Jixiangornis*, detailed data on the morphology of these long-tailed birds, and the description of new *Rahonavis* material will certainly lend clarity to the relationships of these taxa.

The existence of a *Dalianraptor* and *Zhongornis* clade closely related to the confuciusornithids requires further investigation. *Dalianraptor*, which is poorly known, is only ambiguously aligned with *Zhongornis*, the result of uncertain scorings in the former. The poor preservation of both taxa and juvenile status of the only known specimen of *Zhongornis* may be clouding the phylogenetic placement of these taxa. *Dalianraptor* also shows distinct departures from *Jeholornis* and other Jehol long-tailed birds, such as major differences in proportions and a subsequently much shorter wing in *Dalianraptor*. These differences suggest that *Dalianraptor*, like *Jinfengopteryx*, may prove not to be a bird and requires further investigation (Turner *et al.*, 2007).

Confuciusornis is most commonly resolved as the closest sister taxon to Ornithothoraces (Zhou & Zhang, 2005; Clarke *et al.*, 2006; Zhou *et al.*, 2008), however, *Sapeornis* has also been resolved as the more derived pygostylian (O'Connor *et al.*, 2009). These clades are here unresolved in a polytomy with Ornithothoraces. More in-depth study and higher-character sampling, especially within Sapeornithidae, may help to clarify this relationship.

As in previous analyses, Ornithothoraces is resolved as a dichotomy between Enantiornithes and Ornithuromorpha. Enantiornithes, also consistent with previous analyses (Chiappe, 2002a; Zhang *et al.*, 2008b), is largely resolved as a polytomy, with even previously phylogenetically supported clades such as Longipterygidae no longer resolved (Chiappe *et al.*, 2006; O'Connor *et al.*, 2009). This highlights an already apparent problem with this clade, and even enantiornithine specific analyses have failed to resolve the clade (Chiappe & Walker, 2002; Chiappe *et al.*, 2006). While the interrelationships of the clade may prove very difficult to clarify, this polytomy may also be resolved by increased character sampling and new material for existing taxa. The Jehol bird *Liaoningornis* is resolved as part of the enantiornithine clade and closely related to the Spanish *Eoalulavis*. This relationship is supported by the bizarre and similar sternal morphology of these two taxa. The placement of *Liaoningornis* outside Ornithuromorpha is further supported by the absence of derived morphologies such as two cnemial crests on the tibiotarsus or a fully fused tarsometatarsus, as well as the significant departure in sternal morphology from other ornithuromorph taxa, in which the sternum is fairly conservative.

The basal placement of *Archaeorhynchus* within Ornithuromorpha has been supported by previous analyses (Zhou & Zhang, 2006; Zhou *et al.*, 2010). Basal ornithuromorphs are not resolved, and *Apsaravis* is included in this polytomy despite some previous analyses that have suggested this taxon is more derived (Clarke *et al.*, 2006; You *et al.*, 2006; Zhou & Zhang, 2006). A songlingornithid clade formed by *Songlingornis*, *Yixianornis*, and *Yanornis* is resolved, which has been supported in previous analyses (Clarke *et al.*, 2006). *Gansus* falls outside Ornithurae, *contra* You *et al.* (2006), which may reflect new information on this taxon as well as the influence of greater taxonomic sampling of ornithuromorph taxa in the present analysis.

Most past analyses have placed *Ichthyornis* closer to modern birds than *Hesperornis* (Chiappe, 2002a; Clarke, 2004; Zhou & Zhang, 2005; You *et al.*, 2006; Zhou *et al.*, 2010). This study includes a comprehensive review of the anatomy of Hesperornithiformes and thus may be a more accurate assessment of the relationships of these taxa. Hesperornithiformes are here resolved as closer to living birds than *Ichthyornis*, which would then exclude *Ichthyornis* from Ornithurae (this clade is defined as the common ancestor of *Hesperornis* and Neornithes plus all its descendants). The placement of *Vegavis* outside Neornithes is likely the result of very low character sampling in this derived clade.

These results highlight taxa that cannot be resolved further without additional information (i.e. *Dalianraptor*), as well as pre-existing problems (i.e. the large enantiornithine polytomy). The lack of resolution in the basal region of the tree may reflect a trend similar to that observed in the enantiornithine clade, with increasing taxonomic diversity without detailed studies producing a lack of phylogenetic resolution. The relationships between and within avian clades require detailed studies of the several poorly known taxa and greater amounts of available morphological data on all taxa to untangle. An increase in available morphological data can be translated to characters and character states to better reflect the known morphological variation in order to create phylogenetic hypotheses with increasing accuracy.

CONCLUSIONS

When taxa are placed in the frame of a phylogenetic hypothesis, the Mesozoic avian fossil record reveals several interesting hypotheses regarding early evolutionary trajectories. Avian evolution is highly plastic and numerous characters and ecomorphs acquired during the Cenozoic evolution of Neornithes had already been experimented with by multiple lineages of Mesozoic birds. For example, the absence of teeth – a synapomorphy for Neornithes – evolved in several lineages of Mesozoic, pre-modern birds (e.g. Confuciusornithidae, *Gobipteryx*) as well as outside Aves within Theropoda. The postorbital bone likely reduced several times outside Ornithuromorpha where it was lost, and the diapsid condition appears secondarily derived within the confuciusornithids and at least one enantiornithine. Manual reduction also shows convergent reduction within Enantiornithes and Ornithuromorpha although the manus of each group is distinct morphologically from the other. Ecologically, the Mesozoic radiation of birds also achieved and thus pre-dated some of the specializations seen among their living counterparts. For example, flightlessness – typical of numerous lineages of modern birds – is inferred to have evolved independently in at least three lineages of Mesozoic birds (i.e. *Patagopteryx*, Hesperornithiforms, and possibly in the enantiornithine *Elsornis*). Littoral ecomorphs evolved in the form of wading birds in both clades of ornithothoracines (*Lectavis* and the hongshanornithids) as well as in the form of trophically specialized mud-probing taxa (enantiornithines *Longirostravis* and *Rapaxavis*). The flightless hesperornithiforms were highly specialized divers, similar in appearance to the modern grebes, with dense bones and rotating lobed feet.

The Late Jurassic *Archaeopteryx* continues to be the oldest and the most primitive known bird. By around 130–120 Ma, a large number of lineages make their debut in the fossil record in the now celebrated sediments of the Jehol Group in northeastern China. No information, however, is available for the roughly 20 million-year-gap between when *Archaeopteryx* lived and when the Jehol avifauna flourished. Undoubtedly, this time period holds critical clues for understanding the early phases of avian diversification in the Mesozoic. The Jehol avifauna provides an excellent picture of avian diversification in the Early Cretaceous. A number of long-tailed birds, whose precise taxonomy and relationships need to be further studied, are recorded alongside early pygostylians (short-tailed birds) and ornithothoracines (enantiornithines and ornithuromorphs). With the exception of the insular *Rahonavis* (Forster *et al.*, 1998a), which some interpret as nonavian (Makovicky *et al.*, 2005), all pre-ornithothoracine birds disappear from the fossil record after the time recorded by the Jehol avifauna. Later in the Early Cretaceous, both enantiornithines and ornithuromorphs diversified across the globe. Throughout the Cretaceous Period these birds occupied a wide variety of marine, shoreline, and continental habitats using a diversity of morphological features. Early in their divergence, these birds established themselves as strong flyers, possessing skeletal (i.e. narrow furcula, keel) and integumentary modifications for flight (i.e. alula, tail fan). Nested within the ornithuromorph radiation is the radiation of modern birds. The fossil record indicates that this radiation was well established by the close of the Cretaceous although the timing and sequence of this diversification is still unclear.

The inferred evolutionary relationships between pre-modern birds continue to change as new lineages are discovered and new data becomes available for known taxa. The incredible diversity now known reflects a large amount of new morphological information that is not currently reflected in publications. Detailed studies of these fascinating new birds will lend greater detail to our understanding of the earliest avifaunas.

ACKNOWLEDGMENTS

We thank Stephanie Abramowicz for assistance with figures, and Mark Norell, Cathy Forster, Hailu You, and Shu'an Ji for use of specimen photographs. Thanks to Zhonghe Zhou at the

Institute of Vertebrate Paleontology and Paleoanthropology and Chunling Gao at the Dalian Natural History Museum for specimen access. We thank National Science Foundation for funding.

APPENDIX 3A

There are 245 characters total included in this analysis. Each character is described in the list below, which largely follows Gao *et al.* (2008) and O'Connor *et al.* (2009) but with the following modifications. Characters 4, 9, 95, 159, 202, 204, 210, 222, 227, and 234 in the former lists have been reworded or expanded to possess an additional state. Character 5 has been removed. Five new characters are also included (in this list: 80, 108, 202, 237, and 241). Descriptions for some of the characters include clarification of the morphology that is intended to be scored for given characters and states. The scores for each character against each of the 54 taxa (Figure 3.19) are tabulated in Appendix 3B.

Skull and mandible

1. Premaxillae in adults: unfused (0); fused only rostrally (1); completely fused (2). (ORDERED)
2. Maxillary process of the premaxilla: restricted to its rostral portion (0); subequal or longer than the facial contribution of the maxilla (1).
3. Frontal process of the premaxilla: short (0); relatively long, approaching the rostral border of the antorbital fenestra (1); very long, extending caudally near the level of lacrimals (2). (ORDERED)
4. Premaxillary teeth: present throughout (0); present but rostral tip edentulous (1); present but restricted to rostral portion (2); absent (3).
5. Caudal margin of naris: far rostral than the rostral border of the antorbital fossa (0); nearly reaching or overlapping the rostral border of the antorbital fossa (1).
6. Naris longitudinal axis: considerably shorter than the long axis of the antorbital fossa (0);

subequal or longer (1). We are using the longitudinal axis of these structures as a proxy for their relative size. The longitudinal axis is often easier to measure than the actual area enclosed by either the naris or the antorbital fossa.

7. Maxillary teeth: present (0); absent (1).
8. Dorsal (ascending) ramus of the maxilla: present with two fenestra (the promaxilllary and maxillary fenestra) (0); present with one fenestra (1); unfenestrated (2); ramus absent (3). (ORDERED)
9. Caudal margin of choana: located rostrally, not overlapping the region of the orbit (0); displaced caudally, at the same level or overlapping the rostral margin of the orbit (1).
10. Rostral margin of the jugal: away from the caudal margin of the naris (0); or very close to (leveled with) the caudal margin of the naris (1).
11. Contact between palatine and maxilla/premaxilla: palatine contact maxilla only (0); contacts premaxilla and maxilla (1).
12. Vomer and pterygoid articulation: present, well developed (0); reduced, narrow process of pterygoid passes dorsally over palatine to contact vomer (1); absent, pterygoid and vomer do not contact (2).
13. Jugal process of palatine: present (0); absent (1).
14. Contact between palatine and pterygoid: long, craniocaudally overlapping contact (0); short, primarily dorsoventral contact (1).
15. Contact between vomer and premaxilla: present (0); absent (1).
16. Ectopterygoid: present (0); absent (1).
17. Postorbital: present (0); absent (1).
18. Contact between postorbital and jugal: present (0); absent (1).
19. Quadratojugal: sutured to the quadrate (0); joined through a ligamentary articulation (1).
20. Lateral, round cotyla on the mandibular process of the quadrate (quadratojugal articulation): absent (0); present (1).
21. Contact between the quadratojugal and squamosal: present (0); absent (1).
22. Squamosal incorporated into the braincase, forming a zygomatic process: absent (0); present (1).

23 Squamosal, ventral or "zygomatic" process: variably elongate, dorsally enclosing otic process of the quadrate and extending cranioventrally along shaft of this bone, dorsal head of quadrate not visible in lateral view (0); short, head of quadrate exposed in lateral view (1).

24 Frontal/parietal suture in adults: open (0); fused (1).

25 Quadrate orbital process (pterygoid ramus): broad (0); sharp and pointed (1).

26 Quadrate pneumaticity: absent (0); present (1).

27 Quadrate: articulating only with the squamosal (0); articulating with both prootic and squamosal (1).

28 Otic articulation of the quadrate: articulates with a single facet (squamosal) (0); articulates with two distinct facets (prootic and squamostal) (1); articulates with two distinct facets and quadrate differentiated into two heads (2). (ORDERED)

29 Quadrate distal end: with two transversely aligned condyles (0); with a triangular, condylar pattern, usually composed of three distinct condyles (1).

30 Basipterygoid processes: long (0); short (articulation with pterygoid subequal to, or longer than, amount projected from the basisphenoid rostrum) (1).

31 Pterygoid, articular surface for basipterygoid process: concave "socket," or short groove enclosed by dorsal and ventral flanges (0); flat to convex (1); flat to convex facet, stalked, variably projected (2). (ORDERED)

32 Eustachian tubes: paired, lateral, and well-separated from each other (0); paired, close to each other and to cranial mid-line or forming a single cranial opening (1).

33 Osseous interorbital septum (mesethmoid): absent (0); present (1).

34 Dentary teeth: present (0); absent (1).

35 Dentary tooth implantation: teeth in individual sockets (0); teeth in a communal groove (1).

36 Symphysial portion of dentaries: unfused (0); fused (1).

37 Deeply notched rostral end of the mandibular symphysis: absent (0); present (1).

38 Mandibular symphysis, symphyseal foramina: absent (0); single (1); paired (2).

39 Mandibular symphysis, symphyseal foramen/foramina: opening on caudal edge of symphysis (0); opening on dorsal surface of symphysis (1).

40 Small ossification present at the rostral tip of the mandibular symphysis (intersymphysial ossification): absent (0); present (1). Martin (1987:13) refers to this ossification as the "predentary." This term is inappropriate as it implies a homology between this ossification and the predentary bone of ornithischian dinosaurs – a hypothesis that is not supported by parsimony.

41 Caudal margin of dentary strongly forked: unforked, or with a weakly developed dorsal ramus (0); strongly forked with the dorsal and ventral rami approximately equal in caudal extent (1).

42 Mandibular ramus sigmoidal such that the rostral tip is dorsally convex and the caudal end is dorsally concave: absent (0); present (1).

43 Cranial extent of splenial: stops well caudal to mandibular symphysis (0); extending to mandibular symphysis, though noncontacting (1); extending to proximal tip of mandible, contacting on mid-line (2). (ORDERED)

44 Meckel's groove (medial side of mandible): not completely covered by splenial, deep and conspicuous medially (0); covered by splenial, not exposed medially (1).

45 Rostral mandibular fenestra: absent (0); present (1).

46 Caudal mandibular fenestra: present (0); absent (1). We regard the caudal mandibular fenestra of neornithines as homologous to the surangular fenestra of non-avian dinosaurs (Chiappe, 2002b).

47 Articular pneumaticity: absent (0); present (1).

48 Teeth: serrated crowns (0); unserrated crowns (1).

Vertebral column and ribs

49 Atlantal hemiarches in adults: unfused (0); fused, forming a single arch (1).

50 One or more pneumatic foramina piercing the centra of mid-cranial cervicals, caudal to the level of the parapophysis-diapophysis: present (0); absent (1).

51 Cervical vertebrae: variably dorsoventrally compressed, amphicoelous ("biconcave": flat to concave articular surfaces) (0); cranial surface heterocoelous (i.e., mediolaterally concave, dorsoventrally convex), caudal surface flat or slightly concave (1); heterocoelous cranial (i.e., mediolaterally concave, dorsoventrally convex) and caudal (i.e., mediolaterally convex, dorsoventrally concave) surfaces (2). (ORDERED)

52 Prominent carotid processes in the intermediate cervicals: absent (0); present (1).

53 Postaxial cervical epipophyses: prominent, projecting further back from the postzygapophysis (0); weak, not projecting further back from the postzygapophysis, or absent (1).

54 Keel-like ventral surface of cervical centra: absent (0); present (1).

55 Prominent (50% or more the height of the centrum's cranial articular surface) ventral processes of the cervicothoracic vertebrae: absent (0); present (1).

56 Thoracic vertebral count: 13–14 (0); 11–12 (1); fewer than 11 (2). The transition between cervical and thoracic vertebrae is often difficult to identify, which makes counting these vertebrae problematic. Here, thoracic vertebrae are defined as possessing free, ventrally projecting ribs. When inarticulated, vertebral morphology should be used. (ORDERED)

57 Thoracic vertebrae: at least part of series with subround, central articular surfaces (e.g. amphicoelous/opisthocoelous) that lack the dorsoventral compression seen in heterocoelous vertebrae (0); series completely heterocoelous (1).

58 Caudal thoracic vertebrae, centra, length and mid-point width: approximately equal in length and mid-point width (0); length markedly greater than mid-point width (1).

59 Wide vertebral foramen in the mid-caudal thoracic vertebrae, vertebral foramen/articular cranial surface ratio (vertical diameter) larger than 0.40: absent (0); present (1).

60 Hyposphene–hypantrum accessory intervertebral articulations in the thoracic vertebrae: present (0); absent (1).

61 Lateral side of the thoracic centra: weakly or not excavated (0); deeply excavated by a groove (1); excavated by a broad fossa (2).

62 Cranial thoracic vertebrae, parapophyses: located in the cranial part of the centra of the thoracic vertebrae (0); located in the central part of the centra of the thoracic vertebrae (1).

63 Notarium: absent (0); present (1).

64 Sacral vertebrae, number ankylosed (synsacrum): less than 7 (0); 7 (1); 8 (2); 9 (3); 10 (4); 11 or more (5); 15 or more (6). (ORDERED)

65 Synsacrum, procoelous articulation with last thoracic centrum (deeply concave facet of synsacrum receives convex articulation of last thoracic centrum): absent (0); present (1).

66 Cranial vertebral articulation of first sacral vertebra: approximately equal in height and width (0); wider than high (1).

67 Series of short sacral vertebrae with dorsally directed parapophyses just cranial to the acetabulum: absent (0); present, three such vertebrae (1); present, four such vertebrae (2). (ORDERED)

68 Convex caudal articular surface of the synsacrum: absent (0); present (1).

69 Degree of fusion of distal caudal vertebrae: fusion absent (0); few vertebrae partially ankylosed (intervening elements are well-discernible) (1); vertebrae completely fused into a pygostyle (2). (ORDERED)

70 Free caudal vertebral count: more than 35 (0); 35–26 (1); 25–20 (2); 19–9 (3); 8 or less (4). (ORDERED)

71 Procoelous caudals: absent (0); present (1).

72 Distal caudal vertebra prezygapophyses: elongate, exceeding the length of the centrum by more than 25% (0); shorter (1); absent (2). (ORDERED)

73 Free caudals, length of transverse processes: approximately equal to, or greater than, centrum width (0); significantly shorter than centrum width (1).

74 Proximal haemal arches: elongate, at least three times longer than wider (0); shorter (1); absent (2). (ORDERED)

75 Pygostyle: longer than or equal to the combined length of the free caudals (0); shorter (1).

76 Cranial end of pygostyle dorsally forked: absent (0); present (1).

77 Cranial end of pygostyle with a pair of laminar, ventrally projected processes: absent (0); present (1).

78 Distal constriction of pygostyle: absent (0); present (1). In the pygostyles of some enantiornithine taxa, the distalmost mediolateral width is reduced so that the mid-line of the pygostyle projects distally farther than the lateral margins (Chiappe *et al.*, 2002b).

79 Ossified uncinate processes in adults: absent (0); present and free (1); present and fused (2).

80 Uncinate process, orientation: perpendicular to rib (0); angled dorsally defining an acute angle with the rib (1).

81 Gastralia: present (0); absent (1).

Thoracic Girdle and Sternum

82 Coracoid shape: rectangular to trapezoidal in profile (0); strut-like (1).

83 Coracoid and scapula articulation: through a wide, sutured articulation (0); through more localized facets (1).

84 Scapula: articulated at the shoulder (proximal) end of the coracoid (0); well below it (1).

85 Coracoid, humeral articular (glenoid) facet: dorsal to acrocoracoid process/"biceps tubercle" (0); ventral to acrocoracoid process (1).

86 Humeral articular facets of the coracoid and the scapula: placed in the same plane (0); forming a sharp angle (1).

87 Coracoid, acrocoracoid: straight (0); hooked medially (1).

88 Laterally compressed shoulder end of coracoid, with nearly aligned acrocoracoid process, humeral articular surface, and scapular facet, in dorsal view: absent (0); present (1).

89 Procoracoid process on coracoid: absent (0); present (1).

90 Distinctly convex lateral margin of coracoid: absent (0); present (1).

91 Broad, deep fossa on the dorsal surface of the coracoid (dorsal coracoidal fossa): absent (0); present (1).

92 Supracoracoidal nerve foramen of coracoid: centrally located (0); displaced toward (often as an incisure) the medial margin of the coracoid (1); displaced so that its nerve no longer passes through the coracoid (absent) (2). (ORDERED) In some taxa the n. supracoracoideus does not pierce the coracoid, but is assumed to pass medially at the level between the bone's mid-point and its glenoid (humeral articular facet).

93 Coracoid, medial surface, strongly depressed elongate furrow at the level of the passage of n. supracoracoideus: absent (0); present (1).

94 Supracoracoid nerve foramen, location relative to dorsal coracoidal fossa: above fossa (0); inside fossa (1).

95 Coracoid, sternolateral corner: unexpanded (0); expanded (1); well developed squared-off lateral process (sternocoracoidal process) (2); present and with a distinct omal projection (hooked) (3).

96 Scapular shaft: straight (0); sagittally curved (1).

97 Scapula, length: shorter than humerus (0); as long as or longer than humerus (1).

98 Scapular acromion costolaterally wider than deeper: absent (0); present (1).

99 Scapula, acromion process: projected cranially surpassing the articular surface for coracoid (facies articularis coracoidea; Baumel & Witmer, 1993) (0); projected less cranially than the articular surface for coracoid (1).

100 Scapula, acromion process: straight (0); laterally hooked tip (1).

101 Proximal end of scapula, pit between acromion and humeral articular facet (scapular fossa): absent (0); present (1).

102 Costal surface of scapular blade with prominent longitudinal furrow: absent (0); present (1).

103 Scapular caudal end: blunt (may or may not be expanded) (0); sharply tapered (1).

104 Furcular, shape: boomerang-shaped (0); V- to Y-shaped (1); U-shaped (2).

105 Furcula interclavicular angle: approximately 90° (0); less than 70° (1). The interclavicular angle is measured as the angle formed between three points, one at the omal end of each rami and the apex located at the clavicular symphysis.

106 Dorsal and ventral margins of the furcula: subequal in width (0); ventral margin distinctly wider than the dorsal margin so that the furcular ramus appears concave laterally (1).

107 Hypocleideum: absent (0); present as a tubercle or short process (1); present as an elongate process approximately 30% rami length (2); hypertrophied, exceeding 50% rami length (3). ORDERED

108 Sternum: unossified (0); partially ossified, coracoidal facets cartilaginous (1); fully ossified (2).

109 Ossified sternum: two flat plates (0); single flat element (1); single element, with slightly raised mid-line ridge (2); single element, with projected carina (3).

110 Sternal carina: near to, or projecting rostrally from, the cranial border of the sternum (0); not reaching the cranial border of the sternum (1).

111 Sternum, caudal margin, number of paired caudal trabecula: none (0); one (1); two (2). The use of "lateral" and "medial" to identify the specific sternal processes is abandoned here due to the difficulty of identifying trabecula when only one is present. *Eoenantiornis* is scored as a "?" due to the uncertain status of the sternal processes; it is possible that the identified "lateral process" (Zhou *et al.*, 2005) is actually the distal humerus.

112 Sternum, outermost trabecula, shape: tips terminate cranial to caudal end of sternum (0); tips terminate at or approaching caudal

end of sternum (1); tips extend caudally past the termination of the sternal mid-line (2).

113 Prominent distal expansion in the outermost trabecula of the sternum: absent (0); present (1).

114 Rostral margin of the sternum broad and rounded: absent (0); present (1).

115 Sternum, coracoidal sulci spacing on cranial edge: widely separated mediolaterally (0); adjacent (1); crossed on mid-line (2). In taxa such as *Eoalulavis* in which the preserved sternum does not bear actual sulci, the placement of the coracoids can be used to infer their position relative to the sternum.

116 Costal facets of the sternum: absent (0); present (1).

117 Sternal costal processes: three (0); four (1); five (2); six (3); seven (4); eight (5). (ORDERED)

118 Sternal mid-line, caudal end: blunt W-shape (0); V-shape (1); elongate straight projection (xiphoid process) (2); flat (3); rounded (4).

119 Sternum, caudal half, paired enclosed fenestra: absent (0); present (1).

120 Sternum, dorsal surface, pneumatic foramen (or foramina): absent (0); present (1).

Thoracic limb

121 Proximal and distal humeral ends: twisted (0); expanded nearly in the same plane (1).

122 Humeral head: concave cranially and convex caudally (0); globe shaped, craniocaudally convex (1).

123 Proximal margin of the humeral head concave in its central portion, rising ventrally and dorsally: absent (0); present (1).

124 Humerus, proximocranial surface, well-developed circular fossa on mid-line: absent (0); present (1).

125 Humerus with distinct transverse ligamental groove: absent (0); present (1).

126 Humerus, ventral tubercle projected caudally, separated from humeral head by deep capital incision: absent (0); present (1).

127 Pneumatic fossa in the caudoventral corner of the proximal end of the humerus: absent or rudimentary (0); well developed (1).

128 Humerus, deltopectoral crest: projected dorsally (the plane of the crest is coplanar to the cranial surface of the humerus) (0); projected cranially (1).

129 Humerus, deltopectoral crest: less than shaft width (0); approximately same width (1); prominent and subquadrangular (i.e., subequal length and width) (2).

130 Humerus, deltopectoral crest, perforated by a large fenestra: absent (0); present (1).

131 Humerus, bicipital crest: little or no cranial projection (0); developed as a cranial projection relative to shaft surface in ventral view (1); hypertrophied, rounded tumescence (2).

132 Humerus, distal end of bicipital crest, pit-shaped fossa for muscular attachment: absent (0); craniodistal on bicipital crest (1); directly ventrodistal at tip of bicipital crest (2); caudodistal, variably developed as a fossa (3).

133 Distal end of the humerus very compressed craniocaudally: absent (0); present (1).

134 Humerus, demarcation of muscle origins (e. g. m. extensor metacarpi radialis in Aves) on the dorsal edge of the distal humerus: no indication (0); a variably projected scar-bearing tubercle (dorsal supracondylar process) (2).

135 Well-developed brachial depression on the cranial face of the distal end of the humerus: absent (0); present (1). We interpret the brachial fossa not as a depression on the craniodistal end of the humerus but as a distinct scar for muscle attachment.

136 Well-developed olecranon fossa on the caudal face of the distal end of the humerus: absent (0); present (1).

137 Humerus, distal end, caudal surface, groove for passage of m. scapulotriceps: absent (0); present (1).

138 Humerus, m. humerotricipitalis groove: absent (0); present as a well-developed ventral depression contiguous with the olecranon fossa (1).

139 Humerus, distal margin: approximately perpendicular to long axis of humeral shaft (0); ventrodistal margin projected significantly distal to dorsodistal margin, distal margin angled strongly ventrally (sometimes described as a well-projected flexor process) (1).

140 Humeral distal condyles: mainly located on distal aspect (0); on cranial aspect (1).

141 Humerus, long axis of dorsal condyle: at low angle to humeral axis, proximodistally oriented (0); at high angle to humeral axis, almost transversely oriented (1).

142 Humerus, distal condyles: subround, bulbous (0); weakly defined, "strap-like" (1).

143 Humerus, ventral condyle: length of long axis of condyle less than the same measure of the dorsal condyle (0); same or greater (1).

144 Ulna: shorter than humerus (0); nearly equivalent to or longer than humerus (1).

145 Ulnar shaft, radial-shaft/ulnar-shaft ratio: larger than 0.70 (0); smaller than 0.70 (1).

146 Ulna, cotylae: dorsoventrally adjacent (0); widely separated by a deep groove (1).

147 Ulna, dorsal cotyla strongly convex: absent (0); present (1).

148 Ulna, bicipital scar: absent (0); developed as a slightly raised scar (1); developed as a conspicuous tubercle (2).

149 Proximal end of the ulna with a well-defined area for the insertion of m. brachialis anticus: absent (0); present (1).

150 Semilunate ridge on the dorsal condyle of the ulna: absent (0); present (1).

151 Shaft of radius with a long longitudinal groove on its ventrocaudal surface: absent (0); present (1).

152 Ulnare: heart-shaped with little differentiation into short rami (0); U-shaped to V-shaped, well-developed rami (1).

153 Ulnare, ventral ramus (crus longus, Baumel & Witmer, 1993): shorter than dorsal ramus (crus brevis) (0); same length as dorsal ramus (1); longer than dorsal ramus (2).

154 Semilunate carpal and proximal ends of metacarpals in adults: unfused (0); semilunate fused to the alular (I) metacarpal (1); semilunate fused to the major (II) and minor (III) metacarpals (2); fusion of semilunate and all metacarpals (3). Any specimen that is inferred to be a juvenile should be scored as a "?" in order to account for the possibility of ontogenetic change.

155 Semilunate carpal, position relative to the alular metacarpal (I): over entire proximal surface (0); over less than one-half proximal surface or no contact present (1).

156 Carpometacarpus, proximal ventral surface: flat (0); raised ventral projection contiguous with minor metacarpal (1); pisiform process forming a distinct peg-like projection (2).

157 Carpometacarpus, ventral surface, supratrochlear fossa deeply excavating proximal surface of pisiform process: absent (0); present (1).

158 Round-shaped alular metacarpal (I): absent (0); present (1).

159 Alular metacarpal (I), extensor process: absent, no cranioproximally projected muscular process (0); present, tip of extensor process just surpassed the distal articular facet for phalanx 1 in cranial extent (1); tip of extensor process conspicuously surpasses articular facet by approximately half the width of facet, producing a pronounced knob (2); tip of extensor process conspicuously surpasses articular facet by approximately the width of facet, producing a pronounced knob (3). (ORDERED)

160 Alular metacarpal (I), distal articulation with phalanx I: ginglymoid (0); shelf (1); ball-like (2).

161 Metacarpal III, craniocaudal diameter as a percentage of same dimension of metacarpal II: approximately equal or greater than 50% (0); less than 50% (1).

162 Proximal extension of metacarpal III: level with metacarpal II (0); ending distal to proximal surface of metacarpal II (1).

163 Intermetacarpal process or tubercle on metacarpal II: absent (0); present as scar (1); present as tubercle or flange (2).

164 Intermetacarpal space: absent or very narrow (0); at least as wide as the maximum width of minor metacarpal (III) shaft (1).

165 Intermetacarpal space: reaches proximally as far as the distal end of metacarpal I (0); terminates distal to end of metacarpal I (1).

166 Distal end of metacarpals: unfused (0); partially or completely fused (1).

167 Minor metacarpal (III) projecting distally more than the major metacarpal (II): absent (0); present (1).

168 Alular digit (I): long, exceeding the distal end of the major metacarpal (0); subequal (1); short, not surpassing this metacarpal (2). (ORDERED)

169 Proximal phalanx of major digit (II): of normal shape (0); flat and craniocaudally expanded (1).

170 Major digit (II), phalanx 1, "internal index process" (Stegmann, 1978) on caudodistal edge: absent (0); present (1).

171 Second phalanx of major digit (II): longer than proximal phalanx (0); shorter than or equivalent to proximal phalanx (1).

172 Ungual phalanx of major digit (II): present (0); absent (1).

173 Ungual phalanx of major digit (II) much smaller than the unguals of the alular (I) and minor (III) digits: absent (0); present (1).

174 Proximal phalanx of the minor digit (III) much shorter than the remaining nonungual phalanges of this digit: absent (0); present (1).

175 Ungual phalanx of minor digit (III): present (0); absent (1).

176 Length of manus (semilunate carpal + major metacarpal and digit) relative to humerus: longer (0); subequal (1); shorter (2). (ORDERED)

177 Intermembral index = (length of humerus + ulna)/(length of femur + tibiotarsus): less than 0.7, flightless (0); between 0.7 and 0.9 (1); between 0.9 and 1.1 (2); greater than 1.1 (3).

Pelvic girdle

178 Pelvic elements in adults, at the level of the acetabulum: unfused or partial fusion (0); completely fused (1).

179 Ilium/ischium, distal co-ossification to completely enclose the ilioischiadic fenestra: absent (0); present (1).

180 Preacetabular process of ilium twice as long as postacetabular process: absent (0); present (1).

181 Preacetabular ilium: approach on mid-line, open, or cartilaginous connection (0); co-ossified, dorsal closure of "iliosynsacral canals" (1).

182 Ilium, m. cuppedicus fossa as broad, medio-laterally oriented surface directly cranioventral to acetabulum: present (0); surface absent, insertion variably marked by a small entirely lateral fossa cranial to acetabulum (1).

183 Preacetabular pectineal process (Baumel & Witmer, 1993): absent (0); present as a small flange (1); present as a well-projected flange (2). (ORDERED)

184 Small acetabulum, acetabulum/ilium length ratio equal to or smaller than 0.11: absent (0); present (1).

185 Prominent antitrochanter: caudally directed (0); caudodorsally directed (1).

186 Postacetabular process shallow, less than 50% of the depth of the preacetabular wing at the acetabulum: absent (0); present (1).

187 Iliac brevis fossa: present (0); absent (1).

188 Ischium: two-thirds or less the length of the pubis (0); more than two-thirds the length of the pubis (1).

189 Obturator process of ischium: prominent (0); reduced or absent (1). The ischium of *Archaeopteryx* is forked distally; the thicker cranioventrally oriented fork is here interpreted to be the obturator process.

190 Ischium, caudal demarcation of the obturator foramen: absent (0); present, developed as a small flange or raised scar contacting/fused with pubis and demarcating the obturator foramen distally (1).

191 Ischium with a proximodorsal (or proximocaudal) process: absent (0); present (1).

192 Ischiadic terminal processes forming a symphysis: present (0); absent (1).

193 Orientation of proximal portion of pubis: cranially to subvertically oriented (0); retroverted, separated from the main synsacral axis by an angle ranging between 65° and 45° (1); more or less parallel to the ilium and ischium (2). (ORDERED)

194 Pubic pedicel: cranioventrally projected (0); ventrally or caudoventrally projected (1).

195 Pubic pedicel of ilium very compressed laterally and hook-like: absent (0), present (1).

196 Pubic shaft laterally compressed throughout its length: absent (0); present (1).

197 Pubic apron: one-third or more the length of the pubis (0); shorter (1); absent (absence of symphysis) (2). (ORDERED)

198 Pubic foot: present (0); absent (1). This refers to the distinct long, caudodorsal tapered expansion of the distal pubis, as opposed to the gradual expansion of the distal pubes present in taxa such as *Confuciusornis*.

Pelvic limb

199 Femur with distinct fossa for the capital ligament: absent (0); present (1).

200 Femoral neck: present (0); absent (1).

201 Femoral anterior trochanter: separated from the greater trochanter (0); fused to it, forming a trochanteric crest with a laterally curved edge (1); fused to it, forming a trochanteric crest with a flattened edge (2).

202 Femoral trochanteric crest: projects proximally beyond femoral head (0); equal in proximal projection (1); does not project beyond femoral head (2).

203 Femoral posterior trochanter: present, developed as a slightly projected tubercle or flange (0); hypertrophied, "shelf-like" conformation (1) (in combination with development of the trochanteric shelf; see Hutchinson, 2001); absent (2).

204 Femur with prominent patellar groove: absent (0); present as a continuous extension onto the distal shaft (1); present and separated from the shaft by a slight ridge, giving it a pocketed appearance (2).

205 Femur: ectocondylar tubercle and lateral condyle separated by deep notch (0); ectocondylar tubercle and lateral condyle contiguous but without developing a tibiofibular crest (1); tibiofibular crest present, defining laterally

a fibular trochlea (2). Proximal to the lateral condyle in theropod dinosaurs there is a caudal projection known as the ectocondylar tubercle (Welles, 1984). It is hypothesized that this tubercle is homologous to the precursor to the tibiofibular crest, formed through the connection of the ectocondylar tubercle and the lateral condyle (Chiappe, 1996). (ORDERED)

206 Caudal projection of the lateral border of the distal end of the femur, proximal and contiguous to the ectocondylar tubercle/tibiofibular crest: absent (0); present (1).

207 Femoral popliteal fossa distally bounded by a complete transverse ridge: absent (0); present (1).

208 Fossa for the femoral origin of m. tibialis cranialis: absent (0); present (1).

209 Tibia, calcaneum, and astragalus: unfused or poorly co-ossified (sutures still visible) (0); complete fusion of tibia, calcaneum, and astragalus (1).

210 Round proximal articular surface of tibiotarsus: absent (0); present (1).

211 Tibiotarsus, proximal articular surface: flat (0); angled so that the medial margin is elevated with respect to the lateral margin (1).

212 Tibiotarsus, cnemial crests: absent (0); present, one (1); present, two (2).

213 Tibia, caudal extension of articular surface for distal tarsals/tarsometatarsus: absent, articular restricted to distalmost edge of caudal surface (0); well-developed caudal extension, sulcus cartilaginis tibialis of Aves (Baumel & Witmer, 1993), distinct surface extending up the caudal surface of the tibiotarsus (1); with well-developed, caudally projecting medial and lateral crests (2). (ORDERED)

214 Extensor canal on tibiotarsus: absent (0); present as an emarginate groove (1); groove bridged by an ossified supratendinal bridge (2). (ORDERED)

215 Tibia/tarsal-formed condyles: medial condyle projecting farther cranially than lateral condyle (0); equal in cranial projection (1).

216 Tibia/tarsal-formed condyles, mediolateral widths: medial condyle wider (0); approxi-

mately equal (1); lateral condyle wider (2). (ORDERED)

217 Tibia/tarsal-formed condyles: gradual sloping of condyles towards mid-line of tibiotarsus (0); no tapering of either condyle (1).

218 Proximal end of the fibula: prominently excavated by a medial fossa (0); nearly flat (1).

219 Fibula, tubercle for m. iliofibularis: craniolaterally directed (0); laterally directed (1); caudolaterally or caudally directed (2). (ORDERED)

220 Fibula, distal end reaching the proximal tarsals: present (0); absent (1).

221 Distal tarsals in adults: free (0); completely fused to the metatarsals (1). Any specimen that is inferred to be a juvenile should be scored as a "?" in order to account for the possibility of ontogenetic change.

222 Metatarsals II–IV, intermetatarsal fusion: absent or minimal co-ossification (0); partial fusion, sutural contacts easily discernible (1); completely or nearly completely fused, sutural contacts absent or poorly demarcated (2). (ORDERED)

223 Proximal end of metatarsus: plane of articular surface perpendicular to longitudinal axis of metatarsus (0); strongly inclined dorsally (1).

224 Metatarsal V: present (0); absent (1).

225 Proximal end of metatarsal III: in the same plane as metatarsals II and IV (0); plantarly displaced with respect to metatarsals II and IV (1).

226 Tarsometatarsal proximal vascular foramen/foramina: absent (0); one between metatarsals III and IV (1); two (2).

227 Metatarsals, relative mediolateral width: metatarsal IV approximately the same width as metatarsals II and III (0); metatarsal IV narrower than metatarsals II and III (1); metatarsal IV greater in width than either metatarsal II or III (2).

228 Well-developed tarsometatarsal intercotylar eminence: absent (0); present, low, and rounded (1); present, high, and peaked (2).

229 Tarsometatarsus, projected surface and/or grooves on proximocaudal surface (associated

with the passage of tendons of the pes flexors in Aves; hypotarsus): absent (0); developed as caudal projection with flat caudal surface (1); projection, with distinct crests and grooves (2); at least one groove enclosed by bone caudally (3). (ORDERED)

230 Plantar surface of tarsometatarsus excavated: absent (0); present (1).

231 Tarsometatarsal distal vascular foramen completely enclosed by metatarsals III and IV: absent (0); present (1).

232 Metatarsal I: straight (0); J-shaped, the articulation of the hallux is located on the same plane as the attachment surface of the metatarsal I (1); J-shaped; the articulation of the hallux is perpendicular to the attachment surface (2); the distal half of the metatarsal I is laterally deflected so that the laterodistal surface is concave (3).

233 Metatarsal II tubercle (associated with the insertion of the tendon of the m. tibialis cranialis in Aves): absent (0); present, on approximately the center of the proximodorsal surface of metatarsal II (1); present, developed on lateral surface of metatarsal II, at contact with metatarsal III or on lateral edge of metatarsal III (2). (ORDERED)

234 Metatarsal II, distal plantar surface, fossa for metatarsal I (fossa metatarsi I; Baumel & Witmer, 1993): absent (0); shallow notch (1); conspicuous ovoid fossa (2). (ORDERED)

235 Relative position of metatarsal trochleae: trochlea III more distal than trochleae II and IV (0); trochlea III at same level as trochlea IV, both more distal than trochlea II (1); trochlea III at same level as trochleae II and IV (2); distal extent of trochlea III intermediate to trochlea IV and II where trochlea IV projects furthest distally (3).

236 Metatarsal II, distal extent of metatarsal II relative to metatarsal IV: approximately equal in distal extent (0); metatarsal II shorter than metatarsal IV but reaching distally farther than base of metatarsal IV trochlea (1); metatarsal II shorter than metatarsal IV, reaching distally only as far as base of metatarsal IV trochlea (2).

237 Distal tarsometatarsus, trochlea in distal view: aligned in a single plane (0); metatarsal II slightly displaced plantarly with respect to III and IV (1); metatarsal II strongly displaced plantarly in respect to III and IV, such that there is little or no overlap in medial view (2).

238 Trochlea of metatarsal II broader than the trochlea of metatarsal III: absent (0); present (1).

239 Metatarsal III, trochlea in plantar view, proximal extent of lateral and medial edges of trochlea: trochlear edges approximately equal in proximal extent (0); medial edge extends farther (1).

240 Distal end of metatarsal II strongly curved medially: absent (0); present (1).

241 Digit IV phalanges in distal view, medial trochlear rim enlarged with respect to lateral trochlear rim: absent (0); present (1); greatly enlarged with the lateral trochlea reduced to a rounded peg (2).

242 Completely reversed hallux (arch of ungual phalanx of digit I opposing the arch of the unguals of digits II–IV): absent (0); present (1).

243 Size of claw of hallux relative to other pedal claws: shorter, weaker, and smaller (0); similar in size (1); longer, more robust, and larger (2).

Integument

244 Alula: absent (0); present (1).

245 Fan-shaped feathered tail composed of more than two elongate retrices: absent (0); present (1).

APPENDIX 3B

	Characters							
	1	2	3	4	5	6	7	8
Dromaeosauridae	0	0	0	0	0	0	0	0
Archaeopteryx (all ten)	0	0	0	0	0	0	0	0
Jeholornis prima	[01]	?	2	3	1	1	1	?
Rahonavis ostromi	?	?	?	?	?	?	?	?
Shenzhouraptor sinensis	?	?	?	?	?	?	?	?
Jixiangornis orientalis	?	?	?	?	?	?	?	?
Dalianraptor cuhe	?	?	?	?	?	?	?	?
Zhongornis haoae	?	?	?	3	?	?	1	?
Sapeornis chaoyangensis	1	0	1	0	0	1	0	[01]
Didactylornis jii	?	?	?	0	?	?	?	?
Confuciusornis sanctus	1	0	2	3	1	1	1	1
Changchengornis hengdaoziensis	1	?	2	3	1	1	1	1
Eoconfuciusornis zhengi	[12]	?	2	3	1	?	1	[01]
"Dalianornis mengi"	[01]	0	1	0	1	1	0	2
Cathyaornis yandica	[12]	0	[12]	0	?	?	0	[012]
Concornis lacustris	?	?	?	?	?	?	?	?
Elsornis keni	?	?	?	?	?	?	?	?
Eoalulavis hoyasi	?	?	?	?	?	?	?	?
eocathay walkeri from single slab	?	?	?	0	?	?	?	?
Eoenantiornis buhleri	0/1	0	1	0	?	?	0	?
Gobipteryx minuta	1	0	[12]	3	1	0	1	3
Iberomesornis romerali	?	?	?	?	?	?	?	?
Longipteryx chaoyangensis	?	1	2	2	1	1	1	?
Longirostravis hani	[01]	0	2	2	?	?	1	?
Neuquenornis volans	?	?	?	?	?	?	?	?
Otogornis genghisi	?	?	?	?	?	?	?	?
Pengornis houi	0	0	[01]	0	1	1	0	1
Protopteryx fengningensis	[01]	?	[12]	[02]	1	?	?	?
Rapaxavis pani	[01]	0	2	2	?	?	1	[012]
Shanweiniao cooperorum	[01]	?	[12]	[23]	?	?	1	?
Vescornis hebeiensis	?	0	?	?	?	?	?	[12]
Vorona berivotrensis	?	?	?	?	?	?	?	?
Archaeorhynchus spathula	1	0	1	3	1	1	1	[12]
Liaoningornis longidigitrus	?	?	?	?	?	?	?	?
Songlingornis linghensis	?	?	?	[01]	?	?	?	?
Ambiortus dementjevi	?	?	?	?	?	?	?	?
Longicrusavis houi	1	0	[12]	0	1	?	0	?
Apsaravis ukhaana	?	?	?	?	?	?	?	?
Hongshanornis longicresta	?	0	2	0	?	1	0	?
Yanornis martini	1	0	[12]	1	1	1	0	?
Patagopteryx deferrariisi	?	?	?	?	?	?	?	?
Hollanda luceria	?	?	?	?	?	?	?	?
Yixianornis grabaui	1	1	2	1	?	?	?	?
Gansus yumenensis	?	?	?	?	?	?	?	?
Ichthyornis dispar	2	1	[12]	?	?	?	0	?
Hesperornis regalis	1	1	2	3	1	1	0	3
Parahesperornis	1	1	2	3	?	?	0	?
Enaliornis baretti	?	?	?	?	?	?	?	?
Baptornis advenus	1	?	?	3	?	?	?	?
Baptornis varneri	?	?	?	?	?	?	?	?
Limenavis patagonica	?	?	?	?	?	?	?	?
Vegaavis iaai	?	?	?	?	?	?	?	?
Anas	2	1	2	3	0	1	1	3
Gallus	1	1	2	3	1	1	1	3

9	10	11	12	13	14	15	16	17	18	19	20
0	0	0	0	0	0	0	0	0	0	0	0
0	0	0	0	0	0	?	0	0	1	0	0
?	?	?	?	?	?	?	?	?	?	?	0
?	?	?	?	?	?	?	?	?	?	?	?
?	?	?	?	?	?	?	?	?	?	?	?
?	?	?	?	?	?	?	?	0	?	?	?
?	?	?	?	?	?	?	?	?	?	?	?
?	0	?	?	?	?	?	?	0	0	1	0
?	?	?	?	?	?	?	?	?	?	?	?
?	0	?	?	?	?	?	0	0	0	1	0
?	0	?	?	?	?	?	?	?	?	1	?
?	0	?	?	?	?	?	?	0	?	?	?
?	?	?	?	?	?	?	?	0	?	?	?
?	?	?	?	?	?	?	?	?	?	?	?
?	?	?	?	?	?	?	?	?	?	?	?
?	?	?	?	?	?	?	?	?	?	?	?
?	?	?	?	?	?	?	?	?	?	?	?
?	?	?	?	?	?	?	?	?	?	?	?
0	0	0	2	1	?	1	0	?	?	?	?
?	?	?	?	?	?	?	?	?	?	?	?
?	?	?	?	?	?	?	?	?	?	?	?
?	?	?	?	?	?	?	?	?	?	?	?
?	?	?	?	?	?	?	?	?	?	?	?
?	0	?	?	?	?	?	?	0	?	?	?
?	?	?	?	?	?	?	?	?	?	?	?
?	?	?	?	?	?	?	?	?	?	?	?
?	?	?	?	?	?	?	?	?	?	?	?
?	?	?	?	?	?	?	?	?	?	?	?
?	?	?	?	?	?	?	?	?	?	?	?
?	?	?	?	?	?	?	?	?	?	?	?
?	?	?	?	?	?	?	?	?	?	?	?
?	?	?	?	?	?	?	?	?	?	?	?
?	?	?	?	?	?	?	?	?	1	?	?
?	1	?	?	?	?	?	?	?	?	1	?
?	?	?	?	?	?	?	?	1	?	1	1
?	?	?	?	?	?	?	?	?	?	?	?
?	?	?	?	?	?	?	?	1	1	1	1
?	1	0	2	1	0	1	1	?	1	1	1
?	?	0	?	1	0	?	?	1	?	1	1
?	?	?	?	?	?	?	?	?	?	?	?
?	?	?	?	?	?	?	?	?	?	1	1
?	?	?	?	?	?	?	?	?	?	?	?
?	?	?	?	?	?	?	?	?	?	?	?
1	0	1	1	1	1	1	1	1	1	1	1
1	1	1	2	1	1	1	1	1	1	1	1

	Characters							
	21	22	23	24	25	26	27	28
Dromaeosauridae	0	0	0	0	0	0	0	0
Archaeopteryx (all ten)	1	0	?	0	0	?	0	0
Jeholornis prima	?	?	?	?	?	?	?	?
Rahonavis ostromi	?	?	?	?	?	?	?	?
Shenzhouraptor sinensis	?	?	?	?	?	?	?	?
Jixiangornis orientalis	?	?	?	?	?	?	?	?
Dalianraptor cuhe	?	?	?	?	?	?	?	?
Zhongornis haoae	?	?	?	?	0	?	?	?
Sapeornis chaoyangensis	?	1	?	0	?	?	?	?
Didactylornis jii	?	?	?	?	?	?	?	?
Confuciusornis sanctus	1	1	0	0	0	0	1	2
Changchengornis hengdaoziensis	?	?	?	?	?	0	?	?
Eoconfuciusornis zhengi	?	?	?	?	?	?	?	?
"Dalianornis mengi"	?	?	?	0	?	1	?	?
Cathyaornis yandica	?	?	?	?	?	?	?	?
Concornis lacustris	?	?	?	?	?	?	?	?
Elsornis keni	?	?	?	?	?	?	?	?
Eoalulavis hoyasi	?	?	?	?	?	?	?	?
eocathay walkeri from single slab	?	?	?	0	?	?	?	0
Eoenantiornis buhleri	?	?	?	0	?	?	?	?
Gobipteryx minuta	?	?	?	?	0	?	?	?
Iberomesornis romerali	?	?	?	?	?	?	?	?
Longipteryx chaoyangensis	?	?	?	?	?	?	?	?
Longirostravis hani	?	?	?	?	?	?	?	?
Neuquenornis volans	?	?	?	?	?	?	?	?
Otogornis genghisi	?	?	?	?	?	?	?	?
Pengornis houi	?	?	?	0	?	1	?	?
Protopteryx fengningensis	?	?	?	?	?	?	?	?
Rapaxavis pani	?	?	?	?	?	?	?	?
Shanweiniao cooperorum	?	?	?	?	?	?	?	?
Vescornis hebeiensis	?	?	?	?	?	?	?	?
Vorona berivotrensis	?	?	?	?	?	?	?	?
Archaeorhynchus spathula	?	?	?	?	?	?	?	?
Liaoningornis longidigitrus	?	?	?	?	?	?	?	?
Songlingornis linghensis	?	?	?	?	?	?	?	?
Ambiortus dementjevi	?	?	?	?	?	?	?	?
Longicrusavis houi	?	?	?	?	?	?	?	?
Apsaravis ukhaana	?	?	?	?	?	?	1	[01]
Hongshanornis longicresta	?	?	?	?	?	?	?	?
Yanornis martini	?	?	?	?	?	?	?	?
Patagopteryx deferrariisi	1	1	0	0	0	1	1	1
Hollanda luceria	?	?	?	?	?	?	?	?
Yixianornis grabaui	?	?	?	0	?	?	?	1
Gansus yumenensis	?	?	?	?	?	?	?	?
Ichthyornis dispar	1	1	?	?	1	1	1	1
Hesperornis regalis	1	1	1	0	0	0	1	1
Parahesperornis	1	?	?	?	0	0	1	1
Enaliornis baretti	?	?	?	?	?	?	1	1
Baptornis advenus	?	?	?	0	0	?	1	1
Baptornis varneri	?	?	?	?	?	?	?	?
Limenavis patagonica	?	?	?	?	?	?	?	?
Vegaavis iaai	?	?	?	?	?	?	?	?
Anas	1	1	1	1	1	1	1	2
Gallus	1	1	1	1	1	1	1	2

29	30	31	32	33	34	35	36	37	38	39	40
0	0	0	0	0	0	0	0	0	?	?	0
0	?	?	0	?	0	0	0	?	?	?	0
?	?	?	?	?	0	0	?	?	?	?	?
?	?	?	?	?	?	?	?	?	?	?	?
?	?	?	?	?	?	?	?	?	?	?	?
?	?	?	?	?	?	?	0	?	?	?	?
?	?	?	?	?	?	?	?	?	?	?	?
?	?	?	?	?	1	-	?	?	?	?	?
?	?	?	?	?	1	-	0	?	?	?	?
?	?	?	?	?	1	?	0	?	?	?	?
0	?	?	?	1	1	-	1	1	2	1	0
?	?	?	?	?	1	-	1	1	?	?	0
?	?	?	?	?	1	-	0	1	?	?	0
?	?	?	?	?	0	0	0	?	?	?	0
?	?	?	?	?	0	0	0	?	?	?	?
?	?	?	?	?	?	?	?	?	?	?	?
?	?	?	?	?	?	?	?	?	?	?	?
?	?	?	?	?	0	0	0	?	?	?	?
?	?	?	?	?	0	0	?	?	?	?	?
0	?	1	?	?	1	-	0&1	0	?	?	?
?	?	?	?	?	?	?	?	?	?	?	?
?	?	?	?	?	0	0	0	?	?	?	?
?	?	?	?	?	0	0	?	?	?	?	?
?	?	?	?	?	?	?	?	?	?	?	?
?	?	?	?	?	0	0	0	?	?	?	0
?	?	?	?	?	0	0	?	?	?	?	?
?	?	?	?	?	0	0	0	?	?	?	0
?	?	?	?	?	?	?	0	?	?	?	?
?	?	?	?	?	0	0	0	0	?	?	?
?	?	?	?	?	?	?	?	?	?	?	?
?	?	?	?	?	1	n	0	0	?	?	0
?	?	?	?	?	?	?	?	?	?	?	?
?	?	?	?	?	0	0	0	?	?	?	?
?	?	?	?	?	?	?	?	?	?	?	?
?	?	?	?	?	?	0	0	?	?	?	1
?	?	?	?	?	1	-	1	?	?	?	?
?	?	?	?	?	?	0	0	?	?	?	1
?	?	?	?	1	0	0	0	?	?	?	1
1	?	?	?	?	?	?	?	?	?	?	?
?	1	?	?	?	0	?	0	?	?	?	1
?	?	?	0	1	0	0	1	?	?	?	?
1	?	1	?	1	0	1	0	0	?	?	1
1	?	?	?	?	0	1	?	?	?	?	?
?	?	?	?	?	?	?	?	?	?	?	?
1	?	?	?	?	?	?	?	0	?	?	1
?	?	?	?	?	?	?	?	?	?	?	?
?	?	?	?	?	?	?	?	?	?	?	?
1	1	2	1	1	1	-	1	0	2	1	0
1	1	2	1	1	1	-	1	0	2	0	0

	Characters							
	41	42	43	44	45	46	47	48
Dromaeosauridae	0	0	0	0	0	0	0	0
Archaeopteryx (all ten)	0	0	0	0	0	0	?	1
Jeholornis prima	0	0	0	?	0	0	?	1
Rahonavis ostromi	?	?	?	?	?	?	?	?
Shenzhouraptor sinensis	?	?	?	?	?	?	?	?
Jixiangornis orientalis	?	0	?	?	?	?	?	?
Dalianraptor cuhe	?	?	?	?	0	1	?	?
Zhongornis haoae	?	?	?	?	?	?	?	-
Sapeornis chaoyangensis	0	0	?	?	0	?	?	1
Didactylornis jii	0	0	?	?	?	?	?	0
Confuciusornis sanctus	1	0	?	1	1	0	0	-
Changchengornis hengdaoziensis	?	0	?	?	1	0	0	-
Eoconfuciusornis zhengi	0	0	?	?	1	0	0	-
"*Dalianornis mengi*"	0	0	?	?	?	1	?	1
Cathyaornis yandica	0	0	?	?	?	?	?	1
Concornis lacustris	?	?	?	?	?	?	?	?
Elsornis keni	?	?	?	?	?	?	?	?
Eoalulavis hoyasi	?	?	?	?	?	?	?	?
eocathay walkeri from single slab	0	0	?	?	?	?	?	1
Eoenantiornis buhleri	?	?	?	?	?	?	?	1
Gobipteryx minuta	1	0	?	?	?	1	0	-
Iberomesornis romerali	?	?	?	?	?	?	?	?
Longipteryx chaoyangensis	0	0	?	?	0	1	?	1
Longirostravis hani	0	0	?	?	0	1	?	1
Neuquenornis volans	?	?	?	?	?	?	?	?
Otogornis genghisi	?	?	?	?	?	?	?	?
Pengornis houi	?	0	?	?	?	?	?	1
Protopteryx fengningensis	?	0	0	?	?	?	?	1
Rapaxavis pani	?	0	?	?	?	1	?	1
Shanweiniao cooperorum	0	0	?	?	0	1	?	?
Vescornis hebeiensis	0	0	[12]	0	0	1	?	1
Vorona berivotrensis	?	?	?	?	?	?	?	?
Archaeorhynchus spathula	0	?	0	0	?	?	1	n
Liaoningornis longidigitrus	?	?	?	?	?	?	?	?
Songlingornis linghensis	?	0	?	?	?	?	?	1
Ambiortus dementjevi	?	?	?	?	?	?	?	?
Longicrusavis houi	0	1	?	?	0	1	?	?
Apsaravis ukhaana	1	?	?	?	?	?	?	?
Hongshanornis longicresta	0	1	?	?	0	?	?	?
Yanornis martini	?	0	0	?	0	1	?	1
Patagopteryx deferrariisi	?	0	?	?	0	0	0	?
Hollanda luceria	?	?	?	?	?	?	?	?
Yixianornis grabaui	1	0	?	?	?	?	?	1
Gansus yumenensis	?	?	?	?	?	?	?	?
Ichthyornis dispar	0	0	2	1	0	1	1	1
Hesperornis regalis	0	0	0	1	0	1	0	1
Parahesperornis	?	?	?	?	?	?	?	1
Enaliornis baretti	?	?	?	?	?	?	?	?
Baptornis advenus	?	0	?	1	0	1	0	1
Baptornis varneri	?	?	?	?	?	?	?	?
Limenavis patagonica	?	?	?	?	?	?	?	?
Vegaavis iaai	?	?	?	?	?	?	?	?
Anas	0	0	0	1	0	1	1	-
Gallus	0	0	0	1	0	1	1	-

49	50	51	52	53	54	55	56	57	58	59	60
0	0	0	0	0	0	0	0	0	0	0	0
0	0	0	?	1	0	0	0	0	0	?	?
?	?	0	?	?	?	?	0	0	?	?	1
?	1	?	0	?	?	?	?	0	?	1	0
?	?	?	?	?	?	?	?	?	?	?	?
?	?	?	?	?	?	?	?	?	?	?	?
?	?	?	?	?	?	?	0	0	1	?	?
?	?	?	?	1	0	0	[01]	0	1	?	?
?	?	?	?	?	?	?	?	?	1	?	?
1	0	1	?	1	0	1	[01]	0	1	1	1
?	?	[12]	?	?	?	?	?	?	?	?	?
?	?	?	0	?	0	?	0	0	?	?	?
?	?	?	1	1	?	0	?	0	1	?	?
?	?	1	?	?	1	?	?	0	1	?	?
?	?	[01]	?	?	1	0	?	0	?	1	?
?	?	0	0	1	1	1	2	0	0	?	?
?	?	?	?	?	?	?	?	0	0	?	?
?	?	?	?	?	1	?	?	?	?	?	?
?	?	0	?	?	1	1	1	0	?	?	1
?	?	1	0	1	1	?	?	?	?	?	?
?	?	1	?	?	?	?	?	0	?	?	?
?	?	?	?	?	?	?	?	0	?	?	?
?	?	?	?	?	?	?	?	?	?	?	?
1	?	2	?	?	1	?	?	0	?	1	?
?	?	?	?	?	?	?	1	0	?	?	?
?	?	?	?	?	1	?	?	0	1	?	?
?	?	?	?	1	?	?	?	?	?	?	?
?	?	2	1	?	1	?	?	0	?	?	?
?	?	?	?	?	?	?	?	?	?	?	?
?	?	0	?	?	?	?	?	0	?	?	1
?	?	?	?	?	?	?	?	?	?	?	?
?	?	0	1	?	0	?	?	?	?	?	?
?	?	1	1	?	1	1	?	0	?	?	?
?	0	2	1	?	1	?	2	0	1	?	?
?	?	?	?	?	?	?	?	?	?	?	?
?	?	0	?	?	?	?	?	0	1	?	?
1	1	2	1	1	0	1	1	0	0	1	1
?	?	?	?	?	?	?	?	?	?	?	?
?	?	1	?	1	?	?	2	0	1	1	?
?	1	2	1	?	1	?	2	0	1	1	1
1	1	[01]	1	1	0	1	2	0	1	1	1
?	0	2	1	0	0	1	2	1	1	1	1
?	0	2	1	0	0	1	2	1	1	1	1
?	?	2	1	?	0	?	?	1	?	?	?
?	0	2	1	0	0	1	?	1	1	1	1
?	0	2	1	?	0	1	?	1	1	?	1
?	?	?	?	?	?	?	?	?	?	?	?
?	?	2	?	?	?	?	?	?	?	?	?
1	0	2	1	1	0	1	2	1	1	1	1
1	0	2	1	1	0	1	2	1	1	1	1

	Characters							
	61	62	63	64	65	66	67	68
Dromaeosauridae	0	0	0	0	0	0	0	0
Archaeopteryx (all ten)	0	0	0	0	0	?	0	0
Jeholornis prima	0	0	0	0	?	?	?	?
Rahonavis ostromi	2	0	?	0	?	0	?	?
Shenzhouraptor sinensis	[01]	?	?	?	?	?	?	?
Jixiangornis orientalis	?	?	?	?	?	?	?	?
Dalianraptor cuhe	?	?	?	?	?	?	?	?
Zhongornis haoae	0	?	?	[01]	?	?	?	?
Sapeornis chaoyangensis	1	0	0	1	0	?	0	0
Didactylornis jii	0	?	?	?	?	?	?	?
Confuciusornis sanctus	2	0	0	1	0	0	0	0
Changchengornis hengdaoziensis	2	?	?	1	?	?	?	0
Eoconfuciusornis zhengi	0	?	?	1	?	?	?	0
"*Dalianornis mengi*"	1	1	?	?	?	?	?	?
Cathyaornis yandica	1	1	?	2	?	?	?	?
Concornis lacustris	1	1	?	?	?	?	?	?
Elsornis keni	1	0	?	?	?	?	?	?
Eoalulavis hoyasi	2	?	0	?	?	?	?	?
eocathay walkeri from single slab	?	?	?	?	?	?	?	?
Eoenantiornis buhleri	?	?	?	[012]	?	?	?	?
Gobipteryx minuta	?	?	?	[012]	1	0	?	?
Iberomesornis romerali	0	0	0	0	0	1	?	0
Longipteryx chaoyangensis	?	?	?	2	?	?	?	?
Longirostravis hani	?	?	0	1	?	?	?	?
Neuquenornis volans	1	1	?	?	?	?	?	?
Otogornis genghisi	?	?	?	?	?	?	?	?
Pengornis houi	1	0	?	1	?	?	?	?
Protopteryx fengningensis	?	?	0	1	?	?	?	?
Rapaxavis pani	1	1	?	1	?	?	?	0
Shanweiniao cooperorum	?	?	?	?	?	?	?	?
Vescornis hebeiensis	1	?	?	2	?	?	?	?
Vorona berivotrensis	?	?	?	?	?	?	?	?
Archaeorhynchus spathula	?	?	?	2	0	0	0	?
Liaoningornis longidigitrus	?	?	?	?	?	?	?	?
Songlingornis linghensis	2	?	?	?	?	?	?	?
Ambiortus dementjevi	2	0	?	?	?	?	?	?
Longicrusavis houi	2	0	?	?	?	?	?	?
Apsaravis ukhaana	0	0	?	4	?	?	0	?
Hongshanornis longicresta	1	?	?	?	?	?	?	?
Yanornis martini	2	0	0	3	?	?	0	?
Patagopteryx deferrariisi	0	0	0	3	1	1	0	1
Hollanda luceria	?	?	?	?	?	?	?	?
Yixianornis grabaui	1	0	0	3	?	?	0	?
Gansus yumenensis	1	0	0	4	0	0	1	0
Ichthyornis dispar	2	0	?	[34]	?	0	1	0
Hesperornis regalis	2	0	0	4	0	1	0	?
Parahesperornis	2	0	0	?	?	?	?	?
Enaliornis baretti	2	?	?	?	?	?	?	?
Baptornis advenus	2	0	0	4	0	1	0	?
Baptornis varneri	?	?	?	?	?	?	?	?
Limenavis patagonica	?	?	?	?	?	?	?	?
Vegaavis iaai	?	?	?	[56]	?	?	?	?
Anas	0	0	0	6	0	1	2	0
Gallus	0	0	1	6	0	1	2	0

69	70	71	72	73	74	75	76	77	78	79	80
0	0	0	0	0	1	-	-	-	-	1	-
0	2	0	0	0	1	-	-	-	-	0	-
0	1	0	0	0	0	-	-	-	-	?	?
0	[0123]	0	1	0	0	-	-	-	-	?	?
0	[12]	?	?	?	0	n	n	n	n	?	?
0	1	?	?	?	?	n	n	n	n	1	-
[01]	[123]	?	[01]	?	?	n	n	n	n	?	?
1	3	0	?	1	?	-	-	-	-	?	?
2	4	0	0	0	0	0	0	?	0	0	-
2	4	?	?	1	?	?	?	?	?	?	?
2	4	0	0	0	?	0	0	0	0	1	-
2	4	?	?	0	?	0	0	0	0	?	?
2	4	1	?	?	?	0	?	?	0	?	?
?	?	?	?	0	?	?	?	?	?	?	?
2	4	0	?	0	?	0	1	1	1	?	?
?	?	0	?	0	1	?	?	?	?	?	?
?	?	?	?	?	?	?	?	?	?	?	?
?	?	?	?	?	?	?	?	?	?	0	-
?	?	?	?	?	?	?	?	?	?	?	?
2	4	0	2	0	?	?	?	?	?	1	?
[12]	4	0	1	?	?	?	?	?	?	?	?
[12]	4	0	1	0	1	0	?	?	?	0	-
2	4	?	?	0	?	0	1	?	1	1	?
2	4	?	?	0	?	0	1	?	1	0	-
?	?	?	?	?	?	?	?	?	?	?	?
2	4	0	?	0	[12]	?	?	?	?	?	?
1	4	?	?	?	?	0	?	?	1	?	?
2	4	?	?	0	?	0	?	?	1	0	-
2	4	?	?	?	?	?	?	?	1	?	?
2	?	?	?	?	?	?	?	0	0	?	?
?	?	?	?	?	?	?	?	?	?	?	?
?	[34]	0	1	1	1	?	?	?	?	?	?
?	?	?	?	?	?	?	?	?	?	?	?
?	?	?	?	?	?	?	?	?	?	?	?
?	?	?	?	?	?	?	?	?	?	?	?
?	?	?	?	?	?	?	?	?	?	?	?
2	4	?	1	0	?	?	?	?	?	?	?
2	4	?	?	?	?	?	?	?	0	1	-
?	?	?	?	?	?	?	?	?	?	?	?
?	?	1	2	0	?	?	?	?	?	?	?
2	4	?	?	?	?	1	0	0	0	[12]	0
2	4	?	2	0	2	1	0	0	0	0	-
2	4	0	1	0	2	1	0	0	0	?	?
1	3	0	2	0	2	1	0	0	0	1	0
1	?	?	?	?	?	1	0	0	0	1	?
1	3	0	2	0	?	1	0	0	0	1	1
?	?	?	?	?	?	?	?	?	?	?	?
?	?	?	?	?	?	?	?	?	?	?	?
2	4	0	2	0	2	1	0	0	0	2	0
2	4	0	2	0	2	1	0	0	0	2	0

	Characters							
	81	82	83	84	85	86	87	88
Dromaeosauridae	0	0	0	0	0	0	0	0
Archaeopteryx (all ten)	0	0	0	0	0	0	0	0
Jeholornis prima	0	1	1	0	1	?	0	0
Rahonavis ostromi	?	?	1	?	?	?	?	?
Shenzhouraptor sinensis	0	1	1	?	?	?	?	?
Jixiangornis orientalis	0	1	1	?	?	?	?	?
Dalianraptor cuhe	?	?	?	?	?	?	?	?
Zhongornis haoae	0	1	?	?	?	?	0	0
Sapeornis chaoyangensis	0	0	0	0	1	1	0	0
Didactylornis jii	0	1	1	?	?	?	?	?
Confuciusornis sanctus	0	1	0	0	?	0	-	0
Changchengornis hengdaoziensis	0	1	0	0	?	0	-	0
Eoconfuciusornis zhengi	0	0	0	0	?	?	-	0
"*Dalianornis mengi*"	0	1	1	1	1	?	0	?
Cathyaornis yandica	?	1	1	1	1	?	0	?
Concornis lacustris	?	1	1	1	1	1	0	?
Elsornis keni	?	1	1	1	1	1	0	0
Eoalulavis hoyasi	?	1	1	1	1	1	0	1
eocathay walkeri from single slab	?	1	1	?	?	?	?	?
Eoenantiornis buhleri	0	1	1	1	1	1	0	1
Gobipteryx minuta	?	1	1	1	1	?	0	1
Iberomesornis romerali	?	1	1	?	?	?	0	?
Longipteryx chaoyangensis	0	1	1	1	1	1	0	1
Longirostravis hani	0	1	1	1	?	?	?	?
Neuquenornis volans	?	1	1	1	?	?	?	?
Otogornis genghisi	?	1	1	1	1	1	?	?
Pengornis houi	0	1	1	1	1	?	?	?
Protopteryx fengningensis	?	1	1	1	1	?	0	?
Rapaxavis pani	0	1	1	?	?	?	0	?
Shanweiniao cooperorum	0	1	1	1	1	?	?	?
Vescornis hebeiensis	0	1	1	1	?	?	0	?
Vorona berivotrensis	?	?	?	?	?	?	?	?
Archaeorhynchus spathula	0	1	1	1	1	?	0	0
Liaoningornis longidigitrus	?	?	?	?	?	?	?	?
Songlingornis linghensis	?	1	1	?	?	?	?	?
Ambiortus dementjevi	?	1	?	1	1	1	0	0
Longicrusavis houi	?	1	1	1	1	1	0	0
Apsaravis ukhaana	?	1	1	1	1	1	0	0
Hongshanornis longicresta	0	1	1	1	1	1	0	0
Yanornis martini	0	1	1	1	1	1	1	0
Patagopteryx deferrariisi	?	1	1	1	1	1	0	0
Hollanda luceria	?	?	?	?	?	?	?	?
Yixianornis grabaui	0	1	1	1	1	1	0	0
Gansus yumenensis	1	1	1	1	1	1	0	0
Ichthyornis dispar	?	1	1	1	1	1	1	0
Hesperornis regalis	?	0	1	0	1	?	0	0
Parahesperornis	?	0	1	0	1	?	0	?
Enaliornis baretti	?	?	?	?	?	?	?	?
Baptornis advenus	?	0	1	0	1	?	0	0
Baptornis varneri	?	?	?	?	?	?	?	?
Limenavis patagonica	?	?	?	?	?	?	?	?
Vegaavis iaai	?	1	1	1	1	?	0	0
Anas	1	1	1	1	1	1	1	0
Gallus	1	1	1	1	1	1	1	0

89	90	91	92	93	94	95	96	97	98	99	100
0	0	0	0	0	-	0	0	0	0	0	0
0	0	0	0	0	-	0	0	0	0	?	0
0	0	0	1	?	-	1	1	0	?	?	?
?	?	?	?	?	?	?	0	?	0	0	0
?	0	?	?	?	?	?	1	0	?	0	?
0	0	?	?	?	?	?	?	?	?	?	?
?	?	?	?	?	?	?	?	?	?	?	?
0	0	0	?	?	-	0	0	0	?	?	?
0	0	0	0	?	-	0	0	0	0	0	0
?	0	?	?	?	?	?	0	0	?	?	?
0	0	0	?	?	-	0	0	0	0	0	0
0	0	?	?	?	-	0	0	0	0	0	0
0	0	0	0	0	-	0	0	0	0	?	0
0	1	?	[12]	?	?	0	0	0	?	0	?
?	?	0	?	?	?	0	0	0	1	0	0
0	1	?	1	1	?	0	0	?	1	?	?
0	1	1	2	1	?	0	0	0	1	0	0
0	1	1	1	1	0	0	0	0	0	0	0
0	1	?	?	?	?	0	0	0	0	0	0
0	1	?	1	1	0	0	0	?	?	0	0
0	?	1	?	?	?	0	?	?	0	0	0
0	0	?	2	?	?	0	?	?	?	?	?
0	0	?	?	?	?	0	0	0	?	0	0
0	0	?	2	?	?	0	0	0	?	0	0
0	1	1	1	1	1	0	0	0	?	?	?
?	?	?	?	?	?	?	0	?	?	0	0
0	0	?	?	?	?	?	0	?	0	0	1
1	1	?	[12]	?	?	2	0	0	?	0	0
0	0	?	[12]	?	?	0	?	?	?	0	?
0	0	0	[12]	?	-	0	0	0	?	0	0
0	1	?	[12]	?	?	0	0	0	?	0	0
?	?	?	?	?	?	?	?	?	?	?	?
?	0	?	2	0	n	[01]	1	0	0	0	0
?	?	?	?	?	?	?	?	?	?	?	?
1	0	1	?	?	?	2	?	?	?	?	?
1	?	?	?	1	?	[01]	1	?	?	0	1
?	0	?	?	?	?	[12]	1	1	0	0	?
0	0	1	1	1	1	1	1	1	0	0	1
1	0	?	?	?	?	[01]	1	1	?	?	0
1	0	0	?	0	-	2	1	0	0	0	0
0	0	0	?	0	-	1	1	1	0	0	?
?	?	?	?	?	?	?	?	?	?	?	?
1	0	0	1	0	-	2	1	0	0	0	0
1	0	0	2	0	-	3	1	0	0	?	?
1	0	0	1	0	-	2	1	1	0	1	0
?	0	0	1	0	-	0	1	0	?	?	?
?	0	0	1	0	-	0	?	?	?	?	?
?	?	?	?	?	?	?	?	?	?	?	?
1	0	0	1	0	-	0	?	?	?	?	?
?	?	?	?	?	?	?	?	?	?	?	?
?	?	?	?	?	?	?	?	?	?	?	?
?	0	0	1	?	?	[01]	1	0	?	0	0
1	0	0	2	0	-	2	1	0	0	0	0
1	0	0	2	0	-	2	1	1	0	0	0

	Characters							
	101	102	103	104	105	106	107	108
Dromaeosauridae	0	0	0	0	0	0	0	2
Archaeopteryx (all ten)	?	0	0	0	0	0	0	0
Jeholornis prima	?	?	1	?	?	0	0	[12]
Rahonavis ostromi	0	0	0	?	?	?	?	?
Shenzhouraptor sinensis	?	?	1	0	1	?	[01]	?
Jixiangornis orientalis	?	?	?	0	0	?	[01]	[12]
Dalianraptor cuhe	?	?	?	0	? (~70°)	?	[01]	?
Zhongornis haoae	?	0	0	?	0	?	[01]	?
Sapeornis chaoyangensis	?	?	1	0	0	0	2	?
Didactylornis jii	?	?	?	0	0	0	?	?
Confuciusornis sanctus	0	0	0	0	0	0	0	2
Changchengornis hengdaoziensis	?	0	?	0	0	0	0	2
Eoconfuciusornis zhengi	?	0	0	0	0	?	0	2
"*Dalianornis mengi*"	?	?	0	1	1	?	[23]	2
Cathyaornis yandica	?	0	1	1	1	?	3	2
Concornis lacustris	?	?	?	1	1	1	2	2
Elsornis keni	0	1	0	1	1	1	[23]	2
Eoalulavis hoyasi	0	?	1	1	1	?	3	1
eocathay walkeri from single slab	?	?	0	?	?	?	?	2
Eoenantiornis buhleri	?	?	?	1	1	1	2	2
Gobipteryx minuta	0	?	?	1	1	1	[23]	?
Iberomesornis romerali	?	?	?	1	1	?	[23]	[12]
Longipteryx chaoyangensis	0	?	1	1	1	1	3	2
Longirostravis hani	?	?	0	1	1	1	[123]	2
Neuquenornis volans	?	1	?	1	1	1	1	2
Otogornis genghisi	?	?	?	?	?	?	?	?
Pengornis houi	?	?	?	1	1	?	2	?
Protopteryx fengningensis	?	?	0	1	1	1	3	2
Rapaxavis pani	?	?	?	1	1	?	3	2
Shanweiniao cooperorum	?	?	0	1	1	?	2	2
Vescornis hebeiensis	?	?	0	1	1	?	3	2
Vorona berivotrensis	?	?	?	?	?	?	?	?
Archaeorhynchus spathula	?	0	?	2	1	0	0	2
Liaoningornis longidigitrus	?	?	?	?	?	?	?	[12]
Songlingornis linghensis	?	?	?	2	1	?	0	2
Ambiortus dementjevi	?	?	?	2	1	0	?	?
Longicrusavis houi	?	0	1	2	1	0	?	2
Apsaravis ukhaana	0	0	1	?	?	?	?	[12]
Hongshanornis longicresta	?	0	1	2	1	?	1	2
Yanornis martini	?	?	1	2	1	?	0	2
Patagopteryx deferrariisi	0	?	?	?	?	?	?	2
Hollanda luceria	?	?	?	?	?	?	?	?
Yixianornis grabaui	0	0	1	2	1	0	?	2
Gansus yumenensis	0	0	1	2	1	0	0	2
Ichthyornis dispar	0	0	1	2	?	0	0	2
Hesperornis regalis	0	0	0	2	0	0	?	2
Parahesperornis	?	?	?	?	?	?	?	2
Enaliornis baretti	?	?	?	?	?	?	?	?
Baptornis advenus	0	?	?	?	?	?	?	2
Baptornis varneri	?	?	?	?	?	?	?	?
Limenavis patagonica	?	?	?	?	?	?	?	?
Vegaavis iaai	?	?	?	?	?	?	?	?
Anas	0	0	1	2	1	0	0	2
Gallus	0	0	1	2	1	0	1	2

109	110	111	112	113	114	115	116	117	118	119	120
0	-	[01]	-	-	0	0	0	0	0	0	0
?	?	?	?	?	?	?	?	?	?	?	?
?	?	?	?	?	?	?	?	?	?	?	?
?	?	?	?	?	?	?	?	?	?	?	?
[012]	?	?	?	?	?	?	?	?	?	?	?
?	?	?	?	?	?	?	?	?	?	?	?
?	?	?	?	?	?	?	?	?	?	?	?
?	?	?	?	?	?	?	?	?	?	?	?
[12]	1	1	1	-	0	1	1	2	1	0	0
1	-	1	?	?	0	1	?	?	1	0	?
[12]	?	?	?	?	?	?	?	?	?	?	?
[123]	?	[12]	[12]	1	?	1	?	?	2	0	?
2	1	2	2	1	1	1	?	?	2	0	?
2	1	2	2	1	1	1	0	n	2	0	?
2	0	1	0	?	1	1	?	?	[12]	0	?
2	?	0	n	-	0	1	0	-	3	0	?
2	1	2	[12]	1	1	1	?	?	2	0	?
[123]	?	1	n	n	1	1	0	n	2	0	0
?	?	?	?	?	?	?	?	?	?	?	?
[123]	?	?	?	?	?	?	?	?	?	?	?
[23]	?	2	1	1	0	1	?	?	2	0	?
[123]	?	2	2	1	1	1	0	-	2	0	?
3	0	[12]	?	1	?	1	?	?	?	?	?
?	?	?	?	?	?	?	?	?	?	?	?
?	?	?	?	?	?	?	?	?	?	?	?
2	1	1	1	0	?	1	?	?	1	0	?
2	1	2	2	1	1	1	?	?	2	0	?
[123]	?	2	1	1	?	[01]	?	?	2	0	?
2	1	[12]	?	?	1	1	?	?	2	0	?
?	?	?	?	?	?	?	?	?	?	?	?
[23]	1	[12]	?	?	1	?	?	?	?	?	?
2	?	0	n	n	1	[12]	?	?	3	0	?
3	0	1	2	1	1	1	?	?	4	1	?
?	?	?	?	?	?	?	?	?	?	?	?
3	?	1	2	0	1	1	0	-	1	0	?
3	0	?	?	?	1	1	?	?	?	?	?
?	?	2	2	0	1	?	?	?	1	0	?
3	0	1	2	1	1	[12]	?	?	4	1	?
[12]	?	?	?	?	1	0	?	?	?	?	?
?	?	?	?	?	?	?	?	?	?	?	?
3	0	1	1	1	1	?	0	?	4	1	?
3	0	2	1	0	1	0	0	?	1	0	?
3	0	0	?	?	1	?	1	2	?	0	1
1	?	0	-	-	0	1	1	1	?	0	0
?	?	?	?	?	?	?	1	?	?	?	?
[12]	?	?	?	?	0	1	1	1	?	?	0
?	?	?	?	?	?	?	?	?	?	?	?
?	?	?	?	?	?	?	?	?	?	?	?
?	?	?	?	?	?	?	?	?	?	?	?
3	1	1	1	0	0	1	1	3	3	0	1
3	1	1	0	1	0	1	1	2	2	0	1

Taxa\characters	121	122	123	124	125	126	127	128
Dromaeosauridae	0	0	0	0	0	0	0	1
Archaeopteryx (all ten)	0	0	0	?	0	0	0	1
Jeholornis prima	?	0	0	0	0	0	0	0
Rahonavis ostromi	?	?	?	?	?	?	?	?
Shenzhouraptor sinensis	0	?	?	?	?	?	?	0
Jixiangornis orientalis	?	0	0	?	?	?	?	0
Dalianraptor cuhe	?	?	0	?	?	?	?	0
Zhongornis haoae	?	?	0	0	0	0	?	0
Sapeornis chaoyangensis	?	0	0	?	?	0	0	0
Didactylornis jii	?	?	?	?	0	0	?	?
Confuciusornis sanctus	0	0	0	0	1	0	0	0
Changchengornis hengdaoziensis	0	?	0	?	?	?	?	0
Eoconfuciusornis zhengi	0	?	0	0	0	0	0	?
"Dalianornis mengi"	?	0	1	1	0	1	?	0
Cathyaornis yandica	0	0	1	1	?	1	1	0
Concornis lacustris	0	0	1	1	1	?	?	0
Elsornis keni	0	0	1	1	0	0	0	0
Eoalulavis hoyasi	0	0	1	1	1	1	0	0
eocathay walkeri from single slab	?	0	1	?	?	1	?	0
Eoenantiornis buhleri	0	?	1	?	?	1	?	0
Gobipteryx minuta	0	?	?	?	?	?	?	0
Iberomesornis romerali	?	?	?	?	0	?	?	?
Longipteryx chaoyangensis	?	0	1	1	0	?	?	0
Longirostravis hani	?	0	1	?	?	1	?	?
Neuquenornis volans	0	?	?	?	?	1	1	?
Otogornis genghisi	0	?	1	?	?	1	0	0
Pengornis houi	0	0	1	?	0	1	0	0
Protopteryx fengningensis	?	?	1	?	?	?	?	0
Rapaxavis pani	?	0	1	0	0	?	?	0
Shanweiniao cooperorum	?	?	1	?	?	?	?	?
Vescornis hebeiensis	0	0	1	1	0	?	?	0
Vorona berivotrensis	?	?	?	?	?	?	?	?
Archaeorhynchus spathula	?	?	0	0	0	?	?	0
Liaoningornis longidigitrus	?	?	?	?	?	?	?	?
Songlingornis linghensis	?	?	?	?	?	?	?	?
Ambiortus dementjevi	1	0	0	0	1	0	0	?
Longicrusavis houi	1	1	0	?	0	1	?	1
Apsaravis ukhaana	1	1	0	0	1	0	0	0
Hongshanornis longicresta	1	1	0	?	?	?	?	?
Yanornis martini	1	1	0	0	1	?	?	0
Patagopteryx deferrariisi	1	0	0	1	?	?	?	1
Hollanda luceria	?	?	?	?	?	?	?	?
Yixianornis grabaui	?	1	0	?	?	1	1	?
Gansus yumenensis	1	1	0	0	1	?	?	0
Ichthyornis dispar	1	1	0	0	1	1	0	0
Hesperornis regalis	?	?	?	?	0	?	0	?
Parahesperornis	?	?	?	?	?	?	?	?
Enaliornis baretti	?	?	?	?	?	?	?	?
Baptornis advenus	?	?	?	?	0	0	0	?
Baptornis varneri	?	?	?	?	?	?	?	?
Limenavis patagonica	?	?	?	?	?	?	?	?
Vegaavis iaai	1	?	0	?	?	?	?	?
Anas	1	1	0	0	1	1	1	1
Gallus	1	1	0	0	1	1	1	1

129	130	131	132	133	134	135	136	137	138	139	140
0	0	0	0	0	0	0	0	0	0	0	0
1	0	0	0	0	0	0	0	0	0	0	0
0	0	0	?	0	[01]	0	0	?	?	0	1
?	?	?	?	?	?	?	?	?	?	?	?
0	0	?	?	?	?	?	?	?	?	?	?
0	0	?	?	?	?	?	?	?	?	0	?
?	0	?	?	?	?	?	?	?	?	?	?
1	0	0	?	?	?	?	0	0	0	0	?
1	1	0	0	0	0	0	?	?	?	0	1
1	1	?	?	?	?	?	?	?	?	0	1
2	1	0	?	0	0	0	0	0	1	0	1
2	?	?	?	?	?	?	?	?	?	?	?
1	0	0	?	0	?	?	?	0	0	0	?
0	0	?	?	?	?	?	?	?	?	?	1
1	0	2	1	1	[01]	0	1	0	1	1	1
[01]	0	2	1	1	?	?	?	?	?	?	1
0	0	0	[12]	0	[12]	0	0	0	0	1	1
1	0	2	1	1	0	0	1	0	0	1	1
0	0	?	?	?	[01]	?	?	?	?	1	1
0	0	?	?	?	?	?	?	?	?	?	?
?	?	?	?	?	?	0	?	?	?	?	?
?	?	?	?	1	?	0	?	?	?	1	1
0	0	2	?	?	[01]	?	0	?	0	1	1
0	0	?	?	?	[01]	?	?	?	?	?	?
?	?	?	?	1	?	0	1	?	0	1	?
1	0	?	?	?	[01]	?	0	0	0	1	1
1	0	1	?	?	[01]	?	1	0	0	?	1
1	0	?	?	?	?	?	?	?	?	?	?
0	0	0	?	?	[01]	0	?	?	?	1	1
0	0	?	?	?	?	?	?	?	?	1	1
1	0	1	?	1	[01]	0	?	?	?	1	1
?	?	?	?	?	?	?	?	?	?	?	?
1	0	?	?	?	?	0	?	?	?	0	1
?	?	?	?	?	?	?	?	?	?	?	?
1	0	-	-	?	?	?	?	?	?	?	?
1	0	1	2	0	2	1	1	0	0	0	1
1	0	1	1	1	0	0	0	0	1	1	1
1	0	?	?	?	2	?	?	?	?	0	?
1	0	1	1	0	[01]	?	?	?	?	0	1
0	0	0	?	0	?	0	0	0	0	0	1
?	?	?	?	?	?	?	?	?	?	?	?
?	0	?	0	?	0	?	1	1	0	0	1
1	0	1	3	0	?	1	?	?	?	0	1
1	0	1	2	0	2	1	0	?	1	0	1
?	0	?	?	1	0	0	0	?	?	0	-
?	?	?	?	?	?	?	?	?	?	?	?
?	?	?	?	1	0	0	0	?	?	0	-
?	?	?	?	?	2	1	1	1	1	0	1
?	?	?	?	?	?	?	?	?	?	?	?
0	0	0	3	0	1	1	1	1	1	0	1
0	0	0	3	0	1	1	1	1	1	1	1

Taxa\characters	141	142	143	144	145	146	147	148
Dromaeosauridae	0	0	1	0	0	0	0	0
Archaeopteryx (all ten)	?	?	?	0	0	0	0	0
Jeholornis prima	0	1	1	1	1	0	?	0
Rahonavis ostromi	?	?	?	?	1	0	?	0
Shenzhouraptor sinensis	?	?	?	1	0	?	?	?
Jixiangornis orientalis	?	?	?	1	0	?	?	?
Dalianraptor cuhe	?	?	?	0	1	?	?	?
Zhongornis haoae	?	?	?	0	0	?	?	?
Sapeornis chaoyangensis	0	0	1	1	1	?	?	?
Didactylornis jii	?	?	?	1	?	?	?	?
Confuciusornis sanctus	0	0	0	0	1	0	0	2
Changchengornis hengdaoziensis	?	?	?	0	1	?	?	?
Eoconfuciusornis zhengi	?	?	?	0	1	?	?	?
"*Dalianornis mengi*"	?	?	?	1	1	?	?	?
Cathyaornis yandica	1	1	0	1	1	?	?	?
Concornis lacustris	?	?	?	?	1	1	1	2
Elsornis keni	1	?	?	0	1	0	0	[12]
Eoalulavis hoyasi	1	1	1	1	1	1	?	2
eocathay walkeri from single slab	?	?	?	1	0	?	?	?
Eoenantiornis buhleri	?	?	?	1	1	?	?	?
Gobipteryx minuta	?	?	?	1	1	?	?	?
Iberomesornis romerali	1	?	?	1	1	?	?	?
Longipteryx chaoyangensis	?	1	?	1	1	?	?	?
Longirostravis hani	?	?	?	1	0	?	?	?
Neuquenornis volans	?	?	?	1	1	?	1	?
Otogornis genghisi	1	1	?	1	1	?	?	?
Pengornis houi	?	?	?	1	1	?	?	?
Protopteryx fengningensis	?	?	?	1	1	?	?	?
Rapaxavis pani	1	1	?	1	0	?	?	?
Shanweiniao cooperorum	?	?	?	1	1	?	?	?
Vescornis hebeiensis	1	1	1	1	1	?	?	?
Vorona berivotrensis	?	?	?	?	?	?	?	?
Archaeorhynchus spathula	1	0	?	1	0	?	?	?
Liaoningornis longidigitrus	?	?	?	?	1	?	?	?
Songlingornis linghensis	?	?	?	?	?	?	?	?
Ambiortus dementjevi	?	?	?	?	?	?	?	?
Longicrusavis houi	0	0	0	1	1	0	0	?
Apsaravis ukhaana	1	1	1	1	1	0	1	2
Hongshanornis longicresta	?	?	?	1	1	?	?	?
Yanornis martini	0	0	0	1	1	?	0	?
Patagopteryx deferrariisi	0	0	?	0	0	0	0	?
Hollanda luceria	?	?	?	?	?	?	?	?
Yixianornis grabaui	?	0	?	1	0	0	0	2
Gansus yumenensis	0	0	1	1	1	0	0	1
Ichthyornis dispar	0	0	0	1	1	0	0	2
Hesperornis regalis	-	-	-	?	?	?	?	?
Parahesperornis	?	?	?	?	?	?	?	?
Enaliornis baretti	?	?	?	?	?	?	?	?
Baptornis advenus	-	-	-	0	1	0	0	0
Baptornis varneri	?	?	?	?	?	?	?	?
Limenavis patagonica	0	0	0	?	?	?	?	?
Vegavis iaai	?	?	?	?	?	?	?	?
Anas	0	0	0	0	1	0	0	1
Gallus	0	0	0	1	1	0	0	1

149	150	151	152	153	154	155	156	157	158	159	160
0	0	0	-	-	0	0	0	?	0	0	0
0	0	0	?	?	0	0	1	0	0	0	0
0	0	0	0	1	2	1	1	0	0	1	0
0	0	0	?	?	?	?	?	?	?	?	?
?	?	?	?	?	0	1	1	0	0	0	?
?	?	?	?	?	?	?	?	?	?	?	?
?	?	?	?	?	?	?	0	0	0	0	?
?	0	0	?	?	2	1	1	0	0	0	0
?	?	0	?	?	[12]	1	?	?	?	?	?
1	0	0	0	1	2	1	0	?	0	0	0
?	?	0	?	?	?	?	0	?	0	0	0 or ?
?	?	0	?	?	0	1	?	?	0	0	0 or ?
?	1	1	0	?	3	1	1	0	0	0	1
?	?	?	?	?	?	?	?	?	?	?	?
1	?	1	?	?	3	1	?	?	?	?	?
1	?	1	?	?	?	?	?	?	?	?	?
?	?	?	0	n	3	1	?	?	0	0	1
?	?	1	0	n	3	1	?	?	0	0	0
1	1	1	?	?	?	?	?	?	?	?	?
?	?	?	?	?	?	?	?	?	?	?	?
?	?	0	0	-	0	1	1	0	0	0	0
?	?	?	0	-	3	1	?	?	1	0	?
?	1	1	?	?	3	1	?	?	1	0	?
?	?	?	1	[02]	[23]	1	1	1	?	?	?
?	?	?	?	?	?	1	?	?	0	0	0
1	?	?	0	-	?	1	1	?	0	0	?
?	1	?	?	?	?	?	?	?	?	?	?
?	?	?	1	[02]	0	1	1	0	1	0	?
?	?	?	?	?	?	?	?	?	?	?	?
?	1	0	1	?	0	1	?	?	0	0	2
?	?	?	?	?	?	?	?	?	?	?	?
?	?	?	0	?	3	?	?	?	?	[12]	?
?	?	0	?	?	3	1	1	1	0	?	?
1	1	0	1	?	3	1	1	0	0	1	1
?	?	?	?	?	3	1	?	?	0	0	?
1	1	?	0	?	3	1	1	1	0	0	0
0	?	0	?	?	?	?	?	?	?	?	?
?	?	?	?	?	?	?	?	?	?	?	?
1	?	0	?	?	3	1	1	1	0	1	2 see r hand
1	1	?	0	?	3	1	2	?	0	1	1
1	1	0	1	?	3	1	2	1	0	1	2
?	?	?	?	?	?	?	-	?	?	?	?
?	?	?	?	?	?	?	-	?	?	?	?
1	0	?	?	?	?	?	-	?	?	?	?
?	?	?	?	?	?	?	-	?	?	?	?
1	1	1	?	?	3	?	2	1	?	[23]	?
?	?	1	?	?	?	?	?	?	?	?	?
1	1	0	1	2	3	1	2	0	1	3	2
1	1	0	1	1	3	1	2	0	1	3	2

Taxa\characters	161	162	163	164	165	166	167	168
Dromaeosauridae	0	0	0	0	0	0	0	0
Archaeopteryx (all ten)	0	1	0	0	0	0	0	0
Jeholornis prima	0	0	0	1	0	0	0	0
Rahonavis ostromi	?	?	?	?	?	?	?	?
Shenzhouraptor sinensis	0	0	[01]	0	0	0	0	0
Jixiangornis orientalis	?	?	[01]	0	0	?	0	1
Dalianraptor cuhe	1	?	?	0	?	?	?	0
Zhongornis haoae	0	1	0	0	-	0	0	0
Sapeornis chaoyangensis	1	0	0	0	-	0	0	0
Didactylornis jii	?	0	0	0	?	1	0	0
Confuciusornis sanctus	?	1	0	0	-	0	0	0
Changchengornis hengdaoziensis	0	?	0	?	?	0	0	0
Eoconfuciusornis zhengi	0	1	?	0	?	0	0	0
"*Dalianornis mengi*"	0	1	0	0	1	0	1	1
Cathyaornis yandica	0	?	[01]	0	0	0	1	2
Concornis lacustris	0	?	?	0	?	?	?	2
Elsornis keni	0	?	[01]	?	0	1	?	?
Eoalulavis hoyasi	0	?	?	1	?	0	1	1
eocathay walkeri from single slab	0	?	?	0	0	0	1	2
Eoenantiornis buhleri	0	?	0	0	0	0	1	1
Gobipteryx minuta	0	?	?	0	?	0	1	?
Iberomesornis romerali	?	?	?	?	?	?	?	?
Longipteryx chaoyangensis	0	1	0	0	?	0	1	1
Longirostravis hani	0	?	[01]	0	1	0	1	?
Neuquenornis volans	0	?	[01]	0	0	0	1	?
Otogornis genghisi	?	?	?	?	?	?	?	?
Pengornis houi	0	?	0	1	0	0	1	?
Protopteryx fengningensis	?	?	[01]	0	0	0	1	0
Rapaxavis pani	0	1	0	0	0	0	1	2
Shanweiniao cooperorum	?	?	?	?	?	?	?	[01]
Vescornis hebeiensis	0	0	[01]	0	?	0	1	2
Vorona berivotrensis	?	?	?	?	?	?	?	?
Archaeorhynchus spathula	?	?	0	?	?	0	?	2
Liaoningornis longidigitrus	?	?	?	?	?	?	?	?
Songlingornis linghensis	?	?	?	?	?	?	?	?
Ambiortus dementjevi	?	?	?	?	?	?	?	?
Longicrusavis houi	0	0	0	1	0	1	0	0
Apsaravis ukhaana	1	0	1	1	0	?	?	?
Hongshanornis longicresta	0	?	?	0	0	?	0	1
Yanornis martini	0	0	0	0	0	1	0	1
Patagopteryx deferrariisi	0	?	?	1	?	1	?	?
Hollanda luceria	?	?	?	?	?	?	?	?
Yixianornis grabaui	1	0	0	0	0	1	?	1
Gansus yumenensis	1	0	0	0	1	?	1	2
Ichthyornis dispar	1	0	1	0	0	1	0	?
Hesperornis regalis	?	?	?	?	?	?	?	?
Parahesperornis	?	?	?	?	?	?	?	?
Enaliornis baretti	?	?	?	?	?	?	?	?
Baptornis advenus	?	?	?	?	?	?	?	?
Baptornis varneri	?	?	?	?	?	?	?	?
Limenavis patagonica	?	?	2	?	?	1	0	?
Vegaavis iaai	?	?	?	?	?	?	?	?
Anas	1	1	1	1	1	1	0	2
Gallus	1	0	2	1	0	1	1	2

169	170	171	172	173	174	175	176	177	178	179	180
0	0	0	0	0&1	0	0	0	0	0	0	0
0	0	0	0	0	0	0	0	2	0	0	0
0	0	0	0	1	0	0	1	3	0	0	0
?	?	?	?	?	?	?	?	?	0	0	1
0	0	0	0	1	0	0	0	3	?	?	?
0	0	0	0	[01]	?	0	0	3	?	?	?
0	0	?	?	1	?	0	0	1	?	?	?
0	0	0	0	1	0	0	0	1	?	0	0
0	0	1	0	1	0	1	0	3	0	0	0
0	0	1	0	[01]	?	1	[01]	3	0	?	?
0	0	0	0	2	1	0	0	2	0	0	0
0	0	0	0	2	1	0	0	2	?	?	?
0	0	0	0	2	1	0	0	2	0	?	?
0	0	1	0	1	0	1	1	3	0	0	0
0	0	1	0	1	0	1	2	2	0	0	0
0	0	1	0	[01]	0	1	2	2	?	?	?
?	?	?	?	?	?	?	?	?	?	?	?
0	0	1	?	?	?	?	2	?	?	?	0
0	0	1	0	0	0	1	1	?	?	?	?
0	0	1	0	1	0	1	2	2	?	?	?
0	0	?	?	?	?	?	?	2	0	?	?
?	?	?	?	?	?	?	?	?	?	?	?
0	0	1	0	0	0	1	1	3	0	0	0
0	0	1	1	-	0	1	2	2	?	?	0
?	?	?	?	?	?	?	?	?	?	?	?
0	0	1	0	?	0	?	[12]	3	?	?	?
0	0	0	0	0	0	1	0	?	?	?	0
0	0	1	1	-	0	1	2	3	0	0	0
0	?	1	1	-	?	?	2	2	?	?	?
0	0	1	0	[01]	0	1	2	2	0	?	?
?	?	?	?	?	?	?	?	?	?	?	?
1	0	1	0	?	?	?	2	3	0	0	0
?	?	?	?	?	?	?	?	?	?	?	?
1	0	1	0	?	?	?	?	?	?	?	?
1	0	0	0	1	0	1	0	1	?	?	?
1	?	?	?	?	?	?	?	?	1	0	0
1	0	1	0	1	0	1	0	1	?	?	?
1	0	1	0	1	0	1	2	3	1	0	1
0	?	0	0	?	?	?	2	0	?	0	0
?	?	?	?	?	?	?	?	?	?	?	?
1	0	1	0	1	0	1	0	2	1	0	0
1	0	1	0	0	0	1	3	2	1	?	0
1	1	1	1	n	-	?	?	3	1	0	?
?	?	?	?	n	?	?	?	?	1	0	0
?	?	?	?	n	?	?	?	?	1	?	?
?	?	?	?	n	?	?	?	?	1	?	?
?	?	?	?	n	?	?	?	0	1	0	0
?	?	?	?	n	?	?	?	?	1	0	?
1	1	?	?	?	?	?	?	?	?	?	?
?	?	?	?	?	?	?	?	?	0	1	?
1	0	1	1	n	-	1	2	3	1	1	0
1	0	1	1	n	-	1	2	1	1	1	0

Taxa\characters	181	182	183	184	185	186	187	188
Dromaeosauridae	[01]	0	0	0	0	0	0	0
Archaeopteryx (all ten)	0	0	0	0	0	1	?	0
Jeholornis prima	0	0	0	0	0	1	1	?
Rahonavis ostromi	0	0	0	0	-	1	1	0
Shenzhouraptor sinensis	?	?	?	?	?	?	?	?
Jixiangornis orientalis	?	?	?	?	?	?	?	?
Dalianraptor cuhe	?	?	?	?	?	?	?	?
Zhongornis haoae	?	?	?	?	?	?	?	?
Sapeornis chaoyangensis	0	?	?	0	?	1	1	1
Didactylornis jii	?	?	?	?	?	?	?	0
Confuciusornis sanctus	0	0	0	0	0	?	1	0
Changchengornis hengdaoziensis	0	?	?	?	?	?	?	?
Eoconfuciusornis zhengi	0	?	?	0	?	?	?	?
"Dalianornis mengi"	0	?	?	?	?	?	?	0
Cathyaornis yandica	0	0	0	?	?	1	1	?
Concornis lacustris	?	?	?	?	?	?	?	1
Elsornis keni	?	?	?	?	?	?	?	?
Eoalulavis hoyasi	?	?	?	?	?	1	?	?
eocathay walkeri from single slab	?	?	?	?	?	?	?	?
Eoenantiornis buhleri	0	?	?	?	?	?	?	?
Gobipteryx minuta	?	?	?	?	?	?	?	?
Iberomesornis romerali	?	?	?	?	?	?	?	?
Longipteryx chaoyangensis	0	?	?	?	?	1	1	1
Longirostravis hani	?	?	?	?	?	1	?	?
Neuquenornis volans	?	?	?	?	?	?	?	?
Otogornis genghisi	?	?	?	?	?	?	?	?
Pengornis houi	?	?	?	?	?	?	?	?
Protopteryx fengningensis	0	?	?	?	?	1	1	?
Rapaxavis pani	0	?	0	?	?	1	1	1
Shanweiniao cooperorum	?	?	?	?	?	?	1	?
Vescornis hebeiensis	0	?	?	?	?	?	?	?
Vorona berivotrensis	?	?	?	?	?	?	?	?
Archaeorhynchus spathula	0	?	?	?	?	1	0	1
Liaoningornis longidigitrus	?	?	?	?	?	?	?	?
Songlingornis linghensis	?	?	?	?	?	?	?	?
Ambiortus dementjevi	?	?	?	?	?	?	?	?
Longicrusavis houi	?	?	?	?	1	1	?	?
Apsaravis ukhaana	0	1	1 per julia	1	1	1	?	1
Hongshanornis longicresta	?	?	?	?	?	?	?	0
Yanornis martini	0	0	?	1	1	?	1	0
Patagopteryx deferrariisi	0	1	0	0	0	0	0	1
Hollanda luceria	?	?	?	?	?	?	?	?
Yixianornis grabaui	?	1	?	?	0	0	?	0
Gansus yumenensis	0	1	1	1	1	0	1	0
Ichthyornis dispar	0	1	0	1	1	?	?	1
Hesperornis regalis	0	1	2	1	1	0	1	1
Parahesperornis	?	?	?	?	?	?	?	?
Enaliornis baretti	?	?	?	?	1	?	?	?
Baptornis advenus	0	1	2	1	1	0	1	1
Baptornis varneri	?	?	?	?	?	0	?	?
Limenavis patagonica	?	?	?	?	?	?	?	?
Vegaavis iaai	?	?	1	?	?	?	?	?
Anas	1	1	1	1	1	0	1	0
Gallus	1	1	2	1	1	0	1	1

189	190	191	192	193	194	195	196	197	198	199	200
0	0	0	0	0	1	0	0	0	0	[01]	0
0	0	1	?	0	1	0	0	0	0	?	1
1	0	1	?	0	1	0	0	1	0	?	?
0	0	1	1	0	1	0	0	0	0	0	1
?	?	?	?	?	?	?	?	1	?	?	?
?	?	?	?	?	?	?	?	?	?	?	?
1	?	1	?	?	?	?	?	?	?	?	?
1	0	1	1	1	1	?	0	1	0	0	1
?	?	?	0	?	?	?	0	1	0	?	?
1	0	1	1	1	1	0	0	1	0	1	0
?	?	?	?	[12]	?	?	?	1	?	?	?
?	?	?	?	?	1	0	0	1	0	?	?
1	n	1	1	1	?	?	?	1	0	?	?
?	?	?	?	?	1	0	0	?	?	?	?
1	?	1	?	[12]	?	?	0	1	?	0	?
?	?	?	?	?	?	?	?	?	?	?	?
?	?	?	?	?	?	?	?	?	?	?	?
?	?	?	?	[12]	?	?	0	[12]	0	?	?
?	?	?	?	?	1	0	?	1	1	?	?
?	?	?	?	1	?	?	0	?	?	?	?
1	0	1	1	[12]	1	1	0	1	0	?	0
1	0	?	1	?	1	1	?	?	?	?	0
?	?	?	?	?	?	?	?	?	?	?	?
?	?	?	?	?	?	?	?	?	?	?	?
?	?	?	?	?	?	?	?	[01]	0	?	0
?	?	?	?	?	?	?	?	1	0	?	?
1	0	1	?	1	1	1	?	1	0	0	0
?	?	?	?	?	1	?	?	[12]	?	0	?
?	?	?	?	?	?	?	?	?	?	?	?
?	?	?	?	?	?	?	?	?	?	1	0
1	0	0	?	[12]	1	0	0	1	0	?	?
?	?	?	?	?	?	?	?	?	?	?	0
?	?	?	?	?	?	?	?	?	?	?	?
1	?	?	?	[12]	?	?	0	?	?	?	?
1	1	0	1	2	?	?	1	2	1	1	0
?	?	0	1	[12]	?	?	?	[12]	1	?	?
1	0	0	?	1	1	?	0	1	0	?	?
1	1	0	1	2	1	?	0	2	1	?	0
?	?	?	?	?	?	?	?	?	?	?	?
1	?	?	1	1	?	?	0	1	?	?	?
1	0	1	1	2	1	0	0	1	1	0	0
1	0	1	1	2	1	?	1	2	-	1	0
1	0	0	1	2	1	0	1	2	-	1	0
?	?	?	?	?	?	?	?	?	?	1	0
?	?	?	?	2	?	?	?	?	?	1	0
1	1	0	1	2	?	?	1	?	?	1	0
1	1	?	?	2	?	?	1	?	?	?	?
?	1	?	?	2	?	?	?	?	?	?	0
1	1	1	1	2	1	0	1	2	1	1	0
0	1	1	1	2	1	0	1	2	1	1	0

Taxa\characters	201	202	203	204	205	206	207	208
Dromaeosauridae	0	-	0	0	0	0	0	0
Archaeopteryx (all ten)	0	-	0	0	0	0	0	?
Jeholornis prima	?	?	0	?	?	?	0	?
Rahonavis ostromi	1	2	0	0	1	0	0	1
Shenzhouraptor sinensis	?	?	?	?	?	?	?	?
Jixiangornis orientalis	?	?	?	?	?	?	?	?
Dalianraptor cuhe	?	?	?	?	?	?	?	?
Zhongornis haoae	?	?	[02]	0	[01]	?	?	0
Sapeornis chaoyangensis	0	-	?	0	1	0	0	?
Didactylornis jii	?	?	?	?	?	?	?	?
Confuciusornis sanctus	1	1	2	0	1	0	1	?
Changchengornis hengdaoziensis	?	?	?	?	?	?	?	?
Eoconfuciusornis zhengi	?	?	?	?	?	?	?	?
"*Dalianornis mengi*"	?	?	?	?	?	?	?	?
Cathyaornis yandica	1	2	?	?	?	?	1	?
Concornis lacustris	?	?	?	?	?	?	?	?
Elsornis keni	?	?	?	?	?	?	?	?
Eoalulavis hoyasi	1	?	1	?	?	?	?	?
eocathay walkeri from single slab	?	?	?	?	?	?	?	?
Eoenantiornis buhleri	?	?	1	?	?	?	?	?
Gobipteryx minuta	?	?	?	?	?	?	?	?
Iberomesornis romerali	?	?	?	0	?	?	?	?
Longipteryx chaoyangensis	?	[12]	1	?	1	0	?	?
Longirostravis hani	?	1	?	?	?	?	?	?
Neuquenornis volans	1	1	1	0	?	?	?	?
Otogornis genghisi	?	?	?	?	?	?	?	?
Pengornis houi	1	?	?	?	?	?	?	?
Protopteryx fengningensis	?	?	1	?	?	?	?	?
Rapaxavis pani	?	?	?	?	?	?	?	?
Shanweiniao cooperorum	?	?	?	?	[01]	0	?	?
Vescornis hebeiensis	?	?	?	?	?	?	?	?
Vorona berivotrensis	1	?	1	1	[01]	?	1	1
Archaeorhynchus spathula	?	?	?	?	?	?	?	?
Liaoningornis longidigitrus	1	?	?	?	?	?	?	?
Songlingornis linghensis	?	?	?	?	?	?	?	?
Ambiortus dementjevi	?	?	?	?	?	?	?	?
Longicrusavis houi	?	?	?	?	?	0	1	?
Apsaravis ukhaana	1	[12]	?	1	1	0	?	?
Hongshanornis longicresta	?	?	?	?	?	?	?	?
Yanornis martini	1	?	?	?	?	?	?	?
Patagopteryx deferrariisi	?	?	2	0	?	?	?	?
Hollanda luceria	?	?	?	1	2	0	1	1
Yixianornis grabaui	1	?	2	?	1	0	1	?
Gansus yumenensis	1	[12]	?	1	1	0	1	?
Ichthyornis dispar	1	2	2	1	1	0	1	?
Hesperornis regalis	2	1	2	2	1	0	1	1
Parahesperornis	2	1	2	2	1	0	1	1
Enaliornis baretti	1	2	2	1	1	0	1	1
Baptornis advenus	1	2	2	1	1	0	1	1
Baptornis varneri	?	?	?	2	1	0	1	?
Limenavis patagonica	?	?	?	?	?	?	?	?
Vegaavis iaai	1	1	2	1	1	0	?	?
Anas	1	1	2	1	1	0	1	1
Gallus	1	1	2	1	1	0	1	1

209	210	211	212	213	214	215	216	217	218	219	220
0	0	0	1	0	0	0	0	0	0	0	0
0	0	0	1	?	0	?	?	0	?	?	0
0	?	?	0	[01]	0	?	1	0	?	?	1
0	1	1	1	0	0	0	1	0	0	2	1
0	?	0	?	?	?	?	?	?	?	?	?
?	?	?	?	?	?	?	?	?	?	?	?
?	?	?	?	?	?	?	?	?	?	?	?
0	?	0	0	?	0	?	?	0	?	?	1
1	0	0	0	?	0	?	1	1	1	?	1
1	?	0	?	?	?	?	?	?	?	?	?
1	?	0	0	1	0	0	0	0	?	1	1
?	0	?	?	?	?	?	?	?	?	?	1
0	?	?	?	?	?	?	1	0	?	?	?
0	?		?	?	?	?	?	?	?	?	?
?	?	?	[01]	?	?	?	?	?	1	?	1
1	1	1	1	?	?	?	?	?	?	?	1
?	?	?	?	?	?	?	?	?	?	?	?
?	?	?	?	?	?	?	?	?	?	?	?
?	?	?	?	?	?	?	?	?	?	?	?
?	1	0	[01]	1	0	?	0	1	1	?	1
0	1	?	0	?	?	?	?	?	?	?	?
?	?	0	?	?	?	?	0	?	1	?	1
1	?	0	0	?	?	?	?	?	1	?	1
?	?	?	1	?	?	?	?	?	?	?	?
?	?	?	?	?	?	?	?	?	?	?	?
1	?	?	[01]	?	0	1	[12]	1	1	?	1
0	?	?	?	?	?	?	?	?	?	?	?
0	?	0	[01]	?	0	?	1	1	?	?	1
1	?	?	?	?	?	?	?	?	?	?	1
0	?	?	[01]	?	?	?	?	?	?	?	?
0	1	0	1	[12]	1	0	0	1	1	1	1
0	?	0	[12]	0	0	?	?	?	?	2	1
?	?	?	?	?	?	?	?	?	?	?	?
?	?	?	?	?	?	?	?	?	?	?	?
1	0	0	2	2	[01]	1	2	1	?	1	1
1	?	?	?	2	0	1	2	1	?	?	1
?	0	?	[12]	?	?	?	?	?	?	?	1
1	?	0	?	?	0	?	?	0	?	?	1
1	0	0	1	?	0	?	0	1	1	1	1
1	0		1	[12]	0	0	0	0	1	2	1
1	0	0	?	?	?	?	?	?	?	?	1
1	0	0	[12]	2	?	0	?	0	?	2	1
1	0	0	2	2	1	1	1	0	?	2	1
1	0		2	2	1	1	1	1	1	2	1
1	0		2	2	1	1	1	0	?	?	?
1	0		2	2	1	1	1	0	?	?	?
1	0		2	2	1	0	1	0	1	2	1
?	?	?	?	?	?	?	?	?	?	?	?
1	?	0	2	?	2	?	1	?	?	?	?
1	0	0	2	2	2	0	1	1	1	2	1
1	0	0	2	2	2	0	1	1	1	2	1

Taxa\characters	221	222	223	224	225	226	227	228
Dromaeosauridae	0	0	0	0	0	0	0	0
Archaeopteryx (all ten)	0	0	0	0	0	0	0	0
Jeholornis prima	1	0	0	0	0	0	0	0
Rahonavis ostromi	0	0	0	?	0	0	1	0
Shenzhouraptor sinensis	0	0	?	?	0	?	?	?
Jixiangornis orientalis	?	0	?	?	?	?	0	0
Dalianraptor cuhe	?	?	?	?	?	?	?	?
Zhongornis haoae	?	0	0	?	0	?	0	0
Sapeornis chaoyangensis	1	0	0	0	0	0	0	0
Didactylornis jii	1	0	?	0	0	0	0	0
Confuciusornis sanctus	1	0	0	0	0	1	0	0
Changchengornis hengdaoziensis	1	0	0	0	0	?	0	?
Eoconfuciusornis zhengi	?	0	?	0	0	0	0	0
"Dalianornis mengi"	0	0	?	?	0	0	1	0
Cathyaornis yandica	?	?	?	?	?	?	?	?
Concornis lacustris	1	1	0	?	0	0	1	0
Elsornis keni	?	?	?	?	?	?	?	?
Eoalulavis hoyasi	?	?	?	?	?	?	?	?
eocathay walkeri from single slab	?	?	?	?	?	?	?	?
Eoenantiornis buhleri	1	1	?	?	0	?	1	0
Gobipteryx minuta	1	0	0	1	0	?	1	0
Iberomesornis romerali	0	0	?	1	0	0	0	0
Longipteryx chaoyangensis	?	0	?	1	0	0	0	0
Longirostravis hani	1	[01]	?	1	0	0	0	0
Neuquenornis volans	1	0	?	?	0	?	1	?
Otogornis genghisi	?	?	?	?	?	?	?	?
Pengornis houi	1	0	0	?	0	0	1	0
Protopteryx fengningensis	0	0	?	?	0	0	0	0
Rapaxavis pani	0	0	?	1	0	0	0	0
Shanweiniao cooperorum	1	0	?	?	0	0	0	0
Vescornis hebeiensis	0	0	?	?	0	0	1	0
Vorona berivotrensis	1	[12]	0	0	0	1	1	0
Archaeorhynchus spathula	1	[01]	0	1	0	0	0	0
Liaoningornis longidigitrus	1	1	0	?	0	0	[01]	0
Songlingornis linghensis	?	?	?	?	?	?	?	?
Ambiortus dementjevi	?	?	?	?	?	?	?	?
Longicrusavis houi	1	2	0	1	1	1	0	0
Apsaravis ukhaana	1	[12]	?	1	1	1	0	0
Hongshanornis longicresta	1	2	0	?	0	?	0	0
Yanornis martini	1	1	0	1	1	?	0	0
Patagopteryx deferrariisi	1	2	0	1	0	1	0	0
Hollanda luceria	1		0	?	0	0	1	2
Yixianornis grabaui	1	[12]	0	1	?	0	0	0
Gansus yumenensis	1	2	0	1	1	1	0	0
Ichthyornis dispar	1	2	0	1	1	2	0	2
Hesperornis regalis	1	2	0	1	1	2	0	2
Parahesperornis	1	2	0	1	0	2	0	2
Enaliornis baretti	1	2	0	1	0	?	0	1
Baptornis advenus	1	2	0	1	0	2	0	1
Baptornis varneri	1	2	0	1	1	2	0	1
Limenavis patagonica	?	?	?	?	?	?	?	?
Vegaavis iaai	1	2	0	1	1	?	0	?
Anas	1	2	0	1	1	2	0	1
Gallus	1	2	0	1	1	2	0	1

229	230	231	232	233	234	235	236	237	238	239	240
0	0	0	0	[01]	0	0	0	0	0	0	0
0	0	0	0	0	?	0	?	0	0	?	0
0	?	?	0	0	?	0	2	0	0	?	0
0	0	0	0	[12]	0	0	[01]	0	0	0	0
?	0	?	0	?	?	0	?	0	?	?	0
?	?	?	?	?	?	?	[01]	?	?	?	?
?	?	?	?	?	?	?	?	?	?	?	?
0	0	0	0	?	?	0	[12]	0	0	?	0
0	0	0	0	0	?	0	0	0	0	?	0
?	?	0	?	?	?	0	0	0	0	?	0
0	1	?	2	1	0	0	0	0	0	0	0
0	?	?	2	?	?	0	1	0	0	?	0
0	0	0	2	?	0	0	2	0	0	?	0
?	?	?	2	?	?	?	[01]	?	0	?	0
?	?	?	?	?	?	?	?	?	?	?	?
?	?	0	[01]	?	?	0	0	0	0	?	0
?	?	?	?	?	?	?	?	?	?	?	?
?	?	?	?	?	?	?	?	?	?	?	?
?	?	?	?	?	?	0	?	?	1	?	0
0	?	?	1	1	?	0	?	?	1	?	1
0	0	0	1	?	?	0	0	0	?	?	0
0	0	0	0	1	?	[13]	2	0	0	0	0
?	?	0	?	?	?	0	1	0	0	?	0
?	1	?	2	?	0	0	?	0	1	1	0
?	?	?	?	?	?	?	?	?	?	?	?
0	1	0	1	2	0	0	1	0	1	?	0
?	?	0	0	?	?	0	[01]	0	1	?	0
?	?	0	0	2	?	0	1	0	0	?	0
?	?	?	?	1	?	[01]	[01]	?	0	?	0
?	?	0	1	1	?	[01]	2	0	1	?	1
0	1	1	?	0	2	0	1	0	0	0	0
?	?	?	?	0	?	0	1	0	0	?	0
?	?	?	1	[12]	?	0	1	0	0	?	0
?	?	?	?	?	?	?	?	?	?	?	?
?	?	1	0	1	0	0	1	0	0	?	0
1	?	1	?	2	0	0	1	0	0	?	0
?	0	?	0	?	?	0	2	0	0	?	0
[01]	?	?	1	2	?	0	1	0	0	?	0
1	1	1	0	0	0	0	?	0	0	?	0
[123]	1	1	1	2	0	0	3	0	0	0	0
1	1	1	0	?	0	0	1	0	?	?	0
1	0	1	0	2	?	0	2	0	0	0	0
?	0	1	?	2	1	0	2	0	0	0	0
1	0	?	?	1	2	3	2	1	0	1	0
1	0	0	?	1	2	3	2	1	0	0	0
1	0	0	?	?	?	0	1	0	0	0	0
1	0	0	?	1	2	1	1	0	0	0	0
1	0	0	?	1	2	1	2	1	0	0	0
?	?	?	?	?	?	?	?	?	?	?	?
3	?	1	?	?	?	?	1	?	?	?	0
3	0	1	3	2	1	0	2	2	0	0	0
3	0	1	3	2	2	0	1	2	0	1	0

Taxa\characters	241	242	243	244	245
Dromaeosauridae	0	0	0	0	0
Archaeopteryx (all ten)	0	0	1	0	0
Jeholornis prima	0	1	0	?	0
Rahonavis ostromi	0	1	0	?	?
Shenzhouraptor sinensis	0	0	0	?	0
Jixiangornis orientalis	?	1	0	?	0
Dalianraptor cuhe	?	1	1	?	0
Zhongornis haoae	0	1	1	?	?
Sapeornis chaoyangensis	0	1	1	?	?
Didactylornis jii	0	1	1	?	?
Confuciusornis sanctus	0	1	1	0	0
Changchengornis hengdaoziensis	0	1	1	?	0
Eoconfuciusornis zhengi	0	1	0	0	0
"*Dalianornis mengi*"	?	1	?	?	?
Cathyaornis yandica	?	?	?	?	?
Concornis lacustris	0	1	?	?	?
Elsornis keni	?	?	?	?	?
Eoalulavis hoyasi	?	?	?	1	?
eocathay walkeri from single slab	?	?	?	?	?
Eoenantiornis buhleri	?	1	1	1	0
Gobipteryx minuta	?	?	?	?	?
Iberomesornis romerali	0	1	1	?	?
Longipteryx chaoyangensis	0	1	[01]	?	?
Longirostravis hani	0	?	?	?	?
Neuquenornis volans	0	1	[12]	?	?
Otogornis genghisi	?	?	?	?	?
Pengornis houi	0	?	?	?	?
Protopteryx fengningensis	0	1	1	1	0
Rapaxavis pani	0	1	1	?	?
Shanweiniao cooperorum	?	1	[01]	?	1
Vescornis hebeiensis	0	1	1	?	?
Vorona berivotrensis	0	?	?	?	?
Archaeorhynchus spathula	0	?	?	?	?
Liaoningornis longidigitrus	0	1	[12]	?	?
Songlingornis linghensis	?	?	?	?	?
Ambiortus dementjevi	?	?	?	?	?
Longicrusavis houi	0	1	1	?	?
Apsaravis ukhaana	0	?	?	?	?
Hongshanornis longicresta	0	1	1	1	1
Yanornis martini	0	1	1	?	?
Patagopteryx deferrariisi	0	0	1	?	?
Hollanda luceria	0	?	?	?	?
Yixianornis grabaui	0	1	1	?	1
Gansus yumenensis	0	1	1	?	?
Ichthyornis dispar	0	?	?	?	?
Hesperornis regalis	2	?	?	?	?
Parahesperornis	2	?	?	?	?
Enaliornis baretti	0	?	?	?	?
Baptornis advenus	1	?	?	?	?
Baptornis varneri	1	?	?	?	?
Limenavis patagonica	?	?	?	?	?
Vegaavis iaai	?	?	?	?	?
Anas	0	1	0	1	1
Gallus	0	1	0	1	1

REFERENCES

Alonso PD, Milner AC, Ketcham RA, Cookson MJ, RoweTB. 2004. The avian nature of the brain and inner ear of *Archaeopteryx*. *Nature* **430**: 666–669.

Barsbold R, Osmólska H, Watabe M, Currie PJ, Tsogtbataar K. 2000. A new oviraptorosaur (Dinosauria: Theropoda) from Mongolia: the first dinosaur with a pygostyle. *Acta Paleontologica Polonica* **45**: 97–106.

Baumel JJ, Witmer LM. 1993. Osteologia. In *Handbook of Avian Anatomy: Nomina Anatomica Avium*, 2nd Edn, Baumel JJ, King AS, Breazile JE, Evans HE, Vanden Berge JC (eds). Cambridge: Nuttall Ornithological Club; 45–132.

Brett-Surman MK, Paul GS. 1985. A new family of bird-like dinosaurs linking Laurasia and Gondwanaland. *Journal of Vertebrate Paleontology* **5**: 133–138.

Bell A, Chiappe LM, Erickson GM, Suzuki S, Watabe M, Barsbold R, Tsogtbaatar K. 2010. Description and ecologic analysis of a new Late Cretaceous bird from the Gobi Desert (Mongolia). *Cretaceous Research* **31**: 16–26.

Benton MJ. 2004. Origin and relationships of Dinosauria. In *The Dinosauria*, 2nd edn, Weishampel DB, Dodson P, Osmólska H (eds). Berkeley, CA: University of California Press; 7–19.

Burgers P, Chiappe LM. 1999. The wing of *Archaeopteryx* as a primary thrust generator. *Nature* **399**: 60–62.

Burke AC, Feduccia A. 1997. Developmental patterns and the identification of homologies in the avian hand. *Science* **278**: 666–668.

Burnham DA, Derstler KL, Currie PJ, Bakker RT, Zhou Z, Ostrom JH. 2000. Remarkable new birdlike dinosaur (Theropoda: Maniraptora) from the Upper Cretaceous of Montana. *University of Kansas Paleontological Contributions* **13**: 1–12.

Cai Z, Zhao L. 1999. A long tailed bird from the Late-Cretaceous of Zhejiang. *Science in China* **42**: 434–441.

Cambra-Moo O, Buscalioni AD, Cubo J, Castanet J, Loth M-M, de Margerie E, de Ricqlès A. 2006. Histological observations of Enantiornithine bone (Saurischia, Aves) from the Lower Cretaceous of Las Hoyas (Spain). *Comptes Rendus Palevol* **5**: 685–691.

Cau A, Arduini P. 2008. *Enantiophoenix electrophyla* gen. et sp. nov. (Aves, Enantiornithes) from the Upper Cretaceous (Cenomanian) of Lebanon and its phylogenetic relationships. *Atti della Società Italiana di Scienze Naturali e del Museo Civico di Storia Naturale di Milano* **149** (2): 293–324.

Charig AJ, Milner AC. 1986. *Baryonyx*, a remarkable new theropod dinosaur. *Nature* **324**: 359–361.

Chiappe LM. 1993. Enantiornithine (Aves) tarsometatarsi from the Cretaceous Lecho Formation of northwestern Argentina. *American Museum Novitates* **3083**: 1–27.

Chiappe LM. 1995. The phylogenetic position of the Cretaceous birds of Argentina: Enantiornithines and *Patagopteryx deferrarsii*. *Courier Forschungsinstitut Senckenberg* **181**: 55–63.

Chiappe LM. 1996. Late Cretaceous birds of southern South America: anatomy and systematics of Enantiornithes and *Patagopteryx deferrariisi*. *Münchner Geowissenschaftliche Abhandlungen* **30**: 203–244.

Chiappe LM. 2001. Rise of birds. In *Palaeobiology II*, Briggs DEG, Crowther PR (eds). Oxford: Blackwell Science; 102–106.

Chiappe LM. 2002a. Basal bird phylogeny: problems andsolutions. In *Mesozoic Birds: Above the Heads of Dinosaurs*, Chiappe LM, Witmer LM (eds). Berkeley: University of California Press; 448–472.

Chiappe LM. 2002b. Osteology of the flightless *Patagopteryx deferrariisi* from the Late Cretaceous of Patagonia. In *Mesozoic Birds: Above the Heads of Dinosaurs*, Chiappe LM, Witmer LM (eds). Berkeley: University of California Press; 281–316.

Chiappe L.M. 2007. *Glorified Dinosaurs: the Origin and Early Evolution of Birds*. Hoboken: John Wiley & Sons.

Chiappe LM. 2009. Downsized dinosaurs: the evolutionary transition to modern birds. *Evolution: Education and Outreach* **2**: 248–256.

Chiappe LM, Calvo JO. 1994. *Neuquenornis volans*, a new Upper Cretaceous bird (Enantiornithes: Avisauridae) from Patagonia, Argentina. *Journal of Vertebrate Paleontology* **14**: 230–246.

Chiappe LM, Walker CA. 2002. Skeletal morphology and systematics of the Cretaceous Euenantiornithes (Ornithothoraces: Enantiornithes). In *Mesozoic Birds: Above the Heads of Dinosaurs*, Chiappe LM, Witmer LM (eds). Berkeley: University of California Press; 240–267.

Chiappe LM, Norell MA, Clark JM. 1998. The skull of a new relative of the stem-group bird *Mononykus*. *Nature* **392**: 272–278.

Chiappe LM, Ji S, Ji Q, Norell MA. 1999. Anatomy and systematics of the Confuciosornithidae (Aves) from the Late Mesozoic of northeastern China. *Bulletin of the American Mususeum of Natural History* **242**: 1–89.

Chiappe LM, Norell MA, Clark JM. 2001. A new skull of *Gobipteryx minuta* (Aves: Enantiornithines) from the

Cretaceous of the Gobi Desert. *American Museum Novitates* **3346**: 1–17.

Chiappe LM, Norell MA, Clark JM. 2002a. The Cretaceous, short-armed Alvarezsauridae: *Mononykus* and its kin. In *Mesozoic Birds: Above the Heads of Dinosaurs*, Chiappe LM, Witmer LM (eds). Berkeley: University of California Press; 87–120.

Chiappe LM, Lamb JP. Jr, Ericson PG. P. 2002b. New enantiornithine bird from the marine Upper Cretaceous of Alabama. *Journal of Vertebrate Paleontology* **22**: 170–174.

Chiappe LM, Suzuki S, Dyke GJ, Watabe M, Tsogtbaatar K, Barsbold R. 2006. A new enantiornithine bird from the Late Cretaceous of the Gobi Desert. *Journal of Systematic Palaeontology* **5**: 193–208.

Chiappe LM, Ji S, Ji Q. 2007. Juvenile birds from the EarlyCretaceous of China: implications for enantiornithine ontogeny. *American Museum Novitates* **3594**:1–46.

Chiappe L.M, Marugán-Lobón J, Ji S, Zhou Z. 2008. Life history of a basal bird: morphometrics of the Early Cretaceous *Confuciusornis*. *Biology Letters* **4**: 719–723.

Chiappe L.M, Marugán-Lobón J, Chinsamy A. 2010. Paleobiology of the Cretaceous Bird *Confuciusornis*: A Reply to Peters & Peters (2009). *Biology Letters* **6**: 529–530.

Chinsamy A, Elzanowski A. 2001. Evolution of growth pattern in birds. *Nature* **412**: 402–403.

Chinsamy A, Chiappe LM, Dodson P. 1994. Growth rings in Mesozoic birds. *Nature* **368**: 196–197.

Chinsamy A, Chiappe LM, Dodson P. 1995. Mesozoic avian bone microstructure: physiological implications. *Paleobiology* **21**: 561–574.

Choiniere J, Xu X, Clark JM, Forster CA, Guo Y, Han F-L. 2010. A basal alvarezsauroid theropod from the Early Late Jurassic of Xinjiang, China. *Science* **327**: 571–574.

Christiansen P, Bonde N. 2004. Body plumage in *Archaeopteryx*: a review, and new evidence from the Berlin specimen. *Comptes Rendus Palevol* **3**: 99–118.

Chuong C-M, Wu P, Zhang F-C, Xu X, Yu M, Widelitz RB, Jiang T-X, Hou L. 2003. Adaptation to the sky: defining the feather with integument fossils from the Mesozoic of China and experimental evidence frommolecular laboratories. *Journal of Experimental Zoology* **298B**: 42–56.

Clark JM, Norell MA, Barsbold R. 2001. Two new oviraptorids (Theropoda: Oviraptorosauria), Upper Cretaceous Djadokhta Formation, Ukhaa Tolgod, Mongolia. *Journal of Vertebrate Paleontology* **21**: 209–213.

Clark JM, Norell MA, Makovicky P. 2002. Cladistic approaches to the relationships of birds to other theropods. In *Mesozoic Birds: Above the Heads of Dinosaurs*, Chiappe LM, Witmer LM (eds). Berkeley: University of California Press; 31–61.

Clarke JA. 2004. Morphology, phylogenetic taxonomy, and systematics of *Ichthyornis* and *Apatornis* (Avialae: ornithurae). *Bulletin of the American Museum of Natural History* **286**: 1–179.

Clarke JA, Norell MA. 2002. The morphology and phylogenetic position of *Apsaravis ukhaana* from the Late Cretaceous of Mongolia. *American Museum Novitates* **3387**: 1–46.

Clarke JA, Tambussi C, Noriega J, Erickson G, Ketchum R. 2005. Definitive fossil evidence for the extant avian radiation in the Cretaceous. *Nature* **433**: 305–309.

Clarke JA, Zhou Z, Zhang F. 2006. Insight into the evolution of avian flight from a new clade of Early Cretaceous ornithurines from China and the morphology of *Yixianornis grabaui*. *Journal of Anatomy* **208**: 287–308.

Clark JM, Norell MA, Chiappe LM. 1999. An oviraptorid skeleton from the Late Cretaceous of Ukhaa Tolgod, Mongolia, preserved in an avianlike brooding position over an oviraptorid nest. *American Museum Novitates* **3265**: 1–36.

Close RA, Vickers–Rich P, Chiappe LM, O'Connor JK, Rich TH, Kool L. 2009. Earliest Gondwanan bird from the Cretaceous of southeastern Australia. *Journal of Vertebrate Paleontology* **29**: 616–619.

Codd JR, Manning PL, Norell MA, Perry SF. 2007. Avianlike breathing mechanics in maniraptoran dinosaurs. *Proceedings of the Royal Society of London Series B (Biological Sciences)* **275**:157–161.

Cracraft J. 1988. The major clades of birds. In *The Phylogeny and Classification of the Tetrapods, Volume 1: Amphibians, Reptiles, Birds*, Benton MJ (ed.). Oxford: Clarendon Press; 10–12.

Dalla Vecchia FM, Chiappe LM. 2002. First avian skeleton from the Mesozoic of northern Gondwana. *Journal of Vertebrate Paleontology* **22**: 856–860.

Dalsätt J, Zhou Z, Zhang F, Ericson PG. P. 2006. Food remains in *Confuciusornis sanctus* suggest a fish diet. *Naturwissenschaften* **93**: 444–446.

De Beer G. 1954. *Archaeopteryx lithographica, a Study Based upon the British Museum Specimen*. London: British Museum Publication 224;1–68.

De Ricqlès AJ, Padian K, Horner JR, Lamm E-T, Myhrvold N. 2003. Osteohistology of *Confuciusornis sanctus* (Theropoda: Aves). *Journal of Vertebrate Paleontology* **23**: 373–386.

Dyke GJ, Kaiser GW. 2010. Cracking a developmental constraint: egg size and bird evolution. *Records of the Australian Museum*, in press.

Dyke GJ, Nudds RL. 2009. The fossil record and limb disparity of enantiornithines, the dominant flying birds of the Cretaceous. *Lethaia* **42**: 248–254.

Dyke GJ, Malakhov DV, Chiappe LM. 2006. A reanalysis of the marine bird *Asiahesperornis* from northern Kazakhstan. *Cretaceous Research* **27**: 947–953.

Elzanowski A. 1974. Preliminary note on the palaeognathous bird from the Upper Cretaceous of Mongolia. *Acta Palaeontologica Polonica* **30**: 103–109.

Elzanowski A. 1977. Skulls of *Gobipteryx* (Aves) from the Upper Cretaceous of Mongolia. *Acta Palaeontologica Polonica* **37**: 153–165.

Elzanowski A. 1981. Embryonic bird skeletons from the Late Cretaceous of Mongolia. *Acta Paleontologica Polonica* **42**: 147–176.

Elzanowski A. 2002. Archaeopterygidae (Upper Jurassic of Germany). In *Mesozoic Birds: Above the Heads of Dinosaurs*, Chiappe LM, Witmer LM (eds). Berkeley: University of California Press; 129–159.

Erickson GM. 2005. Assessing dinosaur growth patterns: a microscopic revolution. *Trends in Ecology and Evolution* **20**: 677–684.

Erickson GM, Rogers KC, Yerby SA. 2001. Dinosaurian growth patterns and rapid avian growth rates. *Nature* **412**: 429–433.

Erickson GM, Rogers KC, Varricchio DJ, Norell MA, Xu X. 2007. Growth patterns in brooding dinosaurs reveals the timing of sexual maturity in non-avian dinosaurs and genesis of the avian condition. *Biology Letters* **3**: 558–561.

Erickson GM, Rauhut OWM, Zhou Z-H, Turner AH, Inouye BD, Hu D-Y, Norell MA. 2009. Was dinosaurian physiology inherited by birds? Reconciling slow growth in Archaeopteryx. *PLoS ONE* **4**: 1–9.

Feduccia A. 1996. *The Origin and Evolution of Birds*. New Haven: Yale University Press.

Feduccia A. 1999. *The Origin and Evolution of Birds*, 2nd edn. Yale University Press, New Haven.

Feduccia A. 2001. Digit homology of birds and dinosaurs: accomodating the cladogram. *Trends in Ecology and Evolution* **16**: 285–286.

Feduccia A, Nowicki J. 2002. The hand of birds revealed by early ostrich embryos. *Naturwissenchaften* **89**: 391–393.

Fisher HI. 1957. Bony mechanisms of automatic flexionand extension in the pigeon's wing. *Science* **126**: 446.

Forster CA, Chiappe LM, Krause DW, Sampson SD. 1996. The first Cretaceous bird from Madagascar. *Nature* **382**: 532–533.

Forster CA, Sampson SD, Chiappe LM, Krause DW. 1998a. The theropodan ancestry of birds: New evidence from the Late Cretaceous of Madagascar. *Science* **279**: 1915–1919.

Forster CA, Sampson SD, Chiappe LM, Krause DW. 1998b. Genus correction. *Science* **280**: 185.

Galis F, Kundrát M, Metz JAJ. 2005. *Hox* genes, digit identities and the theropod/bird transition. *Journal of Experimental Zoology* **304B** (3):198–205.

Galton P, Martin LD. 2002. *Enaliornis*, an early hesperornithiform bird from England, with comments onother Hesperornithiformes. In *Mesozoic Birds: Above the Heads of Dinosaurs*, Chiappe LM, Witmer LM (eds). Berkeley: University of California Press; 317–338.

Gao C, Liu J. 2005. A new avian taxon from Lower Cretaceous Jiufotang Formation of western Liaoning. *Global Geology* **4**: 313–316.

Gao C, Chiappe LM, Meng Q, O'Connor J.K, Wang X, Cheng X, Liu J. 2008. A new basal lineage of Early Cretaceous birds from China and its implications on the evolution of the avian tail. *Palaeontology* **51**: 775–791.

Gatesy SM, Dial KP. 1996. From frond to fan: *Archaeopteryx* and the evolution of short tailed birds. *Evolution* **50**: 2037–2048.

Gauthier JA. 1986. Saurischian monophyly and the origin of birds. In *The Origin of Birds and the Evolution of Flight*, Padian K (ed.). San Francisco: California Academy of Sciences Memoir; 1–55.

Goloboff PA, Farris JS, Nixon KC. 2008. TNT, a free program for phylogenetic analysis. *Cladistics* **24**: 774–786.

Göhlich UB, Chiappe LM. 2006. A new carnivorous dinosaur from the Late Jurassic Solnhofen archipelago. *Nature* **440**: 329–332.

Grellet-Tinner G, Chiappe LM. 2004. Dinosaur eggs and nesting: implications for understanding the origin of birds. In *Feathered Dragons: Studies on the Transition from Dinosaurs to Birds*, Currie PJ, Koppelhus EB, Shugar MA (eds). Bloomington: Indiana University Press; 185–214.

Grellet-Tinner G, Chiappe LM, Norell MA, Bottjer D. 2006. Dinosaur eggs and nesting behaviors: a paleobiological investigation. *Palaeogeography, Palaeoclimatology, Palaeoecology* **232**: 294–321.

Hackett SJ, Kimball RT, Reddy S, Bowie RCK, Braun EL, Braun MJ, Chojnowski JL, Cox WA, Han K-L,

Harshman J, Huddleston CJ, Marks BD, Miglia KJ, Moore WS, Sheldon FH, Steadman DW, Witt CC, Yuri T. 2008. A phylogenomic study of birds reveals their evolutionary history. *Science* **320**: 1763–1768.

Harris JD, Lamanna MC, You H-L, Ji S-A, Ji Q. 2006. Asecond enantiornithean (Aves: Ornithothoraces) wing from the Early Cretaceous Xiagou Formation near Changma, Gansu Province, People's Republic of China. *Canadian Journal of Earth Sciences* **43**: 547–554.

Holtz TR, Jr. 1994. The phylogenetic position of the Tyrannosauridae: implications for theropod systematics. *Journal of Paleontology* **68**: 1100–1117.

Holtz TR. Jr. 1998. A new phylogeny of the carnivorous dinosaurs. *Gaia* **15**: 5–61.

Holtz TR, Jr. 2001. Arctometatarsalia revisited: the problem of homoplasy in reconstructing theropod phylogeny. In *New Perspectives on the Origin and Evolution of Birds: Proceedings of the International Symposium in Honor of John H. Ostrom*, Gauthier JA, Gall LF (eds). New Haven: Yale University Press; 99–122.

Hopson JA. 2001. Ecomorphology of avian and nonavian theropod phalangeal proportions: implications for the arboreal versus terrestrial origin of bird flight. In *New Perspectives on the Origin and Evolution of Birds: Proceedings of the International Symposium in Honor of John H. Ostrom*, Gauthier JA, Gall LF (eds). New Haven: Yale University Press; 211–235.

Hou L, Liu Z. 1984. A new fossil bird from the Lower Cretaceous of Gansu and the early evolution of birds. *Scientia Sinica (Series B)* **27**: 1296–1302.

Hou L, Martin LD, Zhou Z, Feduccia A. 1996. Early adaptive radiation of birds: evidence from fossils from northeastern China. *Science* **274**: 1164–1167.

Hou L, Martin LD, Zhou Z, Feduccia A, Zhang F. 1999. Adiapsid skull in a new species of the primitive bird *Confuciusornis*. *Nature* **399**: 679–682.

Houck MA, Gauthier JA, Strauss RE. 1990. Allometric scaling in the earliest fossil bird, *Archaeopteryx lithographica*. *Science* **247**: 195–198.

Hu D, Hou L, Zhang L, Xu X. 2009. A pre-*Archaeopteryx* troodontid theropod from China with long feathers on the metatarsus. *Nature* **461**: 640–643.

Hu D-Y, Li L, Hou L-H, Xu X. 2010. A new sapeomithid bird from China and its implication for early avian evolution. *Acta Geologica Sinica*, **84**: 472–482.

Huang SH, Norell MA, Ji Q, Gao K-Q. 2002. New specimens of *Microraptor zhaoianus* (Theropoda: Dromaeosauridae) from northeastern China. *American Museum Novitates* **3381**: 1–44.

Hutchinson JR. 2001. The evolution of pelvic osteology and soft tissues on the line to extant birds (Neornithes). *Zoological Journal of the Linnean Society* **131**: 123–168.

James FC, Pourtless, IVJA. 2009. Cladistics and the origin of birds: a review and two new analyses. *Ornithological Monographs* **66**: 1–78.

Jenkins Jr FA, Dial KP, Goslow Jr G. E. 1988. A cineradiographic analysis of bird flight: the wishbone in starlings is a spring. *Science* **241**: 1495–1498.

Ji Q, Currie P, Norell MA, Ji S-A.. 1998. Two feathered dinosaurs from northeastern China. *Nature* **393**: 753–761.

Ji Q, Chiappe LM, Ji S-A. 1999. A new late Mesozoic confuciusornithid bird from China. *Journal of Vertebrate Paleontology* **19**: 1–7.

Ji Q, Norell MA, Gao K-Q, Ji J-A, Ren D. 2001. The distribution of integumentary structures in a feathered dinosaur. *Nature* **410**: 1084–1088.

Ji Q, Ji S-A, You H, Zhang J, Yuan C, Ji X, Li J, Li Y. 2002a. Discovery of an avialae bird – *Shenzhouraptor sinensis* gen. et sp. nov. – from China. *Geological Bulletin of China* **21**: 363–369.

Ji Q, Ji S-A, Zhang H, You H, Zhang J, Wang L, Yuan C, Ji X. 2002b. A new avialian bird – *Jixiangornis orientalis* gen. et sp. nov. – from the Lower Cretaceous of western Liaoning, NE China. *Journal of Nanjing University* **38**: 723–736.

Ji Q, Ji S-A, Lü J, You H, Chen W, Liu Y, Liu Y. 2005. Firstavialian bird from China: *Jinfengopteryx elegans* gen et sp nov. *Geological Bulletin of China* **24**: 197–210.

Ji S-A, Ji Q, Lü J, Yuan C. 2007. A new giant compsognathid dinosaur with long filamentous integuments from Lower Cretaceous of northeastern China. *Acta Geologica Sinica (English Edition)* **81**: 8–15.

Kurochkin EN. 1982. New order of birds from the Lower Cretaceous in Mongolia. *Paleontological Journal* **1982**: 215–218.

Kurochkin EN. 1985. A true carinate bird from Lower Cretaceous deposits in Mongolia and other evidence of Early Cretaceous birds in Asia. *Cretaceous Research* **6**: 271–278.

Kurochkin EN. 2006. Parallel evolution of theropod dinosaurs and birds. *Entomological Review* **86**: 283–297.

Kurochkin EN, Dyke GJ, Karhu A. 2002. A new presbyornithid bird (Aves: Anseriformes) from the Late Cretaceous of Southern Mongolia. *American Museum Novitates* **3386**: 1–11.

Lambrecht K. 1929. *Neogaeornis wetzeli* n.g.n.sp, der este Kreidevogel der sudlichen Hemisphare. *Palaeontologische Zeitschrift* **11**: 121–129.

Li L, Duan Y, Hu D, Wang L, Cheng S, Hou L. 2006. New eoentantiornithid bird from the Early Cretaceous Jiufotang Formation of western Liaoning, China. *Acta Geologica Sinica (English Edition)* **80** (1): 38–41.

Li Q-G, Gao K-Q, Vinther J, Shawkey MD, Clarke JA, D'Alba L, Meng Q-J, Briggs DEG, Prum RO. 2010. Plumage color patterns of an extinct dinosaur. *Science* **327**: 1369–1372.

Long J, Schouten P. 2008. *Feathered Dinosaurs.* New York: Oxford University Press.

Longrich N. 2006. Structure and function of the hindlimb feathers in *Archaeopteryx lithographica*. *Paleobiology* **32**: 417–431.

Lü J, Hou LH. 2005. A possible long-tailed bird with a pygostyle from the Late Mesozoic Yixian Formation, Western Liaoning, China. *Acta Geologica Sinica* **79**: 7–10.

Makovicky PJ, Apesteguía S, Agnolin FL. 2005. The earliest dromaeosaurid theropod from South America. *Nature* **437**: 1007–1011.

Marsh OC. 1872a. Notice of a new and remarkable fossil bird. *American Journal of Science* **4**: 344.

Marsh OC. 1872b. Preliminary description of *Hesperornis regalis*, with notices of four other new species of Cretaceous birds. *The American Journal of Science and Arts* **3**: 1–7.

Marsh OC. 1880. *Odontornithes: a Monograph on the Extinct Toothed Birds of North America. United States Geological Exploration of the 40th Parallel.* Washington, DC: Government Printing Office.

Martin LD. 1983. The origin and early radiation of birds. In *Perspectives in Ornithology*, Brush AH, Clark JrGA (eds). New York: Cambridge University Press; 291–338.

Martin LD. 1991. Mesozoic birds and the origin of birds. In *Origins of the Higher Groups of Tetrapods*, Brush AH, Clark Jr GA (eds). Ithaca: Comstock Publishing Associates; 485–540.

Martin LD. 1995. The Enantiornithes: terrestial birds of the Cretaceous. *Courier Forschungsinstitut Senckenberg* **181**: 23–26.

Martin LD. 1997. The difference between dinosaurs and birds as applied to *Mononykus*. In *Dinofest International*, Wolberg DL, Stump E, Rosenberg GD (eds). Philadelphia: Academy of Natural Sciences; 337–343.

Martin LD. 2004. A basal archosaurian origin for birds. *Acta Zoologica Sinica* **50**: 978–990.

Martin LD, Stewart JD. 1999. Implantation and replacement of bird teeth. *Smithsonian Contributions to Paleobiology* **89**: 295–300.

Mayr G, Peters DS, Plowdowski G, Vogel O. 2002. Bristle-like integumentary structures at the tail of the horned dinosaur *Psittacosaurus*. *Naturwissenschaften* **89**: 361–365.

McKitrick MC. 1992. Phylogenetic analysis of avian parental care. *The Auk* **109**: 828–846.

Morschhauser E, Varricchio DJ, Gao C-H, Liu J-Y, WangX-R, Cheng X-D, Meng Q-J. 2009. Anatomy of the Early Cretaceous bird *Rapaxavis pani*, a new species from Liaoning Province, China. *Journal of Vertebrate Paleontology* **29**: 545–554.

Nesbitt SJ, Turner AH, Spaulding M, Conrad JL, Norell MA. 2009. The theropod furcula. *Journal of Morphology* **270**: 856–879.

Nessov LA, Borkin L. 1983. New records of bird bonesfrom Cretaceous of Mongolia and Middle Asia. *Trudy Zoologicheskogo Instituta USSR* **116**: 108–110.

Norell MA, Clarke JA. 2001. Fossil that fills a critical gap in avian evolution. *Nature* **409**: 181–184.

Norell MA, Makovicky PJ. 1997. Important features ofthe dromaeosaur skeleton: Information from a new specimen. *American Museum Novitates* **3215**: 1–28.

Norell MA, Makovicky PJ. 1999. Important features of the dromaeosaur skeleton II: information from newly collected specimens of *Velociraptor mongoliensis*. *American Museum Novitates* **3282**: 1–45.

Norell MA, Xu X. 2005. Feathered dinosaurs. *Annual Review of Earth and Planetary Sciences* **33**: 277–299.

Norell MA, Clark JM, Chiappe LM, Dashzeveg D. 1995. A nesting dinosaur. *Nature* **378**: 774–776.

Norell MA, Makovicky PJ, Clark JM. 2001. Relationships among Maniraptora: problems and prospects. In *New Perspectives on the Origin and Evolution of Birds: Proceedings of the International Symposium in Honor of John H. Ostrom*, Gauthier JA, Gall LF (eds). New Haven: Yale University Press; 49–68.

Norell MA, Ji Q, Gao, K-Q, Yuan C, Zhao Y, Wang L. 2002."Modern" feathers on a non-avian dinosaur. *Nature* **416**: 36–37.

Novas FE, Pol D. 2005. New evidence on deinonychosaurian dinosaurs from the Late Cretaceous of Patagonia. *Nature* **433**: 858–861.

O'Connor JK, Wang X-R, Chiappe LM, Gao C-H, Meng Q-J, Cheng X-D, Liu J-Y. 2009. Phylogenetic support for a specialized clade of Cretaceous enantiornithine birds with information from a new species. *Journal of Vertebrate Paleontology* 29: 188–204.

O'Connor JK, Gao K-Q, Chiappe LM. 2010. A new ornithuromorph (Aves: Ornithothoraces) bird from the Jehol Group indicative of higher-level diversity. *Journal of Vertebrate Paleontology* 30 (2): 311–321.

Organ CL, Shedlock AM, Meade A, Pagel M, Edwards SV. 2007. Origin of avian genome size and structure in non-avian dinosaurs. *Nature* 446: 180–184.

Ostrom JH. 1974. *Archaeopteryx* and the origin of flight. *Quarterly Review of Biology* 49: 27–47.

Padian K, Chiappe LM. 1998. The origin and early evolution of birds. *Biological Reviews of the Cambridge Philosophical Society* 73: 1–42.

Padian K, de Ricqlès AJ, Horner JR. 2001. Dinosaurian growth rates and bird origins. *Nature* 412: 405–408.

Provini P, Zhou Z-H, Zhang F-C. 2009. A new species of the basal bird Sapeornis from the Early Cretaceous of Liaoning, China. *Vertebrata PalAsiatica* 47: 194–207.

Prum RO. 2005. Evolution of the morphological innovations of feathers. *Journal of Experimental Zoology* 304B: 570–579.

Prum RO, Brush AH. 2002. The evolutionary origin and diversification of feathers. *Quarterly Review of Biology* 77: 261–295.

Rayner JM. V. 2001. On the origin and evolution of flapping flight aerodynamics in birds. In *New Perspectives on the Origin and Evolution of Birds: Proceedings of the International Symposium in Honor of John H. Ostrom*, Gauthier JA, Gall LF (eds). New Haven: Yale University Press; 363–385.

Ruben J. 1991. Reptilian physiology and the flight capacity of *Archaeopteryx*. *Evolution* 45: 1–17.

Sanz JL, Buscalioni AD. 1992. A new bird from the Early Cretaceous of Las Hoyas, Spain, and the early radiation of birds. *Palaeontology* 35: 829–845.

Sanz JL, Bonaparte JF, Lacasa A. 1988. Unusual Early Cretaceous birds from Spain. *Nature* 331: 433–435.

Sanz JL, Chiappe LM, Buscalioni A. 1995. The osteology of *Concornis lacustris* (Aves: Enantiornithes) from the Lower Cretaceous of Spain and a re-examination of its phylogenetic relationships. *American Museum Novitates* 3133: 1–23.

Sanz JL, Chiappe LM, Perez-Moreno BP, Buscalioni AD, Moratalla J. 1996. A new Lower Cretaceous bird from Spain: implications for the evolution of flight. *Nature* 382: 442–445.

Sanz JL, Chiappe LM, Pérez-Moreno BP, Moratalla J, Hernández-Carrasquilla F, Buscalioni AD, Ortega F, Poyato-Ariza FJ, Rasskin-Gutman D, Martínez Delclos X. 1997. A nestling bird from the Early Cretaceous of Spain: implications for avian skull and neck evolution. *Science* 276: 1543–1546.

Sanz JL, Álvarez JC, Meseguer J, Soriano C, Hernández-Carrasquilla F, Pérez-Moreno BP. 2000. Wing loading in primitive birds. *Vertebrata PalAsiatica* 38: 27.

Sanz JL, Chiappe LM, Fernández-Jalvo Y, Ortega F, Sánchez-Chillon B, Poyato-Ariza FJ, Pérez-Moreno BP. 2001. An Early Cretaceous pellet. *Nature* 409: 998–999.

Schweitzer MH, Wittmeyer JL, Horner J. R. 2005. Gender-specific reproductive tissue in ratites and *Tyrannosaurus rex*. *Science* 308: 1456–1460.

Schweitzer MH, Suo Z, Avci R, Asara JM, Allen MA, Arce FT, Horner JR. 2007. Analyses of soft tissue from *Tyrannosaurus rex* suggest the presence of protein. *Science* 316: 277–280.

Senter P, Robins JH. 2003. Taxonomic status of the specimens of Archaeopteryx. *Journal of Vertebrate Paleontology* 23: 961–965.

Sereno PC. 1993. The pectoral girdle and forelimb of the basal theropod *Herrerasaurus ischigualastensis*. *Journal of Vertebrate Paleontology* 13: 425–450.

Sereno PC. 1997. The origin and evolution of dinosaurs. *Annual Review of Earth and Planetary Sciences* 25: 435–489.

Sereno PC. 1999. The evolution of dinosaurs. *Science* 284: 2137–2147.

Sereno PC, Novas FE. 1992. The complete skull and skeleton of an early dinosaur. *Science* 258: 1137–1140.

Sereno PC, Rao C. 1992. Early evolution of avian flight and perching: new evidence from Lower Cretaceous of China. *Science* 255: 845–848.

Shapiro MD. 2002. Developmental morphology of limb reduction in Hemiergis (Squamata: Scincidae): Chondrogenesis, Osteogenesis, and Heterochrony. *Journal of Morphology* 254: 211–231.

Shubin NH. 1994. The phylogney of development and the origin of homology. In *Interpreting the Hierarchy of Nature*, Grande L, Rieppel O (eds). San Diego: Academic Press; 201–225.

Sibley CG, Ahlquist JE. 1990. *Phylogeny and Classification of Birds: A Study in Molecular Evolution*. New Haven: Yale University Press.

Starck JM, Chinsamy A. 2002. Bone microstructure and developmental plasticity in birds and other dinosaurs. *Journal of Morphology* 254: 232–246.

Starck JM, Ricklefs RE. 1998. Patterns of development: the altricial – precocial spectrum. In *Avian Growth and Development*, Starck JM, Ricklefs RE (eds). Oxford: Oxford University Press; 3–26.

Stegmann BC. 1978. Relationships of the superorders Alectoromorphae and Charadriomorphae (Aves): acomparative study of the avian hand. *Publicationsof the Nuttall Ornithological Club* **17**: 1–118.

Turner AH, Pol D, Clarke JA, Erickson GM, NorellMA.2007. A basal dromaeosaurid and size evolution preceding avian flight. *Science* **317**: 1378–1381.

Varricchio DJ, Jackson FD. 2004a. A phylogenetic assessment of prismatic dinosaur eggs from the Cretaceous Two Medicine Formation of Montana. *Journal of Vertebrate Paleontology* **24**: 931–937.

Varricchio DJ, Jackson FD. 2004b. Two eggs sunny-side up: reproductive physiology in the dinosaur *Troodon formosus*. In *Feathered Dragons: Studies on the Transition from Dinosaurs to Birds*, Currie PJ, Koppelhus EB, Shugar MA (eds). Bloomington: Indiana University Press; 215–233.

Varricchio DJ, Jackson F, Borkowski JJ, Horner, J.R. 1997. Nest and egg clutches of the dinosaur *Troodon formosus* and the evolution of avian reproductive traits. *Nature* **385**: 247–250.

Varricchio DJ, Moore JR, Erickson GM, Norell MA, Jackson FD, Borkowski JJ. 2008. Avian paternal care had dinosaur origin. *Science* **322**: 1826–1828.

Vasquez RJ. 1992. Functional osteology of the avian wrist and the evolution of flapping flight. *Journal of Morphology* **211**: 259–268.

Vasquez RJ. 1994. The automating skeletal and muscular mechanisms of the avian wing (Aves). *Zoomorphology* **114**: 59–71.

Von Meyer H. 1861. *Archaeopteryx litographica* (Vogel-Feder) und *Pterodactylus* von Solenhofen. *Neues Jahrbuch für Mineralogie, Geognosie, Geologie, und Petrefakten-kunde* **1861**: 678–679.

Wagner GP, Gauthier JA. 1999. 1,2,3 = 2,3,4: a solution to the problem of the homology of the digits in the avian hand. *Proceedings of the National Academy of Sciences, USA* **96**: 5111–5116.

Walker AD. 1972. New light on the origin of birds and crocodiles. *Nature* **237**: 257–263.

Walker CA. 1981. New subclass of birds from the Cretaceous of South America. *Nature* **292**: 51–53.

Walker CA, Buffetaut E, Dyke GJ. 2007. Large euenantiornithine birds from the Cretaceous of southern France, North America and Argentina. *Geological Magazine* **144**: 977–986.

Wang X-R, O'Connor JK, Zhao B, Chiappe LM, Gao G-H, Cheng X-D. 2010. A new species of Enantiornithes (Aves: Ornithothoraces) based on a well-preserved specimen from the Qiaotou Formation of northern Hebei, China. *Acta Geologica Sinica* **84**: 247–256.

Welles SP. 1984. *Dilophosaurus wetherilli* (Dinosauria, Theropoda). Osteology and comparisons. *Palaeontographica Abteilung A*, **185**: 85–180.

Wellnhofer P. 2008. *Archaeopteryx. Der Urvogel von Solnhofen.* München: Friedrich Pfeil.

Welman J. 1995. *Euparkeria* and the origin of birds. *South African Journal of Science* **91**: 533–537.

Witmer LM. 1991. Perspectives on avian origins. In *Origins of the Higher Groups of Tetrapods*, Schultze H-P, Trueb L (eds). Ithaca: Comstock Publishing Associates; 427–466.

Witmer LM. 2001. The role of *Protoavis* in the debate on avian origins. In *New Perspectives on the Origin and Evolution of Birds: Proceedings of the International Symposium in Honor of John H. Ostrom*, Gauthier JA, Gall LF (eds). New Haven: Yale University Press; 537–548.

Witmer LM. 2002. The debate on avian ancestry: phylogeny, function, and fossils. In *Mesozoic Birds: Above the Heads of Dinosaurs*, Chiappe LM, Witmer LM (eds). Berkeley: University of California Press; 3–30.

Xu X. 2006. Feathered dinosaurs from China and the evolution of major avian characters. *Integrative Zoology* **1**: 4–11.

Xu X, Norell MA. 2006. Non-avian dinosaur fossils from the Lower Cretaceous Jehol Group of western Liaoning, China. *Geological Journal* **41**: 419–437.

Xu X, Zhang F. 2005. A new maniraptoran dinosaur from China with long feathers on the metatarsus. *Naturwissenschaften* **92**: 173–177.

Xu X, Zhou Z, Wang X. 2000. The smallest known non-avian theropod dinosaur. *Nature* **408**: 705–708.

Xu X, Zhou Z, Prum RO. 2001. Branched integumentary structures in *Sinornithosaurus* and the origin of feathers. *Nature* **410**: 200–204.

Xu X, Zhou Z, Wang X, Kuang X, Du X. 2003. Four-winged dinosaurs from China. *Nature* **421**: 335–340.

Xu X, Norell MA, Kuang X, Wang X, Zhao Q, Jia C. 2004. Basal tyrannosauroids from China and evidence for protofeathers in tyrannosauroids. *Nature* **431**: 680–684.

Xu X, Clark JM, Mo J-Y, Choiniere J, Forster CA, Erickson GM, Hone DW. E, Sullivan C, Eberth DA, Nesbitt SJ, Zhao Q, Hernandez R, Jia C-K, Han F-L, Guo Y. 2009a. A Jurassic ceratosaur from China helps clarify avian digital homologies. *Nature* **459**: 941–944.

Xu X, Zhao Q, Norell MA, Sullivan C, Hone DW. E, Erickson GM, Wang X-L, Han, F-L, Guo Y. 2009b. A new feathered maniraptoran dinosaur fossil that fills a morphological gap in avian origin. *Chinese Science Bulletin* **54**: 430–435.

You HL, Lamanna MC, Harris JD, Chiappe LM, O'Connor JK, Ji SA, Lü JC, Yuan C-X, Li DQ, ZhangX, Lacovara KJ, Dodson P, Ji Q. 2006. A nearly modern amphibious bird from the Early Cretaceous of northwestern China. *Science* **312**: 1640–1643.

Yuan C. 2008. A new genus and species of Sapeornithidae from Lower Cretaceous in Western Liaoning, China. *Acta Geologica Sinica* **82**: 48–55.

Zhang F, Zhou Z-H. 2000. A primitive enantiornithine bird and the origin of feathers. *Science* **290**: 1955–1959.

Zhang F, Zhou Z-H, Xu X, Wang X. 2002. A juvenile coelurosaurian theropod from China indicates arboreal habits. *Naturwissenschaften* **89**: 394–398.

Zhang F, Zhou Z, Hou L-H. 2003. Birds. In The Jehol Biota: the Emergence of Feathered Dinosaurs, Beaked Birds and Flowering Plant, Chang M, Chen P, Wang Y, Wang Y, Miao D (eds). Shanghai: Shanghai Scientific and Technical Publishers; 129–150.

Zhang F, Ericson PG. P, Zhou Z. 2004. Description of a new enantiornithine bird from the Early Cretaceous of Hebei, northern China. *Canadian Journal of Earth Sciences* **41**: 1097–1107.

Zhang F, Zhou Z-H, Dyke GJ. 2006. Feathers and "feather-like" integumentary structures in Liaoning birds and dinosaurs. *Geological Journal* **41**: 395–404.

Zhang F-C, Zhou Z-H, Xu X, Wang X-L, Sullivan C. 2008a. A bizarre Jurassic maniraptoran from China with elongate ribbon-like feathers. *Nature* **455**: 1105–1108.

Zhang F, Zhou ZH, Benton MJ. 2008b. A primitive confuciusornithid bird from China and its implications for early avian flight. *Science in China, Series D – Earth Sciences* **51**: 625–639.

Zhang F-C, Kearns S.L, Orr P.J, Benton MJ, Zhou Z-H, Johnson D, Xu X, Wang X-L. 2010. Fossilized melanosomes and the colour of Cretaceous dinosaurs and birds. *Nature,* **463**: 1075–1078.

Zheng X, Zhang Z, Hou L. 2007. A new enantiornitine bird with four long rectrices from the Early Cretaceous of northern Hebei, China. *Acta Geologica Sinica (English edition)* **81**: 703–708.

Zheng X-T, You H-L, Xu X, Dong Z-M. 2009. An EarlyCretaceous heterodontosaurid dinosaur with filamentous integumentary structures. *Nature* **458**: 333–337.

Zhou ZH. 1995. Discovery of a new enantiornithine bird from the Early Cretaceous of Liaoning, China. *Vertebrata PalAsiatica* **33**: 99–113.

Zhou ZH, Zhang F. 2001. Two new ornithurine birds from the Early Cretaceous of western Liaoning, China. *Chinese Science Bulletin* **46**: 1258–1264.

Zhou ZH, Zhang F. 2002a. A long-tailed, seed-eating bird from the Early Cretaceous of China. *Nature* **418**: 405–409.

Zhou ZH, Zhang F. 2002b. Largest bird from the Early Cretaceous and its implications for the earliest avian ecological diversification. *Naturwissenshaften* **89**: 34–38.

Zhou ZH, Zhang F. 2003a. *Jeholornis* compared to *Archaeopteryx*, with a new understanding of the earliest avian evolution. *Naturwissenshaften* **90**: 220–225.

Zhou ZH, Zhang F. 2003b. Anatomy of the primitive bird *Sapeornis chaoyangensis* from the Early Cretaceous of Liaoning, China. *Canadian Journal of Earth Sciences* **40**: 731–747.

Zhou ZH, Zhang F. 2004. Mesozoic birds of China: an introduction and review. *Acta Zoologica Sinica* **50**: 913–920.

Zhou ZH, Zhang F. 2005. Discovery of an ornithurine bird and its implications for Early Cretaceous avian radiation. *Proceedings of the National Academy of Science* **102**: 18998–19002.

Zhou ZH, Zhang F. 2006. Mesozoic birds of China – a synoptic review. *Vertebrata PalAsiatica* **44**: 74–98.

Zhou ZH, Clarke JA, Zhang F, Wings O. 2004. Gastroliths in *Yanornis*: an indication of the earliest radical diet-switching and gizzard plasticity in the lineage leading to living birds. *Naturwissenschaften* **91**: 571–574.

Zhou Z, Chiappe L. M, Zhang F. 2005. Anatomy of the Early Cretaceous bird *Eoenantiornis buhleri* (Aves: Enantiornithes) from China. *Canadian Journal of Earth Sciences* **42**: 1331–1338.

Zhou Z, Clarke J.A, Zhang F. 2008. Insight into diversity, body size and morphological evolution from the largest Early Cretaceous enantiornithine bird. *Journal of Anatomy* **212**: 565–577.

Zhou Z-H, Zhang, F-C, Li, Z-H. 2009. A new basal orithurine (*Jianchangornis microdonta* gen. et sp. nov.) from the Lower Cretaceous of China. *Vertebrata PalAsiatica* **47**: 299–310.

Zhou Z-H, Zhang F-C, Li Z-H. 2010. A new Lower Cretaceous bird from China and tooth reduction in early avian evolution. *Proceedings of the Royal Society of London Series B (Biological Sciences)* **277**:219–227.

Zusi RL. 1993. Patterns of diversity in the avian skull. In *The Skull: Patterns of Structural and Systematic Diversity*, Hanken J, Hall BK (eds). Chicago: University of Chicago Press; 391–437.

Part 2 "The Contribution of Paleontology to Ornithology": the Diversity of Modern Birds: Fossils and the Avian Tree of Life

4 Progress and Obstacles in the Phylogenetics of Modern Birds

BRADLEY C. LIVEZEY

Carnegie Museum, Pittsburgh, USA

The invitation to participate in this volume provided an unusual opportunity for me to explore the philosophical undercurrents that serve and afflict this dynamic and important period in avian phylogenetics. My contribution is admittedly that of an avian phylogeneticist with principal experience with phenotypic (morphological) data (Livezey & Zusi, 2001, 2006, 2007). Although I endeavor to maintain an essential understanding of molecular methods, the following text clearly includes personal views of some highly contentious issues involving those. As one of the primary objectives here is to differentiate matters of empirical inference from those of personal impression, I expect to fail to some extent.

Although significant advances in methods and inferences are evident in avian phylogenetics, during the latter 20th century the field has been compromised by deep divisions among subgroups of practitioners. Progress that has been made is especially gratifying where diverse approaches reveal comparable patterns, e.g. Galloanserimorphae. However, most inferences either are limited to a single methodological approach or present different pictures of avian phylogeny, and attempts to resolve differences or bring all data to bear on problematic groups are rare to nonexistent. This deficiency reflects a lack of consensus in objectives, standards of investigation, justifications for perspectives, shared philosophies pertaining to phylogenetic hypotheses, and a common nomenclature for critical terminology and concepts that bridge schools.

Perhaps most troubling is the lack of a repeatable protocol to optimize a summary of available data and trees and to approximate nonarbitrarily the status of variably contradictory phylogenetic reconstructions. More specifically, a broadly accepted way to combine groupings across methods and studies, as well as indicate relative support for included groups (Farris & Goloboff, 2008; Grant & Kluge, 2008), and requisite theoretical investment in these urgent priorities' is needed.

In what follows, I attempt to draw attention to the challenges and potential pitfalls presented by the present diversity of data and protocols bearing on avian phylogeny. Moreover, whereas an appreciation and synergism among majority and minority schools seems the most constructive of possible futures, I contend that this diversity instead has become the victim of divisive, unempirical assertion. This domination of cooperative methods by a competition for single-school supremacy has perpetuated the unfortunate politics of systematics of the past 50 years to the present day.

HISTORICAL ORIGINS OF PHYLOGENETICS

For centuries, organisms were grouped into species and higher assemblages purely on the basis of

---

informal assessments of overall phenotypic similarity (Feduccia, 1977, 1995, 1996). Three conceptual events fundamentally revolutionized this practice: (i) Mendelian genetics, the subcellular basis by which phenotypes were rendered heritable; (ii) discovery that the underlying genetic material was DNA, a double helix comprising genes, linear triplets of nucleotides (four possible alternatives), the informative sites being limited to the first two of three nucleotides, and a redundancy of translative equivalence of 20 amino acids; and (iii) the pivotal recognition by Hennig (1966) that historical hierarchies reflected by shared derived characters (synapomorphies) are proper reflections of phylogenetic descent and natural groups. A glance at these three revolutionary leaps reveals by its absence what is perhaps the greatest remaining obstacle to progress in phylogenetics, that is the multitude of factors that influence the translation of genotype into phenotype. The early roots and applications of cladistics were in phenotypic contexts most frequently referred to as morphological characters, in part because such data were most familiar and were those with which Hennig (1966) and other proponents (Wiley, 1981) primarily exemplified cladistic methodology. However, a growing interest in molecular biology – e.g. paternity, identity by finger-printing, and insights into the structural nuances of the genome and chromosomes – assumed new priority before either an understanding of ontogeny was accomplished or applications of phenotypic cladistics had matured.

DNA HYBRIDIZATION AND THE DEMISE OF MOLECULAR PHENETICS

The first decades following the introduction of cladistics were largely consumed by a theoretical debate concerning the comparative propriety of overall similarity (phenetic school) and special or apomorphic similarity (cladistic or phylogenetic school) for the reconstruction of phylogenetic relationships, with some confusion including essences of both by the eclectic school (Raikow, 1985). The final stages of this current debate marked an increasing investment by the National Science Foundation (NSF) and other funding opportunities into study of molecular biology.

Following digressions into the phylogenetic implications of interspecific hybridization (Sibley, 1957) and distances based on compositions of amino acids through electrophoresis (Sibley & Ahlquist, 1970, 1972), the coincident interests of molecular structure and phylogenetic relationships led Sibley & Ahlquist (1990) to the quintessential phenetics of DNA–DNA hybridization. Widely acclaimed by many and awaited with unrestrained zeal by a community frustrated by the absence of reliable phylogenies, it was surprising to many that this massive effort generally was judged to be a failure within a few years of publication (Cracraft, 1987; Houde, 1987; Lanyon, 1992; Mindell, 1992; Harshman, 1994). Subsequently, most systematists enamored of molecular evidence turned to nucleotide sequences for phylogenetic purposes, progressing from mitochondial to nuclear genomes and from parsimony to maximum-likelihood (Felsenstein & Sober, 1987; Felsenstein, 2004) and Bayesian estimators (Harney, 2003; Ronquist, 2004) as theory and computational power advanced.

During this period, cladistic exploration of phenotypic features of modern Aves augmented a traditional paleontological enterprise (Padian & Chiappe, 1998; Fountaine et al., 2005), and despite lower profiles and publicity than corresponding molecular works, these efforts came to include increasingly large taxonomic groups and anatomical data sets of greater dimension (Livezey, 1986, 1991, 1995a, 1996, 1997, 1998; Bertelli et al., 2002; Ksepka et al., 2006). However, these concurrent programs failed to achieve mutual respect or collaborative synergy, and phenotypic methods were regular subjects of condescension or derision by some (McCracken et al., 1999; van Tuinen, 2005; Scotland et al., 2003), an unfortunate and ironic perpetuation of the polemics extolled in the ill-fated enterprise by Sibley & Ahlquist (1990).

A reanalysis of the sparse data matrix of Sibley & Ahlquist (1990) by a few stalwarts (Harshman,

1994; Bleiweiss *et al.*, 1995) salvaged little signal. Infrequent use of the method persisted (e.g. Houde, 1987) but failed to alter the marked course of the field away from phenetic methods. Remarkably, a minority still employ DNA–DNA hybridization (Hedges & Sibley, 1994; van Tuinen *et al.*, 2001), and some bolster claims of molecular superiority through comparisons of the phenograms by Sibley & Ahlquist (1990) with traditional classifications (van Tuinen, 2005) or morphological phylogenetics (Fain & Houde, 2004). Despite fundamental shortcomings, large studies can show unmerited longevity in some professional circles.

As much as the genome captures increasing interest among the upcoming cohort of systematists, the phenotype remains essential to numerous evolutionary topics – e.g. vestigialization (Fong *et al.*, 1995; Livezey, 2003), heterochrony and appendicular growth (Oster *et al.*, 1988; Livezey, 1995b) – for which phylogenetic insights are critical. Despite some condescending calls for limiting phenotypic data to attributes mapped *a posteriori* on molecular phylogenies (e.g. Scotland *et al.*, 2003), confident insights into distributions, transformations, rates, and correlations of genic and phenotypic patterns require detailed study of both genotypes and phenotypes by comparable means (Omland, 1997). Intuitively, it is imprudent to limit investigation of a massive and complex historical problem to the weaknesses and strengths of a single tool, a narrowness of practice no longer necessary (Eldredge & Cracraft, 1980; Wiley, 1981; Felsenstein, 2004). Also, intuitively bizarre unions grounded solely on molecular evidence and fossil-based estimates of evolutionary rates (Bromham *et al.*, 2002) will remain tenuous in the absence of a plausible explanation for grossly contradictory characteristics of phenotypes. Where disagreements prove intransigent, such tandem investigation is essential to the identification of those characters holding most promise for gaining insights into the ontogenetic mechanisms and the processes influential in the history of the controversial lineage.

During recent years, diversification of methods has led to a substantive debate concerning philo-sophical implications (Sober, 1985, 1987, 1988, 1996, 2002, 2004, 2005; Felsenstein & Sober, 1987; Patterson, 1987; Patterson *et al.*, 1993; Steel & Penny, 2000, 2004b; Kluge, 1997a,b; 2001a,b 2007; van Tuinen, 2005), reliability of various sources of signal (Baker *et al.*, 1998; Jenner, 2004a; Ray *et al.*, 2006; Revell *et al.*, 2008), methods for refinement of models upon which inferences are based (Penny *et al.*, 1994; Kim, 1998a,b; Burnham & Anderson, 2002; Durrett, 2002; Konishi & Anderson, 2002; Kelchner & Thomas, 2006), objective comparisons of studies (Farris *et al.*, 1994; Jenner, 2004b; Pisani *et al.*, 2007), and exploration or synthesis of multiple empirical pathways (Raup & Gould, 1974; Hecht, 1989; Geeta, 2003). Nevertheless, overtly prejudicial views on morphological phylogenetics persist, a notable number of titles include the archetypical term of adversarial points of view – vs. or versus (e.g. Hedges & Maxson, 1996; Givnish & Systsma, 1997; Hillis & Wiens, 2000). Fortunately, systematists of more even-handed perspectives also are in evidence (Lee, 1997; Hillis & Wiens, 2000; Baker & Gatesy, 2002; Cracraft *et al.*, 2003), preserving hopes for a phylogenetics of genotypic and phenotypic bases (Fitch, 1979; Davidson, 1991; Simmons *et al.*, 2004a) and objectivism regarding signal (Penny & Hendy, 1986).

At present, it is undeniable that phenotypic phylogenetics remains in the minority, although explicit endorsements persist (e.g. Sereno, 2007). Ease of data collection almost certainly contributes to the popularity of molecular investigations, but commitment to phenotypic evolution endures, despite blatant assaults to the contrary (e.g. van Tuinen, 2005) and in the absence of empirical justification or a single, widely accepted phylogeny by which to judge relative accuracy. The increasing economy of direct sequencing of DNA represents an advantage, but also necessitates alternative modes of analysis (maximum-likelihood estimates, Bayesian) and increasingly complex evolutionary models. Some consider quantitative comparisons between morphological and molecular inferences meritorious and of evolutionary interest (Patterson, 1987; Patterson *et al.*, 1993; Omland, 1997;

Benton, 1999; Bromham *et al.*, 2002; Pisani *et al.*, 2007), spawning increased formality of comparison (Huelsenbeck & Bull, 1996; Goldman *et al.*, 2000). Only where molecular data are not available, notably in the case of analyses of fossil taxa, does the balance of effort broadly favor phenotypic evidence among the majority persuaded by signal indicated by molecular patterns.

A TYPICAL YEAR IN AVIAN PHYLOGENETICS

A year of research on the systematics of modern birds generally witnesses the publication of several dozen analyses at family level or lower, and fewer than one-half dozen analyses of higher-order relationships. Studies of higher-order phylogenetics by phenotypic means currently are dominated by those for fossil taxa. Of neontological projects, a substantial majority are based on DNA sequences, most of which are analyzed by Bayesian or maximum-likelihood methods and use, at least in part, sequences available from the internet (GenBank).

Among these diverse studies, most share only a minority of ingroup taxa and nodes, a circumstance making identification of specific points of disagreement and advantage problematic. The single avian order Pelecaniformes (or subparts thereof) has been the subject of phylogenetic investigation diverse in method and findings (Table 4.1) at a semi-annual pace beginning with Cracraft (1985) to the present. Unfortunately, this annual harvest almost never is subjected to any comparative assessment or synthesis of findings, leaving readers to face the increasingly confusing published record without informed oversight or reliably objective input by colleagues.

Consensus regarding nodes is attained at present only when a substantial majority of works including the relevant taxa recover the same or very similar topologies, an agreement bolstered where substantial support statistics and robustness grace most or all sources. Realistically, however, hypotheses seldom are proposed *a priori*, and no explicit means of falsifiability is recognized, a deficiency that has led to a common, if unempirical adoption of methodological bias by which variably plausible scenarios and unsupported

Table 4.1 Chronology of recent analyses of phylogeny of modern Pelecaniformes with critical analytical criteria.

Reference	Modern taxa	Type of data	Method	Notable findings
Cracraft (1985)	6 families	Phenotypic	Parsimony	Ordinal monophyly
Siegel-Causey (1988)	2 families	Morphology	Parsimony	Familial monophyly
Sibley & Ahlquist (1990)	7 families	Nuclear DNA	DNA hybridization	Ordinal polyphyly
McKitrick (1991)	4 families	Pelvic myology	Parsimony	Polyphyly, several trees
Hedges & Sibley (1994)	6 families	Mtdna	DNA hybridization	Polyphyly
Siegel-Causey (1997)	7 families	Mtdna, morphology	Parsimony	Ordinal polyphyly
Farris *et al.* (1999)	2 families	*Contra* Hedges	Parsimony	Inadequate support
Kennedy *et al.* (2000)	2 families	Sequence data	Parsimony	Familial monophyly
Van Tuinen *et al.* (2001)	3 families	Mtdna	Phenetics	Ordinal monophyly
Mayr (2003)	7 families	Osteology	Parsimony	Ordinal monophyly
Mayr & Clarke (2003)	3 families	Morphology	Parsimony	Ordinal monophyly
Cracraft *et al.* (2004)	7 families	Sequence data	Parsimony	Ordinal polyphyly
Kennedy & Spencer (2004)	5 families	Mtdna sequence	Likelihood	Novel topology
Bourdon (2005)	5 families	Osteology	Parsimony	Ordinal monophyly
Bourdon *et al.* (2005)	7 families	Osteology	Parsimony	Ordinal polyphyly
Kennedy *et al.* (2005)	4 families	Sequence data	Neighbor networks	Familial monophyly
Livezey & Zusi (2007)	7 families	Morphology	Parsimony	Ordinal monophyly

speculation serve as surrogate conciliations of competing inferences. In general, the enterprise of phylogenetic systematics typically comprises multiple, methodologically diverse exploratory exercises of different taxonomic groups. Unfortunately, there is little in the way of overarching planning with respect to taxonomic groups of highest priority for study, no statistical standards for optimizing inferences within studies, and no accepted protocol for judging among inferences drawn from independent investigations.

SOCIOLOGICAL ASPECTS OF CONTEMPORARY SYSTEMATICS

Marketplace of science and publication

In systematics, a triplet of multidisciplinary periodicals are considered by many to be of special influence in the scientific community. These few favorites are the for-profit, multidisciplinary weekly journals *Nature* (UK) and *Science* (USA), and the membership-constrained, governmentally subsidized *Proceedings of the National Academy of Sciences* (USA). As in most professional journals, editorial boards typically are dominated by peers whose specialties favor topics of comparatively widespread interest, which for systematics have favored discoveries ancient and unexpected (notably fossils) and methods considered to be sophisticated (notably molecular techniques), and reflect as well the level of interest shown by readers of the wider scientific community. Together, the market-driven editorial filters pose significant obstacles to the publication of phenotypic phylogenetics of modern taxa (Jenner, 2004a,b).

These challenges fortunately are not typical of the systematic community *per se*, although trends are not encouraging of methodological diversity, a circumstance unlikely to be balanced by comparisons based on dubious criteria such as estimates of homoplasy (Sanderson & Donaghue, 1989). This imbalance is worsened by the format shared by many outlets, one that severely limits length of works and included figures, and relies heavily on associated web-based supplementary files. The editorial and stylistic hurdles posed by these outlets are conducive neither to traditionally descriptive, morphological works nor analytical detail.

With the spread of technology bearing on sequencing of DNA, the ease of automated collection of data, and multiple means for establishment of alignment and inferences among taxa, a comparative tide of molecular studies inundated limited outlets of publication. In the instance of birds and other "higher" vertebrates, this led to several new, well funded, and professionally produced monthly or quarterly journals dedicated to molecular phylogenetics and associated topics (e.g. *Molecular Phylogenetics and Evolution*, *Journal of Molecular Biology*, and *Molecular Biology and Evolution*). The advantage of publication that these afford molecular works over phenotypic explorations is substantial. This change is especially notable in view of the centuries of morphological work and specimens that preceded molecular techniques, much of the former having remained unincorporated and the latter unstudied in a phylogenetic context during subsequent decades.

The comparative sizes of molecular data matrices precluded from the start the publication of matrices in molecular studies. Consequently, an early web-based archive of sequence data was introduced (GENBANK), rendering the associated phylogenetic analyses free of the cost of publication of data and a considerable increase in economy. By contrast, use of descriptive morphology for phylogenetic analysis was delayed by an absence of similar web-based outlets, despite the fact that such matrices tended to be much smaller than their counterparts for molecular works. Fortunately, comparable distribution of character matrices via websites by authors, journals, or digital archives (e.g. MORPHOBANK) is rapidly closing this gap between disciplines.

Influence and ease of plug-and-play analyses

It is both undeniable and regrettable that few of the systematists toiling in the discipline – interests avian or otherwise – have in-depth training in mathematics let alone interests in the special areas of critical importance in phylogenetic theory. It appears from the literature that systematists, like specialists in other biological disciplines, tend to suffer the traditional phobia concerning numerical skills. The restriction of mathematical skills to a small minority of the field has imposed two negative effects on the diversity and magnitude of progess in systematics, both of which derive in part from the uninformed or fashion-driven selection of approach, analytical methods, numerical models, and graduate experience.

The first is that those few fortunates having mathematical facilities have profound theoretical and methodological influence on the practitioners comprising the majority in the field: most of the latter are intimidated into uninformed conformance. The skewed distribution of mathematical skills also has led to a largely cookbook approach to the implementation of software by those amassing the evidence upon which trees and consensuses ultimately are built. Most practitioners of Bayesian inference have had no formal training in the statistical foundations of this subtly but fundamentally distinct framework, and insight is limited to that fortunately gleaned from the writings of a favored few. Predictably, flawed analyses often are introduced to the literature by substandard methods and lead to the predictable fruits of haste (Erixon *et al.*, 2003; Simmons *et al.*, 2004a). The combination of untrained implementation of easy-use software with limited training and unconscious bias regarding inferences and alternative approaches justifiably encourages a suspicion concerning the empirical content of an indeterminate proportion of phylogenetic works. Initial use of Bayesian methods in phylogenetics already has been called into question regarding realism of support statistics (Simmons *et al.*, 2004b) and dubiously justified adoption of priors (Hartigan, 1998; Felsenstein, 2004).

Phylogenetics by committee

Enduring uncertainty can foster an imposition of opinion, even in science. Unfortunately for ornithology, a recent third-party adjudication, the 49th

supplement of the American Ornithologists' Union *Check-list* (Banks *et al.*, 2008), underscored an important but unintended issue in avian systematics. Although the *Check-list* neither has legalistic standing or represents the views of societal membership, it has the appearance of such, and arguably serves as a taxonomic resource. In light of these real and apparent roles, the strangely hurried, biased, and needless adoption by Banks *et al.* (2008) of the interordinal affinity of the grebes (Podicipediformes) and flamingos (Phoenicopterigiformes) overtly injected politics into science. A sympathetic but equally unempirical view was expressed in the British sister periodical, *The Ibis* (Sangster, 2005).

Wording in Banks et al. (2008) ranges from biased to incorrect: under the flamingos (p. 760), the proposal is adopted "to *recognize* the close relationship to the order Podicipediformes *shown* by several genetic studies ... they were *formerly* considered more closely related to the Ciconiiformes [emphasis added]." Given the implication of certainty, nonspecialists may assume the classification to be a reliable basis for design of phylogenetic contrasts, whereas informed readers justifiably might doubt the veracity of the entire work, useful instead as an example of the low standards too frequently employed in higher-order systematics of birds.

The recent addendum to the *Check-list* cited only studies consistent with the revision, and neither molecular (Brown *et al.*, 2008) nor morphological counterevidence (Livezey & Zusi, 2006, 2007) or other dissensions (Storer, 2006) were noted; even the review of supportive works was superficial (Mayr & Clarke, 2003). The absence of a single credible morphological synapomorphy evidently also was deemed unworthy of mention (Livezey & Zusi, 2007; *contra* Mayr, 2004). Mayr (2004) pointedly searched for morphological characters quasi-consistent with the alliance, but nevertheless the few noted also were shared with other taxa relevant to the question (e.g. some Ciconiiformes). As a paleontologist, Mayr (2004) lamented that fossil grebes were so modern in form as to be of little use in deciphering ordinal origins. Evidently, the qualities sought in fossils for such cases include taxa (O'Keefe & Sander, 1999): (i) synapomorphically diagnosable as members of the Podicipediformes, but (ii) sufficiently plesiomorphic so as not to be too "grebe-like" to hint at a possible role as "linking" the two modern orders, and (iii) apomorphies to unite this intermediate fossil with nonpodicipediforms.

The four molecular papers that were cited in support of the revision followed no protocol of demonstrated rigor: one was based on few taxa and the defunct method of DNA hybridization (van Tuinen *et al.*, 2001); another combined nuclear sequences of five genes with selected fossils (Ericson *et al.*, 2006); and the remaining two merged multiple molecular data sets and either yielded poor resolution (Cracraft *et al.*, 2004) or only modest support (Chubb, 2004). Chubb (2004) rationalized this counterintuitive union of distinctly different orders by citing conditions normally marshaled to support a claim of *convergence* instead of leading to obfuscation of close relationships by mere specialization in *phenotypically disparate* groups of divergent aquatic habit.

At best the action by the *Check-list* committee reflects a traditional discomfort with "unattached" orders (appearance of ignorance), and at worst a concerted attempt to impose the false clarity of opinion where data fail to resolve. Lacking any original contribution or a balanced presentation, are such classificatory votes actually constructive? Obviously, systematists will continue empirical pursuit of evidence bearing on hypotheses, but derivative edicts such as the *Check-list* would profit taxonomic summaries if based on the entire empirical basis rather than a single prejudical vote by nonspecialists. Also notable is an arbitrariness to higher-order groupings that are recognized in the *Check-list*, e.g. the monophyly of the orders Galliformes and Anseriformes, supported both by a growing body of morphological (e.g. Livezey, 1997) and molecular data (e.g. Groth & Barrowclough, 1999) for more than a decade, remains unnoted. The *Check-list* committee would do well to recall the earlier conservatism that rejected the groupings by Sibley & Ahlquist (1990), one in spite of strong advocacy for adoption *in toto* by then-chairperson and collaborator Monroe (1989), who considered wholesale adoption inevitable and piecemeal incorporation to be an unjustified delay of modern enlightenment.

Perils of the novel but untested

The aforementioned divisions between molecular and phenotypic schools, in part, have led to poor understanding of methods between schools and a mutual mistrust of findings. Perhaps most troubling are applications comprising multiple methods of poor familiarity and limited usage, a novel complexity best illustrated by the stated methods. For example, Kennedy *et al.* (2000, p. 348) performed a sequence-based analysis in which the Pelecaniformes comprised both the ingroup and outgroups, and in which data " ... were aligned by eye ... with reference to the seabird data ... and used the secondary structure model and conserved motifs approach .... All gaps of more than one base were removed ... [and] trees were constructed with both maximum-parsimony and maximum-likelihood. ... For the [latter] heuristic analysis, the TIM substitution model with invariable sites and among-site rate heterogeneity was selected with the Akaike information criterion .... The nucleotide frequencies, gamma shape parameter for rate heterogeneity (with four rate categories), substitution rate matrix (with four substitution types)...and proportion of invariable sites were all estimated by maximum-likelihood." Whereas the detail was comparatively impressive, justification of these options was not, and a subsequent work (Kennedy *et al.*, 2005) adopted a significantly more complex sequence of analyses.

While it is undeniable that new methods should be explored and tested, it is doubtful that a fair reception of such methods and resultant inferences is forthcoming unless the authors accompany this ingenuity with a justification for unique mathods over the suite of established methods. Until unfamiliar methods are subjected to careful review, studies involving new approaches will be viewed with caution, whereas if properly described and justified, the new method will be accorded the appropriate scrutiny, oversight, and acceptance. Without such insights, a choice to devise new methods in molecular systematics, a field already populated with a number of known variations, will understandably engender suspicions on the part of readers, especially those primarily concerned with phenotypes.

In parallel, colleagues in molecular systematics reading a morphological work such as that by Livezey & Zusi (2006, 2007) are likely to be distracted by anatomical detail and discouraged by an absence of first-hand implementation of the concepts of characters and states, let alone be in a position to judge the propriety of ordering, weighting, noncomparability, or methods of search. Indeed, the advisability of inverse weighting of morphological characters on the basis of relative homoplasy – like that of third-position nucleotides in sequence data – is far from decided (Goloboff et al., 2008). This increasing specialization among subdisciplines currently relies on an editorial system for professional and uniform scientific standards, i.e. publication resides in the hands of a very few readers and participants.

PERSISTENT PROBLEMS OF AVIAN SYSTEMATICS

Homology: foundation of phylogenetic inference

The traditional criteria for homology of phenotypic characters by Remane (1954) comprised observational qualities (Ingilis, 1966) and few fundamental steps (Brower & Zchawaroch, 1996), i.e. similarity of form, proximity of position, and phenotypic intermediates as constituents of "primary" homology (Agnarsson & Coddington, 2008). Two additional diagnostic components – similarity of function and ontogenetic pathways – also were included in the diagnosis of homology (Wagner, 1989; Hall 1994, 1995), with recent reviews of counterexamples and theory (Hall, 2003). A utility for topological optimization in affirmation of homology a posteriori (de Pinna, 1991) also was recognized. Although the process of comparative study leading to hypotheses of phenotypic homol-

ogy (condition of characters) and variants thereof (state or homologue) often is inadequately treated in morphological phylogenetics, identification and description of homologues is critical to the strength of morphological study (Arthur, 1984, 1988; Wilkinson, 1991), and fundamental also to consideration of constraints on phenotype (Arnold, 1992) and development (Arthur, 1997, 2004).

In practice, one seldom is treated to details concerning the parallel procedures in sequence analyses, and the problems of alignment arguably are at least as contentious and potent as more traditional assessments of evolutionary history. Any application involving hypotheses of homology and their role in the reconstruction of phylogeny is deserving of concern if only the analytical options leading to the "final" or "best" inferences are provided (Wheeler, 2003a,b). The potential for mutual suspicions is aggravated further by the absence of entities comparable to phenotypic characters in molecular analyses.

Not uncommonly, higher-order phylogenetic studies encounter phenotypic variation within and among related characters sufficient to render assessments of homology and assignments of states problematic (Poe & Wiens, 2000; Wiens, 2001), although the experience of the author finds such difficulties to be the small minority (Livezey & Zusi, 2006). Extreme cases in phenotypic comparisons are named "missing characters" (contra "missing data") as a result of the undiagnosability of such superimposition of extreme apomorphy (Maddison, 1993). Comparable to the analytical challenges of "gaps" in sequence data (Giribet & Wheeler, 1999; Simmons & Ochotereno, 2000; Ogden & Rosenberg, 2007), in practice the only viable treatment is to code such noncomparable conditions as the absence of diagnosed character-states. In the latter ruling, unknown entries and those undiagnosable by coincidence of hyperapomorphy are analyzed identically, allowing for parsimony-optimal substitutions to be made for both. As for a number of practical problems (and corresponding options) in phylogenetics of phenotypes – delimitation of characters and states, value of taxa vs. characters, binary vs. multistatic characters, ordering of states

(Graybeal, 1989; Hauser & Presch, 1991; Hauser, 1992; Kim, 1996; Wiens, 1998; Siddall & Jensen, 2003; Steel & Penny, 2004a, 2005) – superior alternatives for the treatment of noncomparability seem likely to be available in the near future.

Procedures by which sequences are aligned are of enormous influence, and this class of homology presents the analyst with the variably combined problems of indels, transposed segments, silent substitutions, bias in compositon of nucleotides, and genomic segments of noncomparability among taxa. These issues are subjected (singly or in combination), at the disgression of analysts, to alignment by hand, a parsimony of assumed change imposed by software, multiple alignments of genes by criteria for simultaneous fits, and "dynamic" alignments iteratively revised ("optimized") on the basis of favored topological outcomes or superior summary statistics (Hein, 1989).

The latter class of alignments – which explicitly confounds hypotheses of sequence homologies with phylogenetic inferences of relationships – represents a worrisome circularity; a principle recognized in lesser contexts (de Queiroz, 2000). Distinct from manipulation of included phenotypic characters toward a preferred topology (de Queiroz, 2000), or recursive weighting thereof (Goloboff, 1993; Kluge, 1997), both share the danger of considering topologies in the inferences of homology (Fleissner *et al.*, 2005). The challenge of homology and the dangers of some proposals have led to multifactorial approaches to alignment (Minelli, 1998).

Substandard discourse, campaigns, and premature confidence

Scientific endeavors are replete with disagreements of method and message, but systematics historically has been peopled with specialists more vociferous and less collaborative than those of many other biological endeavors. Despite notable periodical reviews of certain issues (Hull, 1970; Sober, 1983a,b), differences of opinion in systematics seldom engender an intensification of study or syntheses, but instead are often simply ignored, mischaracterized, or misunderstood. Even progress as recognized by the broad community simply can be dismissed and perpetuated by authors where considered censure is deemed advantageous, thereby adding to the conflagration of real and exaggerated arguments about methods.

Perhaps most disturbing is the proliferation of confidence in method and inference where there is a undeniable lack of empirical clarity, as if the competition of theory and evidence is at times supplanted by head counts and bravado. Also distressing is the apparently insatiable hunger for new and previously unproposed groupings (Livezey & Zusi, 2007), an inclination paralleled by the relative disinterest shown by some where new evidence or analysis are merely confirmatory of accepted groups (e.g. Mayr, 2007), as if any aspect of avian phylogeny is beyond question. Critical importance must be afforded to the independence among alignment, topological search (Simmons, 2004), and choice of model (Anderson, 2008).

Long-branch attraction and combinatorial implications

"Long-branch attraction" principally is a problem for molecular data – wherein states are limited to the *same four* unordered alternatives per character – chance coincidence of states between lineages inferred to have undergone protracted terminal augmentation being a realistic expectation. Chance repeats of sequences in "long branches" are considered to lead to the homoplasious attraction of lineages in the repeats in the serial throws of the four-sided die of nucleotides, although concerns vary (Siddall & Whiting, 1999). Sequence data differ from phenotypic characters in this likelihood in that phenotypic characters do not share the *same* four alternatives (i.e. "attraction" by repeated series resulting from sequential trials of a four-bead urn with replacement). Circumstances both conducive and robust with respect to long-branch attraction exist (Siddall & Whiting, 1999), but a growing consensus among those working with sequences advocates minimizing long branches through more intensive sampling of taxa (Wilson, 1999; Slack *et al.*, 2007).

In the case of "long branches" of phenotypic changes optimized as autapomorphies, the feasibility of attraction or homoplasy can in some instances be dismissed *a priori*. For example, in a prior analysis of gruiform birds (Livezey, 1998), a truly remarkable gruiform genus *Aptornis* was resolved to be extremely apomorphic, reflected by a very long terminal branch in the global tree. However, a criticism of its placement in the tree was leveled on the basis of the notorious potential of "long branches." An examination of the "long branch" in question revealed that it almost entirely comprised apomorphies unique to the ingroup and (in some cases) all of Neornithes, thereby precluding a pseudo-affinity between lineages on combinatoric grounds.

The importance of autapomorphy differs between molecular and morphological contexts. In morphological studies, genuinely autapomorphic states can be deemed so *a priori*, and in many cases are determinable (within error of identification or diagnosis of homology) prior to analysis. In the molecular context, given the limitation of possible states to four, *a priori* uniqueness is seldom if ever demonstrable, and autapomorphies are limited to changes in state inferred by assignment to terminal branches *a posteriori* considered to be optimal. Hence while it is not uncommon for molecular systematists to advocate complete exclusion of autapomorphies (under parsimony), especially in light of a coincident apprehension concerning "long branches," it is justifiably treated as topologically uninformative (neutral) while being evolutionarily insightful in many cases (e.g. Livezey, 2003), and improve comparability of trees and evolutionary rates (Leman, 1965; Omland, 1997).

Laxity concerning outgroups and rooting

In recent decades, especially among those practicing molecular methods, it has become fashionable to offer variably plausible stories of common life-historical characteristics, shared selection regimes, and convergent form in those instances where a dismissal of a contrary morphological

inference is more convenient than even a temporary admission of uncertainty for ones own findings. Notable examples in the ornithological literature pertain to waterfowl, one of the first avian orders subjected to intensive phylogenetic study (e.g. McCracken *et al.*, 1999). An hypothesis of polyphyly of ratites indicated by sequence data from at least one nuclear gene (Harshman *et al.*, 2008; but see Omland, 1994, 1997) – essentially limited to linking Tinamiformes with sympatric Rheiformes among other, morphologically diverse ratites – apparently is contradicted with evidence from virtually homoplasy-free molecular short interspersed repetitive elements (SINEs; Ray *et al.*, 2006) as well as morphological data (Livezey & Zusi, 2006, 2007). This dispute is likely to remain contentious, in part because it is conditional on specific genes and the potentially white-noise provided by a crocodylian outgroup. An insight into the effect of ancient outgroups might be provided by comparing topologies of the adjacent ingroup (Galloanserimorphae) using both *Caiman* and a tinamou as alternative outgroups.

Molecular systematics of higher order, especially that of basalmost relationships of birds, are compromised by a common, critical impediment to informative polarities – the closest extant relative of modern birds being the Crocodylia (pre-Theropodan origins among Archosauromorpha), with nodes separated by approximately 25 Myr in the Middle Triassic (Carroll, 1988). During this protracted period of evolutionary change, substantial, largely undetectable transition and transformation of nucleotides and concatenation of indels in the dinosaurian branches leading to Neornithes are virtually certain, and silent and multiple substitutions may asymptotically approach the white noise of throwing four-sided (nucleotide) dice. Most systematists reliant on this dubious rooting standard typically fail to mention this shortcoming or plea that as the only available means, acquiesconce is justified.

An outgroup of unsuitable antiquity is likely to be historically uniformative and subsequent assemblages of ingroup members are essentially phenetic; optimal solutions with respect to the effectively random initial "signal" have ramifications for an indeterminate span of the resultant tree and are at least as important as other analytical options (e.g. evolutionary models, parameters). To decry such concerns as exaggerated is tantamout to the claim that outgroups are of little importance, or requires the unfounded proposal that the genes used were so conservative for sequence data to manifest extreme convergence since the divergence of the outgroup (e.g. Crocodylia) only to begin to evolve informative changes ca. the K–Pg boundary, at the base of the ingroup (e.g. Neornithes or Neognathae). Although the Crocodylia are considered the only outgroup remotely suited for rooting Neornithes by molecular means, it is sufficiently defective so as to prompt the proposal of the turtles (Testudines) as a better outgroup for birds (García-Moreno *et al.*, 2003). It is obvious, however, that resorting to the Crocodylia would not serve well for purposes of morphological investigation. Inasmuch as polarities and directionality of change is central to phylogenetic (as opposed to phenetic) techniques, without a reliable outgroup, what inference can be trusted where outgroups require *assumptions* of optimal heterogeneity of evolutionary rates? The absence of an alternative outgroup for sequence data is undeniable, but has no logical bearing on the problem of reliability of initial conditions employed in the analytical process.

Avian systematists working on molecular data are not alone in the challenge of finding a sound root. Although fortunately in the minority, some paleornithological systematists depart from the use of outgroups for purposes of rooting in the analysis of fossil and modern taxa, and some do so for reasons suggestive of prior preferences. For example, Mayr & Clarke (2003) repeatedly offered local argumentation of polarites instead of direct rooting by outgroups. Of course, use of a hypothetical ancestor represents an opportunity for wholesale replacement polarities of outgroups with those of intuition.

"Form follows function" trumps "phenotype follows genotype"

Amidst the current politics and empirical dissension of modern systematics, two aphorisms typically are assumed but expressed singly, but nevertheless both are central to natural selection and evolutionary change (Endler, 1986). Moreover, both processes are supported empirically, i.e. both genes (instructions) and phenotypes (products) probably deserve serious consideration with respect to departures from neutrality of change by selection, convergence, and interdependence. This state of affairs is surprising in that the great majority of currently active systematists were educated within a fundamental evolutionary paradigm that can be simplified as: (i) phenotype being the joint product of genotype and environment, (ii) natural selection acting on the resultant phenotypic characters, (iii) selection acting on comparative success of phenotypes and their genotypic bases, (iv) differential transmittal of the genes through the stochastics of this phenotypic competition (and chance), (v) phylogenetic lineages undergoing evolutionary change through generational repetitions of this process, and (vi) these lineages reflect these trends through time in variation tractable for purposes of historical reconstruction.

At what point in this process is it implied that an immunity of genes and their constituents to selection driving directional change, stabilization of frequencies, reversals, and convergence among lineages pertains, a notion widely assumed in many phylogenetic circles? The implicit contradiction attending much of molecular phylogenetics – that selection for functional optima (and to a lesser degree environmental interactions during ontogeny) alter modal form while not noticeably tainting the genome involved with such refinements for historical reconstructions – has not been logically reconciled or empirically demonstrated.

This disparity is confounded by analytical assumptions required for methods employed by systematists. For example, independence of evolution in (i) different lineages and (ii) at different sites on a tree is assumed for numerical solutions of MLE (Felsenstein, 2004), while under the criterion of parsimony no such assumption is made. The latter does not make parsimony more vulnerable to the bias of inderdependence among sites and lineages, but simply frees the criterion from arguably unrealistic assumptions concerning underlying evolutionary patterns. In MLE of sequence data, realism of these assumptions is jeopardized by hierarchical relationships among lineages, multigenic selection, and constraints of genomic architecture.

Most critical is the recognition that choice between products of duplication (DNA) and translation (phenotype) is essentially one of scale and reductionism of the historical signal sought. Natural selection and mutation exert change in both currencies, and the unjustified pardon granted molecular methods from the obsession with convergence seems more a reflection of our inadequate knowledge of selection acting on the genome, one precluding easy speculation and contrivance of evolutionary scenarios. Although complexity of characters may pose challenges of diagnosis, such complexity likely holds a greater richness of signal and potential continuity of transformation among states, both of special importance to recovery of deep history.

Rejection of selection-related (convergence) scenarios as legitimate argumentation is mistaken by some as a dismissal of natural selection and its role in evolutionary change. Instead, the criticism simply turns on the implication that such selective change does not justify the quasi-Lamarckian assertion that "form follows function" or confirm an omnipresence of "convergence" and "parallelism" as the basis for similarity of form in lineages sharing aspects of life history (Coddington, 1994; Arendt & Reznick, 2007). Systematists whose fear of being misled by convergence approaches paranoia, often merely extol the obvious, that common circumstance vies with common ancestry as the basis of shared characters of the phenotype.

Fossils: contributions, folklore, and peculiarities of perspective

Paleontology has been considered as an indespensible source of evolutionary insight, the most celebrated objective being the discovery of fossil "links" among taxonomic (especially higher-order) groups of otherwise obscure origins (e.g. Olson & Feduccia, 1980). An important realization concerning the role of fossils in phylogenetics was that geological age does not correlate necessarily with the primitive status of the taxon, but only an estimate of the *earliest* date of its extinction – characters, not strata, inform polarities (O'Keefe & Sander, 1999). Also of renewed concern are the analytical implications of the abundance of missing data typical of fossils. Although fossils can augment modern groups with long-extinct members and can shed light on the ancestry of modern taxa through inclusion in phylogenetic study, the provision of critical "linkages" between problematic modern groups is at best uncommon and of vastly different importance among taxonomic groups. Of lesser phylogenetic importance but significant evolutionary interest are fossil members of groups completely unknown by neontological means.

The phylogenetic relationships of birds have been illuminated greatly by a burgeoning wealth of Mesozoic fossils – taxa largely bearing on pre-Neornithine roots. Until the early 1990s, *Archaeopteryx* was the only fossil taxon that provided any understanding of the phylogenetic transition between basal Archosauria and Aves, whereas other breakthroughs – e.g. among Anhimidae, *Anseranas*, Anseriformes, and Galliformes – were achieved through study of modern taxa, both using morphological (Livezey, 1997) and molecular data (Groth & Barrowclough, 1999). At ordinal scale, the Neornithes are undergoing substantial resolution by both means, although points of disagreement remain (Livezey & Zusi, 2007).

An important source of confusion in paleophylogenetics derives from the different views of the diagnosability and likelihood of recovery of direct ancestors to one or more modern taxa among available fossils in a given ingroup (Wiley, 1981). Avoiding for the present "stem" and "crown" groups in vogue among paleornithologists, ancestral status is equivalent to diagnosed inclusion in the paraphyletic group (grade) subtending the corresponding, often modern "crown" (terminal) clade (Kemp, 1999). Fossils are indeed extinct members of clades enlarged thereby, but the phylogenetic significance of these taxa ranges from negligible to critical, and has no claim of superiority over neotological evidence.

The assumption that fossil members are plesiomorphic relative to modern relatives is now recognized to be possible in some cases but unfounded in general *a priori* (Kemp, 1999). The expectation of plesiomorphy seldom is demonstrated empirically to the extent that fossil status reliably informs decisions of local polarity or ordering of states. However, the latter two notions persist in the literature of avian systematics, and to the extent that the latter assumptions influence analyses, the price of such oversimplicity of age may be significantly higher than any of the intended benefits.

An unnecessary and unfounded practice too commonly employed in phylogenetic analyses of fossils is the culling of characters not represented in most or all of the fossil taxa that are the principal focus of a given analysis. Recent examples of the truncation of character information to those features present in the fossils, the latter typically the lowest common denominator for characterization, include many analyses involving descriptions of Mesozoic taxa. The practice is especially obvious where the same author analyzes different fossils from the same locality and age based on matrices of significantly different reductions from those typical of modern relatives (Bourdon, 2005; Bourdon *et al.*, 2005; Chapter 8, this volume). Whereas this approach often is advocated as a means of reducing the burden of missing data, such *a priori* deletions weaken the inferred relationships among modern taxa included in the analyses, thereby undermining the entire study for fossil and modern taxa alike. Such deficiencies are worsened where resultant inferences are extrapolated to recommendations regarding the modern taxa and related clades thereof.

The traditional folklore that elevates the perceived importance of fossils in establishing higher-order relationships is intimately related to the rarely deserved appelation "missing links" (Cracraft, 1972). The modifier "missing" conjures a perception of discovery, and "link" implies a critical role in resolving relationships between groups and novel evolutionary leaps (Cracraft, 1979), a combination first popularized with respect to the pongid affinities of hominids. Also, there is an unjustified fascination attached to fossil taxa considered to be "mosaics" of plesiomorphies and apomorphies. However, such mixtures of apomorphy and plesiomorphy are logically unavoidable for *all* taxa under independent evolution of characters and heterogeneity of evolutionary rates among phylogenetic lineages, and fossils provide no new insights or confirmation of evolutionary processes of the kind (Livezey & Zusi, 2007). This criticism does not pertain to the different emphasis on the "mosaicism" or functional modularity of characters, which instead refers to organization of data matrices with respect to evolutionary interrelationships of characters that evidently holds promise for better fits of models for phylogenetic trends in character complexes (von Dassow & Munro, 1999; Kim & Kim, 2001; Clarke & Middleton, 2008).

In sum, fossils enrich our knowledge of evolutionary diversity and in a minority of cases provide critical information where modern taxa fail to reflect phylogenetic episodes or transformation of characters. Enrichment of taxa by fossils creates new problems challenging to phylogenetic resolution (e.g. Dromornithiformes, Phorhusraciformes, Gastornithiformes, Diatrymiformes, and *Aptornis*), and uses of fossils to narrow gaps in our phylogenetic understanding (e.g. *Sinosauropteryx, Archaeopteryx, Confuciusornis*) are critical but rare. On the other hand, modern taxa can provide critical phylogenetic links where a considerable fossil record for the group(s) either was uninformative or misleading (Livezey, 1997), and the morphofunctional diversity represented by modern taxa is beyond dispute. The typical negative characteristics of fossils are: (i) comparatively poor preservation; (ii) inaccessibility to DNA sampling; and (iii) extinction. The single advantage unique to fossils is that, if stratigraphically known, they provide point estimates of *minimal* ages of divergence for the lineages represented (Foote & Raup, 1996), and under some circumstances this can contribute to estimates of confidence intervals (Burbrink & Pyron, 2008).

Paleomorphological and neomolecular marriage of convenience

Amidst this proliferation of methods and theory, a peculiar marriage of methods was born. In a discipline to which cooperation among schools should qualify as a godsend, this union is rooted in the less desirable of motives. Among those molecular systematists who hold that the signals encoded in gene products (phenotypes) are unreliable, an "exception" is often accorded fossils, most recently for calibration of age (Scotland *et al.*, 2003; Hedges & Maxson, 1996; Givnish & Sytsma, 1997). Even the most egregious of critics of morphological evidence (e.g. van Tuinen, 2005) grudgingly admit to a utility of those applications permitting the use of fossils to calibrate absolute time in molecular trees (Benton & Donoghue, 2006).

The perception that morphological and molecular data are of predictably different reliability, however, has encouraged some systematists to seek a niche that purports to serve the best of both classes of evidence. Although broadly foundational to the limited collaboration in which paleomorphological inferences are applied to neomolecular trees (see below), this partial coincidence of aims also has motivated biased surveys with the objective of pseudoconfirmation of molecular inferences. In this new protocol of "ambulance-chasing," an explicitly biased survey of characters among select groups of taxa is performed to discover morphological features apparently consistent with the molecular inference in favor (e.g. Mayr, 2002, 2004). Such prejudial surveys intended to fit favored molecular inferences often depend predominately or solely on fossils (Mayr, 2007), a perpetuation of the supposed importance of fossils and an apparent belief that anatomical characters improve with geological

age and rarity. Unfortunately, in terms of reasonably objective analysis of characters and acceptably broad sampling of taxa, such exercises are not phylogenetic analyses and do not constitute independent affirmation of any group. This process is especially intractable where modern and dubiously related fossil taxa are assembled for the purpose of finding characters suggestive of the favored taxonomic group or a higher-order "linkage" (e.g. Mayr, 2002, 2003, 2004).

Perhaps least appreciated among narrowly defined disagreements of large-scale works is that each dichotomy in a tree is a virtually inextractable part of a highly interrelated, mutually constrained whole, and the small topological point of comparison typically imposes ramifications throughout much of the total tree. This seldom-conceded issue can be ameliorated by use of optimization procedures that hold the remainder of the tree unchanged while the point of interest is manipulated (shift of a branch in otherwise unchallenged topology, e.g. by movement of a single branch in MacClade; Maddison & Maddison, 1992). This differs greatly from the exercise in which the entire tree is subjected to further searches while examining the "cost" of a given dichotomy of contention (e.g. constrained searches in PAUP – Phylogenetic Analysis using Parsimony).

Slow, uniform rates of morphological change – the "straw man" of punctuationalists (Eldredge & Gould, 1972; Stanley, 1979) – are not consistent with mechanisms known to influence morphological macroevolution. In the face of growing counterevidence, however, assumptions of "clock-like" molecular change persists. At the other extreme, where molecular inference fails to resolve relationships, the literature is rife with *ad hoc* pleas of rapid radiations for short internodes or pulses of evolutionary change related to diversification, assumptions especially dangerous given the imprecision of fossils as benchmarks of evolutionary rates.

Perhaps the most confusing and potentially misleading of this clan are those trusting virtually all molecular inferences accompanied *a posteriori* by searches for morphological characters argued to be consistent with the groupings. These efforts appear to seek co-evidentiary groupings by acceptable empirical means and to please almost everyone. Unfortunately, the original bias concerning molecular veracity typically leads to biased sampling of taxa and characters, the most egregious being one-sided lists of characters that appear to be congruent (phenetically) with the molecular inferences already assumed valid, a one-sided pretense that is especially deficient where fossils are involved. Mayr (2004) performed a biased search for agreeable characters among selected taxa that were marshaled in combination with fossil-based scenarios to corroborate the recent, tenuous grouping of grebes (Podicipediformes) with flamingos (Phoenicopterygiformes) on molecular grounds. Where the purportedly uniting characters are accurately described, these exercises constitute attempts to reassure opinion where evidence is contradictory, and fail even the most minimal of investigational standards.

This approach is not a tandem study of taxa by different means, but instead morphological methods are limited to fossils whereas modern taxa are analyzed by molecular means. A subtle misfortune of this duplex of methods is that modern taxa are far more amenable to complete morphological characterization, and use of morphological criteria for fossils depends on a sound, independent or combined morphological analysis of modern relatives. The offspring of this methodological marriage comprises molecular systematists reluctantly salvaging what is revealed by fossils for purposes of calibrating dates on otherwise molecular phylogenies, or morphologists plying a trade in which they have dubious commitment and doing so with the most difficult of taxa, those known solely from fossils.

Unintuitive phylogenetic groups and the limits of credulity

Despite a minority status of formal morphological phylogenetics, it is virtually indisputable that systematists admit to morphological homology of some features and certain phylogenetic relationships *a priori*. Formal analysis of characters and inferences of homology and relationships aside, it is worthwhile for systematists to consider their confidence or doubts in particular traditional groupings. While it seems unlikely that there is a phylogenetic systematist alive who doubts the broad homology or informativeness of feathers, beaks, and wings among extant birds, and the spatulate bills of ducks appear unquestioned as homologs among Anseriformes (*contra* Olson & Feduccia, 1980), confidence in the proposed homologies of Neornithes and subgroups thereof is distinctly less common and subject to increasing doubts.

Some characters of historical importance – acarinate sterna, abarbulate pennae, and lateral processes of the sternum – stand at the margin of general reliability. Characters for which the constituency in support of homology and historical veracity merit objective evaluation are more numerous, but the critical nature of many of these to phylogenetic hypotheses calls for their continued evaluation from the perspectives of ontogeny, comparative anatomy, and phylogenetic pattern. As explicit contradictions between inferences made by morphological and molecular means arise (e.g. Fain & Houde, 2004) and withstand definitive evaluation, the importance of objective comparison and bases for trust in characters, both molecular and morphological, is paramount. The relevance of such introspection is exemplified by the surprising advocacy won by the biased phenetics of Lignon (1967) that weakly suggested a close relationship between storks (Ciconiiformes) and New World vultures (Cathartidae).

The horizon of confidence and disbelief becomes less clear when entire data sets stand in the balance, especially where classes of evidence or methods of analysis differ in concert. An understanding of the inclinations of researchers toward inferences of diverse implication cannot but facilitate the reasoned integration of multiple views toward a convergence on the single, true history of avian phylogeny. At the pinnacle of the work, the proposals by Sibley & Ahlquist (1990) tested the stands of systematists differing in prior information and personal investment in phylogenetics. Such points of controversy are too numerous to list here, but the positions of the Turnicidae, Piciformes, and eclectic Ciconiiformes fueled many early doubts regarding the work in the absence of definitive and formal analyses to the contrary at the time.

Properties of estimators: importance, confusion, and tactics

A welcome addition to the literature of phylogenetics has been an increased attention to the statistical qualities of estimators and corresponding parameters, and the relation between these on analytical approach (Gascuel, 2005). Accuracy (proximity of expected value of estimator to parameter, difference being the bias), precision (dispersion of estimates about parameter among samples), consistency (asymptotic convergence of estimator to parameter with increasing sample size), efficiency (minimal error of estimate of parameter with increasing sample size), robustness (resilience of estimator to departures from assumptions), and sufficiency (use of information relevant to parameter by feasible estimator) are generally considered most relevant to systematists. A property of estimators less frequently noted but virtually synonymous with efficiency is power (precision relative to sample size). Clearly, most of these qualities of estimators are interrelated to varying degrees in practice, e.g. precision, efficiency, and power tend to be directly related. However, confusion between distinct properties also persists, e.g. accuracy (related to mean estimates) and precision (variance of estimates) can be virtually independent but are often confused. These qualities are relatively straightforward and critical to understand, but confusion persists in real-world applications.

A subtly complex examples one germane to both the terminology of estimator and to the lackluster record of analyses of combined data, attempted a consensus regarding phylogeny of modern Charadriiformes. Thomas *et al.* (2004, p. 28) noted "that the supertree is generally more *consistent* with molecular data (both recent sequence studies and DNA–DNA hybridization)

than analyses based on morphology. However, it is of course possible that this reflects the greater *number* of molecular source trees [29 of 51] *available* rather than indicating that molecular data are [*sic*] actually better at *resolving* shorebird phylogeny [emphasis added]." The inclusion of both the phenogram of Sibley & Ahlquist (1990) and the related classification by Monroe & Sibley (1993) twice tainted the source trees by its exclusion as phenetic, and these were the only sources virtually complete to species level. The unfortunate use of the term "consistent" *sensu* similarity, the bias in "available" source trees (worsened by genes shared among studies), and the vague expression of "better at resolving," shows that the authors have confused the closer resemblance of the supertree to their molecular source trees with the likely accuracy of molecular methods. The most favorable proposal suggested regarding molecular methods by this outcome is that the estimators employed produced estimates more similar to each other (especially with indeterminate sharing of data) than the estimators used for morphological data. Nothing can be inferred regarding relative accuracies of data types based on this exercise. In addition, if the same suite of estimators were used for both data types, one might be able to gain a vague sense of the comparability of the estimators with respect to the two classes of data.

With respect to consistency *sensu* statistical estimators, cases have been made that a given estimator or criterion may not qualify (Faith & Cranston, 1991, 1992; Forster & Sober, 1994), and some have argued that this alone relegates the estimator substandard to an alternative that is consistent. Given the inaccessibility of truth in real-world applications, such assertions generally turn on simulations and philosophical exchanges, which for the parsimony criterion has proved inconclusive (e.g. Farris, 1999; Huelsenbeck & Lander, 2000). Unfortunately, it remains to be demonstrated that in practical contexts and sample sizes, consistency is a critical property of topological estimators and alone may preclude its reliable use (Steel *et al.*, 1993). For example, the performance of estimators (and magnitudes of differences) under realistic sample sizes with comparable data remains an essential but unknown issue among phylogenetic estimators, whereas approach to truth at the limit as sample sizes approach infinity is of lesser relevance.

Parsimony: philosophical pedigree and simplicity vs. realism

The philosphical germ of parsimony is simplicity (Kluge, 2005), and in cladistic contexts it has principally served as a criterion of optimization in which trees of maximal simplicity (minimal lengths) were considered optimal, although homoplasy is recognized as comprising both evidence of mistaken homology and a source of topological stability (Givnish & Systsma, 1997; Källersjo *et al.*, 1999) and patterns of homoplasy pose problems of comparison across studies (Sanderson & Donoghue, 1989, 1996). Most cladists working with phenotypic characters viewed parsimony to be a criterion for minimization of tree lengths and the additional steps or *ad hoc* changes that such augmentation entails (Goloboff, 2003). Attacks in recent years against the parsimony criterion – including consistency, Popperian perspective on evidence, and the purported importance for *a priori* evolutionary models – have been advanced largely by molecular systematists (e.g. De Laet, 2005; Faith 2006). Defenders of parsimony, however, have countered all criticisms, in the contexts of both molecular and phenotypic data (Farris, 2000, 2008).

In practice, searches for minimal tree length invariably approximate those for minimal homoplasy, and the transformations of characters in such optimal trees imply a post-analytical hypothesis of homology and evolution; this capacity can be employed with pre-analytical interpretations in a dynamic approach to the determination of homology (Ramírez, 2007). Whereas most applications of parsimony in phenotypic contexts have performed to the expectations of morphological practitioners, the criterion met with less acceptance among those analyzing DNA sequences. However, most in both camps employ criteria of parsimony in various ways, e.g. adoption of the least complex model of comparable explanatory power or the parsimony implicit in sequence alignment (Anderson, 2008). Still others defend the comparative explanatory power of parsimony (Farris, 2008).

Some detractors of parsimony consider that it does not assume a probabilistic model of evolution or that the criterion is not intrinsically related to the evolutionary process, some consider parsimony equivalent to a model of "minimal evolution" (Rzhetsky & Nei, 1993; Penny *et al.*, 1994; Gascuel, 2000; Steel & Penny, 2000; Denis & Gascuel, 2003), and still others condemn it by the claim that parsimony actually is a model of evolution but is burdened by an exorbitant number of implicit parameters (Felsenstein, 2004). Although for a given data-set the search for minimal steps or change is consistent with this view, the terminology incorrectly implies that those using the parsimony criterion are limited by a preconception that evolutionary rates and changes are rare, whereas the method only seeks a minimalist explanation for data sets characterized by any of a range of evolutionary rates. Moreover, although parsimony favors topological solutions at least as *efficient* (i.e. of minimal homoplasy) in the explanation of characters among lineages (other parameters being equal), it is virtually routine to consider "suboptimal" or implications of longer topologies in such studies (Sober, 1983a).

Sequence data – in which numbers of base-pairs, a serial (sub)structure of codons, and limited concerns about selection – were relatively suited to probabilistic modeling. This soon led to a favoring of alternative methods to optimize topologies, notably maximum-likelihood (Severini, 2000) and Bayesian models (Felsenstein, 2004; Ronquist, 2004). Nevertheless, systematists persuaded of the philosophical and practical advantages of maximum parsimony continue to hold a virtual monopoly on the phylogenetics of phenotypic (discrete) characters (Felsenstein, 1973; Goloboff, 2003), and this schizm in methods has intensified the nonconstructive estrangement between many morphological and molecular systematists.

Such comparative use of different methods – e.g. philosophical perspective of parsimony vs. statistical models of maximum likelihood – is a natural and generally constructive process in scientific fields (Felsenstein & Sober, 1987; Sober, 2004). However, it seems likely that probabilistic models refined for discrete (including morphological) characters will be available soon, permitting more direct comparisons of classes of data. For example, evolutionary models for phenotypes have been the topic of informal discussion in the systematic literature for decades. Refinement of such ideas into statistical language (Johnson et al., 2005) – e.g. null models for numbers of state-changes (Geeta, 2003), elucidation of underlying processes (Fusco, 2001), combined with empirical estimates of invariance and ordered states in morphological characters – seem well within the feasiblity of present-day systematists should such methods prove preferable and are critical to phylogenetic perspectives on development and allometry (Klingenberg, 1998). The paucity of such details is not an unavoidable characteristic of morphological data, but instead reflects the view that such topics were not vital to phenotypic phylogenetics.

The preference for model-based analyses instead of those based on global parsimony is not a simple difference of interpretation. Most importantly, the parsimony criterion does not entail the supposition and numerical comparison of a specific model of evolution, a probabilistic framework by which assumed (null) and empirical estimates of parameters permit comparisons among models and optimization of parameters so as to fit a tree to a data set by means of branch lengths (Felsenstein, 2004; Gascuel, 2005). The key worry in the latter approaches is the assumption of realism or requirement of such models employed to parametrically fit a tree to the data (Brower, 2000), although there is concern regarding the use of likelihood ratios for selection of models in general. As for the predicted use of such methods for phenotypic inferences, the optimality of phylogenetic estimates is constrained by the realism of the model chosen (Cunningham, 1998), and criteria of choice do not enjoy confident consensus. Unfortunately, certainty of the underlying model is as unattainable as the real or true phylogeny.

Interpretability is a strong advantage of parsimony-optimized phylogenetics, at least with respect to patterns and plasticity of individual morphological characters among branches and clades. The groupings defined by maximum-likelihood or Bayesian optima are thought to represent natural (monophyletic) groups, but may not meaningfully be equivalent to clades. For example, there is an inability to identify specific characters and changes thereof on branches in probabilistic models, a significant disadvantage for most morphologists, especially those with a strong interest in evolutionary patterns of specific characters within phylogenies. For the present, however, this inscrutability of trees optimized by likelihoods is of limited concern in that most such analyses pertain to sequence data, the latter seldom subjected to patterns of change in single nucleotides or codons, the latter an aspect of the same disinterest that protects most sequence data from speculations of convergence.

These two basic approaches to phylogenetic reconstruction are employed in tandem by some attempting calibrations of geological ages on phylogenetic hypotheses. Phenotypic phylogenies, typically inferred within the parsimony paradigm, form the basis for referral of fossil taxa to modern groups. The fossils, for which an estimate of geological age is available, then serve as points of calibration in trees typically based on likelihood models and molecular data and for which some hazard assumptions concerning the passage of phylogenetic time.

Evolutionary models: criteria of selection and influence

In addition to variable scale of primary structure (e.g. sequences of nucleotides collectively and by position, codons, and amino-acid equivalents), additional attributes of the genome have been deemed phylogenetically informative. Intuitively cause for optimism, this diversity of signal is too frequently treated in terms of the favorability with which the phylogenetic inferences are viewed by authors. There is a marked paucity of explorations of the effects of models, parameters of models, and class of data on resultant inferences in real contexts. This glib simplification of method represents an extension of a tradition of terse statements of methods in which presented trees were inferred, while explorations of the effects of alternative models, selection of genes, method(s) of alignment, treatment of third-position nucleotides and gaps, relative influence accorded indels and SINEs, and the class and parametric specificity of models employed were not outlined or discussed.

In the face of criticism of the comparatively direct optimization of phenotypic change under parsimony, such brevity not only contributes to confusion of nonspecialists but has raised suspicions on the part of morphological cladists. It seems inescapable that some such cases involve less than optimal analytical design, and at worst this area of homology and modeling, especially where standards of detail and justification are lowered, provides an ominous opportunity for error and abuse. These concerns are worsened by the easy use of software and varying fashions of analysis without obvious justification in the literature.

ARE COMMON GOALS AND CRITERIA FEASIBLE?

For those committed to the reconstruction of phylogenetic history, it is apparent that these systematists are convinced not only of the feasibility of reaching the goal, but that these committed systematists typically favor one class of evidence over others in pursuit of this common goal. As to feasibility, consistency and detail of inferences to date fall short of resounding affirmation of this hope. Perhaps more sinister is the possibility that the various classes of evidence or signal bearing on phylogenetic history – nucleotides or products thereof – may not encode reliable history or provide adequate geological ages for other evolutionary aims to be achieved (Ziman, 1978; Lake, 1997; Yang, 1997). There is no logical basis for the assumption that the mode of genetic inheritance or its expression necessarily evolved in a manner permitting accurate retracing of biotic history. Moreover, whatever the potential for the genome and its phenotypic expressions for phylogenetic inferences, it is painfully clear that the

falsifiability described by Popper (1963, 1968, 1972) and associated concepts of corroboration (de Queiroz & Poe, 2001, 2003) represent the best to be hoped for phylogenetic hypotheses (Ruse, 1979). The truth of biotic history, for reasons of philosophy and practicality (e.g. unrecovered fossils), remains a goal toward which we toil but we will never knowingly achieve.

Consequently, a mathematical exposition of the feasibility of this goal, and the relative and absolute richness of recoverable signal of available evidence, is overdue. This exercise is not straightforward, but involves statistical properties and historical signal by class of evidence. Moreover, comparative assessments would entail defensibly objective tests between the potential signal of phenotypic signatures and of sequence data (variably long and unpredictably edited "sentences" composed of tri-nucleotide "words" coding semiredundantly for amino acids), and for which neither class (phenotypic and molecular) is adequately understood regarding mutability, alteration of content, or antiquity.

As was discovered at the onset of statistical modeling for systematics, sequence data were numerically better suited for probabilistic analysis and fitting of models. Morphological characters and states are not amenable to direct integral tallies and combinatorial inferences, but instead require the development of a common method of characterization and symbolism before probabilistic modeling can begin. Although this primary phase may seem prohibitive, distributional generalities certainly can be estimated by a sample of character analyses performed by common and accepted methods. Attention to frequencies of phenotypic invariance is especially critical for comparable uses of combinatorics and probabilistic theory in molecular and phenotypic realms.

Unfulfilled promise of "total" evidence and supertrees

"Just how difficult will it be to build a comprehensive avian Tree of Life (ATOL)? Several observations suggest that it will be extremely so. First, there are about 20,000 nodes on the extant avian Tree of Life. Fossil taxa only add to that number." – Cracraft *et al.* (2004, p. 484).

As illustrated by the well-intentioned effort by Thomas *et al.* (2004), hopes held out for a genuinely holistic, "total-evidence" reconstruction, one robust by diversity and testing of methods, have not been met (Baum, 2004; Kluge, 2004). Reasons for this failure are several. Perhaps most pervasive of the impediments among some investigators is a disinterest, or unwillingness, to compromise their findings and contaminate their data with those of fellow systematists, and accepted practices are lacking even where the combined data are limited to different genes. A viable alternative with comparable aims to "total-evidence" methods is a formal protocol for comparing inferences while maintaining individuality of studies, or "meta-trees" (Nye, 2008).

Implicitly unreasonable prejudice against the "dilution" of one data-set assumed superior by another assumed to be deficient (in the absence of any objective criterion of performance), burdens the empirical endeavor of phylogenetics with self-serving and subjective perspectives. Less obviously, balking or explicit refusal to attempt total-evidence methods reveals pessism so profound as to view use of multiple lines of evidence as dilution of the truth or an unwise empirical compromise. The latter belief that a single data set or approach is sufficient for the enormous reconstruction of the "Tree of Life" (TOL) is premature at least and antithetical to the objective of the program at worst. Feasibility of the goals of TOL remains to be demonstrated, an uncertainty that is far more serious than discarding the information to be found in fossils, but perpetuates a view that fossils and extant lineages are adequate for the recovery of signal for omniresolutory analysis. Unfortunately, there is no basis for such optimism.

TEN TALKING POINTS FOR PHYLOGENETICS

1 Even if an assumption of accuracy is made uniquely for molecular methods, given the failure of reproducibility of trees and low precision of estimators, is it likely that reliable assessment of accuracy is forthcoming? If so, in what timescale is this expectation likely to be realized?

2 What credence can be placed on a single clade or couplet of interest if little else in the tree, or at least nodes immediately subtending the clade, is strongly supported?

3 What, if any, are the justifications for the overwhelming reliance on nucleotide sequences instead of codons, amino-acid equivalents, or more inclusive genomic segments, especially given the increased richness of more complex characters but reduced informativeness of the low-scale signals for problems of deep history?

4 Do specialists react similarly to disagreements between a morphological and a molecular study of comparable scale as they do regarding two molecular studies of similar concordance, scale, and feasible methods? Are disagreements between and among methods or data-types treated by comparable criteria?

5 What, if any, are the widely recognized criteria underlying choice of parsimony, maximum likelihood, Bayesian estimators, or indices of information-content for a given analysis?

6 What, if anything, can be inferred from wildly different phylogenetic reconstructions deriving from two different molecular data-sets, analyzed by different means, and sharing only a portion of sampled taxa – e.g. Fain & Houde (2004) vs. Hackett *et al.* (2008)?

7 By what logical frameworks and evidentiary equivalences are the various currencies of phylogeny – e.g. characters, states, steps, nucleotides, codons, SINEs, indels – to be reconciled? At a larger scale, what quantity of morphological characters has an expected signal equal to 1 kb of sequence, say, for a particular nuclear gene?

8 In explorations for new sources of signal, what units are comparable in phenotypic and molecular contexts – organ-systems and genes, coding options and alternative metrics (e.g. SINEs)?

9 If afflicted by poor resolution within a phylogeny, is an increase in taxa or characters *sensu lato* to be sought most assiduously? Is the generality the same for genotypic and phenotypic analyses?

10 What are the essential properties and empirical bases for a critically needed measure of congruence of independently derived topologies (and inherent strengths and weaknesses) for a taxonomic group, a metric that minimally would permit assessments of: (i) practical unanimity, (ii) significant majority, (iii) virtually split opinion, (iv) significant minority, and (v) absence of progress?

CONCLUSIONS

Realism of near-term goals

I took a vivid memory from an informal conversation at the International Ornithological Congress in Vienna, Austria, in 1994. A colleague, with the greatest and most introspective understanding of molecular systematics of which I am aware, predicted that the avian phylogeny would be resolved by phylogenetic inference from DNA sequences within the *next five years* (i.e. by the change of millenium). This prediction was accompanied by doubts in the value of morphological insights into avian phylogeny. Both views were reminiscent of attending the appearance of the "tapestry" from DNA–DNA hybidization a few years before. An abiding interest in morphological evolution, however, nurtured continuing efforts to resolve avian phylogeny by way of phenotypic signal during the next decade (Livezey & Zusi, 2006, 2007).

Unfortunately, despite undeniable progress in the pursuit of phylogenetic history in multiple taxonomic groups, neither the mindset nor a means for synthesizing this progress exists. In some practical instances, disagreement of inferences among datasets and analytical protocols may even raise doubts that some or all of the currently used phylogenetic methods may not be estimators of the true phylogeny of interest. Until a philosophically sound, empirical foundation for a uniform synthesis of analyses among schools is implemented, one based on the resolution of mechanisms that unite genotype with phenotype in reality (Arthur, 1984, 1988, 1997;

Krumlauf, 1994; Hall, 1996; Sommer, 1999; Wray, 1999; Telford, 2000; Steppan *et al.*, 2002; Sarkar & Fuller, 2003; Reid, 2004; Moens & Selleri, 2006) and constraints thereon (Wagner, 1984, 2000; van Tienderen & Koelewijn, 1994; Wade, 2000), problems distinct from the historically important view of recapitulation and the ontogenetic criterion for determination of polarities (Kraus, 1988), it is my contention that much of the resources and apparent advances accomplished in systematics will be misdirected, misunderstood, or wasted. Comparison of partially congruent findings is only a limited means of fostering confident consensus. Furthermore, unless methodological partisanship is overcome by empirically directed, mutually respectful syngergy among schools, an essential and dynamic discipline will remain mired in the fog of bias, distortion, intuition, and political spin (Ebach *et al.* 2008). The critical currency in this enterprise is information (Farris, 1979; Wägele, 1995; Wägele & Rödding, 1998; Shpak and Churchill, 2000), and unraveting the translation of genotype into phenotype through ontogeny will broader the potential of all sources of signal (Mayo, 1983; Hall, 1996) and expectations for phylogenetics (Lecointre *et al.*, 1994; Lake, 1997).

A comparative approach to inferences and associated infighting is important but limited in potential for a broad understanding, whereas an empirical understanding of the mechanisms of translation from genes to phenotype (including environmental effects on phenotype and reaction norms) will permit an informed transformation from circumstantial evidence to the informed unification of historical signal from multiple sources. A renewed commitment to objective, empirical reconciliation of differing inferences deriving from fundamentally distinct, arguably noncomparable methodologies, as opposed to a confused blend of theoretical presumption and data-based findings, is essential to broadly supported advances in avian phylogenetics in the coming decades.

ACKNOWLEDGMENTS

This research was supported by National Science Foundation (NSF) grant BSR-9396249 to Livezey,

NSF grant DEB-9815248 to Livezey and Zusi, and TOL-0228604 to Livezey (among other co-PIs). The author appreciated the many editorial improvements to the manuscript, but assumes sole responsibility for the opinions expressed herein.

REFERENCES

Agnarsson I, Coddington JA. 2008. Quantitative tests of primary homology. *Cladistics* **24**: 51–61.

Anderson DR. 2008. *Model-based Inference in the Life Sciences: A Primer on Evidence.* New York: Springer-Verlag.

Arendt J, Reznick D. 2007. Convergence and parallelism reconsidered: What thave we learned about the genetics of adaptation? *Trends in Ecology and Evolution* **23**: 26–32.

Arnold SJ, 1992. Constraints on phenotypic evolution. *American Naturalist* **140**: 85–107.

Arthur W. 1984. *Mechanisms of Morphological Evolution: A Combined Genetic, Developmental and Ecological Approach.* Chichester: John Wiley & Sons.

Arthur W. 1988. *A Theory of the Evolution of Development.* Chichester: John Wiley & Sons.

Arthur W. 1997. *The Origin of Animal Body Plans: A Study in Evolutionary Developmental Biology.* Cambridge: Cambridge University Press.

Arthur W. 2004. *Biased Embryos and Evolution.* Cambridge: Cambridge University Press.

Baker, A, X. Yu, and DeSalle R. 1998. Assessing the relative contribution of molecular and morphological characters in simultaneous analysis trees. *Molecular Phylogenetics and Evolution* **9**: 427–436.

Baker RH, Gatesy J. 2002. Is morphology still relevant? In *Molecular Systematics and Evolution: Theory and Practice*, DeSalle R, Giribet G, Wheeler W (eds). Basel: Birkhäuser; 163–174.

Banks RC, Chesser RT, Cicero C, Dunn JL, Kratter AW, Lovette IJ, Rasmussen PC, RemsenJrJV, Rising JD, Stotz DF, Winker K. 2008. Forty-ninth supplement to the American Ornithologists' Union check-list of North American birds. *The Auk* **125**: 758–768.

Baum BR. 2004. The MRP method. In *Phylogenetic Supertrees: Combining Information to Reveal the Tree of Life*, Bininda-Emonds ORP (ed.). Dordrecht: Kluwer; 17–34.

Benton MJ. 1999. Early origins of modern birds and mammals: molecules vs. morphology. *Bioessays* **21**: 314–321.

Benton MJ, Donoghue PCJ. 2006. Paleontological evidence to date the tree of life. *Molecular Biology and Evolution* **24**: 26–53.

Bertelli S, Giannini NP, Goloboff PA. 2002. A phylogeny of the tinamous (Aves: Palaeognathiformes) based on integumentary characters. *Systematic Biology* **51**: 959–979.

Bleiweiss R, Kirsch JAW, Shafi N. 1995. Confirmation of a portion of the Sibley–Ahlquist "tapestry." *The Auk* **112**: 87–97.

Bourdon E. 2005. Osteological evidence for sister group relationship between pseudo-toothed birds (Aves: Odontopterygiformes) and waterfowls [sic]. *Naturwissenschaften* **92**: 586–591.

Bourdon E, Boya B, Iarochène M. 2005. Earliest African neornithine bird: a new species of Prophaethontidae (Aves) from the Paleocene of Morocco *Journal of Vertebrate Paleontology* **25**: 157–170.

Bromham L, Woolfit M, Lee MS, Rambaut YA. 2002. Testing the relationship between morphological and molecular rates of change along phylogenies. *Evolution* **56**: 1921–1930.

Brower A. 2000. Evolution is not a necessary assumption of cladistics. *Cladistics* **16**: 143–154.

Brower A, Schawaroch VZ. 1996. Three steps of homology assessment. *Cladistics* **12**: 265–272.

Brown JW, Rest JS, García-Moreno J, Sorenson MD, Mindell DP. 2008. Strong mitochondrial DNA support for a Cretaceous origin of modern avian lineages. *BMC Biology* **6**: 1–18.

Burbrink FT, Pyron RA. 2008. The taming of the skew: estimating proper confidence intervals for divergence dates. *Systematic Biology* **57**: 317–328.

Burnham KP. Anderson DR. 2002. *Model Selection and Multimodel Inference: A Practical Information-Theoretic Approach*, 2nd edn. New York: Springer-Verlag.

Carroll RL. 1988. *Vertebrate Paleontology and Evolution.* New York: W. H. Freeman.

Chubb A. 2004. New nuclear evidence for the oldest divergence among neognath birds: the phylogenetic utility of ZENK (i). *Molecular Phylogenetics and Evolution* **30**: 140–151.

Clarke JA, Middleton KM. 2008. Mosaicism in the evolution of birds: an evaluation of quantitative approaches for the use of discrete cladistic character data in the study of morphological evolution. *Systematic Biology* **57**: 185–201.

Coddington J.A. 1994. The roles of homology and convergence in studies of adaptation. In *Phylogenetics and Ecology* Eggleton P, Vane-Wright R (eds). London: Academic Press; 53–78.

Cracraft J. 1972. The relationships of the higher taxa of birds: problems in phylogenetic reasoning. *Condor* **74**: 379–392.

Cracraft J. 1979. Phylogenetic analysis, evolutionary models, and paleontology. In *Phylogenetic Analysis and Paleontology*, Cracraft J Eldredge N (eds). New York: Columbia University Press; 7–39.

Cracraft J. 1985. Monophyly and phylogenetic relationships of the Pelecaniformes: a numerical cladistic analysis. *The Auk* **102**: 834–853.

Cracraft J. 1987. DNA hybridization and avian phylogenetics. *Evolutionary Biology* **21**: 47–96.

Cracraft J, Barker FK, Cibois A. 2003. *Avian Higher-Level Phylogenetics and the Howard and Moore Complete Checklist of Birds of the World*. New Haven: Yale University Press.

Cracraft J, Barker FK, Braun MJ, Harshman J, Dyke GJ, Feinstein J, Stanley S, Cibois A, Schikler P, Beresford P, García-Morena J, Sorenson MD, Yuri T, Mindell DP. 2004. Phylogenetic relationships among modern birds (Neornithes): toward an avian tree of life. In *Assembling the Tree of Life*, Cracraft J Donoghue MJ (eds). New York: Oxford University Press; 468–489.

Davidson D. 1991. On the individuation of events. *Synthése* **86**: 229–254.

De Laet JE. 2005. Parsimony and the problem of inapplicables in sequence data. In *Parsimony, Phylogeny, and Genomics*, Bert VAA Oxford: Oxford University Press; 81–116.

Denis F, Gascuel O. 2003. On the consistency of the minimum evolution principle of phylogenetic inference. *Discrete Applied Mathematics* **127**: 63–77.

De Pinna MCC. 1991. Concepts and tests of homology in the cladistic paradigm. *Cladistics* **7**: 367–394.

De Queiroz K. 2000. Logical problems associated with including and excluding characters during tree reconstruction and theeir implications for the study of morphological character evolution. In *Phylogenetic Analysis of Morphological Data*, (ed.) Wiens JJ Washington, DC: Smithsonian Institution; 192–212.

De Queiroz K, Poe S. 2001. Philosophy and phylogenetic inference: a comparison of likelihood and parsimony methods in the context of Karl Popper's writings on corroboration *Systematic Biology* **50**: 305–321.

De Queiroz K, Poe S. 2003. Failed refutations: further comments on parsimony and likelihood methods and their relationship to Popper's degree of corroboration. *Systematic Biology* **52**: 352–367.

Durrett R. 2002. *Probability Models for DNA Sequence Evolution*. Berlin: Springer-Verlag.

Ebach MC, Williams DM, Gill AC. 2008. O cladistics, where are thou? *Cladistics* **24**: 851–852.

Eldredge N, Cracraft J. 1980. *Phylogenetic Patterns and the Evolutionary Process*. New York: Columbia University Press.

Eldredge N, Gould SJ. 1972. Punctuated equilibria: an alternative to phyletic gradualism. In *Models in Paleobiology*. Schopf TJM Freeman Cooper: San Francisco: 82–115.

Endler JA. 1986. *Natural Selection in the Wild*. Princeton, NJ: Princeton University Press.

Ericson PGP, Anderson CL, Britton T, Elzanowski A, Johansson US, Källersö M, Olson JI, Parsons TJ, Zuccon D, Mayr G. 2006. Diversification of Neoaves: integration of molecular sequence data and fossils. *Proceedings of the Royal Society, Biology Letters* **2**: 543–547.

Erixon P, Svennblad B, Britton T, Oxelman B. 2003. Reliability of Bayesian posterior probabilities and bootstrap frequencies in phylogenetics. *Systematic Biology* **52**: 665–673.

Fain MG, Houde P. 2004, Parallel radiations in the primary clades of birds. *Evolution* **58**: 2558–2573.

Faith D.P. 2006. Science and philosophy for molecular systematists: Which is the cart and which is the horse? *Molecular Phylogenetics and Evolution* **38**: 533–557.

Faith DP, Cranston PS. 1991. Could a cladogram this short have arisen by chance alone?: on permutation tests for cladistic structure *Cladistics* **7**: 1–28.

Faith DP, Cranston PS. 1992. Probability, parsimony, and Popper. *Systematic Biology* **41**: 252–257.

Farris JS. 1979. The information content of the phylogenetic system. *Systematic Zoology* **28**: 483–519.

Farris JS. 1999. Likelihood and inconsistency. *Cladistics* **15**: 199–204.

Farris JS. 2000. Corroboration and "strongest evidence." *Cladistics* **16**: 385–393.

Farris JS. 2008. Parsimony and explanatory power. *Cladistics* **24**: 825–847.

Farris JS, Goloboff PA. 2008. Is REP a measure of "objective support"? *Cladistics* **24**: 1065–1069.

Farris JS, Källersjö M, Kluge AG, Bull C. 1994. Testing significance of incongruence. *Cladistics* **10**: 315–319.

Farris JS, Källersjö M, Crowe TM, Lipscomb D, Johansson U. 1999. Frigatebirds, tropicbirds, and Ciconiida: excesses of confidence probability. *Cladistics* **15**: 1–7.

Feduccia A. 1977. A model for the evolution of perching birds. *Systematic Zoology* **26**: 19–31.

Feduccia A. 1995. Explosive evolution in Tertiary birds and mammals. *Science* **267**: 637–638.

Feduccia A. 1996. *The Origin and Evolution of Birds.* New Haven: Yale University Press.

Felsenstein JS. 2004. *Inferring Phylogenies.* Sunderland: Sinauer.

Felsenstein J, Sober E. 1987. Parsimony and likelihood: an exchange. *Systematic Zoology* **35**: 617–626.

Fitch WM. 1979. Cautionary remarks on using gene expression events in parsimony procedures. *Systematic Zoology* **28**: 375–379.

Fleissner R, Metzler D, Haeseler A. 2005. Simultaneous statistical multiple alignment and phylogeny reconstruction. *Systematic Biology* **54**: 548–561.

Fong DW, Kane TC, Culver DC. 1995. Vestigialization and loss of nonfunctional characters. *Annual Review of Ecology and Systematics* **26**: 249–268.

Foote M, Raup DM. 1996. Fossil preservation and the stratigraphic ranges of taxa. *Paleobiology* **22**: 121–140.

Forster MR, Sober E. 1994. How to tell when simpler, more unified, or less *ad hoc* theories will provide more accurate predictions. *British Journal of Philosophical Science* **45**: 1–35.

Fountaine TMR, Benton MJ, Dyke GJ, Nudds RL. 2005. The quality of the fossil record of Mesozoic birds. *Proceedings of the Royal Society, Series* **B272**: 289–294.

Fusco G. 2001. How many processes are responsible for phenotypic evolution? *Evolution and Development* **3**: 279–286.

García-Moreno J, Sorenson MD, Mindell DP. 2003. Congruent avian phylogenies inferred from mitochondrial and nuclear DNA sequences. *Journal of Molecular Evolution* **57**: 27–37.

Gascuel O. 2000. On the optimization principle in phylogenetic analysis and the minimum-evolution criterion. *Molecular Biology and Evolution* **17**: 401–405.

Gascuel O. 2005. *Mathematics of Evolution and Phylogeny.* Oxford: Oxford University Press.

Geeta R. 2003. Structure trees and species trees: what they say about morphological development and evolution. *Evolution and Development* **5**: 609–621.

Giribet G, Wheeler G. 1999. On gaps. *Molecular Phylogenetics and Evolution* **13**: 132–143.

Givnish TJ, Sytsma KJ. 1997. Homoplasy in molecular vs. morphological data: the likelihood of correct phylogenetic inference. In *Molecular Evolution and Adaptive Radiation*, Givnish TJ, Sytsma KJ (eds). Cambridge: Cambridge University Press; 55–101.

Goldman N, Anderson JP, Rodrigo AG. 2000. Likelihood-based tests of topologies in phylogenetics *Systematic Biology* **49**: (4) 652–670.

Goloboff PA. 1993. Estimating character weights during tree search. *Cladistics* **9**: 83–91.

Goloboff PA. 2003. Parsimony, likelihood, and simplicity. *Cladistics* **19**: 91–103.

Goloboff PA, Carpenter JM, Salvador Arias J, Miranda Esquivel DR. 2008. Weighting against homoplasy improves phylogenetic analysis of morphological data sets. *Cladistics* **24**: 758–773.

Grant T, Kluge AG. 2008. Clade support measures and their adequacy. *Cladistics* **24**: 1051–1064.

Graybeal A. 1998. Is it better to add taxa or characters to a difficult phylogenetic problem? *Systematic Biology* **47**: 9–17.

Groth JG, Barrowclough GF. 1999. Basal divergences in birds and the phylogenetic utility of the nuclear RAG-1 gene. *Molecular Phylogenetics and Evolution* **12**: 115–123.

Hackett SJ, Kimball RT, Reddy S, Bowie RCK, Braun EL, Braun MJ, Chojnowski JL, Cox WA, Han K-L, Harshman J, Huddleston CJ, Marks BD, Miglia KJ, Moore WS, Sheldon FH, Steadman DW, Witt CC, Yuri T. 2008. A phylogenomic study of birds reveals their evolutionary history. *Science* **320**: 1763–1768.

Hall BK. 1994. *Homology: The Hierarchical Basis of Comparative Biology.* San Diego: Academic Press.

Hall BK. 1995. Homology and embryonic development. *Evolutionary Biology* **28**: 1–37.

Hall BK. 1996. *Baupläne,* phylotypic stages, and constraint: why are there so few types of animals? *Evolutionary Biology* **29**: 215–261.

Hall BK. 2003. Descent with modification: the unity underlying homology and homoplasy as seen through an analysis of development and evolution. *Biological Reviews of the Cambridge Philosophical Society* **78**: 409–433.

Harney HL. 2003. *Bayesian Inference: Parameter Estimation and Decisions.* Springer-Verlag, Berlin.

Harshman J. 1994. Reweaving the tapestry: what can we learn from Sibley and Ahlquist (1990)? *The Auk* **111**: 377–388.

Harshman J, Braun EL, Braun MJ, Huddleston CJ, Bowie RCK, Chojnowski JL, Hackett SJ, Han K-L, Kimball RT, Marks BD, Miglia KJ, Moore WS, Reddy S, Sheldon FH, Steppan SJ, Witt CC, Yuri T. 2008. Phylogenomic evidence for multiple losses of flight in ratite birds. *Proceedings of the National Academy of Sciences* **105**: 13462–13467.

Hartigan JA. 1998. The maximum likelihood prior. *Annals of Statistics* **26**: 2083–2103.

Hauser DL. 1992. Similarity, falsification and character state order – a reply to Wilkinson. *Cladistics* **8**: 339–344.

Hauser DL, Presch W. 1991. The effect of ordered characters on phylogenetic reconstructions. *Cladistics* 7: 243–265.

Hecht MK. 1989. Development, trends and phylogenetic inference. *Geobios* (Supplement 2) 12: 207–216.

Hedges SB, Maxson LR. 1996. Molecules and morphology in amniote phylogeny. *Molecular Phylogenetics and Evolution* 6: 312–319.

Hedges SB, Sibley CG. 1994. Molecules vs morphology in avian evolution: the case of the "pelecaniform" birds. *Proceedings of the National Academy of Sciences* 91: 9861–9865.

Hennig W. 1966. *Phylogenetic Systematics*. Urbana: University of Illinois Press.

Hein J. 1989. A new method that simultaneously aligns and reconstructs ancestral sequences for any number of homologous sequence, when the phylogeny is given. *Molecular Biology and Evolution* 6: 649–668.

Hillis DM, Wiens JJ. 2000. Molecules versus morphology in systematics: conflicts, artifacts, and misconceptions. In *Phylogenetic Analysis of Morphological Data*, Wiens JJ (ed.). Washington DC: Smithsonian Instution; 1–19.

Houde PW. 1987. Critical evaluation of DNA hybridization studies in avian systematics. *The Auk* 104: 17–32.

Huelsenbeck JP, Bull JJ. 1996. A likelihood ratio test for detection of conflicting phylogenetic signal. *Systematic Biology* 45: 92–98.

Huelsenbeck JP, Lander KM. 2000. Frequent inconsistency of parsimony under a simple model of cladogenesis. *Systematic Biology* 52: 641–648.

Hull DL. 1970. Contemporary systematic philosophies. *Annual Reviews of Ecology and Systematics* 1: 19–54.

Ingilis WG. 1966. The observational basis of homology. *Systematic Zoology* 15: 219–228.

Jenner RA. 2004a. Accepting partnership by submission? Morphological phylogenetics in a molecular millenium. *Systematic Biology* 53: 333–342.

Jenner RA. 2004b. When molecules and morphology clash: reconciling conflicting phylogenies of the Metazoa by considering secondary character loss. *Evolution and Development* 6: 372–378.

Johnson NL, Kemp AW, Katz S. 2005. *Univariate Discrete Distributions*, 3rd edn. Hoboken: John Wiley & Sons.

Källersjö M, Albert VA, Farris JS. 1999. Homoplasy increases phylogenetic structure. *Cladistics* 15: 91–93.

Kelchner SA, Thomas MA. 2006. Model use in phylogenetics: nine key questions. *Trends in Ecology and Evolution* 22: 87–94.

Kemp TS. 1999. *Fossils and Evolution*. Oxford: Oxford University Press.

Kennedy M, Spencer HG. 2004. Phylogenies of the frigatebirds (Fregatidae) and tropicbirds (Phaethontidae), two divergent groups of the traditional order Pelecaniformes, inferred from mitochondrial DNA sequences. *Molecular Phylogenetics and Evolution* 31: 31–38.

Kennedy M, Gray RD, Spencer HG. 2000. The phylogenetic relationships of the shags and cormorants: can sequence data resolve a disagreement between behavior and morphology? *Molecular Phylogenetics and Evolution* 17: 345–359.

Kennedy M, Holland BR, Gray RD, Spencer H. G. 2005. Untangling long branches: identifying conflicting phylogenetic signals using spectral analysis, neighbor-net, and consensus networks. *Systematic Biology* 54: 620–633.

Kim J. 1996. General inconsistency conditions for maximum parsimony: effects of branch lengths and increasing numbers of taxa. *Systematic Biology* 45: 363–374.

Kim J. 1998a. Large-scale phylogenies and measuring the performance of phylogenetic estimators. *Systematic Biology* 47: 43–60.

Kim J. 1998b. What do we know about the performance of estimators for large phylogenies? *Trends in Ecology and Evolution* 13: 2526.

Kim J, Kim M. 2001. The mathematical structure of characters and modularity. In *The Character Concept in Evolutionary Biology*, Wagner GP (ed.). San Diego: Academic Press; 215–236.

Klingenberg CP. 1998. Heterochrony and allometry: the analysis of evolutionary change in ontogeny. *Biological Reviews of the Cambridge Philosophical Society* 73: 79–123.

Kluge AG. 1997a. Testability and the refutation and corroboration of cladistic hypotheses. *Cladistics* 13: 81–96.

Kluge AG. 1997b. Sophisticated falsification and research cycles: consequences for differential character weighting in phylogenetic systematics. *Zoologica Scripta* 26: 349–360.

Kluge AG. 2001a. Parsimony with and without scientific justification. *Cladistics* 17: 199–210.

Kluge AG. 2001b. Philosophical conjectures and their refutation. *Systematic Biology* 50: 322–330.

Kluge AG. 2004. On total evidence: for the record. *Cladistics* 20: 205–207.

Kluge AG. 2005. What is the rationale for "Ockam's razor" (a.k.a. parsimony) in phylogenetic inference?

In *Parsimony, Phylogeny and Genomics*, Albert VA (ed.). Oxford: Oxford University Press; 15–42.

Kluge AG. 2007. Completing the neo-Darwinian synthesis with an event criterion. *Cladistics* **23**: 613–633.

Konishi KP, Anderson DR. 2002. *Model Selection and Multimodel Inference: A Practical Information-Theoretic Approach*, 2nd edn. New York: Springer-Verlag.

Kraus F. 1988. An empirical evaluation of the use of the ontogeny polarization criterion in phylogenetic inference. *Systematic Zoology* **37**: 106–141.

Krumlauf R. 1994. *Hox* genes in vertebrate development. *Cell* **78**: 191–201.

Ksepka DT, Bertelli S, Giannini NP. 2006. The phylogeny of the living and fossil Sphenisciformes (penguins). *Cladistics* **22**: 412–441.

Lake JA. 1997. Phylogenetic inference: how much evolutionary history is knowable? *Molecular Biology and Evolution* **14**: 213–219.

Lanyon SM. 1992. Review: Phylogeny and classification of birds: a study in molecular evolution. *Condor* **94**: 304–307.

Lecointre G, Philippe H, van Le HL, Guyader H. 1994. How many nucleotides are required to resolve a phylogenetic problem? *Molecular Phylogenetics and Evolution* **3**: 292–309.

Lee MSY. 1997. Molecules, morphology, and phylogeny: a response to Hedges and Maxson. *Molecular Phylogenetics and Evolution* **7**: 394–395.

Leman A. 1965. On rates of evolution and unit characters and character complexes. *Evolution* **19**: 16–25.

Ligon JD. 1967. Relationships of the cathartid vultures. *University of Michigan Museum of Zoology Occasional Papers* **651**: 1–26.

Livezey BC. 1986. A phylogenetic analysis of Recent anseriform genera using morphological characters. *The Auk* **105**: 681–698.

Livezey BC. 1991. A phylogenetic analysis and classification of Recent dabbling ducks (Tribe Anatini) based on comparative morphology. *The Auk* **108**: 471–508.

Livezey BC. 1995a. Phylogeny and evolutionary ecology of modern seaducks (Anatidae: Mergini). *Condor* **97**: 233–255.

Livezey BC. 1995b. Heterochrony and the evolution of avian flightlessness. In *Evolutionary Change and Heterochrony*, McNamara KJ (ed.). Chichester: John Wiley & Sons; 169–193.

Livezey BC. 1996. A phylogenetic analysis of the geese and swans (Anseriformes: Anserinae), including selected fossil species. *Systematic Biology* **45**: 415–450.

Livezey BC. 1997. A phylogenetic analysis of basal Anseriformes, the fossil *Presbyornis*, and the interordinal relationships of waterfowl. *Zoological Journal of the Linnean Society* **121**: 361–428.

Livezey BC. 1998. A phylogenetic analysis of the Gruiformes (Aves) based on morphological characters, with an emphasis on the rails (Rallidae). *Proceedings of the Royal Society of London Series B (Biological Sciences)* **353**: 2077–2151.

Livezey BC. 2003. Evolution of flightlessness in rails (Gruiformes: Rallidae): phylogenetic, ecomorphological, and ontogenetic perspectives. *Ornithological Monographs* **53**: 1–654.

Livezey BC, Zusi RL. 2001. Higher-order phylogenetics of modern Aves based on comparative anatomy. *Netherlands Journal of Zoology* **51**: 179–206.

Livezey BC, Zusi RL. 2006. Higher-order phylogeny of modern birds (Theropoda, Aves: Neornithes) based on comparative anatomy: I. – Methods and characters. *Bulletin of the Carnegie Museum of Natural History* **37**: 1–556.

Livezey BC, Zusi RL. 2007. Higher-order phylogeny of modern birds (Theropoda, Aves: Neornithes) based on comparative anatomy: II. – Analysis and discussion. *Zoological Journal of the Linnean Society* **149**: 1–95.

Maddison WP. 1993. Missing data versus missing characters in phylogenetic analysis. *Systematic Biology* **42**: 576–581.

Maddison WP, Maddison DR. 1992. MAC*CLADE*©: *Analysis of Phylogeny and Character Evolution, Version 3*. Sunderland: Sinauer.

Mayo O. 1983. *Natural Selection and its Constraints*. London: Academic Press.

Mayr G. 2002. Osteological evidence for paraphyly of the avian order Caprimulgiformes (nightjars and allies). *Journal of Ornithology* **143**: 82–97.

Mayr G. 2003. The phylogenetic affinities of the shoebill (*Balaeniceps rex*). *Journal of Ornithology* **144**: 507–512.

Mayr G. 2004. Morphological evidence for a sister-group relationship between flamingos (Aves: Phoenicopteridae) and grebes (Podicipedidae). *Zoological Journal of the Linnean Society* **140**: 157–169.

Mayr G. 2007. The renaissance of avian paleontology and its bearing on the higher-level phylogeny of birds. *Journal of Ornithology* **148**: 455–458.

Mayr G, Clarke JA. 2003. The deep divergences of neornithine birds: a phylogenetic analysis of morphological characters. *Cladistics* **19**: 527–553.

McCracken KG, Harshman J, McClellan DA, Afton AD. 1999. Data set incongruence and correlated character evolution: an example of functional convergence in the hind-limbs of stifftail diving ducks. *Systematic Biology* 48: 683–714.

Mindell DP. 1992. Review: Phylogeny and classification of birds: a study in molecular evolution. *Systematic Biology* 41: 126–134.

Minelli A. 1998. Molecules, developmental modules, and phenotypes: a combinatorial approach to homology. *Molecular Phylogenetics and Evolution* 9: 340–347.

Moens CB, Selleri L. 2006. *Hox* cofactors in vertebrate development. *Developmental Biology* 291: 193–206.

Monroe BL. 1989. Review: Response to E. Mayr. *The Auk* 106: 515–516.

Monroe BL Jr, Sibley CG. 1993. *A World Checklist of Birds.* Yale University Press.

Nye TMW. 2008. Trees of trees: an approach to comparing multiple alternative phylogenies. *Systematic Biology* 57: 785–794.

Ogden TH, Rosenberg MS. 2007. How should gaps be treated in parsimony? A comparison of approaches using simulation. *Molecular Phylogenetics and Evolution* 42: 817–826.

O'Keefe FR, Sander PM. 1999. Paleontological paradigms and inferences of phylogenetic pattern: a case study. *Paleobiology* 25: 518–533.

Olson SL, Feduccia A. 1980. *Presbyornis* and the origin of the Anseriformes (Aves: Charadriomorphae). *Smithsonian Contributions to Zoology* 323: 1–24.

Omland KE. 1994. Character convergence between a molecular and a morphological phylogeny for dabbling ducks. *Systematic Biology* 43: 369–386.

Omland KE. 1997. Correlated rates of molecular and morphological evolution. *Evolution* 51: 1381–1393.

Oster GF, Shubin N, Murray JD, Alberch P. 1988. Evolution and morphogenetic rules: the shape of the vertebrate limb in ontogeny and phylogeny. *Evolution* 42: 862–884.

Padian K, Chiappe LM. 1998. The origin and early evolution of birds. *Biological Reviews of the Cambridge Philosophical Society* 73: 1–42.

Patterson C. 1987. *Molecules and Morphology in Evolution: Conflict or Compromise?* Cambridge: Cambridge University Press.

Patterson C, Williams DM, Humphries CJ. 1993. Congruence between molecular and morphological phylogenies. *Annual Reviews of Ecology and Systematics* 24: 153–188.

Penny D, Hendy MD. 1986. Estimating the reliability of evolutionary trees. *Molecular Biology and Evolution* 3: 403–417.

Penny D, Lockhart PJ, Steel MA, Hendy MD. 1994. The role of models in reconstructing evolutionary trees. In *Models in Phylogeny Reconstruction*, Scotland RW, Siebert DJ, Williams DM (eds). Oxford: Oxford University Press; 211–230.

Pisani D, Benton MJ, Wilkinson M. 2007. Congruence of morphological and molecular phylogenies. *Acta Biotheoretica* 55: 269–281.

Poe S, Wiens JJ. 2000. Character selection and the methodology of morphological phylogenetics. In *Phylogenetic Analysis of Morphological Data*, Wiens JJ (ed.). Washington, DC: Smithsonian Institution; 20–36.

Popper KR. 1963. *Conjectures and Refutations: The Growth of Scientific Knowledge.* New York: Harper and Row.

Popper KR. 1968. *The Logic of Scientific Discovery.* New York. Harper and Row.

Popper KR. 1972. *Objective Knowledge: An Evolutionary Approach.* Oxford: Clarendon Press.

Raikow RJ. 1985. Problems in avian classification. *Current Ornithology* 2: 187–212.

Ramírez MJ. 2007. Homology as a parsimony problem: a dynamic homology approach for morphological data. *Cladistics* 23: 588–612.

Raup DM, Gould SJ. 1974. Stochastic simulation and evolution of morphology – towards a nomothetic paleontology. *Systematic Zoology* 23: 305–322.

Ray DA, Xing J, Salem A-H, Batzer MA. 2006. SINEs of a nearly perfect character. *Systematic Biology* 55: 928–935.

Reid RGB. 2004. Epigenetics and environment: the historical matrix of Matsuda's panenvironmentalism. In *Environment, Development, and Evolution: Toward a Synthesis*, Hall BK, Pearson RD, Müller GB (eds). Cambridge, MA: MIT Press; 7–36.

Remane A. 1954. Morphologie als Homologienforschung. *Verhandlungen der Deutschen zoologischen Gesellschaft* 1954: 159–183.

Revell LJ, Harmon LJ, Collar DC. 2008. Phylogenetic signal, evolutionary process, and rate. *Systematic Biology* 57: 591–601.

Ronquist F. 2004. Bayesian inference of character evolution. *Trends in Ecology and Evolution* 19: 475–481.

Ruse M. 1979. Falsifiability, consilience, and systematics. *Systematic Zoology* 28: 530–536.

Rzhetsky A, Nei M. 1993. Theoretical foundation of the minimum-evolution method of phylogenetic inference. *Molecular Biology and Evolution* 10: 1073–1095.

Sanderson MJ, Donoghue MJ. 1989. Patterns of variation in levels of homoplasy. *Evolution* 44: 1673–1684.

Sanderson MJ, Donoghue MJ. 1996. The relationship between homoplasy and confidence in a phylogenetic tree. In *Homoplasy: The Recurrence of Similarity in Evolution*, Sanderson MJ, Hufford L (eds). San Diego: Academic Press; 67–89.

Sangster G. 2005. A name for the flamingo–grebe clade. *Ibis*. **147**: 612–615.

Sarkar S, Fuller T. 2003. Generalized norms of reaction for ecological developmental biology. *Evolution and Development* **5**: 106–115.

Scotland RW, Olmstead RG, Bennett JR. 2003. Phylogeny reconstruction: the role of morphology. *Systematic Biology* **52**: 539–548.

Sereno PC. 2007. Logical basis for morphological characters in phylogenetics. *Cladistics* **23**: 565–587.

Severini TA. 2000. *Likelihood Methods in Statistics*. Oxford: Oxford University Press.

Shpak M, Churchill GA. 2000. The information content of a character under a Markov model of evolution. *Molecular Phylogenetics and Evolution* **17**: 231–243.

Sibley CG. 1957. The evolutionary and taxonomic significance of sexual dimorphism and hybridization in birds. *Condor* **59**: 166–191.

Sibley CG, Ahlquist JE. 1970. A comparative study of the egg white proteins of passerine birds. *Peabody Museum of Natural History Bulletin* **32**: 1–131.

Sibley CG, Ahlquist JE. 1972. A comparative study of the egg white proteins of non-passerine birds. *Peabody Museum of Natural History Bulletin* **39**: 1–276.

Sibley CG, Ahlquist JE. 1990. *Phylogeny and Classification of Birds: A Study in Molecular Evolution*. New Haven: Yale University Press.

Siddall ME, Jensen K. 2003. Incorrect evaluation of the information content of multistate characters. *Cladistics* **19**: 269–272.

Siddall ME, Whiting M. 1999. Long-branch abstractions. *Cladistics* **15**: 9–24.

Siegel-Causey D. 1988. Phylogeny of the Phalacrocoracidae. *Condor* **90**: 885–905.

Siegel-Causey D. 1997. Phylogeny of the Pelecaniformes: molecular systematics of a primitive group. In *Avian Molecular Evolution and Systematics*, Mindell DP (ed.). San Diego: Academic Press; 159–172.

Simmons MP. 2004. Independence of alignment and tree search. *Molecular Phylogenetics and Evolution* **31**: 874–879.

Simmons MP, Ochoterena H. 2000. Gaps as characters in sequence-based phylogenetic analyses. *Systematic Biology* **49**: 369–381.

Simmons MP, Pickett KM, Miya M. 2004a. How meaningful are Bayesian support values? *Molecular Phylogenetics and Evolution* **21**: 188–199.

Simmons MP, Reeves A, Davis JI. 2004b. Character-state space versus rate of evolution in phylogenetic inference. *Cladistics* **20**: 191–204.

Slack KE, Delsuc F, Mclenachan PA, Arnason U, Penny D. 2007. Resolving the root of the avian mitogenomic tree by breaking up long branches. *Molecular Phylogenetics and Evolution* **42**: 1–13.

Sober E. 1983a. Parsimony in systematics: philosophical issues. *Annual Reviews of Ecology and Systematics* **14**: 335–357.

Sober E. 1983b. Parsimony methods in systematics. In *Advances in Cladistics, Volume 2*, Platnick NI, Funk VA (eds). New York: Columbia University Press; 37–47.

Sober E. 1985. A likelihood justification for parsimony. *Cladistics* **1**: 209–233.

Sober E. 1987. Likelihood and convergence. *Philosophical Science* **55**: 228–237.

Sober E. 1988. *Reconstructing the Past: Parsimony, Evolution and Inference*. Cambridge, MA: MIT Press.

Sober E. 1996. Parsimony and predictive equivalence. *Erkenntnis* **44**: 167–197.

Sober E. 2002. Reconstructing ancestral character states – a likelihood perspective on cladistic parsimony. *The Monist* **85**: 156–176.

Sober E. 2004. The contest between likelihood and parsimony. *Systematic Biology* **53**: 644–653.

Sober E. 2005. Parsimony and its presuppositions. In *Parsimony, Phylogeny and Genomics*, Albert VA (ed.). Oxford: Oxford University Press; 43–53.

Sommer RJ. 1999. Convergence and the interplay of evolution and development. *Evolution and Development* **1**: 8–10.

Stanley SM. 1979. *Macroevolution, Pattern and Process*. W. H. Freeman: San Francisco.

Steel MA, Penny D. 2000. Parsimony, likelihood, and the role of models in molecular phylogenetics. *Molecular Biology and Evolution* **17**: 839–850.

Steel MA, Penny D. 2004a. Maximum parsimony and the phylogenetic information in multistate characters. In *Parsimony, Phylogeny and Genomics*, Albert VA (ed.). Oxford: Oxford University Press; 163–180.

Steel MA, Penny D. 2004b. Two links between maximum parsimony and maximum likelihood under the Poisson model. *Applied Mathematics Letters* **17**: 785–790.

Steel MA, Penny D. 2005. Maximum parsimony and the phylogenetic information in multistate characters. In

Parsimony, Phylogeny and Genomics, Albert VA (ed.). Oxford: Oxford University Press; 163–178.

Steel MA, Hendy MD, Penny D. 1993. Parsimony can be consistent! Systematic Zoology **42**: 581–587.

Steppan SJ, Phillips PC, Houle D. 2002. Comparative quantitative genetics: evolution of the G matrix. *Trends in Ecology and Evolution* **17**: 320–327.

Storer R. 2006. The grebe–flamingo connection: a rebuttal. *The Auk* **123**: 1183–1184.

Telford MJ. 2000. Turning *Hox* "signatures" into synapomorphies. *Evolution and Development* **2**: 360–364.

Thomas GH, Wills MA, Székely T. 2004. A supertree approach to shorebird phylogeny. *BMC Evolutionary Biology* **4**: 1–18.

Van Tuinen M. 2005. Relationships of birds – molecules versus morphology. In *Electronic Encyclopedia of Life Sciences*. Chichester: John Wiley & Sons. doi: xlink: href="10.1038/npg.els.0004163.

Van Tuinen M, Futvill DB, Kirsch JAW, Hedges SB. 2001. Convergence and divergence in the evolution of aquatic birds. *Proceedings of the Royal Society of London, Series B (Biological Sciences)* **268**: 1–6.

Van Tienderen PH, Koelewijn HP. 1994. Selection on reaction norms, genetic correlations and constraints. *Genetics Research* **64**: 115–125.

Von Dassow G, Munro E. 1999. Modularity in animal development and evolution: elements of a conceptual framework for evo-devo. *Journal of Experimental Biology, Series B* **285**: 307–325.

Wade MJ. 2000. Epistasis as a genetic constraint within populations and an accelerant of adaptive divergence among them. In *Epistasis and the Evolutionary Process*, Wolf JB, BrodieIIIED, Wade MJ (eds). Oxford: Oxford University Press; 213–231.

Wägele JW. 1995. On the information content of characters in comparative morphology and molecular systematics. *Journal of Zological Systematics and Evolutionary Research* **33**: 42–47.

Wägele JW, Rödding F. 1998. *A priori* estimation of phylogenetic information conserved in aligned sequences. *Molecular Phylogenetics and Evolution* **9**: 358–365.

Wagner GP. 1984. Coevolution of functionally constrained characters: prerequisites of adaptive versatility. *BioSystems* **17**: 51–55.

Wagner GP. 1989. The origin of morphological characters and the biological basis of homology. *Evolution* **43**: 1157–1171.

Wagner GP. 2000. Evolutionarily stable configurations: functional integration and the evolution of phenotypic stability. *Evolutionary Biology* **31**: 155–217.

Wheeler WC. 2003a. Iterative pass optimization of sequence data. *Cladistics* **19**: 254–260.

Wheeler WC. 2003b. Implied alignment: a synapomorphy-based multiple-sequence alignment method and its use in cladogram search. *Cladistics* **19**: 261–268.

Wiens JJ. 1998. Does adding characters with missing data increase or decrease phylogenetic accuracy? *Systematic Biology* **47**: 625–640.

Wiens JJ. 2001. Character analysis in morphological phylogenetics: problems and solutions. *Systematic Biology* **50**: 689–699.

Wiley EO. 1981. *Phylogenetics: The Theory and Practice of Phylogenetic Systematics*. New York: John Wiley & Sons.

Wilkinson M. 1991. Homoplasy and parsimony analysis. *Systematic Biology* **40**: 105–109.

Wilson SJ. 1999. A higher-order parsimony method to reduce long-branch attraction. *Molecular Biology and Evolution* **16**: 694–705.

Wray GA. 1999. Evolutionary dissociations between homologous genes and homologous structures. In *Homology*, Bock GR, Cardew G (eds). Chichester: John Wiley & Sons; 189–203.

Yang ZH. 1997. How often do wrong models produce better phylogenies? *Molecular Biology and Evolution* **14**: 105–108.

Ziman J. 1978. *Reliable Knowledge: An Explanation of the Grounds for Belief in Science*. Cambridge: Cambridge University Press.

5 The Utility of Fossil Taxa and the Evolution of Modern Birds: Commentary and Analysis

GARETH DYKE[1] AND EOIN GARDINER[2]

[1]University College Dublin, Dublin, Ireland
[2]University of Bristol, Bristol, UK

As discussed in other chapters, the tempo and mode of the evolutionary radiation of modern birds (Neornithes) remains debated. On the one hand, estimates for lineage divergences that have been based on the interpretation and modeling of molecular sequence data strongly suggest a deep Cretaceous origination for the bulk of Neornithes. On the other, the fossil record of modern birds remains dominated by specimens from the Cenozoic, especially from the Eocene (e.g. Mayr, 2009). In fact for decades the vast majority of the recorded earliest occurrences for the modern clades (irrespective of whether one refers to groupings as "stem-groups," "crown-group," "orders," or "families;" see below) have come from rocks of primarily Early and Middle Eocene age with just a handful of European (e.g. London Clay, Messel, Quercy) and North American (e.g. Green River, Willwood) formations accounting for the bulk of diversity. Extensive compendia of early fossil occurrences, such as the comprehensive work of Mayr (2009), only serve to confirm what avian paleontologists have known since the 1940s: lots of fossil taxa demonstrably referable to modern lineages are known from the Eocene and Oligocene, yet are largely absent from the Late Cretaceous record (Feduccia, 1999).

Most famously, this clear numerical pattern in the fossil record of Neornithes led Feduccia (1995) to propose his "explosive" model for the modern avian evolutionary radiation (Lindow, Chapter 14, this volume). In this view modern birds diversified rapidly in the aftermath of the end-Cretaceous extinction event that marks the boundary between this period and that of the Tertiary (now referred to as the Paleogene by geologists; the boundary, then, is now referred to as the K–Pg). Paleogene birds – and mammals too – according to Feduccia (1995), took over the ecological niches left unoccupied by the demise of their earlier-diverging counterparts at the K–Pg.

This "rapid radiation model" (Lindow, Chapter 14, this volume) for the diversification of modern birds can, by its nature, also explain the apparent paucity of fossil records from the Cretaceous, which indeed has stood up to the natural test provided by continued "fossil collectorship" since the 1990s (Fountaine et al., 2005). In other words, if anatomically modern birds were present in abundance in the Cretaceous then one would expect that paleontologists would have found their fossils by now; after all, other vertebrates of similar body sizes – mammals, lizards, and amphibians – are well-known from the Cretaceous. This argument, that the vertebrate fossil record accurately represents the broad pattern of evolutionary events has been championed forcibly by Benton and co-workers (e.g. Benton, 1999).

More than nagging doubts over this view of the modern bird evolutionary radiation remain, however. Why do molecular estimates for divergences continue to demand significant range extensions into the Cretaceous, in many cases to ages tens of millions before the oldest known fossils (e.g. Cooper & Penny, 1997; Paton *et al.*, 2002; Hackett *et al.*, 2008)? Our understanding of the neornithine radiation is different to the prevailing world-view when it comes to mammals; early proposals for the bird radiation, including that of Feduccia (1995), have not been corroborated by continued fossil discoveries. In contrast, divergences of major "modern" mammalian lineages are now known to *have* been deep in the Cretaceous as more fossils have been discovered and described.

The fossil record, as currently known, has also been brushed under the carpet by some; perhaps neornithine birds were present in the Cretaceous but we have not found them yet because they were somehow cryptic (and thus hard to determine) in morphology (Cooper & Fortey, 1998), or lived in areas of the world (like the Southern Hemisphere) that have not been explored to any great extent by paleontologists (Cracraft, 2001). We classify these arguments, forcing a correspondence between "fossils" and "molecules," as simply "bullet dodging." As with evolutionary studies dealing with the other major groups of vertebrates, one has no choice but to take the fossil record at face value, deal with its inherent problems as efficiently as possible, and analyze it quantitatively to extract patterns. Feduccia (1995) did not discuss any quantitative trends inherent to the modern avian fossil record, he merely noted (quite fairly as it turns out) that we have far more identifiable fossils from the Paleogene than we do from the Cretaceous; this is still the case (Mayr, 2009).

In this chapter, we move on from the work of Feduccia (1995) taking as our starting point the fact that the fossil record is our only direct source of information about the history of life. We reiterate a number of clear issues that have, and continue to plague, the fossil record of birds and ask: what can the record tell us about the shape of neornithine evolution? We end by making, in agreement with Livezey (Chapter 4, this volume)

and Lindow (Chapter 14, this volume), a number of pleas for future students of fossil birds. For more than 50 years we have actually made little real progress in our understanding of this important evolutionary radiation in the history of vertebrates.

THE BIG PROBLEM WITH THE MODERN BIRD FOSSIL RECORD

Although no-one would argue that the fossils of modern birds are not abundant (Mayr, 2009), interpreting these records has always been problematic almost completely because of the lack of a clear phylogenetic topology of living birds based on anatomy. Thus throughout the history of avian paleontology, when faced with a bird fossil from the Paleogene, workers have largely resorted to making direct comparisons with skeletal collections of living birds. This works very well, and with increasing accuracy (as would be expected), as the fossils being dealt with get younger in age; one can make direct comparisons between Miocene-aged (ca. 15 million years old [Ma]) fossils and living skeletons and identify groups and clades with accuracy. However, obviously, the deeper in time you go, the less accurate this approach will be (Livezey, Chapter 4, this volume). If one has no idea of the characters, the synapomorphies, that characterize neornithine lineages, then the "comparative approach" to dealing with the fossil record is doomed to failure. We argue that the best we can hope for, when addressing the earliest modern avian fossils, is a very broad brush approach to taxonomy, identifying representatives of major lineages in the fossil record.

Over the past 15 years we have seen almost the opposite effect as a characteristic of the fossil record of non-modern birds from the Mesozoic. Until relatively recently very few bird fossils from the whole of this huge time period were known – just *Archaeopteryx* from the Jurassic of Germany (140 Ma) and an array of toothed, marine taxa (e.g. *Hesperornis*, *Ichthyornis*) from the Late Cretaceous (90 Ma). As has been widely discussed, an explosion of fossil discoveries over the past two

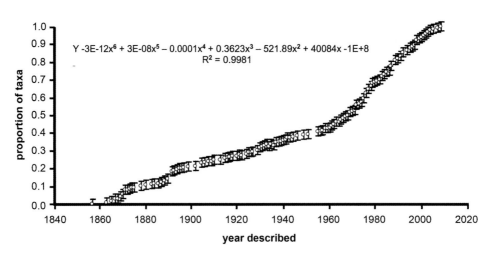

Fig. 5.1   Collectorship of birds from the Paleocene and Eocene (Cenozoic, our sample). Data pooled from subdivided analyses (see text for details).

decades has completely altered this picture; we now know about a range of lineages of "non-modern" taxa that diverged earlier than Neornithes (Chiappe, 2007) and in some cases (Enantiornithes) were likely just as diverse (Chiappe, 2007). However, because little was known about the diagnostic anatomical characteristics of successive clades of birds in the Mesozoic as recently as the mid-1980s, our understanding of taxonomic diversity has grown hand-in-hand with our understanding of their phylogenetics, based on osteology. This of course means that now when new fossil Mesozoic bird fossils are discovered one can assess their relationships with relative ease, at least compared to fossil Neornithes.

A number of commentators (including Livezey, Chapter 4, this volume) have remarked over the past few years that even though the history of ornithological classification dates back more than a century and a half, we still lack clear consensus on the interrelationships of the major clades of living birds. Indeed, many of the 23 or so traditionally recognized "orders" (depending on source) are of questionable monophyly; no clear, phylogenetically tested, synapomorphies have been documented in a number of cases. The phrase "phylogenetically tested" is the key one in this context; although some have argued to the con-

trary (Mayr, 2009), characters they propose, and use, for the classification of fossil taxa have often never been tested by global analysis (Livezey, Chapter 4, this volume).

Nevertheless, this potential drawback has never deterred students of avian paleontology who have merrily carried on naming and describing hundreds of taxa of fossil birds from the earliest Paleogene with little noticeable slow-down since the 1960s (Figure 5.1), indeed ideas about phylogenetic systematics based on demonstrably shared-derived characters did not fully solidify until the 1980s (Schuh, 2000). Nevertheless, its been written before and we write it again, the situation remains dire – large numbers of new taxa of Paleocene and Eocene birds have been named and described, yet their placement within specific lineages of Neornithes has yet to be tested via phylogenetic analysis.

WHAT CAN BE SAID BASED ON THE FOSSIL RECORD?

If we do not know precise phylogenetic placements for the bulk of the described neornithine fossil record then how can we use them to address the shape of the divergence?

Since a comprehensive review of the avian fossil record remains a goal (Mayr, 2009), but is beyond the scope of this chapter, analysis of a compendium of records offers one, meaningful approach. It is at least possible for us to ask the question: what does the published fossil record for modern birds suggest about the shape of their evolutionary radiation. However, this kind of data-based approach must be done in a statistically meaningful way; simply quoting the published, putative first occurrences for groups does not provide a measure of confidence. For example, the first described swift (Apodiformes) could very well be from the Paleocene, but is this an isolated fossil, tens-of-millions of years older than its next youngest published counterpart? Fossil records described far out of age-context from their supposed group of membership have never inspired confidence in their likely accuracy.

Bleiweiss (1998) was the first to go further than simply listing the fossils of modern birds, by quantitatively analyzing a relatively large and (at the time) comprehensive set of data for three clades. By collating published fossil records for Apodiformes (swifts and hummingbirds), Strigiformes (owls) and Caprimulgiformes (goatsuckers), arranging them according to their age and performing a gap analysis, Bleiweiss (1998) argued that the divergence of these three clades occurred in the early Tertiary. In other words, these three lineages of Neornithes diversified in the aftermath of the K–Pg extinction, in agreement with the hypothesis of Feduccia (1995). In 1998, as now, relatively little was known for sure about avian higher-level relationships and choice of these three taxa was deliberate; i.e. lineages of living birds that are comprised within the so-called "higher land bird" assemblage, certainly towards the crown of the neornithine phylogeny. Owls, as it turns out, were at one point described – incorrectly – from rocks of Late Cretaceous age (Harrison & Walker, 1977); the other two living groups have never been recorded from sediments older than Paleocene (Mayr, 2009).

Bleiweiss's (1998) approach is commendable because by taking a statistical view of the modern avian fossil record, one can determine the

likelihood that known ranges extend older than the 65 Ma marker, the end of the Cretaceous. Bleiweiss (1998) did this at two signficance levels, 99% and 95%, and reported no support for pre-Pg divergence based on fossil stratigraphic distributions. In 2009, we updated and expanded this approach, building a database that comprises more than 1000 fossil records encompassing seven clades of living birds. Clades were chosen (i.e. Anseriformes, Apodiformes, Pelecaniformes, Procellariiformes, Piciformes, and Strigiformes) to: (i) span the entirety of the neornithine phylogeny (as currently understood), from early diverging (Anseriformes) to crown-ward (Piciformes); (ii) to have an abundant fossil record in terms of numbers of described specimens; and (iii) to be of relatively certain monophyly. Of course, all three of these starting points represent significant assumptions (as discussed above) that will feed into any analysis.

In any case, few workers would question the current hypothesis that clade Anseriformes likely diverged significantly earlier in time than did Piciformes: fossils placed in the former lineage have been described from the Late Cretaceous (Clarke *et al.*, 2005) and, along with Galliformes, Anseriformes is most often considered to be a basal divergence within Neornithes (Livezey, Chapter 4, this volume). Piciformes, on the other hand, are often grouped together with the perching songbirds (Passeriformes) at the very tip of the neornithine tree. One would expect such a difference in phylogenetic position to be reflected in the shape of their fossil record, but this has never been tested before using a large database of occurrences.

Our results (Figures 5.1 and 5.2) show that collectorship for birds from the Paleocene and Eocene (Cenozoic, our sample) increases constantly over time and appears to reach a plateau after the year 2000. Overall data suggest a relatively constant increase over time, with a slightly sharper increase between 1960 and 1990. When compared with collectorship curves for Mesozoic taxa for both the complete Mesozoic fossil record (Fountaine *et al.*, 2005) and just enantiornithine taxa (O'Connor & Dyke, 2010), it is clear that

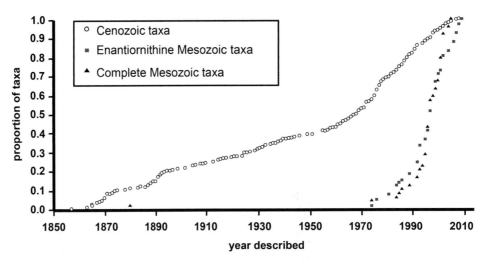

Fig. 5.2 Comparison of Cenozoic collectorship curve with Mesozoic collectorship curves of complete Mesozoic taxa and enantiornithine Mesozoic taxa (see text for data sources).

while Cenozoic collectorship has been relatively constant throughout time and has now reached a plateau, Mesozoic collectorship is in the middle of a sharp increase and has been increasing sharply since 1980. Based on this simple analysis, then, our sample of the Cenozoic avian fossil record appears more complete than does its Mesozoic counterpart. There is also no bias evident in this data sample, between year of description and age of fossils: it seems that paleornithologists working in the Cenozoic have described fossils of all ages

(Figure 5.3). This adds further evidence to the argument that patterns can be extracted from the known fossil record.

Results of the gap analysis can be seen in Table 5.1 and Figure 5.4. The known fossil record for both the Anseriformes (Hope, 2002; Clarke *et al.*, 2005) and Procellariiformes (Hope, 2002) already extends into the Late Cretaceous (~68 Ma) with the confidence intervals (possible range extensions) extending no further than 74 Ma for the Anseriformes and 76 Ma for the

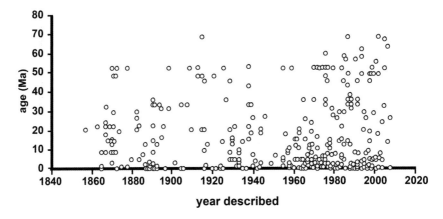

Fig. 5.3 Scatter plot showing no relationship between age of specimen and year of description.

Table 5.1 Results of the gap analysis for each individual order; 95% confidence interval given in million years.

Order	Clade rank	Number of fossil taxa	Maximum age (Ma)	Confidence interval (CI)	Maximum age with CI (Ma)
Anseriformes	2	135	68.2	5.6	73.8
Apodiformes	6	24	55.5	13.2	68.7
Columbiformes	6	17	32.9	13	45.9
Piciformes	6	42	51.9	10	61.9
Pelecaniformes	3	113	57.9	3.7	61.6
Procellariiformes	3	56	68.2	7.7	75.9
Strigiformes	5	93	63.3	5.3	68.6

Procellariiformes. The known fossil records for the Pelecaniformes, Strigiformes, Apodiformes, and Piciformes extend only as far as the early Tertiary (50–65 Ma) with the known fossil record for the Columbiformes extending only as far as the mid-Tertiary (late Oligocene; ~33 Ma). The confidence intervals of the Apodiformes and Strigiformes, however, extend as far as the latest

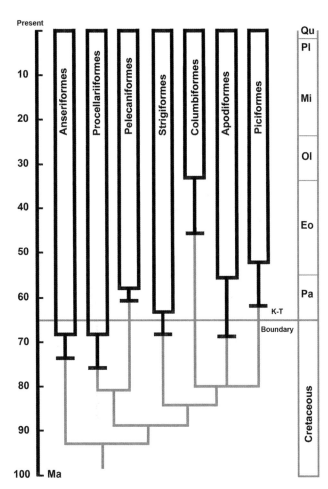

Fig. 5.4 Graphical depiction of simple gap analysis results (see text for discussion).

Cretaceous (up to 69 Ma). The confidence intervals for the Pelecaniformes and Piciformes extend only as far as the earliest Tertiary (early Paleocene; ~62 Ma). The confidence interval for the Columbiformes extends to the mid-Eocene (~46 Ma).

The results of the gap analysis fail to support the hypothesis that all modern bird orders diverged and diversified either before (Cooper & Penny, 1997) or after (Feduccia, 2003) the K–Pg boundary. The confidence intervals for the Columbiformes, Pelecaniformes, and Piciformes are restricted to the Tertiary and do not project past the K–Pg boundary, while the confidence intervals for the Anseriformes, Apodiformes, Procellariiformes, and Strigiformes extend into the Late Cretaceous. In order to provide any support for the hypothesis that all modern bird orders diverged and diversified in the Cretaceous, these confidence intervals would all need to extend past the K–Pg boundary. Likewise, to support the theory that all modern bird orders diverged in the Tertiary, none of the confidence intervals could extend past the K–Pg boundary. The results of the gap analysis also fail to support Feduccia's "Transitional Shorebird Hypothesis" (Feduccia, 2003), i.e. that a single lineage of "transitional shorebirds" originated in the Cretaceous, survived the K–Pg extinction event, and diversified rapidly into the modern avian orders alive today. Feduccia's hypothesis is supported by the fact that fragmentary remains from the Cretaceous are often assigned to "shorebird" or "waterbird" modern orders (Dyke & van Tuinen, 2004) and that such short diversification times have also been seen in the evolution of whales from terrestrial ungulates in less than 10 million years (Feduccia, 2003). However, the extension of the strigiform and apodiform confidence intervals into the Late Cretaceous dispute the "Transitional Shorebird Hypothesis" since these orders are much more derived than those of the "shorebirds" or "waterbirds" (according to the Mayr & Clarke (2003), Livezey & Zusi (2007) and Hackett et al. (2008) phylogenetic trees).

The results of our gap analysis do, however, support the hypothesis that the basal lineages of neornithine birds and a number of basal neoavian lineages diverged in the Cretaceous with the remaining lineages diversifying in the Tertiary (Dyke, 2001). The known fossil records of the Anseriformes (Hope, 2002; Clarke et al., 2005) and Procellariiformes (Hope, 2002) already extend into the Late Cretaceous and, with the confidence intervals included, extend as far as 76 Ma (still in the Late Cretaceous). It is widely accepted that the Anseriformes, as well as the Galliformes (making up the Galloanserae), are the second most basal group of neornithines, after the Paleognathae (Dyke & van Tuinen, 2004). According to the three phylogenetic bird trees used in this study, the Procellariiformes are a basal group among the Neoaves and also relatively basal among neornithine birds as a whole (Mayr & Clarke, 2003; Livezey & Zusi, 2007; Hackett et al., 2008). Despite the fossil records of the Strigiformes and Apodiformes only extending to the early Tertiary, the confidence intervals for these orders extend into the Late Cretaceous. While the morphological trees of Mayr & Clarke (2003) and Livezey & Zusi (2007) place the Apodiformes as a more derived order among the Neoaves, Hackett et al.'s (2008) molecular tree places the Apodiformes as a more basal order among the Neoaves. The extension of the apodiform confidence interval into the Late Cretaceous supports Hackett et al.'s (2008) placement of the Apodiformes as a basal order rather than a derived one. However, the confidence interval for the Apodiformes is the longest in the gap analysis since the Apodiformes had the poorest fossil record out of the seven orders used in this study. If more Tertiary fossil occurrences are found for this order, the confidence interval will shorten and will likely leave the Late Cretaceous and move into the early Tertiary. The confidence interval for the Strigiformes is more reliable since the fossil record for this order was quite comprehensive. Any further Tertiary fossil occurrences are not likely to have a huge affect on this confidence interval. According to Livezey & Zusi (2007) and Hackett et al. (2008), the Strigiformes are in the middle between the most basal and most derived orders of neoavian birds. Mayr & Clarke (2003), however, suggest that the Strigiformes are more basal than even the Procellariiformes. The extension of the strigiform confidence interval into the late Cretaceous supports Mayr & Clarke's (2003) evaluation of the Strigiformes. Regardless of this phylogenetic uncertainty, the results of the gap analysis support the hypothesis that the basal orders of neornithine birds and a number of basal neoavian orders diverged in the Cretaceous (Dyke, 2001).

It has been argued that the "classic confidence intervals" calculated in this study and also calculated by Bleiweiss (1998) underestimate divergence dates (Marshall, 1999). This is because they assume that the probability of finding fossils is constant through time. Marshall suggests that modern birds had a long initial history in the Cretaceous but were rare and had a cryptic diversity, thus resulting in a drastically lower probability of finding fossils from that time. According to Benton (1999), however, this suggestion fails on a probability argument. How could over 20 orders of modern birds have existed undetected through the Cretaceous and all uniformly remained cryptic until the Tertiary? As an example of fossils being found despite their rarity before the major radiation, Benton notes the finding of sharks and bony fishes from the Harding Sandstone, 50–60 million years before abundant and more complete fossils appear. Thus, while the theory that modern birds were cryptic in the Cretaceous is plausible, it is also improbable. Therefore, without clear evidence for the contrary, the appropriate null hypothesis is to assume a constant probability of fossil recovery through time for gap analysis studies on the fossil records of modern birds (Bleiweiss, 1999).

While our collectorship curves show that the fossil records of each order appear complete enough to be used to make inferences about the history of the Neornithes with real confidence, the differing divergence dates given by fossil and molecular evidence is one of the most highly debated topics in evolutionary biology. While this gap analysis study has not bridged the gap between these two lines of evidence, there remains room for improvement on both sides of the debate, e.g. fossil evidence can account for

"ghost lineages" in the fossil record and molecular evidence can develop more realistic models of rate evolution for genetic sequences. The only foreseeable end to the "rocks and clocks" debate would be the discovery of Early Cretaceous neornithine fossils. The single addition of a Cretaceous neornithine bird to the fossil record of any order used in this study could dramatically increase the confidence interval obtained for that order, especially the orders with the least comprehensive fossil records. However, Benton (1999) notes that such fossils "will not be found because they do not exist".

ACKNOWLEDGMENTS

We thank Evgeny Kurochkin for advice on compiling fossil databases. This work was supported by University College Dublin. The fossil databases used in this Chapter are available on request from Gareth Dyke (gareth.dyke@ucd.ie).

REFERENCES

Benton MJ. 1999. Early origins of modern birds and mammals: molecules vs. morphology. *Bioessays* **21**: 1043–1051.

Bleiweiss R. 1998. Fossil gap analysis supports early Tertiary origin of trophically diverse avian orders. *Geology* **26**: 323–326.

Chiappe LM. 2007. *Glorified Dinosaurs: the Origin and Early Evolution of Birds.* Hoboken: John Wiley & Sons.

Clarke JA, Tambussi CP, Noriega JI, Erickson GM, Ketcham RA. 2005. Definitive fossil evidence for the extant avian radiation in the Cretaceous. *Nature* **433**: 305–308.

Cooper A, Penny D. 1997. Mass survival of birds across the KT boundary: molecular evidence. *Science* **275**: 1109–1113.

Cooper A, Fortey R. 1998. Evolutionary explosions and the phylogenetic fuse. *Trends in Ecology and Evolution* **13**: 151–156.

Cracraft J. 2001. Avian evolution, Gondwana biogeography and the Cretaceous-Tertiary mass extinction event. *Proceedings of the Royal Society of London, Series B (Biological Sciences)* **268**: 459–469.

Dyke GJ. 2001. The evolution of birds in the early Tertiary: systematics and patterns of diversification. *Geological Journal* **36**: 305–315.

Dyke GJ, van Tuinen M. 2004. The evolutionary radiation of modern birds Neornithes: reconciling molecules, morphology and the fossil record. *Zoological Journal of the Linnean Society* **141**: 153–177.

Feduccia A. 1995. Explosive evolution in Tertiary birds and mammals. *Science* **267**: 637–638.

Feduccia A. 1999. *The Origin and Evolution of Birds*, 2nd edn. New Haven: Yale University Press.

Feduccia A. 2003. "Big bang" for Tertiary birds? *Trends in Ecology and Evolution*, **18**: 172–176.

Fountaine TMR, Benton MJ, Dyke GJ, Nudds RL. 2005. The quality of the fossil record of Mesozoic birds. *Proceedings of the Royal Society of London, Series B (Biological Sciences)* **272**: 289–294.

Hackett SJ, Kimball RT, Reddy S, Bowie RCK, Braun EL, Braun MJ, Chojnowski JL, Cox WA, Han K-L, Harshman J, Huddleston CJ, Marks BD, Miglia KJ, Moore WS, Sheldon FH, Steadman DW, Witt CC, Yuri T. 2008. A phylogenomic study of birds reveals their evolutionary history. *Science* **320**: 1763–1768.

Harrison CJO, Walker CA. 1975. The Bradycnemidae: a new family of owls from the Upper Cretaceous of Romania. *Palaeontology* **18**: 563–570.

Hope S. 2002. The Mesozoic radiation of Neornithes. In: *Mesozoic Birds: Above the Heads of Dinosaurs*, Chiappe LM, Witmer LM (eds). Berkeley: University of California Press; 339–388.

Livezey BC, Zusi RL. 2007. Higher-order phylogeny of modern birds Theropoda, Aves: Neornithes based on comparative anatomy. II. Analysis and discussion. *Zoological Journal of the Linnean Society* **149**: 1–95.

Marshall CR. 1999. Fossil gap analysis supports early Tertiary origin of trophically diverse avian orders: Comment. *Geology* **27**: 95.

Mayr G. 2009. Paleogene Fossil Birds. Springer, Heidelberg.

Mayr G, Clarke JA. 2003. The deep divergences of neornithine birds: a phylogenetic analysis of morphological characters. *Cladistics* **19**: 527–553.

O'Connor JK, Dyke GJ. 2010. A re-assessment of *Sinornis santensis* and *Cathayornis yandica* (Aves: Enantiornithes). In *Proceedings of the VII International Meeting of the Society of Avian Paleontology and Evolution*, Boles WE, Worthy TH (eds). *Records of the Australian Museum* **62**(1): 7–20.

Paton T, Haddrath O, Baker AJ. 2002. Complete mitochondrial DNA genome sequences show that modern birds are not descended from transitional shorebirds. *Proceedings of the Royal Society of London, Series B (Biological Sciences)* **269**: 839–846.

Schuh RT. 2000. *Biological Systematics: Principles and Applications*. Ithaca: Cornell University Press.

6   Penguins Past, Present, and Future: Trends in the Evolution of the Sphenisciformes

DANIEL T. KSEPKA[1] AND TATSURO ANDO[2]

[1]North Carolina State University, Raleigh, USA
[2]Ashoro Museum of Paleontology, Hokkaido, Japan

Sphenisciformes (penguins) are flightless sea birds widely distributed in the Southern Hemisphere. These birds have completely lost the capacity for aerial flight. Instead, they employ modified flipper-like wings in wing-propelled diving or underwater flight. Penguins are truly marine animals, though as birds they remain tied to land for molting and breeding. The highly specialized morphology and remarkable life histories of penguins have long attracted interest not only from scientists but also from the general public. In the popular imagination, penguins are commonly associated with icy Antarctic environments, however, these birds are by no means restricted to polar waters. Representatives of the group inhabit a diverse array of environments as far north as the Equator, including coastal deserts, sea-ice shelves, barren sub-Antarctic islands and coastal forests. The remarkable breeding cycle of *Aptenodytes forsteri* (emperor penguin) in Antarctica has been both widely popularized and intensely studied (e.g. Stonehouse, 1953; Jouventin *et al.*, 1995; Barbraud & Weimerskirch, 2001). This species and the closely related *Aptenodytes patagonicus* (king penguin) both possess brood pouches that allow them to incubate their eggs and warm their chicks during long periods of subfreezing temperatures. In contrast, *Spheniscus mendiculus* (Galápagos penguin) breeds at the Equator and must protect its eggs from the sun to keep them from overheating (Boersma, 1975).

Penguins have undergone a suite of morphological, physiological, and behavioral modifications in the course of their transition to a primarily underwater existence. These include development of a unique integument characterized by densely packed, scale-like feathers that both insulate and waterproof (Watson, 1883), modifications to the eye lens and shifts in visual sensitivity towards parts of the spectrum that increase the efficiency of prey detection underwater (Sivak 1976; Sivak & Millodot, 1977; Bowmaker & Martin, 1985), stiffening of the wing joints (Raikow *et al.*, 1988), wholesale reduction of the distal wing musculature (Gervais & Alix, 1828; Schoepss, 1829; Schreiweis, 1982), dense bones to counteract buoyancy (Meister, 1962), rete mirabile systems in the head, flipper, and legs for enhanced thermoregulation (Frost *et al.*, 1975; Thomas & Fordyce, 2007), thickened eggshell to reduce the risk of breakage on hard nesting substrates (Boersma *et al.*, 2004), and incubation strategies that include brood pouches and nesting in burrows.

Many of the aquatic specializations exhibited by penguins also increase their fossilization potential. Penguin limb bones have a greatly reduced marrow cavity and are much more robust than those of volant birds. This makes them more likely to withstand pre-burial damage by scavengers and wave action. Penguins frequent nearshore marine habitats, which also increases the likelihood of their bones being buried compared to those of birds inhabiting inland environments. These factors have led to an abundance of fossil penguin material. To date, over 4000 fossil penguin specimens have been deposited in museum collections (see Table 6.1).

Extant penguins occur throughout the Southern Hemisphere, with only the Galápagos penguin (*Spheniscus mendiculus*) ranging slightly north of the Equator. The distribution of fossil penguins closely follows their current range (Figure 6.1). In addition to the abundance of fossil material, numerous subfossil penguin remains have been obtained from localities that include abandoned colonies and middens (McEvey & Vestjens, 1973; van Tets & O'Conner, 1983; Worthy, 1997; Lambert *et al.*, 2002; Emslie & Woehler, 2005; Emslie *et al.*, 2007; Emslie & Patterson, 2007). While subfossil penguin bones and eggshells have been the subject of many interesting studies, Pleistocene materials are not included in Table 6.1 because we lack accurate estimates of the amount of material known.

In this chapter, we attempt to highlight the ways in which our knowledge of extant penguins guides interpretation of the penguin fossil record, and the ways in which fossil penguins expand our understanding of major evolutionary trends in the clade. Due to the sheer volume of work, it is beyond the scope of this chapter to touch on all aspects of extant penguin biology. Excellent collections and summaries of such work are readily available (e.g. Stonehouse, 1975; Davis & Darby, 1990; Williams, 1995; Davis & Renner, 2003).

TAXONOMY

A clearly defined taxonomy is important for facilitating communication between neontologists and paleontologists, and has implications for conservation status of living species. Historically, the total number of species recognized for living penguins has fluctuated. Penguins moult through successive distinct plumages before reaching sexual maturity, and individuals appear markedly different in their natal down, immature plumage, and adult plumage. Early accounts misclassified birds of different age classes as separate species. For example, the downy brown juveniles of the king penguin were originally considered a distinct (and presumably non-aquatic!) taxon, known as the "woolly penguin" (Latham, 1821). Once penguins became better understood, the generally accepted total remained stable at 17 species for many years (Sibley & Monroe, 1990; Martínez, 1992), though as discussed below the true total may be 19 or more.

Molecular studies are currently modifying our understanding of the limits of extant penguin species. Most recently, the taxonomy of the widespread rockhopper penguins (previously recognized as three subspecies of *Eudyptes chrysocome*) was re-evaluated. Banks *et al.* (2006) found that genetic differences support species status for three groups formerly considered subspecies of *Eudyptes chrysocome*: *Eudyptes chrysocome* (southern rockhopper penguin), *Eudyptes moseleyi* (northern rockhopper penguin), and *Eudyptes filholi* (eastern rockhopper penguin). This conclusion is also supported by morphological differences and the allopatric distribution of the three groups (Banks *et al.*, 2006). Also at issue is the status of *Eudyptula minor* (little blue penguin). This taxon was in the past split into two species, *Eudyptula minor* and *Eudyptula albosignata* (Peters, 1931; Gruson, 1976). However, the most recent revision by Kinsky & Falla (1976) recognized just one species, with six subspecies. Unexpected molecular divergence patterns between different geographic populations of *Eudyptula minor* have been detected (Banks *et al.*, 2002). Although these patterns suggest that two distinct lineages are present (Banks *et al.*, 2002), whether these should be treated as species or populations is debatable. In this chapter, we recognize 19 species of extant

Table 6.1 Summary of Fossil Penguin Distribution. For body size: S is smaller than the extant king penguin, M is intermediate in size between the king penguin and emperor penguin, L is larger than the emperor penguin. Because all valid holotypes from Seymour Island are tarsometatarsi, it is difficult to assign additional elements to these species. We therefore consider such elements tentatively referable for purposes of this table (e.g. c.f. *Anthropornis grandis*). We agree with Jadwiszczak (2006b) that the presence of *Archaeospheniscus lopdelli* in that Formation is not supportable (specimens referred to *Archaeospheniscus lopdelli* are placed in Sphenisciformes indet)

AFRICA

REGION	TAXON	MATERIAL	SAMPLE	AGE	SIZE	CITATIONS
South Africa	*Nucleornis insolitus*	isolated elements	2	early Pliocene	S	Simpson, 1979b
South Africa	Sphenisciformes indet. ("*Palaeospheniscus huxleyorum*")	isolated elements	4	early Pliocene	S	Simpson, 1973
South Africa	*Dege hendeyi*	isolated elements	13	early Pliocene (Quartzose Sand Member, Varswater Fm.)	S	Simpson, 1979a
South Africa	Sphenisciformes indet (distinctly smaller than *Dege hendeyi* from same site)	isolated elements	4	early Pliocene	S	Simpson, 1979a
South Africa	*Inguza predemersus*	isolated elements	12	early Pliocene (Quartzose Sand Member, Varswater Fm.)	S	Simpson, 1971c, 1975a
South Africa	cf. *Inguza predemersus*	fragmentary bones	>10	early Pliocene (Quartzose Sand Member, Varswater Fm.)	S	Simpson, 1975a
South Africa	cf. *Inguza predemersus*	partial femur	1	early Pliocene (Gravel Member, Varswater Fm.)	S	Simpson, 1975a

ANTARCTICA

REGION	TAXON	MATERIAL	SAMPLE	AGE	SIZE	CITATIONS
Seymour Island	*Crossvallia unienwillia*	3 partial limb bones	1	late Paleocene (Cross Valley Fm.)	L	Tambussi et al., 2005
Seymour Island	Sphenisciformes indet.	rare partially articulated individual elements	18	early Eocene (Telm 1-2, La Meseta Fm.)	-	Jadwiszczak, 2006b
Seymour Island	*Anthropornis grandis*	tarsometatarsi	10	middle-late Eocene (Telm 4-7, La Meseta Fm.)	L	Wiman, 1905a,b; Myrcha et al., 2002
Seymour Island	cf. *Anthropornis grandis*	isolated elements	12	middle-late Eocene (Telm 5-7, La Meseta Fm.)	L	Jadwiszczak, 2006a; Tambussi et al., 2006
Seymour Island	*Anthropornis nordenskjoeldi*	tarsometatarsi	8	late Eocene (Telm 7, La Meseta Fm.)	L	Wiman, 1905a,b; Marples, 1953; Myrcha et al., 2002
Seymour Island	cf. *Anthropornis nordenskjoeldi*	isolated elements	52	middle and/or late Eocene (Telm 4,5 or 6 and Telm 7, La Meseta Fm.)	L	Marples, 1953; Jadwiszczak, 2006a; Tambussi et al., 2006
Seymour Island	*Anthropornis* sp.(gracile species)	tarsometatarsus	1	late Eocene (La Meseta Fm.)	L	Ksepka, 2007
Seymour Island	cf. *Anthropornis* sp.	isolated bones	13	middle-late Eocene (Telm 5-7, La Meseta Fm.)	L	Myrcha et al., 2002; Tambussi et al., 2006

(continued)

Table 6.1 (*Continued*)

ANTARCTICA

REGION	TAXON	MATERIAL	SAMPLE	AGE	SIZE	CITATIONS
Seymour Island	*Delphinornis arctowskii*	tarsometatarsi	2	late Eocene (Telm 7, La Meseta Fm.)	S	Myrcha et al., 2002
Seymour Island	cf. *Delphinornis arctowskii*	isolated elements	3	late Eocene (Telm 7, La Meseta Fm.)	S	Jadwiszczak, 2006a
Seymour Island	*Delphinornis gracilis*	tarsometatarsi	2	late Eocene (Telm 7, La Meseta Fm.)	S	Myrcha et al., 2002
Seymour Island	cf. *Delphinornis gracilis*	isolated elements	2	late Eocene (Telm 7, La Meseta Fm.)	S	Jadwiszczak, 2006a
Seymour Island	*Delphinornis larseni*	tarsometatarsi	9	middle-late Eocene (Telm 5-7, La Meseta Fm.)	S	Wiman, 1905a,b; Myrcha et al., 2002
Seymour Island	cf. *Delphinornis larseni*	isolated elements	19	late Eocene (Telm 7, La Meseta Fm.)	S	Marples, 1953; Jadwiszczak, 2006a
Seymour Island	*Delphinornis wimani*	tarsometatarsi	6	middle-late Eocene (Telm 5-7, La Meseta Fm.)	S	Marples, 1953; Myrcha et al., 2002
Seymour Island	cf. *Delphinornis wimani*	isolated elements	26	middle and/or late Eocene (Telm 4,5 or 6 and Telm 7, La Meseta Fm.)	S	Jadwiszczak, 2006a
Seymour Island	cf. *Delphinornis* sp.	isolated elements	8	middle-late Eocene (Telm 5-7, La Meseta Fm.)	S	Myrcha et al., 2002; Jadwiszczak, 2006a
Seymour Island	*Marambiornis exilis*	tarsometatarsi	2	late Eocene (Telm 7, La Meseta Fm.)	S	Myrcha et al., 2002
Seymour Island	cf. *Marambiornis exilis*	isolated elements	2	late Eocene (Telm 7, La Meseta Fm.)	S	Jadwiszczak, 2006a
Seymour Island	cf. *Marambiornis* sp.	isolated elements	1	late Eocene (Telm 7, La Meseta Fm.)	S	Jadwiszczak, 2006a
Seymour Island	*Mesetaornis polaris*	tarsometatarsus	1	late Eocene (Telm 7, La Meseta Fm.)	S	Myrcha et al., 2002
Seymour Island	cf. *Mesetaornis polaris*	isolated elements	3	late Eocene (Telm 7, La Meseta Fm.)	S	Jadwiszczak, 2006a
Seymour Island	cf. *Mesetaornis* sp.	tarsometatarsi	3	late Eocene (Telm 7, La Meseta Fm.)	S	Myrcha et al., 2002; Jadwiszczak, 2006a
Seymour Island	*Palaeeudyptes gunnari*	tarsometatarsi	44	late Eocene (Telm 7, La Meseta Fm.)	L	Wiman, 1905a,b; Marples, 1953; Myrcha et al., 2002; Tambussi et al., 2006
Seymour Island	cf. *Palaeeudyptes gunnari*	isolated elements	62	middle and/or late Eocene (Telm 4,5 or 6 and Telm 7, La Meseta Fm.)	L	Marples, 1953; Jadwiszczak, 2006a
Seymour Island	*Palaeeudyptes klekowskii*	tarsometatarsi	32	late Eocene (Telm 7, La Meseta Fm.)	L	Myrcha et al., 2002
Seymour Island	cf. *Palaeeudyptes klekowskii*	isolated elements	45	middle-late Eocene (Telm 5-7, La Meseta Fm.)	L	Jadwiszczak, 2006a; Tambussi et al., 2006
Seymour Island	cf. *Palaeeudyptes* sp.	tarsometatarsi	107	middle-late Eocene (Telm 4/5-7, La Meseta Fm.)	L	Myrcha et al., 2002; Jadwiszczak, 2006a
Seymour Island	Sphenisciformes indet.("*Ichtyopteryx gracilis*")	partial tarsometatarsus	1	Eocene (La Meseta Fm.)	S	Wiman, 1905a,b
Seymour Island	Sphenisciformes indet.("*Tonniornis mesetaensis*")	humerus	1	Eocene (La Meseta Fm.)	S	Tambussi et al., 2006; Jadwiszczak, 2006b
Seymour Island	Sphenisciformes indet.("*Tonniornis minimum*")	humeri	2	Eocene (La Meseta Fm.)	S	Tambussi et al., 2006; Jadwiszczak, 2006b

REGION	TAXON	MATERIAL	SAMPLE	AGE	SIZE	CITATIONS
Seymour Island	Sphenisciformes indet.	isolated bones, very rarely articulated	508	Eocene (La Meseta Fm.)	–	Jadwiszczak, 2006a; Tambussi et al., 2006
Seymour Island	Sphenisciformes indet.	isolated bones, very rarely articulated	>2000	Eocene (La Meseta Fm.)	–	unpublished MLP specimens
Seymour Island	Sphenisciformes indet.	isolated bones, very rarely articulated	>300	Eocene (La Meseta Fm.)	–	unpublished USNM specimens
Seymour Island	Sphenisciformes indet.	isolated bones, very rarely articulated	>300	Eocene (La Meseta Fm.)	–	unpublished UCMP specimens
Seymour Island	Sphenisciformes indet.	isolated bones	38	Eocene (La Meseta Fm.)	–	Wiman, 1905a,b

AUSTRALIA

REGION	TAXON	MATERIAL	SAMPLE	AGE	SIZE	CITATIONS
South Coast	Sphenisciformes indet.("*Pachydyptes simpsoni*")	partial skeleton, isolated bones	3	late Eocene (Blanche Point Marls)	L	Jenkins, 1974
South Coast	Sphenisciformes indet.(cf. *Palaeeudyptes*)	humerus, tibiotarsus	2	late Eocene	L	Finlayson, 1938; Simpson, 1957
South Coast	Sphenisciformes indet.	femur	1	late Oligocene (Gambier Limestone)	S	Simpson 1957
South Coast	Sphenisciformes indet.	humerus	1	late Oligocene (Gambier Limestone)	L	Simpson, 1957
South Coast	*Anthropodytes gilli*	humerus	1	early Miocene? (see Gill, 1959; Jenkins, 1974)	L	Gill, 1959; Simpson, 1959
South Coast	*Pseudaptenodytes macraei*	partial humerus	1	late Miocene (Black Rock Sandstone)	M	Simpson, 1970
South Coast	cf. *Pseudaptenodytes macraei*	carpometacarpi	2	late Miocene (Black Rock Sandstone)	M	Simpson, 1970
South Coast	Sphenisciformes indet. ("*Pseudaptenodytes minor*")	isolated elements	8	late Miocene (Black Rock Sandstone)	S	Simpson, 1970
South Coast	Sphenisciformes indet.	isolated fragmentary elements	10	late Miocene (Black Rock Sandstone)	S	Simpson, 1970
South Coast	Sphenisciformes indet.	partial coracoid	1	late Miocene (Sandringham Sands)	S	Simpson, 1965

NEW ZEALAND

REGION	TAXON	MATERIAL	SAMPLE	AGE	SIZE	CITATIONS
South Island	*Waimanu manneringi*	partial skeleton	1	late early Paleocene (Waipara Greensand)	M	Slack et al., 2006
South Island	*Waimanu tuatahi*	skull, partial skeletons	3	late Paleocene (Waipara Greensand)	S	Slack et al., 2006
South Island	Sphenisciformes indet. (Gore Bay penguin)	partial femur	1	? early middle Eocene (uncertain horizon, see Simpson, 1972a)	L	Marples, 1952
South Island	*Palaeeudyptes marplesi*	tarsometatarsus, partial skeleton	2	late Eocene (Burnside Mudstone)	L	Brodkorb, 1963

(continued)

Table 6.1 (*Continued*)

NEW ZEALAND

REGION	TAXON	MATERIAL	SAMPLE	AGE	SIZE	CITATIONS
South Island	New Burnside Species (Burnside "*Palaeeudyptes*")	partial skeleton	1	late Eocene (Burnside Mudstone)	L	Marples, 1952; Ando, 2007
South Island	cf. *Palaeeudyptes*	skull and carpometacarpus	1	late Eocene (Burnside Mudstone)	L	Ando, 2007
South Island	*Pachydyptes ponderosus*	partial postcranial skeletons	7	latest Eocene (Ototara Limestone)	L	Oliver 1930; Marples, 1952; Simpson, 1970; Ando, 2007
South Island	*Palaeeudyptes antarcticus*	tarsometatarsi	2	latest Eocene – early Oligocene (Ototara Limestone?)	L	Huxley, 1859; Ando, 2007
South Island	cf. *Palaeeudyptes antarcticus*	partial postcranial skeleton	1	latest Eocene – early Oligocene (Ototara Limestone)	L	Ando, 2007
South Island	cf. *Palaeeudyptes marplesi*	partial postcranial skeleton	1	latest Eocene – early Oligocene (Ototara Limestone)	L	Hector, 1871; Ando, 2007
South Island	cf. Waihao penguin species A	partial skeleton	1	latest Eocene – early Oligocene (Ototara Limestone)	L	Ando, 2007
South Island	Sphenisciformes indet. (Waimate penguin)	partial hindlimbs	1	late Oligocene (Ototara Limestone, [= Waihao Stone of Marples, 1952])	–	Marples, 1952
South Island	Sphenisciformes indet.(Kakanui penguin)	partial limb bones	1	latest Eocene – early Oligocene (Ototara Limestone)	–	Marples, 1952
North Island	Sphenisciformes indet.(Glen Murray penguin)	partial hind limb	1	early Oligocene (Glen Massey Sandstone)	L	Grant-Mackie and Simpson, 1973
North Island	Sphenisciformes indet. (Motutara Point penguin)	partial femur	1	late Oligocene (Aotea Sandstone or Whaingaroa Siltstone)	L	Marples and Fleming, 1963; Grant-Mackie and Simpson, 1973
South Island	*Archaeospheniscus lopdelli*	fragmentary mandible, partial skeleton	1	late Oligocene (Kokoamu Greensand)	L	Marples, 1952
South Island	*Archaeospheniscus lowei*	partial mandible, partial skeleton	3	late Oligocene (Kokoamu Greensand)	M	Marples, 1952; Ando, 2007
South Island	Waihao penguin species A (Duntroon "*Palaeeudyptes*")	partial skull and partial skeleton	1	late Oligocene (Kokoamu Greensand)	L	Marples, 1952; Ando, 2007
South Island	Waihao penguin species B (Duntroon "*Palaeeudyptes*")	partial skull and partial skeletons	2	late Oligocene (Kokoamu Greensand)	L	Marples, 1952; Ando, 2007
South Island	Waihao penguin species A/ B (Duntroon "*Palaeeudyptes*")	partial skeletons	6	late Oligocene (Kokoamu Greensand)	L	Marples, 1952; Ando, 2007
South Island	cf. Waihao species A/ B(Woodpecker Bay/ Seal Rock penguin)	partial skeleton	1	late Oligocene (uncertain horizon, see Simpson 1972a)	L	Hector, 1871; Ando, 2007
South Island	Sphenisciformes indet. (Duntroon specimens)	isolated elements	4	late Oligocene (Kokoamu Greensand)	–	Marples, 1952

REGION	TAXON	SAMPLE	MATERIAL	AGE	SIZE	CITATIONS
South Island	Sphenisciformes indet. (Omihi penguin)	1	femur	late Oligocene? (uncertain horizon, see Simpson, 1972a)	S	Marples, 1952
North Island	Sphenisciformes indet. (Te Kauri penguin)	1	associated limb bones and fragments	late Oligocene (Aotea Sandstone)	M	Grant-Mackie and Simpson, 1973
South Island	Duntroonornis parvus	4	isolated limb bones, associated limb bones	late Oligocene (Kokoamu Greensand, Otekaike Limestone)	S	Marples, 195
South Island	Korora oliveri	1	tarsometatarsus	late Oligocene (Otekaike Limestone)	S	Marples, 1952
South Island	Platydyptes novaezealandiae	4	associated hindlimbs; mandible, partial skeletons	late Oligocene (Otekaike Limestone)	M	Oliver, 1930; Marples, 1952; Ando, 2007
South Island	New Hakataramea Species	3	associated and isolated limb bones	late Oligocene (Otekaike Limestone)	S	Ando, 2007
South Island	Platydyptes amiesi	4	isolated and associated limb bones; partial skeletons	late Oligocene and/or earliest Miocene (Otekaike Limestone)	M	Marples, 1952; Ando, 2007
South Island	Platydyptes marplesi	7	isolated limb bones; partial skeletons	late Oligocene and/or earliest Miocene (Kokoamu Greensand; Otekaike Limestone, Gee Greensand?)	M	Marples, 1952; Simpson, 1972b
South Island	Aptenodytes ridgeni	1	partial hindlimbs	middle-late Miocene? (Greta Sandstone?)	L	Simpson, 1972b
South Island	Marplesornis novaezealandiae	1	skull, partial skeleton	middle-late Miocene? (Greta Sandstone?)	M	Marples, 1960
South Island	Pygoscelis tyreei	1	partial postcranial skeleton	middle-late Miocene? (Greta Sandstone?)	S	Simpson, 1972b
North Island	Tereingaornis moisleyi	2	partial skeleton; associated limb bones	Pliocene (unnamed calcareous sandstone)	S	Scarlett, 1983; McKee, 1987

SOUTH AMERICA

REGION	TAXON	SAMPLE	MATERIAL	AGE	SIZE	CITATIONS
Tierra del Fuego	Sphenisciformes indet.	1	partial hindlimb and pelvis	late middle Eocene (Leticia Fm.)	M	Clarke et al., 2003
Peru (Pacific Coast)	Perudyptes devriesi	1	partial skull and skeleton	late middle Eocene (Paracas Fm.)	M	Clarke et al., 2007

(continued)

Table 6.1 (*Continued*)

SOUTH AMERICA

REGION	TAXON	MATERIAL	SAMPLE	AGE	SIZE	CITATIONS
Peru (Pacific Coast)	Sphenisciformes (unpublished MUSM specimens)	partial tarsometatarsus and tibiotarsus	1	late middle Eocene (Paracas Fm.)	S	Ksepka and Clarke, 2010
Peru (Pacific Coast)	Sphenisciformes (unpublished MUSM specimens)	partial skeleton, isolated bones	3	late middle Eocene (Paracas Fm.)	L	Ksepka and Clarke, 2010
Peru (Pacific Coast)	*Icadyptes salasi*	skull and partial skeleton	1	late Eocene (Otuma Fm.)	L	Clarke et al., 2007
Peru (Pacific Coast)	*Inkayacu paracasensis*	skull and partial skeleton	1	late Eocene	L	Clarke et al., 2010
Peru (Pacific Coast)	Sphenisciformes indet.	tibiotarsus	1	late Eocene (Otuma Fm.)	L	Acosta Hospitaleche and Stucchi, 2005
Patagonia (Atlantic Coast)	*Arthrodytes andrewsi*	scapula, coracoid, humerus	1	late Oligocene (San Julian Fm.)	L	Ameghino 1901, 1905; Acosta Hospitaleche 2005
Patagonia (Atlantic Coast)	*Paraptenodytes brodkorbi*	humerus	1	late Oligocene (San Julian Fm.)	S	Simpson, 1972b
Patagonia (Atlantic Coast)	*Paraptenodytes robustus*	humerus	1	late Oligocene (San Julian Fm.)	S	Ameghino, 1895; Simpson, 1972b
Patagonia (Atlantic Coast)	cf. *Paraptenodytes robustus*	isolated bones	4	late Oligocene (San Julian Fm.)	S	Ameghino, 1895; Simpson, 1972b
Patagonia (Atlantic Coast)	*Paraptenodytes antarcticus*	tarsometatarsus, nearly complete skeleton	2	early Miocene (Monte Leon Fm.)	M	Moreno and Mercerat, 1891; Simpson, 1946; Bertelli et al., 2006
Patagonia (Atlantic Coast)	*Eretiscus tonnii*	tarsometatarsus, 3 humeri	4	early/middle Miocene (Gaiman Fm.)	S	Simpson, 1981; Acosta Hospitaleche et al., 2004
Patagonia (Atlantic Coast)	*Palaeospheniscus bergi*(including "*P. gracilis*")	isolated bones	23	early/middle Miocene (Gaiman Fm.)	S	Simpson, 1972b
Patagonia (Atlantic Coast)	*Palaeospheniscus biloculata*(including "*P. wimani*")	partial skeleton, isolated bones	14	early/middle Miocene (Gaiman Fm.)	S	Simpson, 1970; Acosta Hospitaleche, 2007

Location	Taxon	Material	Number	Age (Formation)	S/M	Reference
Patagonia (Atlantic Coast)	*Palaeospheniscus patagonicus*	partial associated skeleton; isolated bones	42	early/middle Miocene (Gaiman Fm)	S	Simpson, 1972b; Acosta Hospitaleche et al., 2008; Simpson, 1972b
Peru (Pacific Coast)	*Palaeospheniscus patagonicus*	tarsometatarsus	1	early/middle Miocene (Chilcatay Fm.)	S	Acosta Hospitaleche and Stucchi, 2005; Ksepka, 2007
Patagonia (Atlantic Coast)	Small penguin species indet.	humerus	1	early/middle Miocene (Gaiman Fm)	S	Ksepka, 2007
Peru (Pacific Coast)	*Spheniscus muizoni*	partial skeleton, isolated bones	7	middle/late Miocene (Pisco Fm.)	S	Göhlich, 2007
Patagonia (Atlantic Coast)	*Madrynornis mirandus*	nearly complete individual	1	late Miocene (Puerto Madryn Fm.)	S	Acosta Hospitaleche et al. (2007)
Patagonia (Atlantic Coast)	cf. *Paraptenodytes*	partial humerus and femur	2	late Miocene (Puerto Madryn Fm.)	M	Acosta Hospitaleche (2003)
Peru (Pacific Coast)	*Spheniscus megaramphus*	skulls, nearly complete individual	6	late Miocene (Pisco Fm.)	M	Stucchi et al., 2003; Stucchi, 2007
Peru (Pacific Coast)	*Spheniscus urbinai*	skulls, partial and complete skeletons	6	late Miocene - early Pliocene (Pisco Fm.)	M	Stucchi 2002; 2007
Chile (Pacific Coast)	Sphenisciformes indet. (cf. *Spheniscus/Palaeospheniscus*)	skulls	5	middle/late Miocene or Pliocene (Bahia Inglesa Fm.)	S	Acosta Hospitaleche and Canto, 2005; Chávez, 2007
Chile (Pacific Coast)	"*Pygoscelis*" *calderensis*	skulls	3	middle/late Miocene or Pliocene (Bahia Inglesa Fm.)	S	Acosta Hospitaleche et al. (2006)
Chile (Pacific Coast)	*Pygoscelis grandis*	partial associated skeleton, individual elements	6	late Miocene, possibly also early Pliocene (Bahia Inglesa Fm.)	M	Walsh and Suárez, 2006
Chile (Pacific Coast)	*Spheniscus chilensis*	isolated bones	72	late Pliocene (Caleta Herradura de Mejillones Fm.)	S	Emslie and Guerra Correa, 2003
Peru (Pacific Coast)	Sphenisciformes (undescribed MUSM specimens)	various	~500	Miocene/Pliocene (Pisco Fm.)	–	Stucchi, 2007
Argentina	Sphenisciformes indet.	various	>150	various	–	Acosta Hospitaleche, 2002

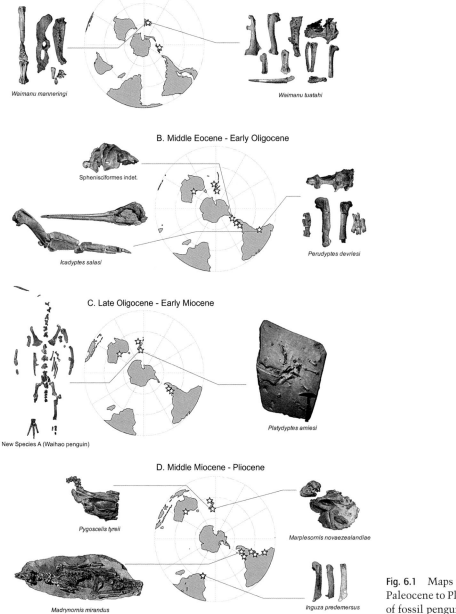

A. Paleocene - Early Eocene

Waimanu manneringi

Waimanu tuatahi

B. Middle Eocene - Early Oligocene

Sphenisciformes indet.

Icadyptes salasi

Perudyptes devriesi

C. Late Oligocene - Early Miocene

New Species A (Waihao penguin)

Platydyptes amiesi

D. Middle Miocene - Pliocene

Pygoscelis tyreii

Marplesornis novaezealandiae

Madrynornis mirandus

Inguza predemersus

Fig. 6.1 Maps showing the Paleocene to Pliocene distribution of fossil penguin localities.

penguins (including three rockhopper species and a single little blue penguin species) and acknowledge that our understanding of *Eudyptula minor* taxonomy is evolving.

Contrary to most avian groups, the recognized diversity of extinct penguins clearly exceeds current diversity. At present, at least 49 species are known from clearly diagnosable remains (see Table 6.1) and several distinct taxa await formal naming. A total of 74 species of fossil penguin have been proposed historically, but many species named in the late 19th and early 20th centuries were based on undiagnostic or pathological remains, or remains of birds other than penguins. Simpson (1946) completed a revision of all species known at the time and eliminated many invalid taxa. Since that time, periodic revisions have weeded out additional invalid species (Simpson, 1971a, 1972a; Jenkins, 1985; Myrcha *et al.*, 2002; Acosta Hospitaleche, 2004; Jadwiszczak, 2006b; Ando, 2007; Ksepka, 2007). Even by the most conservative tallies, the large majority of penguin species that have ever existed are now extinct. The number of known fossil penguin species is growing rapidly, with 10 new species named since 2005.

Recent classifications have placed penguins within the order Sphenisciformes and family Spheniscidae. Clarke *et al.* (2003) proposed phylogenetic definitions for these names and proposed the new name Pansphenisciformes. Under these definitions, Pansphenisciformes is applied to the clade including all taxa more closely related to Spheniscidae than any other extant avian lineage. Sphenisciformes is applied to the clade including all penguins that share the apomorphic loss of aerial flight. These names currently denote the same set of known taxa. However, if volant basal members of the penguin lineage were to be discovered, they would be placed within Pansphenisciformes but excluded from Sphenisciformes. Spheniscidae is restricted to the crown clade of penguins, comprising the most recent common ancestor of all living penguin species and its descendants.

Attempts to formulate a more detailed taxonomy inclusive of stem diversity have thus far met with mixed results. Simpson (1946)

proposed a subfamily level classification of fossil penguins including five subfamilies: Palaeospheniscinae (including *Palaeospheniscus*), Paraptenodytinae (including *Paraptenodytes*), Anthropornithinae (including *Anthropornis*, *Eospheniscus* [later synonymized with *Palaeeudyptes*], *Delphinornis* and *Pachydyptes*), Palaeeudyptinae (including *Palaeeudyptes*), and Spheniscinae (extant penguins). Later authors (Marples, 1952; Brodkorb, 1963; Acosta Hospitaleche, 2004; Tambussi *et al.*, 2005) updated the contents or definitions of these subfamilies. However, following several decades of new discoveries, Simpson (1971a) himself abandoned this classification system because he felt it did not reflect true evolutionary relationships. Only Palaeospheniscinae appears to represent a monophyletic group (Ksepka *et al.*, 2006). A stable phylogenetic taxonomy of stem fossil penguins is desirable, though at present a poor understanding of the relationships of many fragmentary taxa and lack of resolution in many parts of the penguin tree complicates the clear definition of higher taxa. However, we believe that as more complete materials of multiple taxa currently under study (e.g. *Archaeospheniscus*, "*Palaeeudyptes*," *Platydyptes*, *Palaeospheniscus*) are fully described and incorporated into phylogenetic analyses, our understanding of fossil penguin relationships will solidify enough for a new taxonomy to be erected around well-placed taxa.

HISTORICAL OVERVIEW OF FOSSIL PENGUINS

Thomas Henry Huxley (1859) made the first report of a fossil penguin. Huxley (1859) described this find, a partial tarsometatarsus, as *Palaeeudyptes antarcticus*. Although the fossil was fragmentary, it revealed both the antiquity of penguins and the prior existence of very large forms. For nearly a century following Huxley's report, fossil penguins remained known only from highly incomplete material. During this interval, a handful of additional elements were reported from New Zealand

(Hector, 1872) and a large collection of isolated bones was reported from South America by Ameghino (1891, 1895, 1899, 1901, 1905). Accounts of truly "giant" penguins by the Swedish Polar Expedition of 1901–1903 (Wiman, 1905a, 1905b) aroused popular interest in penguin fossils.

George Gaylord Simpson, while best known for his contributions to mammalian paleontology and evolutionary theory, also wrote extensively on fossil penguins. His description of an exquisitely preserved partial skull and skeleton of *Paraptenodytes antarcticus* (Simpson, 1946) marks a major milestone in penguin paleontology. This fossil provided the foundation for Simpson's argument that penguins evolved directly from a volant ancestor, without undergoing a flightless terrestrial interval. Simpson's subsequent work includes comprehensive reviews of regional penguin faunas (Simpson, 1957, 1971a, 1971b, 1972a), descriptions of new species (Simpson, 1972b, 1973, 1979a,b, 1981), and the popular book *"Penguins Past and Present, Here and There"* (Simpson, 1976). Brian J. Marples also generated extensive early work on fossil penguins, describing in detail a diverse penguin fauna from New Zealand and new finds from Antarctica, and providing a foundation for fossil penguin comparative anatomy (Marples, 1952, 1953, 1960, 1962, 1974; Marples & Fleming, 1963).

A lull in the study of fossil penguins followed the era of Simpson and Marples, during which a few accounts of new finds saw publication (Scarlett, 1983; McKee, 1987; Myrcha *et al.*, 1990) and some updated overviews appeared (Olson, 1985; Fordyce & Jones, 1990; Fordyce, 1991; Cozzuol *et al.*, 1991, 1993). In recent years the pace of fossil penguin discoveries has accelerated to unprecedented levels. Since 2005, 10 new species have been named and large volumes of material have been described (Jadwiszczak, 2001, 2003, 2006a,b; Myrcha *et al.*, 2002; Stucchi, 2002; Emslie & Guerra Correa, 2003; Stucchi *et al.*, 2003; Acosta Hospitaleche, 2004; Acosta Hospitaleche *et al.*, 2004, 2006, 2008; Acosta Hospitaleche & Canto, 2005; Slack *et al.*, 2006; Clarke *et al.*, 2007, 2010; Ksepka *et al.*, 2008). These include the oldest and most basal penguin (*Wai-*manu manneringi), a unique spear-billed giant penguin (*Icadyptes salasi*) and the oldest crown penguin fossil (*Spheniscus muizoni*). Figure 6.2 illustrates a sampling of fossil penguin diversity.

PHYLOGENIC RELATIONSHIPS

Disparities between penguins and other birds prompted some bizarre evolutionary scenarios for their higher level relationships in the late 19th and early 20th centuries. Menzbier (1887) went as far as to suggest that penguins have a reptilian origin separately from other birds. Beddard (1898) and Chandler (1916) suggested that the ancestor of penguins was related to the extinct *Hesperornis*, a hypothesis that would place penguins distant from other extant birds. Misinterpretation of fossil morphologies likewise led Lowe (1933, 1939) to propose that penguins evolved from a flightless ancestor separate from all other birds.

Aside from these easily discredited hypotheses, most pre-cladistic taxonomists allied Sphenisciformes with Procellariiformes (albatrosses, petrels, and allies) and Gaviiformes (loons). Seebohm (1888), Sharpe (1891), and Gadow (1889) favored the hypothesis that penguins were closest to a group comprising both Procellariiformes and Gaviiformes, while Fürbringer (1888), Gadow (1893), and Pycraft (1898) considered Procellariiformes to be most closely related to penguins. Fossil evidence has historically been considered consistent with these hypotheses. Wiman (1905a) noted morphological similarities between Eocene Antarctic fossil Sphenisciformes and extant Procellariiformes. Simpson (1946) commented specifically on the similarities of the pterygoid of *Paraptenodytes antarcticus* to the same bone in Procellariiformes, while Olson (1985) noted similarities between the partial skulls of Eocene penguins and those of Gaviiformes.

Recent studies including both morphology-based and molecular-based analyses support a general framework in which penguins are part of a large seabird clade including Procellariiformes,

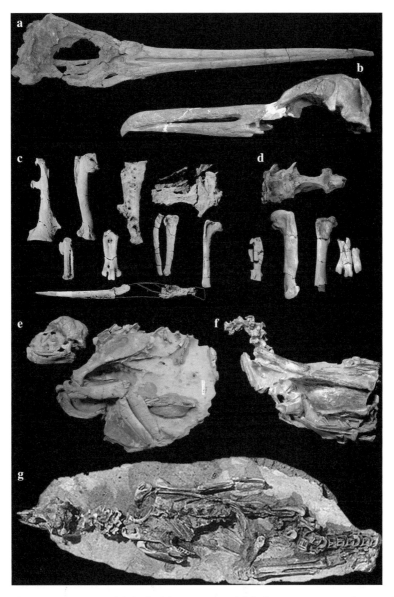

Fig. 6.2 Select fossil penguin specimens. (a) skull of the giant spear-beaked stem penguin *Icadyptes salasi* (MUSM 897, after Ksepka *et al.*, 2008). (b) Skull of the crown penguin *Spheniscus megaramphus* (MUSM 175, after Stucchi *et al.* 2003). (c) Select elements of the basal penguin *Waimanu tuatahi* (OU 12651 and CM zfa 33-34, after Slack *et al.*, 2006). (d) Skull and limb elements of the earliest equatorial penguin *Perudyptes devriesi* (MUSM 889, after Clarke *et al.*, 2007). (e) Skull and skeleton of *Marplesornis novaezealandiae* (CM zfa16527), the potential sister taxon of crown Spheniscidae. (f) Articulated postcranial skeleton of *Pygoscelis tyreei* (CM zfa 22631). (g) *Madrynornis mirandus* (MEF-PV 100, after Acosta Hospitaleche *et al.*, 2007), one of the oldest crown penguins. Not to scale.

Gaviiformes, Pelecaniformes, and Ciconiiformes (Ho et al., 1976; Cracraft, 1988; Sibley & Ahlquist, 1990; McKitrick, 1991; Cooper & Penny, 1997; Siegel-Causey, 1997; Groth & Barrowclough, 1999; van Tuinen et al., 2000; Mey et al., 2002; Mayr & Clarke, 2003; Fain & Houde, 2004; Simon et al., 2004; Mayr, 2005; Slack et al., 2006; Livezey & Zusi, 2006, 2007; Hackett et al., 2008). Within this large clade, the balance of evidence suggests Sphenisciformes and Procellariiformes are sister taxa. While the diving petrel has been considered as an example of the ancestral penguin body plan, all evidence indicates penguins share a relationship with the clade Procellariiformes as a whole and are not descended from any particular lineage within that group.

Extant penguin phylogeny has been thoroughly investigated and consensus appears to be at hand for their relationships, save for the issue of rooting. Molecular data strongly support a topology in which Aptenodytes is the basalmost genus. Sibley & Ahlquist (1990) recovered this pattern in DNA hybridization studies (though they did not include Eudyptes). More recently, Baker et al. (2006) included 18 extant penguin taxa and sequence data from five genes to rigorously test the interrelationships of penguins and found strong support for placing Aptenodytes as the basalmost genus. Morphological analyses consistently support Spheniscus or Spheniscus + Eudyptula as the basal divergence in the crown clade (O'Hara, 1989; Giannini & Bertelli, 2004; Bertelli & Giannini, 2005; Ksepka et al., 2006; Acosta Hospitaleche et al., 2007; Ando, 2007). Although morphology-based and sequence-based phylogenies for Spheniscidae appear very different, the unrooted networks of relationships derived from these data are largely congruent (Figure 6.3). Rooting appears to account for almost all of the disparity between results from the two types of data (Bertelli & Giannini, 2005). Because the unrooted networks are so similar, concentrating efforts on more complete sequence representation for penguin outgroups and further study of fossil taxa that represent proximal outgroups to the crown clade could potentially provide the signal to reroot the morphological or molecular trees. Notably, in combined analyses of morphological and molecular data, the molecular rooting is preferred. The signal in the molecular data also appears to be very robust to choice of optimality criterion: parsimony, maximum likelihood, and Bayesian analyses of molecular data (Baker et al., 2006) all recovered the same topology, while analyses of combined morphological and molecular data using direct optimization (Ksepka et al., 2006; Clarke et al., 2007) and parsimony (Ksepka & Clarke, 2010) produced an extremely similar topology, with differences only for the placement of some species within Eudyptes.

Most recently, attention has been focused on incorporating fossil diversity into the framework of penguin phylogeny (Bertelli et al., 2006; Ksepka et al., 2006; Walsh & Suarez, 2006; Slack et al., 2006; Ando, 2007; Clarke et al., 2007; Acosta Hospitaleche et al., 2007, 2008). These studies have generally agreed well where taxa overlap. We present a summary of our current understanding of the phylogeny of Sphenisciformes in Figure 6.4. This tree reconciles the results of recent phylogenetic analyses (Ksepka et al., 2006; Clarke et al., 2007) with data from new fossil specimens from New Zealand (Ando, 2007). The implications of this phylogeny are discussed further below.

MORPHOLOGICAL EVOLUTION

Bannasch (1994) provided a comprehensive discussion of the functional anatomy of the flight apparatus in extant penguins. Essentially, the locomotion of penguins is similar enough to aerial flight that it is commonly referred to as aquatic or underwater flight. There are, of course, major differences between aerial and aquatic flight due primarily to the density of water, which is approximately 800 times greater than the density of air at sea level. Unlike volant birds, penguins do not need to generate lift with the flight stroke, producing some key differences in the wingbeat cycle. Volant birds produce thrust

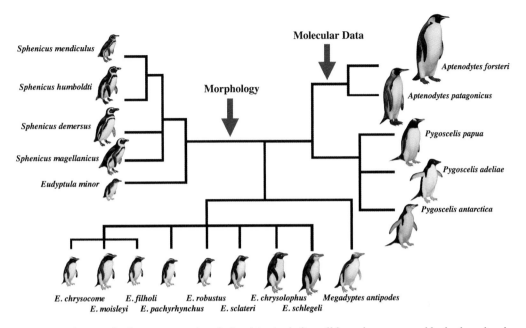

Fig. 6.3 Unrooted network of extant penguin relationships including all branches supported by both molecular and morphological data. While the different preferred rooting results in very different trees, the networks are highly congruent.

during the downstroke; the upstroke is important in repositioning the wing, but does not produce thrust in most birds. Penguins produce thrust both during the upstroke and the downstroke (Clark & Bemis, 1979). There is no folding or bending of the wing during the upstroke, in contrast to the cycle in typical birds and even in volant wing-propelled divers such as alcids (Kaiser, 2007). Instead, the wing remains stiff, increasing the efficiency of thrust generation. An active upstroke also occurs during dives in the wing-propelled Alcidae (Johansson & Wetterholm Aldrin, 2002) as well as in the aerial flight of hummingbirds (Trochilidae; Weis-Fogh, 1972).

It is well agreed upon that penguins evolved from a volant ancestor. However, the earliest stage in penguin evolution, during which the penguin lineage had diverged from its sister group but still retained aerial flight, remains unknown. A few characteristics of the basalmost pansphenisciform can nonetheless be inferred. Stonehouse (1967) estimated that volant members of Panspheniciformes would be constrained in size by the competing needs of the high muscle mass required for underwater propulsion and the upper weight threshold for aerial flight. Stonehouse (1967) predicted that the loss of aerial flight in the penguin lineage occurred in a form with a body mass of approximately 1 kg, the same body mass as the smallest modern species, *Eudyptula minor*. Empirical evidence suggests that this mass is near the upper threshold for maintaining both aerial and underwater flight, and indeed the body masses of the largest extant wing-propelled divers which still maintain aerial flight (e.g. the common murre [*Uria aalge*] and thick-billed murre [*Uria lomvia*]) are near this value (Dunning, 1993). Thus, early Panspheniciformes were probably broadly similar to alcids in their ecology, and no more massive than the largest extant auks and murres in body size.

In many ways, the loss of aerial flight can be viewed as removing an evolutionary constraint on penguins, especially as regards wing structure and body size. Many of the features that make the

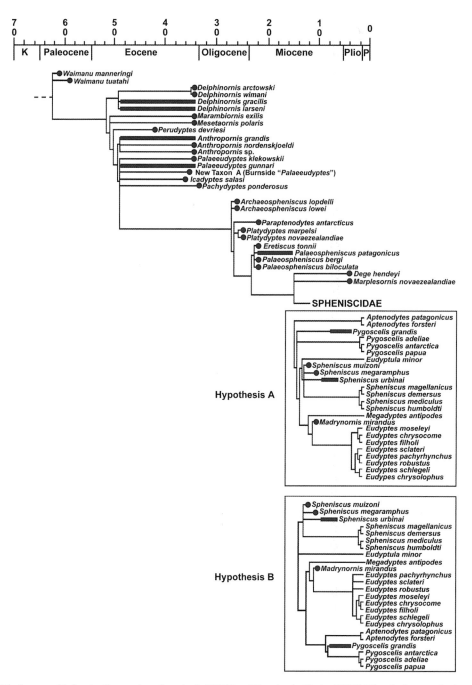

Fig. 6.4   Phylogeny of Sphenisciformes based on Ando (2007) and Ksepka & Clarke (2010). The cladogram depicts areas of agreement between these two studies as fully resolved, and collapses branches where results of these two studies conflict (e.g. placement of *Platydyptes* and *Paraptenodytes*). Hypothesis A reflects the topology for Spheniscidae supported by molecular and combined analyses, while Hypothesis B reflects the topology supported by morphology alone.

extant penguin flipper so efficient in underwater propulsion are incompatible with aerial flight, and thus must have evolved after the loss of aerial flight. These include the high degree of stiffening of the wing joints (Raikow *et al.* 1988), simplification of the intrinsic musculature (Schreiweis, 1982), and marked osteosclerosis of the long bones (Meister, 1962). Freed from the restraints imposed on birds by aerial flight, early Sphenisciformes were able to explore new body plans and attain larger sizes, leading to more efficient diving.

Penguins also exhibit a radically reorganized integument characterized by the development of scale-like feathers with flattened rachi, homogenization of the wing feathers, and loss of apterygia (e.g. Livezey, 1989;Giannini & Bertelli, 2004) that may contribute to efficient underwater flight by reducing drag (Culik *et al.*, 1994). Recently, the first fossil penguin feathers were reported from *Inkayacu paracasensis*, revealing that most of the key features of penguin wing feathers evolved deep within the stem lineage and were already in place by the Eocene (Clarke *et al.*, 2010). These fossil feathers also preserve melanosomes, providing evidence for unique reddish-brown and gray color patterns in *Inkayacu paracasensis* (Clarke *et al.*, 2010).

While the remarkable preservation of feathers offers a glimpse at stem penguin integument, we must rely on bony evidence for insight into the musculature. The origins and insertions of many of the most important flight muscles possess clearly identifiable osteological correlates, allowing changes in myology to be traced through phylogeny. Below, we highlight some of the major changes in the penguin wing apparatus in a phylogenetic context.

Waimanu reveals presumably one of the earliest stages in penguin evolution following the loss of flight. High wing loading is indicated by the short wing elements, though *Waimanu* retains a flipper relatively longer (proportional to body size) than in more crownward penguins. The major wing bones are in almost every way intermediate between outgroup taxa and more crownward penguins (Figure 6.5). The humerus is flattened dorso-

ventrally, whereas the ulna and radius retain a subcircular cross-section (as in modern auks); in more crownward penguins flattening of the humerus is more pronounced and the distal limb bones are also compressed. The coracoid is elongate compared to outgroups, but remains shorter than the humerus. Pronounced elongation of the coracoid displaces the triosseal canal relative to the sternum, increasing space for the pectoralis major and minor muscles and increasing the leverage of the tendon of *m. supracoracoideus*, the major muscle for the upbeat of the wing (Bannasch, 1994). The scapula retains a blade-like shape, as opposed to the expanded paddle-like shape of extant penguins. *Waimanu* also differs in having a more gracile tarsometatarsus, with a posteriorly directed trochlea II. Modern penguins employ the foot as a rudder in underwater flight, but do not use the hindlimb to provide thrust. Differences in the foot of *Waimanu* suggests that early penguins may have utilized the foot more actively in propulsion as do some extant Procellariiformes (Warham, 1996, p. 394; Ando, 2007), but because many of the similarities in the tarsometatarsus are optimized as plesiomorphic caution must be taken when inferring function.

A large morphological gap occurs between *Waimanu* and more crownward penguins. These penguins show further shortening of the wing (relative to body size) and more pronounced flattening of wing elements. Ksepka *et al.* (2006) hypothesized that the more elliptical cross-section of the long bones would be more hydrodynamically efficient, but slightly less resistant to shear forces. However, Kaiser (2007) noted that the flattened humerus of penguins may actually be more resistant to torsion than a typical bird humerus because a flattened shaft has the potential to twist to a certain degree without breaking. In all penguins except *Waimanu*, the coracoid exceeds the humerus in length, further increasing space available to and leverage of the flight muscles. The scapular blade is notably expanded in stem taxa including the undescribed "Waihao penguin species A" and *Platydyptes* (Ando, 2007), and greatly expanded into a paddle-like shape in crown Spheniscidae. This

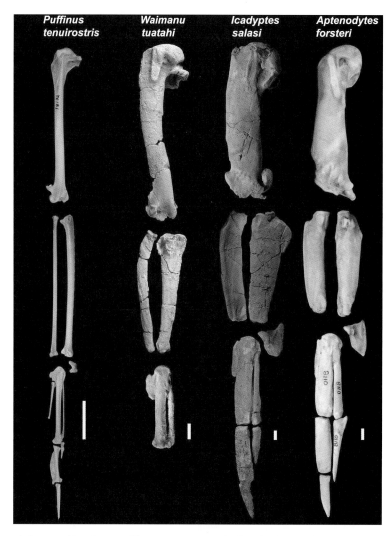

Fig. 6.5 The wing of the procellariiform *Puffinus tenuirostris*, the basal sphenisciform *Waimanu tuatahi*, the stem sphenisciform *Icadyptes salasi* and the extant spheniscid *Aptenodytes forsteri* illustrating changes in the morphology of the wing over the transition to crown Spheniscidae (radiale and sesamoids not illustrated). In order to show proportions, elements are scaled so that the total wing lengths of the four taxa are equal. Because the phalanges are not known for *Waimanu tuatahi*, the lengths of these elements were estimated. Note that a free alular phalanx was probably present in *Waimanu* and *Icadyptes* though this element is not yet reported for these taxa. Scale bars = 1 cm.

expanded blade forms an enlarged area that accommodates the strongly developed *m. scapulohumeralis caudalis*, a muscle important in isometrically transferring force during the downstroke (Bannasch, 1994). A free alular phalanx appears to have been present in many stem taxa such as *Waimanu* and *Icadyptes*, but, at later stages of penguin evolution, it becomes incorporated into the carpometacarpus (Ksepka *et al.*, 2008). Because the alular phalanx has not yet been discovered intact in a fossil penguin, it remains unclear what role, if any, the alula may have played in early penguin locomotion. While the proportions of digits II and III are unspecialized in basal penguins, crown penguins are unique among modern birds in that phalanx III-1 is longer than phalanx II-1 and possesses a proximally directed process. This arrangement of the digits gives the wingtip a less abruptly tapering shape than in basal penguins and, if a similar integument is assumed, correlates to a decrease in aspect ratio. Because of preservation of the phalanges is rare in fossil penguins, we do not yet know at which point in phylogeny this shift occurred, though *Icadyptes* exhibits primitive proportions.

PALEOECOLOGY

One of the most interesting facets of early penguin evolution is the rise of giant penguins. The largest species were estimated to have reached masses of 81–97 kg (Livezey, 1989; Jadwiszczak, 2001; Ando, 2007) and heights of 1.66–1.99 m (Jadwiszczak, 2001) based on extrapolations from limb bones. These values are now recognized as slight overestimates, as more complete skeletons reveal proportional differences between stem and crown penguins that preclude direct scaling of estimates from single bones. Penguins seem to have achieved large body sizes soon after shifting from aerial to underwater flight. Even the oldest fossil penguin (*Waimanu manneringi*) exhibits large body size (ca. 20 kg; Ando, 2007) compared to typical extant birds. We present scaled reconstructions of fossil penguins of various sizes in Figure 6.6, and rough estimates of the sizes of all fossil taxa in Table 6.1. By the late Paleocene, taxa exceeding the living Emperor penguin in size are known from Antarctica (Tambussi *et al.*, 2005). Giant penguins appear in the fossil record by the middle Eocene in South America and late Eocene in Australia and

Fig. 6.6 Scaled reconstructions of fossil and living penguins. Fossils, in black from left to right: *Paraptenodytes antarcticus*, *Icadyptes salasi*, and *Waimanu tuatahi*. Extant penguin silhouettes, in grey, from left to right: *Eudyptes pachyrhynchus* (Fiordland penguin), *Eudyptula minor* (little blue penguin), and *Aptenodytes forsteri* (emperor penguin). Scale bar = 10 cm.

New Zealand. Giant forms may have dispersed throughout the Southern Hemisphere even earlier, but appropriate deposits of Paleocene to middle Eocene age are wanting at many localities. Clearly, giant penguins had a somewhat different ecology than extant penguins as necessitated by their size and indicated by the highly derived morphology of known skulls (Olson, 1985; Myrcha *et al.*, 1990; Jadwiszczak, 2003; Ando, 2007; Ksepka *et al.*, 2008).

Even the oldest fossil penguins are clearly adapted for diving (Slack *et al.*, 2006). However, diving capabilities of early penguins have not been widely discussed. Tambussi *et al.* (2006) concluded the giant penguin *Anthropornis nordenskjoeldi* was not specialized for diving based on the angled (as opposed to straightened) flipper. However, this morphology does not necessarily rule out deep diving capacities for *Anthropornis* or other stem penguins, because there is no demonstrated correlation between the angle of the wing bones in articulation and diving ability in extant diving birds. Extant alcids hold their wings in a partially folded posture throughout the flight stroke during underwater propulsion (Stettenheim, 1959; Lovvorn, 2001). These birds are accomplished divers: depth gauge data show that the rhinoceros auklet (*Cerorhinca monocerata*) reaches depths of 57 m (Kuroki *et al.*, 2003) and data from gill nets indicate the thick-billed murre (*Uria aalge*) reaches depths of up to 180 m (Piatt & Nettleship, 1985). Alcids achieve these depths while retaining the capacity for aerial flight. Extant penguins attain even greater depths (see Kooyman *et al.*, 1971, 1982, 1992), reaching beyond 500 m in the case of the emperor penguins (summarized in Williams, 1995). In extant marine birds and mammals, a strong correlation between body size and diving depth/duration has been demonstrated (Halsey *et al.*, 2006; Kooyman, 1989; Watanuki & Burger, 1999). Thus, the balance of evidence suggests giant penguins were capable of deep diving rather than restricted to near-surface waters.

The feeding ecology of extinct penguins has also received little attention until recently, due to the previous scarcity of described cranial remains.

Though all living penguins are somewhat opportunistic in diet, consuming fish, squid, and crustaceans as available, some broad patterns in cranial morphology related to primary prey type can be identified. Zusi (1975) related cranial osteology, myology, and tongue morphology of living penguins to diet, separating extant penguins into species specializing on small shoaling prey, species specializing on fish, and species with a more generalized diet. Livezey (1989) also found support for distinguishing these three foraging groups using morphometric patterns. Among the features Zusi (1975) identified in penguins specializing on fish are well-developed jaw adductor musculature, stoutly constructed upper jaws and smaller buccal papillae. In contrast, planktonic specialists (*Eudyptes* penguins, *Pygoscelis adeliae*, and *Pygoscelis antarctica*) showed reduced adductor musculature, deeper mandibles, and stronger papillae. Fossils and phylogeny suggest that these specializations were acquired late in the evolutionary history of penguins. Though fossil penguins preserving the beak are not very abundant, all such specimens possess a narrow, slender bill (Myrcha *et al.*, 1990; Slack *et al.*, 2006; Ando, 2007; Clarke *et al.*, 2007). Olson (1985) and Myrcha *et al.* 1990, 2002 noted that such long spear-like beaks seem most suitable for spearing large prey (e.g. fish or squids). An elongate beak can now be optimized as present for a long phylogenetic interval in penguins (Ando, 2007; Clarke *et al.*, 2007), suggesting many or even all stem forms hunted relatively large prey items. Based on the distribution of cranial features related to feeding strategy in living and fossil penguins, Ksepka & Bertelli (2006) hypothesized that specializations for capturing small, shoaling prey arose close to or within the crown clade. The exploitation of planktonic crustaceans by some *Eudyptes* and *Pygoscelis* penguins could be a relatively recent innovation, possibly arising in conjunction with expansion of Antarctic sea-ice. However, recent research (Emslie & Patterson, 2007) suggests that a major shift towards planktonic crustaceans as prey may have occurred even later – within historical times – in some extant penguin lineages (see below).

PENGUIN EVOLUTION, DISTRIBUTION AND CENOZOIC GLOBAL CHANGE

Sphenisciform fossils represent some of the oldest undisputed remains of crown clade birds. *Waimanu manneringi*, the oldest reported fossil penguin, is dated to 60.5–61.6 ma (Slack *et al.*, 2006), just a few million years after the Cretaceous–Paleogene boundary. This age provides a minimum estimate of the divergence of Pansphenisciformes from its extant sister taxon. Molecular estimates suggest that this divergence occurred during the Cretaceous (Baker *et al.*, 2006; Brown *et al.*, 2008), consistent with the large degree of morphological disparity between penguins and their sister taxon Procellariiformes. The divergence of penguins from Procellariiformes was almost certainly a separate event from the loss of aerial flight, which probably occurred some time afterwards. Because all known fossil penguins were clearly incapable of aerial flight, the length of the interval in which basal pansphenisciforms retained aerial flight is uncertain. One interesting hypothesis is that penguins first became flightless immediately after the K–P mass extinction wiped out marine reptiles, leaving a new aquatic predator niche open for exploitation by these birds (Fordyce & Jones, 1990; Kriwet & Benton, 2004; Ando, 2007). In this scenario, basal volant Pansphenisciformes survived the K–P extinction in the Southern Hemisphere and rapidly entered diving predator niches vacated by small mosasaurs and plesiosaurs. Aside from removal of competition, release from predatory pressure by large sharks and marine reptiles may also have been important (Ando, 2007). This hypothesis is consistent with the inferred time frame of early penguin evolution.

A separate question is when the crown clade Spheniscidae arose. Molecular divergence estimates suggest the basal divergence within Spheniscidae occurred in the Eocene (ca. 41 Ma) (Baker *et al.*, 2006). In contrast, the fossil record supports a much later origin, as the oldest crown penguin fossils are only mid–late Miocene (11–13 Ma) in age (Göhlich, 2007). Although hundreds of stem penguin fossils are known from multiple continents during the Eocene–Oligocene interval, all pre-Miocene taxa can be excluded from the crown clade based on retention of plesiomorphies (Ksepka & Clarke, 2010). This suggests that the late appearance of crown penguins in the fossil record approximates their true time of origin, rather than reflecting large gaps in the penguin fossil record (Clarke *et al.*, 2007). Clearly, a straight reading of either the molecular results or of the fossil record will lead to a dramatically different interpretation of the interplay between crown penguin evolution and major biotic, tectonic, and climactic events. The true age of the crown clade may lie somewhere between the extremes of 11 Ma and 41 Ma. Due to the nature of the fossil record, it can only provide a minimum estimate of the origin of any given group, and the discovery of older fossils can always push this estimate further back in time. However, the 41 Ma divergence estimate should now also be considered unreliable, because it was obtained using only external calibration points from previous dating studies. The apparent conflict between the molecular and fossil data may be best resolved by incorporating recently discovered crown penguin fossils as internal calibration points. The oldest crown penguins are the middle Miocene *Spheniscus muizoni* (Göhlich, 2007) and late Miocene *Madrynornis mirandus* (Acosta Hospitaleche *et al.*, 2007), and provide internal calibration points that can replace less reliable external calibrations for future investigations into the timing of penguin divergences.

Biogeographic studies suggest that dispersal played a major role in speciation and distribution for extant penguins (Bertelli & Giannini, 2005; Baker *et al.*, 2006; Ksepka *et al.*, 2006). Many living penguins are capable of traversing large swaths of open water. Several extant species range over thousands of miles throughout the year pursuing food sources or traveling to their breeding grounds (summarized in Williams, 1995). Reports of vagrant penguins arriving on continents distant from their home range also occur regularly (e.g. Condon, 1975; Williams, 1995). Thus, it is unsurprising that long-range dispersal seems to have occurred regularly throughout the Cenozoic. Major ocean currents and the emergence of new island land masses during the Cenozoic influenced the distribution patterns of penguins and are inferred to have provided new areas for vagrants to colonize, leading to allopatric speciation (Bertelli & Giannini, 2005). This capacity for dispersal appears to extend deep into the phylogenetic history of penguins. In the early Paleocene, penguins inhabited at least New Zealand. By the late Paleocene, penguins reached Antarctica, and further extended their range into South America by the middle Eocene and Australia by the late Eocene (Figure 6.7). Although extinct penguins likely inhabited many remote minor islands, only subfossil records have been reported from such localities. However, these islands can still reinforce our understanding of penguin biogeography. For example, geologically young volcanic islands inhabited by penguins appear in terminal branches in biogeographical reconstructions, rather than being reconstructed as breeding grounds for deep nodes (Bertelli & Giannini, 2005; Ksepka et al., 2006).

Vicariance also needs to be considered as a driver of early penguin evolution. The breakup of Gondwana was nearly complete by the time the first penguins appear, but two major plate tectonic events related to the final breakup frame Cenozoic penguin biogeography. First, the opening of the Tasman Passage between Antarctica and Australia near the Eocene–Oligocene boundary completely separated these continents and allowed formation of a deepwater gateway. Second, the opening of the Drake Passage between South American and Antarctica in the Oligocene finally severed the "stepping stone" connection of islands between the continents. These events are significant to penguin biogeography not only in breaking the continuity of the coastlines, but in creation of new major ocean currents. The final separation of Australia and Antarctica allowed cold water currents to begin circling Antarctica. As South America drifted further north, the opening of the Drake Passage allowed the Antarctic circumpolar current to encircle Antarctica entirely (Lawyer & Gahagan, 1998). These events may be expected to have caused increasing provincialism of penguin faunas, and increasingly less frequent dispersal between Antarctica and other landmasses as Antarctica became more isolated. Unfortunately, the Antarctic record of penguin fossils is almost entirely restricted to a single locality sampling the late Eocene (La Meseta Formation of Seymour Island), and the record on Australia consists of a handful of fragmentary specimens. Thus, we do not yet have sufficient temporal and geographic sampling to reliably infer the effects of these major vicariance events on penguins.

In terms of diversity, regional species densities in the fossil record mirror present-day levels to a large degree. New Zealand, South America, and Antarctica are the richest areas in terms of fossil species diversity and these three areas all host numerous breeding taxa today. At present, the highest diversity for any single locality and horizon is in the late Eocene at Seymour Island. Ten described species and at least one distinct undescribed taxon are known from a single unit of the La Meseta Formation at Seymour Island (see Myrcha et al., 2002; Jadwiszczak, 2008). The Duntroonian (late Oligocene) penguin fauna of New Zealand is likewise very diverse, with eight known species (two awaiting formal description). In contrast, few fossil penguin species are known from Australia and Africa, which today each host only a single native species of penguin. In Africa, several fossil species have been described (Simpson, 1971a, 1973, 1975a, 1979a, 1979b), but because of poor stratigraphic constraint it remains uncertain if all overlapped in time. In Australia, no more than two species can be demonstrated to occur at any time interval.

Although penguins have maintained a widespread distribution and at least moderate species diversity in the face of Cenozoic change, the once diverse giant penguin fauna vanished near the opening of the Neogene. Giant forms are last recorded in the late Oligocene at most localities, but appear to persist into the early Miocene in Australia (Gill, 1959; Simpson, 1959). Causes for their ultimate demise remain elusive (Fordyce & Jones, 1990; Ando, 2007). Such taxa persisted beyond the initial onset of Cenozoic global cooling during the early Eocene as well as an episode of

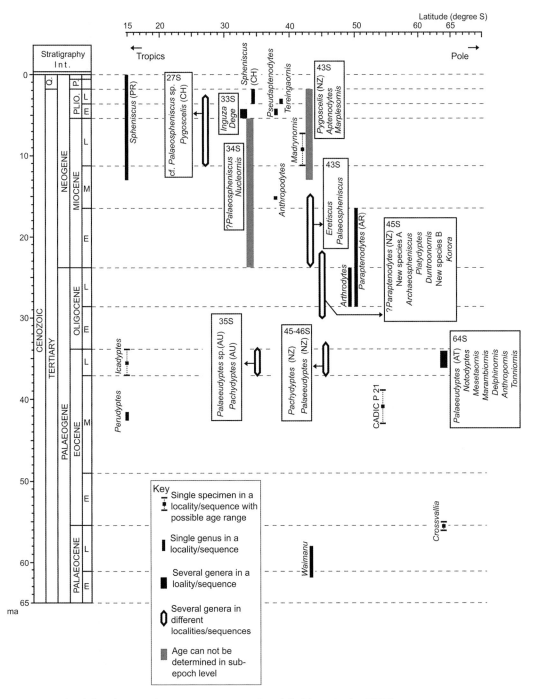

Fig. 6.7 Latitudinal distribution of penguins over time. (Modified from Ando, 2007.)

abrupt cooling at the Oligocene–Eocene (Zachos et al., 2001), suggesting climate change did not drive their extinction (Clarke et al., 2007). One interesting hypothesis is that the competition from marine mammals (cetaceans and pinnipeds) was responsible for the extinction of giant penguin taxa (Stonehouse, 1969; Simpson, 1976; Olson, 1985). Livezey (1989) suggested that decreased maneuverability and increased vulnerability to marine mammalian predators could account for the disappearance of large penguin taxa. Timing is also an issue. Archaic pinnipeds are known from the late Oligocene of the Northern Hemisphere (Bardet, 1994; Démére et al., 2003) but their invasion of Southern Hemisphere waters appears not to have occurred until the middle Miocene at low latitudes and the latest Miocene at high latitudes (Cozzuol, 2001; de Muizon & Bond, 1982; Fordyce, 1989). Perhaps a more likely competition scenario posits the divergence of new clades of whales during the late Oligocene as the leading factor in the demise of giant penguins (Ando, 2007). The global record of marine vertebrates is poor in the Oligocene, hampering our understanding of the timing of arrival of marine mammals and extinction of giant penguins in different regions. Because inferring competition from the fossil record remains difficult, the competition hypothesis yet to be tested in a quantitative manner (Olson, 1985).

TOWARDS THE RECENT

No extinctions among penguins have been documented in historical records. However, recent work suggests a close relative of the yellow-eyed penguin was extirpated by humans some time around AD 1500 (Boessenkool et al., 2009). Molecular and morphometric data support the presence of a previously unrecognized species, *Megadyptes waitaha*, in pre-settlement New Zealand (Boessenkool et al., 2009). Given the timing of extinction and the association of *Megadyptes waitaha* bones with archeological sites, the case for anthropogenic extinction seems strong.

Another possible instance of anthropogenic extinction exists. Four penguin bones recovered

from a midden on Hunter Island, Tasmania and dated to 760 ± 60 yr BP were referred to a new species, *Tasidyptes hunteri* (van Tets & O'Conner, 1983). If this fossil does represent a distinct species, it would represent a second case of anthropogenic extinction in penguins. However, whether *Tasidyptes hunteri* is adequately diagnosable has been considered debatable (Fordyce & Jones, 1990). There is no clear association between the referred specimens and some or all elements may belong to *Eudyptes chrysocome*, a species known to occur in Australia (e.g. Condon, 1975). The referred tarsometatarsus and coracoid of *Tasidyptes hunteri* are not distinct from extant *Eudyptes* (Ksepka, 2007), leaving only the posterior sacral "vertebrae with long slender lateral processes" as a possible diagnostic character, certainly in need of quantitative evaluation.

Pressure from human hunting has fortunately had less of an impact on penguins than it has on other flightless birds. Famously, the closest ecological equivalent of penguins, the great auk, was wiped out by overhunting. While less persecuted, penguins have nonetheless weathered major pressures from human activities. In historical times, vast numbers of penguin eggs were taken for food (Frost et al., 1976) and significant numbers of adult and immature birds were harvested for the rendering of their oil at some colonies (Stokes & Soper, 1987). Populations of penguins were thus exterminated from several islands. Today, hunting of penguins is essentially nonexistent and a greater threat is posed by introduced predators (Stahel & Gales, 1987; Dann, 1992; Massaro & Blair, 2003). Habitat alteration, including modifications of breeding beaches for agriculture (Seddon & Davies, 1989), disruptive harvesting of guano at active colonies (Stokes & Boersma, 1991), and degradation of marine habitats from oil spills (Gandini et al., 1994; Garcìa-Borboroglu et al., 2006), also imperil penguins in many parts of the world.

Another major threat to living penguins is the overharvesting of many preferred prey species (Boersma et al., 1994; Bingham, 2002). Evidence exists for anthropogenic activity driving shifts in penguin ecology in the past as well. Recent investigations into the diets of Pleistocene–Holocene penguins reveals that a shift from a primarily fish to a primarily krill-based diet occurred within historic times in the extant Adélie penguin. Emslie & Patterson (2007) compared stable isotope values from fossil eggshells collected at abandoned Adélie penguin colonies to values from historical and modern eggshells and found pronounced shifts in $\delta^{13}C$ and $\delta^{15}N$ values. These shifts indicate an abrupt shift in diet, consistent with a move from a fish-based to a krill-based diet. Emslie & Patterson (2007) hypothesized that penguins only recently began to rely on krill as a major component of their diets after baleen whale populations were decimated during the historic whaling era. They further conclude that with Antarctic fish stocks currently depleted, increased exploitation of krill for aquiculture could significantly strain food resources for Adélie penguins. Interestingly, Emslie & Patterson's (2007) research suggests that overharvesting of prey species could have a much more profound effect on penguins than direct hunting of the birds themselves has had in the past.

A final serious threat facing all penguins today is climate change. Radiocarbon dating of remains from abandoned Adélie penguin colonies shows that penguin distribution in the Antarctica has been dynamic over the past few tens of thousands of years (Emslie *et al.*, 2007). Climate shifts appear to have driven the colonization and abandonment of localities (Baroni & Orombelli, 1994) and also to have influenced shifts in diet (Emslie & McDaniel, 2002). Species that exist at the extremes of the latitudinal range of Spheniscidae are particularly vulnerable to rapid warming. Emperor penguins breed only on sea-ice, and thus some colonies risk physical loss of their breeding grounds to receding ice shelves. Population decreases for *Aptenodytes forsteri* and other ice-adapted species such as *Pygoscelis adeliae* have been linked to receding sea-ice (Barbraud, 2001; Forcada *et al.*, 2006). At the other extreme, *Spheniscus mendiculus* (the rarest living species) suffered large population losses during recent El Niño events due to a decrease in food availability caused by water temperature increase (Boersma, 1998: Vargas

et al., 2006). The restricted range of these penguins leaves them vulnerable to any major local disruption or to global warming generally: isolated on the Galápagos Islands, they have nowhere to go to escape unfavorable environmental shifts. Baker *et al.* (2006) warned that based on ancestral distribution patterns, global warming might drive species towards higher latitudes, resulting in multiple extinctions. This threat remains very real.

ACKNOWLEDGMENTS

We thank Gareth Dyke and Gary Kaiser for the opportunity to contribute to this volume. R. Ewan Fordyce, Julia Clarke, and Kristin Lamm provided valuable feedback on portions of this manuscript. Barbara Harmon and the New England Aquarium kindly permitted use of the penguin artwork in Figure 6.3. This project is based largely on collections studies and we thank R. Ewan Fordyce, Sue Heath, Alan Tennyson, Paul Scofield, Norton Hiller, Mark Norell, Carl Mehling, Joel Cracraft, Paul Sweet, Peter Capainolo, Mark Florence, Storrs Olson, and Pat Holroyd for access to fossil and extant materials.

REFERENCES

Acosta Hospitaleche C. 2003. *Paraptenodytes antarcticus* (Aves: Sphenisciformes) en la Formación Puerto Madryn (Mioceno Tardío Temprano), Provincia de Chubut, Argentina. *Revista Española de Paleontología* **18**: 179–183.

Acosta Hospitaleche C. 2004. *Los pinguinos (Aves, Sphenisciformes) fosiles de Argentina. Sistematica, biogeografia y evolution.* PhD Dissertation, Universidad Nacional de La Plata.

Acosta Hospitaleche C. 2005. Systematic revision of *Arthrodytes* Ameghino, 1905 (Aves, Spheniscidae) and its assignment to the Paraptenodytinae. *Neues Jahrbuch für Geologie und Paläontologie - Abhandlungen* **7**: 404–414.

Acosta Hospitaleche C. 2007. Revísion sistemática de *Palaeospheniscus biloculata* (Simpson) nov. comb. (Aves, Spheniscidae) de la Formación Gaiman (Mioceno Temprano), Chubut, Argentina. *Ameghiniana* **44**: 417–426.

Acosta Hospitaleche C, Canto J. 2005. Primer registro de cráneos asignados a *Palaeospheniscus* (Aves, Spheniscidae) procedentes de la Formación Bahía Inglesa (Mioceno Medío-Tardio), Chile. *Revista Chilena de Historia Natural* **78**: 219–231.

Acosta Hospitaleche C, Stucchi M. 2005. Nuevos restos terciarios de Spheniscidae (Aves, Sphenisciformes) procedentes de la costa del Perú. *Revista Española de Paleontología* **20**: 1–5.

Acosta Hospitaleche C, Tambussi C, Cozzuol M. 2004. *Eretiscus tonnii* (Simpson) (Aves, Sphenisciformes): materiales adicionales, status taxonómico y distribución geográfica. *Revista del Museo Argentino de Ciencias Naturales* **6**: 233–237.

Acosta Hospitaleche C, Chávez M, Fritisn O. 2006. Pingüinos fósiles (*Pygoscelis calderensis* nov. sp.) en la Formación Bahía Inglesa (Mioceno Medio-Plioceno), Chile. *Revista Geológica de Chile* **33**: 327–338.

Acosta Hospitaleche C, Tambussi C, Donato M, Cozzuol M. 2007. A new Miocene penguin from Patagonia and its phylogenetic relationships. *Acta Palaeontologica Polonica* **52**: 299–314.

Acosta Hospitaleche C, Castro L, Tambussi C, Scasso RA. 2008. *Palaeospheniscus patagonicus* (Aves, Sphenisciformes): new discoveries from the early Miocene of Argentina. *Journal of Paleontology* **82**: 565–575.

Ameghino F. 1891. Enumeracion de las aves fosiles de la republica Argentina. *Revista Argentina de Historia Natural* **1**: 441–445.

Ameghino F. 1895. Sur les oiseaux fossiles de Patagonie. *Boletin del Instituto Geografico de Argentina* **15** (501–602).

Ameghino F. 1899. *Sinópsis geológico-paleontológica.* La Plata: Suplemento (adiciones y correciones).

Ameghino F. 1901. L'Age des formations sedimentaires de Patagonie. *Anales de la Scoiedad Cientifica Argentina* **51**: 20–29, 65–91.

Ameghino F. 1905. Enumeración de los impennes fosiles de Patagonia y de la Isla Seymour. *Anales del Museum Nacional de Buenos Aires (Series 36)* **6**: 97–167.

Ando T. 2007. *New Zealand fossil penguins: origin, pattern, and process.* PhD Dissertation, University of Otago.

Baker AJ, Pereira SL, Haddrath OP, Edge K-A. 2006. Multiple gene evidence for expansion of extant penguins out of Antarctica due to global cooling. *Proceedings of the Royal Society of London Series B (Biological Sciences)* **217**: 11–17.

Banks J, van Buren A, Cherel Y, Whitfield JB. 2006. Genetic evidence for three species of rockhopper penguins, Eudyptes chrysocome. *Polar Biology* **30**: 61–67.

Banks JC, Mitchell AD, Waas JR, Paterson AM. 2002. An unexpected pattern of molecular divergence within the blue penguin (*Eudyptula minor*) complex. *Notornis* **49**: 29–38.

Bannasch R. 1994. Functional anatomy of the "flight" apparatus in penguins. In *Mechanics and Physiology of Animal Swimming*, Maddock L, Bone Q, Rayner, JMV (eds). Cambridge: Cambridge University Press; 163–192.

Barbraud C, Weimerskirch H. 2001. Emperor penguins and climate change. *Nature* **411**: 183–186.

Bardet N. 1994. Extinction events among Mesozoic marine reptiles. *Historical Biology* **7**: 313–324.

Baroni C, Orombelli G. 1994. Abandoned penguin rookeries as Holocene paleoclimatic indicators in Antarctica. *Geology* **22**: 23–26.

Beddard FE. 1898. *The Structure and Classification of Birds.* London: Longmans, Green & Co.

Bertelli S, Giannini NP. 2005. A phylogeny of extant penguins (Aves: Sphenisciformes) combining morphology and mitochondrial sequences. *Cladistics* **21**: 209–239.

Bertelli S, Giannini NP, Ksepka DT. 2006. Redescription and phylogenetic position of the early Miocene penguin *Paraptenodytes antarcticus* from Patagonia. *American Museum Novitates* **3525**: 1–36.

Bingham M. 2002. The decline of Falkland Islands penguins in the presence of a commercial fishing industry. *Revista Chilena de Historia Natural* **75**: 805–818.

Boersma PD. 1975. Adaptations of Galapagos penguins for life in two different environments. In *The Biology of Penguins*, Stonehouse B (ed.). London: Macmillan; 101–114.

Boersma PD. 1998. Population trends of Galápagos penguins: impacts of El Niño and La Niña. *Condor* **110**: 245–253.

Boersma PD, Rebstock GA, Stokes DL. 2004. Why penguin eggshells are thick. *The Auk* **121**: 148–155.

Boessenkool S, Austin JJ, Worthy TH, Scofield P, Cooper A, Seddon PJ, Waters JM. 2009. Relict or colonizer? Extinction and range expansion of penguins in southern New Zealand. *Proceedings of the Royal Society of London Series B (Biological Sciences)* **276**: 815–821.

Bourdon E, Bouya B, Iarochene M. 2005. Earliest African neornithine bird: a new species of Prophaethontidae (Aves) from the Paleocene of Morocco. *Journal of Vertebrate Paleontology* **25**: 157–170.

Bowmaker JK, Martin GR. 1985. Visual pigments and oil droplets in the penguin, Spheniscus humboldti. *Journal of Comparative Physiology A: Neuroethology, Sensory, Neural, and Behavioral Physiology* **156**: 71–77.

Brodkorb P. 1963. Catalogue of fossil birds, part 1 (Archaeopterygiformes through Ardeiformes). *Bulletin of the Florida State Museum* **7**: 179–293.

Brown JW, Rest JS, García-Moreno J, Sorenson MD, Mindell DP. 2008. Strong mitochondrial DNA support for a Cretaceous origin of modern avian lineages. *BMC Biology* **6**: 6.

Chandler AC. 1916. A study of the structure of feathers with reference to their taxonomic significance. *University of California Publications in Zoology* **13**: 243–446.

Chávez M. 2007. Sobre la presencia de Paraptenodytes y Palaeospheniscus (Aves: Sphenisciformes) en la Formación Bahía Inglesa, Chile. *Revista Chilena de Historia Natural* **80**: 255–259.

Chubb AL. 2004. New nuclear evidence for the oldest divergences among neognath birds: phylogenetic utility of ZENK (I). *Molecular Phylogenetics and Evolution* **30**: 140–151.

Clark BD, Bemis W. 1979. Kinematics of swimming penguins at the Detroit Zoo. *Journal of Zoology* **188**: 411–428.

Clarke JA, Ksepka DT, Salas-Gismondi R, Altamirano AJ, Shawkey MD, D'Alba L, Vinther J, DeVries TJ, Baby P. 2010. Fossil evidence for evolution of the shape and color of penguin feathers. *Science* **330**: 954-957.

Clarke JA, Olivero EB, Puerta P. 2003. Description of the earliest fossil penguin from South America and first Paleogene vertebrate locality of Tierra del Fuego, Argentina. *American Museum Novitates* **3423**: 1–18.

Clarke JA, Ksepka DT, Stucchi M, Urbina M, Giannini N, Bertelli S, Narváez Y, Boyd CA. 2007. Paleogene equatorial penguins challenge the proposed relationship between biogeography, diversity, and Cenozoic climate change. *Proceedings of the National Academy of Sciences* **104**: 11545–11550.

Condon HT. 1975. *Checklist of the Birds of Australia, I. Non-passerines.* Royal Australasian Ornithologists Union: Melbourne.

Cooper A, Penny D. 1997. Mass survival of birds across the Cretaceous–Tertiary boundary: molecular evidence. *Science* **275**: 1109–1113.

Cozzuol MA. 2001. A "northern" seal from the Miocene of Argentina: Implications for phocid phylogeny and biogeography. *Journal of Vertebrate Paleontology* **21**: 415–421.

Cozzuo MA, Fordyce RE, Jones CM. 1991. La presencia de *Eretiscus tonnii* (Aves, Spheniscidae) en el Mioceno Temprano de Nueva Zealandia y Patagonia. *Ameghiniana* **28**: 406.

Cozzuol MA, Tambussi CP, Noriega JI. 1993. Un pigüino (Aves, Spheniscidae) de la Formación Puerto Madryn (Mioceno medio) en Península Valdés, Chubut, Argentina, con importantes implicancias filogenéticas. *Ameghiniana* **30**: 327–328.

Cracraft J. 1985. Monophyly and phylogenetic relationships of the Pelecaniformes: a numerical cladistic analysis. *The Auk* **102**: 834–853.

Cracraft J. 1988. The major clades of birds. In *The Phylogeny and Classification of the Tetrapods, Volume 1: Amphibians, Reptiles, Birds*, Benton MJ (ed.). Oxford: Clarendon Press; 339–361.

Cracraft J, Barker FK, Braun J, Harshman J, Dyke GJ, Feinstein J, Stanley S, Cibois A, Schikler P, Beresford P, García-Moreno J, Sorenson MD, Yuri T, Mindell DP. 2004. Phylogenetic relationships among modern birds (Neornithes): towards an avian tree of life. In *Assembling the Tee of Life*, Cracraft J, Donoghue MJ (eds). New York: Oxford Press; 468–489.

Culik B, Wilson R, Bannasch R. 1994. Underwater swimming at low energetic cost by pygoscelid penguins. *Journal of Experimental Biology* **197**: 65–78.

Dann P. 1992. Distribution, population trends and factors influencing the population size of Little Penguins *Eudyptula minor* on Phillip Island, Victoria. *Emu* **91**: 263–272.

Davis LS, Darby JT. 1990. *Penguin Biology.* San Diego: Academic Press.

Davis LS, Renner M. 2003. *Penguins.* New Haven: Yale University Press.

Démére TA, Berta A, Adam PJ. 2003. Pinnipedimorph evolutionary biogeography. *Bulletin of the American Museum of Natural History* **279**: 32–76.

De Muizon C, Bond M. 1982. Le Phocidae (Mammalia) Miocene de la formation Parana (Entre Rios, Argentina). *Bulletin de Museum National d'Historie Naturelle, section C, 4eme serie* **4**: 165–207.

Dunning Jr. JB (ed.). 1993. *CRC Handbook of Avian Body Masses.* Boca Raton, FL: CRC Press.

Emslie SD, Guerra Correa C. 2003. A new species of penguin (Spheniscidae: *Spheniscus*) and other birds from the late Pliocene of Chile. *Proceedings of the Biological Society of Washington* **116**: 308–316.

Emslie SD, McDaniel JD. 2002. Adélie penguin diet and climate change during the middle to late Holocene in northern Marguerite Bay, Antarctic Peninsula. *Polar Biology* **25**: 222–229.

Emslie SD, Patterson WP. 2007. Abrupt recent shift in δC and δN values in Adélie penguin eggshell in Antarctica. *Proceedings of the National Academy of Sciences* **104**: 11666–11669.

Emslie SD, Woehler EJ. 2005. A 9000-year record of Adélie penguin occupation and diet in the Windmill Islands, East Antarctica. *Antarctic Science* **17**: 57–66.

Emslie SD, Coats L, Licht K. 2007. A 45, 000 yr record of Adélie penguins and climate change in the Ross Sea, Antarctica. *Geology* **35**: 61–64.

Ericson PGP, Anderson CL, Britton T, Elzanowski A, Johansson US, Källersjö M, Ohlson JI, Parsons TJ, Zuccon D, Mayr G. 2006. Diversification of Neoaves: integration of molecular sequence data and fossils. *Biology Letters* **4**: 543–547.

Fain MG, Houde P. 2004. Parallel radiations in the primary clades of birds. *Evolution* **58**: 2558–2573.

Finlayson HH. 1938. On the occurrence of a fossil penguin in Miocene beds in South America. *Transactions of the Royal Society, South Australia* **62**: 14–17.

Forcada J, Trathan PN, Reid K, Murphy EJ, Croxall JP. 2006. Contrasting population changes in sympatric penguin species in association with climate warming. *Global Change Biology* **12**: 411–423.

Fordyce RE. 1989. Origins and evolution of Antarctic marine mammals. In *Origins and Evolution of Antarctic Biota*, Crame JA (ed.). London: The Geological Society, Special Publication 47; 269–281.

Fordyce RE. 1991. A new look at the Fossil Vertebrate Record of New Zealand. In *Vertebrate Palaeontology of Australasia*, Vickers-Rich P (ed.). Melbourne: Pioneer Design Studio and Monash University; 1191–1316.

Fordyce RE. 2003. Cetacean evolution and Eocene-Oligocene oceans revisited. In *From Greenhouse to Icehouse: The Marine Eocene–Oligocene Transition*, Prothero DR, Ivany LC, Nesbitt EA (eds), New York: Columbia University Press; 154–170.

Fordyce RE, Jones CM. 1990. Penguin history and new fossil material from New Zealand. In *Penguin Biology*, Davis LS, Darby JT (eds). San Diego: Academic Press; 419–446.

Frost PGH, Siegfried WR, Greenwood PJ. 1975. Arteriovenous heat exchange systems in the Jackass penguin Spheniscus demersus. *Journal of Zoology* **175**: 231–241.

Frost PGH, Siegfried WR, Cooper J. 1976. Conservation of the Jackass Penguin (*Spheniscus demersus* (L.)). *Biological Conservation* **9**: 79–99.

Fürbringer M. 1888. *Untersuchungen Morphologie und Systematick der Vögel, Volumes I/II.* Amsterdam: Verlag von TJ van Holkema.

Gadow, H. 1889. On the taxonomic value of the intenstinal convolutions in birds. *Proceedings of the Zoological Society of London* **1889**: 303–316.

Gadow M. 1893. *Bronn's Klassen und Ordnungen des Thier-Reichs, Volume 6.* Leipzig: CF Winter.

Gandini P, Boersma D, Frere E, Gandini M, Holik T, Lichtschein V. 1994. Magellanic penguins (*Spheniscus magellanicus*) affected by chronic petroleum pollution along coast of Chubut, Argentina. *The Auk* **111**: 20–27.

Garcìa-Borboroglu P, Boersma PD, Ruoppolo V, Reyes L, Rebstock GA, Griot K, Heredia SR, Adornes AC, da Silva RP. 2006. Chronic oil pollution harms Magellanic penguins in the Southwest Atlantic. *Marine Pollution Bulletin* **52**: 193–198.

Gervais P, Alix E. 1877. Ostéologie et myologie des manchots ou Sphéniscidés. *Journal de Zoologie (Paris)* **6**: 424–472.

Giannini NP, Bertelli S. 2004. Phylogeny of extant penguins based on integumentary and breeding characters. *The Auk* **121**: 422–434.

Gill ED. 1959. Provenance of a fossil penguin from western Victoria. *Proceedings of the Royal Society of Victoria* **71**: 121–123.

Göhlich UB. 2007. The oldest fossil record of the extant penguin genus *Spheniscus* – a new species from the Miocene of Peru. *Acta Paleontologica Polonica* **52**: 285–298.

Grant-Mackie JA, Simpson GG. 1973. Tertiary penguins from the North Island of New Zealand. *Journal of the Royal Society of New Zealand* **3**: 441–452.

Groth JG, Barrowclough GF. 1999. Basal divergences in birds and the phylogenetic utility of the nuclear RAG–1 gene. *Molecular Phylogenetics and Evolution* **12**: 115–123.

Gruson ES. 1976. *A Checklist of the Birds of the World.* London: Collins.

Hackett SJ, Kimball RT, Reddy S, Bowie RCK, Braun EL, Braun MJ, Chojnowski JL, Cox WA, Han K-L, Harshman J, Huddleston CJ, Marks BD, Miglia KJ, Moore WS, Sheldon FH, Steadman DW, Witt CC, Yuri. T. 2008. A phylogenomic study of birds reveals their evolutionary history. *Science* **320**: 1763–1768.

Halsey LG, Butler PJ, Blackburn TM. 2006. A phylogenetic analysis of allometry of diving. *American Naturalist* **167**: 267–287.

Harrison GL, McLenachan PA, Phillips MJ, Slack KE, Cooper A, Penny D. 2004. Four new avian mitochondrial genomes help get to basic evolutionary questions in the Late Cretaceous. *Molecular Biology and Evolution* 21: 974–983.

Hector J. 1872. On the remains of a gigantic penguin (*Palaeeudyptes antarcticus*, Huxley) from the Tertiary rocks on the west coast of Nelson. *Transactions and Proceedings of the New Zealand Institute* 4: 341–346.

Ho CY-K, Prager EM, Wilson AC, Osuga DT, Feeney RE. 1976. Penguin evolution: protein comparisons demonstrate phylogenetic relationships to flying aquatic birds. *Journal of Molecular Evolution* 8: 271–282.

Huxley TH. 1859. On a fossil bird and a cetacean from New Zealand. *Quarterly Journal of the Geological Society* 15: 670–677.

Jadwiszczak P. 2001. Body size of Eocene Antarctic penguins. *Polish Polar Research* 22: 147–158.

Jadwiszczak P. 2003. The early evolution of Antarctic penguins. In *Antarctic Biology in a Global Context*, Huiskes AHL, Gieskes WWC, Rozema J, Schorno RML, van der Vies SM, Wolff WJ (eds). Leiden: Backhuys Publishers; 148–151.

Jadwiszczak P. 2006a. Eocene penguins of Seymour Island, Antarctica: Taxonomy. *Polish Polar Research* 27: 3–62.

Jadwiszczak P. 2006b. Eocene penguins of Seymour Island, Antarctica: the earliest record, taxonomic problems and some evolutionary considerations. *Polish Polar Research* 27: 287–302.

Jadwiszczak P. 2008. An intriguing penguin bone from the late Eocene of Seymour Island, Antarctica. *Antarctic Science* 20: 589–590. doi: 10.1017/S0954102008001405.

Jenkins RJF. 1974. A new giant penguin from the Eocene of Australia. *Palaeontology* 17: 291–310.

Jenkins R.J.F. 1985. *Anthropornis nordenskjoeldi* Wiman, 1905: Nordenskjoeld's giant penguin. In *Kadimakara, Extinct Vertebrates of Australia*, Rich PV, van Tets GF (eds). Victoria: Pioneer Design Studio; 183–187.

Johansson LC, Wetterholm Aldrin BS. 2002. Kinematics of diving Atlantic puffins (Fratercula arctica L.): evidence for an active upstroke. *Journal of Experimental Biology* 205: 371–378.

Jouventin P, Barbraud C, Rubin M. 1995. Adoption in the emperor penguin, Aptenodytes forsteri. *Animal Behaviour* 50: 1023–1029.

Kaiser GK. 2007. *The Inner Bird: Anatomy and Evolution.* Vancouver: UBC Press.

Kinsky FC, Falla RA. 1976. A subspecific revision of the Australian Blue Penguin (*Eudyptula minor*) in the New Zealand area. *Records of the Natural History Museum of New Zealand* 1: 105–126.

Kooyman GL. 1989. *Diverse Divers: Pysiology and Behavior (Zoophysiology)* Berlin: Springer-Verlag.

Kooyman GL, Drabek CM, Elsner R, Campbell WB. 1971. Diving behaviours of the Emperor penguin *Aptenodytes forsteri*. *The Auk* 88: 775–795.

Kooyman GL, Davis RW, Croxall JP, Costa DP. 1982. Diving depths and energy requirements of King penguins. *Science* 217: 726–727.

Kooyman GL, Cherel Y, Le Maho Y, Croxall JP, Thorson PH, Ridoux V. 1992. Diving patterns and energetics during foraging cycles in King penguins. *Ecological Monographs* 62: 143–163.

Kriwet J, Benton MJ. 2004. Neoselachian (Chondrichthyes, Elasmobranchii) diversity across the Cretaceous–Tertiary boundary. *Palaeogeography, Palaeoclimatology, Palaeoecology* 214: 181–194.

Ksepka DT. 2007. Phylogeny, histology and functional morphology of fossil penguins (Aves: Sphenisciformes). PhD Dissertation, Columbia University.

Ksepka DT, Bertelli S. 2006. Fossil penguin (Aves: Sphenisciformes) cranial material from the Eocene of Seymour Island (Antarctica). *Historical Biology* 18: 389–395.

Ksepka DT, Clarke JA. 2010. The basal penguin (Aves: Sphenisciformes) *Perudyptes devriesi* and a phylogenetic evaluation of the penguin fossil record. *Bulletin of the American Museum of Natural History* 337: 1–77.

Ksepka DT, Bertelli S, Giannini NP. 2006. The phylogeny of the living and fossil Sphenisciformes (penguins). *Cladistics* 22: 412–441.

Ksepka DT, Clarke JA, DeVries TJ, Urbina M. 2008. Osteology of *Icadyptes salasi*, a giant penguin from the Eocene of Peru. *Journal of Anatomy* 213: 131–147.

Kuroki M, Kato A, Watanuki Y, Niizuma Y, Takahashi A, Naito Y. 2003. Diving behavior of an epipelagically feeding alcid, the Rhinoceros Auklet (*Cerorhinca monocerata*). *Canadian Journal of Zoology* 81: 1249–1256.

Lambert DM, Ritchie PA, Millar C.D. Holland B, Drummond AJ, Baroni C. 2002. Rates of evolution in ancient DNA from Adélie penguins. *Science* 295: 2270–2273.

Latham J. 1821. *General History of Birds.* London: Winchester, Jacob and Johnson.

Lawyer LA, Gahagan LM. 1998. Opening of the Drake Passage and its impact on Cenozoic ocean circulation. In *Tectonic Boundary Conditions for Climate*

Reconstructions, Crowley TJ, Burkes KG (eds). New York: Oxford University Press; 212–226.

Livezey BC. 1989. Morphometric patterns in recent and fossil penguins (Aves, Sphenisciformes). *Journal of the Zoological Society of London* 219: 269–307.

Livezey BC, Zusi RL. 2006. Higher-order phylogenetics of modern birds (Theropoda, Aves: Neornithes) based on comparative anatomy. I. Methods and characters. *Bulletin of the Carnegie Museum of Natural History* 37: 1–544.

Livezey BC, Zusi RL. 2007. Higher-order phylogeny of modern birds (Theropoda, Aves: Neornithes) based on comparative anatomy: II. Analysis and discussion. *Zoological Journal of the Linnean Society* 149: 1–95.

Lovvorn JR, 2001. Upstroke thrust, drag effects, and stroke glide cycles in wing-propelled swimming by birds. *American Zoologist* 41: 154–165.

Lowe PR. 1933. On the primitive characters of the penguins, and their bearing on the phyogeny of birds. *Proceedings of the Zoological Society of London* 2: 483–538.

Lowe PR. 1939. Some additional notes on Miocene penguins in relation to their origin and systematics. *Ibis* 81: 281–296.

Marples BJ. 1952. Early Tertiary penguins of New Zealand. *New Zealand Geological Survey Paleontological Bulletin* 20: 1–66.

Marples BJ. 1953. Fossil penguins from the mid-Tertiary of Seymour Island. *Falkland Islands Dependencies Survey Scientific Reports* 5: 1–15.

Marples BJ. 1960. A fossil penguin from the Late Tertiary of North Canterbury. *Records of the Canterbury Museum* 7: 185–195.

Marples BJ. 1962. Observations on the history of penguins. In *The Evolution of Living Organisms*, Leeper GW (ed.). London: Cambridge University Press; 408–416.

Marples BJ. 1974. Fossil penguins. *New Zealand Nature Heritage* 1: 142–144.

Marples BJ, Flemming CA. 1963. A fossil penguin bone from Kawhia, New Zealand. *New Zealand Journal of Geology and Geophysics* 6: 189–192.

Martínez I. 1992. Order Sphenisciformes. In *Handbook of the Birds of the World, Volume 1 Ostrich to Ducks*, del Hoyo J, Elliott A, Sargatal J (eds). Barcelona: Lynx Edicions; 140–161.

Massaro M, Blair D. 2003. Comparison of population numbers of yellow-eyed penguins, *Megadyptes antipodes*, on Stewart Island and adjacent cat-free islands. *New Zealand Journal of Ecology* 27: 107–113.

Mayr G. 2005. Tertiary plotopterids (Aves, Plotopteridae) and a novel hypothesis on the phylogenetic relationships of penguins (Spheniscidae). *Journal of Zoological Systematics and Evolutionary Research* 43: 61–71.

Mayr G, Clarke J. 2003. The deep divergences of neornithine birds: a phylogenetic analysis of morphological characters. *Cladistics* 19: 527–553.

McEvey AR, Vestjens WJM. 1973. Fossil penguin bones from Macquarie Island, Southern Ocean. *Proceedings of the Royal Society of Victoria* 86: 151–174.

McKee JW. 1987. The occurrence of the Pliocene penguin *Tereingaornis moisleyi* (Sphenisciformes: Spheniscidae) at Hawera, Taranaki, New Zealand. *New Zealand Journal of Zoology* 14: 557–561.

McKitrick MC. 1991. Phylogenetic analysis of avian hindlimb musculature. *University of Michigan Museum of Zoology Miscellaneous Publications* 179: 1–85.

Meister W. 1962. Histological structure of the long bones of penguins. *The Anatomical Record* 143: 377–387.

Menzbier M. 1887. Vergleichende osteologie der pinguine in anwendung zur haupteintheilung der vogel. *Bulletin de la Societe Imperiale des Naturalistes de Moscou* 1: 483–587.

Mey E, Chastel O, Beaucournu JC. 2002. A "penguin" chewing louse *Nesiotinus* on a Kergelen diving-petrel (*Pelecanoides urinatrix exsul*): an indication of a phylogenetic relationship? *Journal of Ornithology* 143: 472–476.

Moreno FP, Mercerat A. 1891. Paleontologia Argentina I. *Catalogo de los Pajaros de la Republica Argentina Conservados en el Museo de la Plata* 1: 8–71.

Myrcha A, Tatur A, del Valle R. 1990. A new species of fossil penguin from Seymour Island, West Antarctica. *Alcheringa* 14: 195–205.

Myrcha A, Jadwiszczak P, Tambussi CP, Noriega JI, Gazdzicki A, Tatur A, del Valle RA. 2002. Taxonomic revision of Eocene Antarctic penguins based on tarsometatarsal morphology. *Polish Polar Research* 23: 5–46.

Norell MA. 1992. The effect of phylogeny on temporal diversity and evolutionary tempo. In *Extinction and Phylogeny*, Novacek MJ, Wheeler QD (eds), New York: Columbia University Press; 89–118.

O'Hara RJ. 1989. Systematics and the study of natural history, with an estimate of the phylogeny of penguins (Aves: Spheniscidae). PhD dissertation, Harvard University.

Oliver WRB. 1930. *New Zealand Birds*. Fine Arts: Wellington.

Olson SL. 1985. The fossil record of birds. In *Avian Biology*, Framer DS, King JR, Parkes KC (eds). New York: Academic Press; 79–238.

Peters JE. 1931. *Check-list of Birds of the World, Volume 1*. Cambridge, MA: Harvard University Press.

Piatt JF, Nettleship DN. 1985. Diving depths of four aclids. *The Auk* **102**: 293–297.

Pycraft, W.P. 1898. Contributions to the osteology of birds. Part II. Impennes. *Proceedings of the Zoological Society of London* **1898**: 958–989.

Raikow RJ, Bicanovsky L, Bledsoe AH. 1988. Forelimb joint mobility and the evolution of wing-propelled diving in birds. *The Auk* **105**: 446–451.

Scarlett RJ. 1983. *Tereingaornis moisleyi* – a new Pliocene penguin. *New Zealand Journal of Geology and Geophysics* **26**: 419–428.

Schoepss CG. 1829. Bescheibung der flugelmuskeln der vögel. In *Archiv fur Anatomie und Physiologie*, Meckel JF (ed.). Leipzig; 72–176.

Schreiweis DO. 1982. A comparative study of the appendicular musculature of penguins (Aves: Sphenisciformes). *Smithsonian Contributions to Zoology* **341**: 1–46.

Seddon PJ, Davis LS. 1989. Nest-site selection by yellow-eyed penguins. *Condor* **91**: 653–659.

Seebohm H. 1888. An attempt to diagnose the suborders of the great Galino-Gralline group of birds, by the aid of osteological characters alone. *The Ibis* **30**: 415–435.

Sharpe R. 1891. A review of recent attempts to classify birds. Proceedings of the 2nd International Ornithological Congress, Budapest

Sibley CG, Ahlquist JE. 1990. *Phylogeny and Classification of Birds: A Study in Molecular Evolution*. New Haven: Yale University Press.

Sibley CG. MonroeJrBL. 1990. *Distribution and Taxonomy of Birds of the World*. New Haven: Yale University Press.

Siegel-Causey D. 1997. Phylogeny of the Pelecaniformes: Molecular systematics of a privative group. In *Avian Molecular Evolution and Systematics*. Names of editor(s) & location. Academic Press; 159–171.

Simon J, Laurent S, Grolleau G, Thoraval P, Soubieux D, Rasschaert D. 2004. Evolution of preproinsulin gene in birds. *Molecular Phylogenetics and Evolution* **30**: 755–766.

Simpson GG. 1959. A new fossil penguin from Australia. *Proceedings of the Royal Society of Victoria* **71**: 113–119.

Simpson GG. 1946. Fossil penguins. *Bulletin of the American Museum of Natural History* **87**: 7–99.

Simpson GG. 1957. Australian fossil penguins, with remarks on penguin evolution and distribution. *Records of the South Australian Museum* **13**: 51–70.

Simpson GG. 1965. New record of a fossil penguin in Australia. *Proceedings of the Royal Society of Victoria* **79**: 91–93.

Simpson GG. 1970. Miocene penguins from Victoria, Australia, Chubut, Argentina. *Memoirs of the National Museum, Victoria* **31**: 17–24.

Simpson GG. 1971a. A review of the pre-Pliocene penguins of New Zealand. *Bulletin of the American Museum of Natural History* **144**: 319–378.

Simpson GG. 1971b. Review of Fossil Penguins from Seymour Island. *Proceedings of the Royal Society of London Series B (Biological Sciences)* **178** (1053): 357–387.

Simpson GG. 1971c. Fossil Penguin from the Late Cenozoic of South Africa. *Science* **171** (3976): 1144–1145.

Simpson GG. 1972a. Conspectus of Patagonian fossil penguins. *American Museum Novitates* **2488**: 1–37.

Simpson GG. 1972b. Pliocene penguins from North Canterbury. *Records of the Canterbury Museum* **9** (2): 159–182.

Simpson GG. 1973. Tertiary penguins (Sphenisciformes, Spheniscidae) from Ysterplaats, Cape Town, South Africa. *South African Journal of Science* **69**: 342–344.

Simpson GG. 1975a. Notes on variation in penguins and on fossil penguins from the Pliocene of Langebaanweg, Cape Province, South Africa. *Annals of the South African Museum* **69**: 59–72.

Simpson GG. 1975b. Fossil penguins. In *The Biology of Penguins*, Stonehouse B (ed.). London: MacMillan Press; 19–41.

Simpson GG. 1976. *Penguins Past and Present, Here and There*. New Haven: Yale University Press.

Simpson GG. 1979a. Tertiary penguins from the Duinefontein site, Cape Province, South Africa. *Annals of the South African Museum* **79**: 1–17.

Simpson GG. 1979b. A new genus of Late Tertiary penguin from Langebaanweg, South Africa. *Annals of the South African Museum* **78**: 1–9.

Simpson GG. 1981. Notes on some fossil penguins, including a new genus from Patagonia. *Ameghiniana* **18**: 266–272.

Sivak JG. 1976. The role of a flat cornea in the amphibious behaviour of the blackfoot penguin *Spheniscus demersus*. *Canadian Journal of Zoology* **54**: 1341–1345.

Sivak JG, Millodot M. 1977. Optical performance of the penguin eye in air and water. *Journal of Comparative Physiology A: Neuroethology, Sensory, Neural, and Behavioral Physiology* **119**: 241–247.

Slack KE, Jones CM, Ando T, Harrison GL, Fordyce RE, Arnason U, Penny D. 2006. Early penguin fossils, plus mitochondrial genomes, calibrate avian evolution. *Molecular Biology and Evolution* **23**: 1144–1155.

Sparks J, Soper T. 1987. *Penguins*. New York: Facts on Files Publications.

Stahel C, Gales R. 1987. *Little Penguin: Fairy Penguins in Australia*. Kensington: New South Wales University Press.

Stettenheim P. 1959. Adaptations for underwater swimming in the Common murre (Uria aalge). Unpublished PhD thesis, University of Michigan Ann Arbor.

Stokes DL, Boersma PD. 1991. Efects of substrate on the distribution of Magellanic penguin (*Spheniscus magellanicus*) burrows. *The Auk* **4**: 923–933.

Stonehouse B. 1953. The Emperor Penguin *Aptenodytes forsteri* Gray. I. Breeding behaviour and development. *Falkland Islands Dependencies Survey Scientific Reports* **6**: 1–33.

Stonehouse B. 1967. The general biology and thermal balances of penguins. *Advances in Ecological Research* **4**: 131–196.

Stonehouse B. 1969. Environmental temperatures of Tertiary penguins. *Nature* [Vol. No.] 673–675.

Stonehouse B. 1975. *The Biology of Penguins*. London: MacMillan Press.

Stucchi M. 2002. Una nueva especie de *Spheniscus* (Aves: Spheniscidae) de la Formación Pisco, Perú. *Boletín de la Sociedad Geológica del Perú* **94**: 17–24.

Stucchi M. 2007. Los pingüinos de la Formación Pisco (Neógeno), Perú. In, *4th European Meeting on the Palaeontology and Stratigraphy of Latin America Cuadernos del Museo Geominero*, Díaz-Martínez E, Rábano I. (eds). Madrid: Instituto Geológico y Minero de España; No. 8: 367–373.

Stucchi M, Urbina M, Giraldo A. 2003. Una nueva specie de Spheniscidae del Mioceno Tardío de la Formacíon Pisco, Perú. *Bulletin de l'Institut Français d'Études Andines* **32**: 361–375.

Tambussi CP, Reguero MA, Marenssi SA, Santillana SN. 2005. *Crossvalia unienwillia*, a new Spheniscidae (Sphenisciformes, Aves) from the late Paleocene of Antarctica. *Geóbios* **38**: 667–675.

Tambussi CP, Acosta Hospitaleche CI, Reguero MA, Marenssi SA. 2006. Late Eocene penguins from West Antarctica: systematics and biostratigraphy. *Geological Society, London, Special Publication* **258**: 145–161.

Thomas DB, Fordyce RE. 2007. The heterothermic loophole exploited by penguins. Australian. *Journal of Zoology* **55**: 317–321.

Van Tets GF, O'Conner S. 1983. The Hunter Island penguin, an extinct new genus and species from a Tasmanian midden. *Records of the Queen Victoria Museum* **81**: 1–12.

Van Tuinen M, Sibley CG, Hedges SB. 2000. The early history of modern birds inferred from DNA sequences of nuclear and mitochondrial genes. *Molecular Biology and Evolution* **17**: 451–457.

Vargas FH, Harrison S, Rea S, Macdonald DW. 2006. Biological effects of El Niño on the Galápagos penguin. *Biological Conservation* **127**: 107–114.

Walsh SA, Suárez ME. 2006. New penguin remains from the Pliocene of Northern Chile. *Historical Biology* **18**: 115–126.

Warham J. 1996. *The Behaviour, Population Biology and Physiology of the Petrels*. London: Academic Press.

Watanuki Y, Burger AE. 1999. Body mass and dive duration in alcids and penguins. *Canadian Journal of Zoology – Revue Canadienne De Zoologie* **77**: 1838–1842.

Watson M. 1883. Report on the anatomy of the Spheniscidae collected during the voyage of H.M.S. Challenger. In *Report of the scientific results of the voyage of H.M.S. Challenger during the years 1973–1876, Zoology, Volume 7*, Murray J (ed.). Edinburgh: Neill and Company; 1–244.

Weis-Fogh T. 1972. Energetics of hovering flight in hummingbirds and in *Drosphila*. *Journal of Experimental Biology* **56**: 79–104.

Williams TD. 1995. *The Penguins*. Oxford: Oxford University Press.

Wiman C. 1905a. Vorfläufige Mitteilung über die alttertiären Vertebraten der Seymourinsel. *Bulletin of the Geological Institute of Uppsala* **6**: 247–253.

Wiman C. 1905b. Vorläufige mitteilung über vertebraten der Seymourinsel. *Wissenschaftliche Ergebnisse der Schwedischen Südpolar-Expedition 1901–1903* **3**: 1–38.

Worthy TH. 1997. The identification of fossil *Eudyptes* and *Megadyptes* bones at Marfels Beach, Marlborough, South Island. *New Zealand Natural Science* **23**: 71–85.

Zachos J, Pagani M, Sloan L, Thomas E, Billups K. 2001. Trends, rhythms, and aberrations in global climate 65 Ma to present. *Science* **292**: 686–693.

Zusi RL. 1975. An interpretation of skull structure in penguins. In *The Biology of Penguins*, Stonehouse B (ed.). Baltimore: University Park Press; 55–84.

7 Phorusrhacids: the Terror Birds

HERCULANO ALVARENGA,[1] LUIS CHIAPPE,[2]
AND SARA BERTELLI[3]

[1]Museu de História Natural de Taubaté, Taubaté, Brazil
[2]Los Angeles County Museum, Los Angeles, USA
[3]Museum für Naturkunde, Berlin, Germany

At the beginning of the Tertiary, while mammals were undergoing an evolutionary explosion, several groups of birds developed a tendency to gigantism. These groups were scattered across almost the whole planet. The Gastornithidae (also known as Diatrymidae), with possible affinity with the Anseriformes (Andors, 1992), have been recovered from Paleocene and Eocene deposits in North America, Europe and Asia (Matthew & Granger, 1917; Martin, 1992; Hou, 1980). In Australia, another group of giant birds, the Dromornithidae, with possible affinity to the Anseriformes (Wroe, 1998; Murray & Vickers-Rich, 2004) had a broad diversity in the mid-Tertiary. Among these, *Dromornis stirtoni* would have been comparable in size to the largest birds ever found. The Ratitae, also present since the Paleocene in South America and possibly in Europe (Alvarenga, 1983; Martin, 1992), is another group remarkable for the large size reached in several species. Among the giant birds of the Tertiary, they are the only present-day survivors. The elephant-bird, *Aepyornis maximus*, a ratite from Madagascar that died out around 700 years ago, is probably the biggest bird ever found (Amadon, 1947; Wetmore, 1967).

The Phorusrhacidae, another group of giants was present in South America from the Paleocene (Alvarenga, 1985) and survived until the end of the Pleistocene (Alvarenga *et al.*, 2010). Ratites and phorusrhacids apparently lived together, perhaps competing for similar terrestrial habitats, during the whole of the Tertiary. At that time South America was an isolated island, and only the ratites, perhaps the most vulnerable among these birds, have survived to present times.

Scientific investigations on the Phorusrhacidae started at the end of the 19th century with the work of Ameghino (1887), who described a mandible of a "probable toothless mammal", which he named *Phorusrhacus longissimus*. However, it was Moreno (1889) who first called attention to the huge bones that he identified as giant birds from the Tertiary period in Argentina. Later, Moreno & Mercerat (1891) and two other publications of Ameghino (1891a,b) recognized *Phorusrhacus* as a bird and named several genera and species for the Phorusrhacidae. A number of bones and fragments of bones were described as a new species or genera, resulting in a complicated and extended synonymy within the family; currently 14 genera and 18 species are recognized (Table 7.1). A more detailed history of the first investigations of this family, as well as of the first classifications proposed, is set out by Alvarenga & Höfling (2003).

In this chapter, our main objective is to present a phylogenetic analysis of the Phorusrhacidae, and at the same time call attention to anatomic details

Table 7.1 List of the Phorusrhacidae species, including the junior synonymous, the geographical and stratigraphical distribution.

Taxon/reference	Geographic locality	Horizon/Age	Additional references
Mesembriornis milneedwardsi Moreno, 1889 Paleociconia australis Moreno, 1889; Moreno & Mercerat, 1891 Driornis pampeanus Moreno & Mercerat, 1891 (part: only femur) Hermosiornis milneedwardsi Rovereto, 1914 Hermosiornis rapax Kraglievich, 1946 Prophororhacos australis Brodkorb, 1967	Argentina, Province, Buenos Aires: Monte Hermoso, Rio Loberia	Upper Pliocene (Montehermosan)	Alvarenga & Höfling, 2003
Mesembriornis incertus (Rovereto, 1914) Phororhacos incertus Rovereto, 1914	Argentina, Province, Catamarca: Andalgalá, Corral Quemado; Province, Buenos Aires: P. de Villarino	Upper Miocene to Lower Pliocene (Huayquerian)	Acosta Hospitaleche, 2002; Alvarenga & Höfling, 2003
Procariama simplex Rovereto,1914	Argentina, Province, Catamarca, Andalgalá,Corral Quemado, Belém, Chiquimil, Rio Santa Maria	Upper Miocene to Lower Pliocene (Huayquerian, Chasicoan)	Tonni, 1980; Alvarenga & Höfling, 2003
Paleopsilopterus itaboraiensis, Alvarenga,1985	Brazil, Est. Rio de Janeiro: São José de Itaboraí	Upper Paleocene (Itaboraian)	Alvarenga & Höfling, 2003
Psilopterus bachmanni (Moreno & Mercerat, 1891) Patagornis bachmanni Moreno & Mercerat, 1891 Psilopterus communis Moreno & Mercerat, 1891 Psilopterus intermedius Moreno & Mercerat, 1891 Phororhacus delicatus Ameghino, 1891 Pelecyornis minutus Ameghino, 1891 Pelecyornis pueyrredonensis Sinclair & Farr, 1932	Argentina, Province, Santa Cruz: Santa Cruz, Lago Pueyrredon, Monte Observación, La Cueva	Middle Miocene (Santacrucian)	Alvarenga & Höfling, 2003
Psilopterus affinis (Ameghino, 1899) Phororhacus affinis Ameghino, 1899	Argentina, Province, Chubut: Cabeça Blanca	Middle Oligocene (Deseadan)	Alvarenga & Höfling, 2003; Agnolin, 2006
Psilopterus colzecus Tonni & Tambussi, 1988	Argentina, Province, Buenos Aires: Partido de Villarino	Upper Miocene (Chasicoan)	Alvarenga & Höfling, 2003
Psilopterus lemoinei (Moreno & Mercerat, 1891) Patagornis lemoinei Moreno & Mercerat, 1891 Psilopterus australis Moreno & Mercerat, 1891; Brodkorb, 1967 Pelecyornis tubulatus Ameghino, 1895 Phororhacus modicus Ameghino, 1895 Staphylornis gallardoi Mercerat, 1897 Staphylornis erythacus Mercerat, 1897 Pelecyornis tenuirostris Sinclair & Farr, 1932	Argentina, Province, Santa Cruz: Santa Cruz, Killik Aike, Monte Observación, Take Harvey, La Cueva, Corriguen Kaik, Taguaquemada	Middle Miocene (Santacrucian)	Alvarenga & Höfling, 2003; Degrange & Tambussi, 2008

Species	Locality	Age	References
Patagornis marshi Moreno & Mercerat, 1891 Tolmodus inflatus Ameghino, 1891 Phororhacos inflatus Ameghino, 1891; Andrews, 1899; Brodkorb, 1967 Paleociconia cristata Brodkorb, 1967	Argentina, Province, Santa Cruz;Monte Observación, Tagua Quemada, La Cueva	Lower to Middle Miocene. (Santacrucian)	Alvarenga & Höfling, 2003
Andrewsornis abbotti Patterson, 1941	Argentina, Province,Chubut (Cabeça Blanca) and Santa Cruz(Pico Truncado)	Middle to Upper Oligocene (Deseadan)	Alvarenga & Höfling, 2003
Andalgalornis steulleti (Kraglievich, 1931) Phororhacos steulleti Kraglievich, 1931 Phororhacos deautieri Kraglievich, 1931 Andalgalornis ferox Paterson & Kraglievich, 1960 Andalgalornis steulleti Brodkorb, 1967	Argentina, Province, Entre Rios and Catamarca (Chiquimil)	Upper Miocene to Lower Pliocene. Fm Andalgalá, (Huayquerian)	Tonni, 1980; Alvarenga & Höfling, 2003; Noriega & Agnolin, 2008
Phorusrhacus longissimus Ameghino, 1887 Phororhacos longissimus, Ameghino, 1889 Stereornis rollieri Moreno & Mercerat, 1891 Stereornis gaudryi Moreno & Mercerat, 1891 Mesembriornis studeri Moreno & Mercerat, 1891 Mesembriornis quatrefragesi Moreno & Mercerat, 1891 Darwinornis copei Moreno & Mercerat, 1891 Darwinornis zittelli Moreno & Mercerat, 1891 Darwinornis socialis Moreno & Mercerat, 1891 Owenornis affinis Moreno & Mercerat, 1891 Owenornis lydekkeri Moreno & Mercerat, 1891 Phororhacos sehuensis Ameghino, 1891 Phororhacos platygnathus Ameghino, 1891 Titanornis mirabilis, Moreno & Mercerat, 1891 Callornis giganteus, Ameghino, 1895 Eucallornis giganteus Ameghino, 1901 Liornis floweri Ameghino, 1895	Argentina, Province, Santa Cruz, La Cueva, Tagua/Quemada, Monte Observación, Rio Sehuén	Lower and Middle Miocene (Santacrucian)	Alvarenga & Höfling, 2003
Titanis walleri Brodkorb, 1963	USA, Florida (Inglish) and Texas (Baskin)	Pliocene (Hemphillian to Late Blancan)	Gould & Quitmer, 2005; MacFadden et al., 2006

(continued)

Table 7.1 (*Continued*)

Taxon/reference	Geographic locality	Horizon/Age	Additional references
Devincenzia pozzi (Kraglievich, 1931) *Phororhacos pozzi* Kraglievich, 1931 *Phororhacus longissimus mendocinus* Kraglievich, 1931 *Devincenzia gallinali* Kraglievich, 1932 *Onactornis depressus* Cabrera, 1939 *Onactornis pozzi* Brodkorb, 1967 *Onactornis mendocinus* Brodkorb, 1967	Argentina, Province, Buenos Aires and Entre Rios; Uruguay	Upper Miocene to Upper Pliocene (Huayquerian and Mesopotamien); Fm Raigon (Uruguay).	Tambussi *et al.*, 1999; Alvarenga & Höfling, 2003; Alfaro & Perea, 2004; Agnolin, 2006; Noriega & Agnolin, 2008
Kelenken guillermoi Bertelli, Chiappe & Tambussi, 2007	Argentina, Province, Rio Negro, Comallo.	Midle Miocene (Fm Collón Curá)	Chiappe & Bertelli, 2007
Physornis fortis Ameghino, 1895 *Aucornis euryrhynchus* Ameghino, 1898	Argentina, Province, Santa Cruz, Puerto Deseado, Punta Nova, La Flexa.	Middle Oligocene (Deseadan)	Alvarenga, 1993; Alvarenga & Höfling, 2003
Paraphysornis brasiliensis (Alvarenga, 1982) *Physornis brasiliensis* Alvarenga, 1982	Brazil, Est. São Paulo, Tremembé.	Upper Oligocene or Lower Miocene ("Upper" Deseadan)	Alvarenga, 1993
Brontornis burmeisteri Moreno and Mercerat, 1891 *Rostrornis floweri* Moreno & Mercerat, 1891 *Brontornis platyonyx* Ameghino, 1895	Argentina, Province, Santa Cruz (Lago Argentina, Monte León, Monte Observación, Karaiken, La Cueva, Rio Gallegos.	Lower to Middle Miocene (Santacrucian)	Alvarenga & Höfling, 2003; Agnolin, 2007

that need further clarification in these birds. The following institutional abbreviations are used in this chapter: DGM, Divisão de Geologia e Mineralogia do Departamento Nacional da Produção Mineral, Rio de Janeiro, Brazil; FMNH, Field Museum of Natural History, Chicago, USA; MHNT, Museu de História Natural de Taubaté, Taubaté, Brazil; UF, University of Florida, Gainesville, USA.

GEOLOGICAL SETTING

Remains of Phorusrhacidae have been found in continental beds of a great variety of Cenozoic strata in South America, from all epochs of the Tertiary. No species has been described for the Eocene although a few fragmentary specimens for this epoch have been attributed to Psilopterinae (Acosta Hospitaleche & Tambussi, 2005). Most of the fossil remains and described taxa of phorusrhacids have been found in Argentina but they are also recorded in Uruguay and Brazil, and recently an ungual phalanx was recovered from Oligo-Miocene deposits in Peru (Shockey *et al.*, 2006). There is also an important record of a beak fragment from Seymour Island, Antarctica (Case *et al.*, 1987) that is certainly a mandibular symphysis of a huge phorusrhacid, closely related to the *Brontornis* genus (Alvarenga & Höfling, 2003) from the La Meseta Formation (Upper Eocene). In North America, likely as a result of the Great American Biotic Interchange (GABI) that occurred during the Tertiary, we have a single species, *Titanis walleri* Brodkorb, 1963, from the Pliocene of Florida and Texas (Chandler, 1994; Baskin, 1995; Gould & Quitmyer, 2005; MacFadden *et al.*, 2006). Claims supporting the existence of European Phorusrhacidae have been discarded by Alvarenga & Höfling (2003). In Table 7.1, a brief summary of the chronology of each site for each species of Phorusrhacidae is given, as well as the bibliographic references that may add details of the paleoenvironment and associated fauna for each species.

SYSTEMATIC PALEONTOLOGY

Phylogenetic analysis

Here we present a cladistic analysis based on 61 characters (see Appendix 7A); three are multistate, two of them (i.e., 4 and 14) can be arranged as a morphological series and were treated as ordered. Damaged or absent structures are coded as missing-data, and all characters are weighted equally. The character matrix (see Appendix 7B) was constructed using the software NDE 0.5.0 and submitted to character optimization and parsimony analysis in PAUP* 4.0b (Swofford, 2001). A heuristic search was conducted using the Tree Bisection Reconnection (TBR) algorithm with 1000 replicates in the branch-swapping cycles with random addition of the taxa. The maximum saved trees (maxtress) were automatically increased by 100 when necessary, and branches were collapsed if the minimum branch length was equal to zero. Bremer support was also calculated in PAUP* based on creation of constrains for the clades result in the previous analysis; this process was conducted with 10 replicates in order to optimize computer time. The cladistic analysis was rooted using the taxon *Anseranas semipalmata*. A total of 16 out of the18 species traditionally included in Phorusrhacidae (Alvarenga & Höfling, 2003; Bertelli *et al.*, 2007) were included in this analysis and only the taxa *Paleopsilopterus itaboaiensis* Alvarenga, 1985 and *Psilopterus affinis* (Ameghino, 1899) were excluded due to the relatively meager quality of the known material.

Paleopsilopterus itaboraiensis is similar in size to *Procariama simplex* and the morphology of its tibiotarsus and tarsometatarsus is very similar to that of other small Phorusrhacidae with long tarsometatarsi (Figure 7.2). In addition to material previously described (Alvarenga, 1985), we can attribute five ungual phalanges to this species (MHNT:5316–5320) (Figure 7.2J–N). R. Silva Santos collected this material during the 1970s, in the Itaboraí Basin, RJ, Brazil, which is the type locality of the species. The specimens of *Psilopterus affinis* suggest affinities with other species of Psilopterinae (Alvarenga & Höfling,

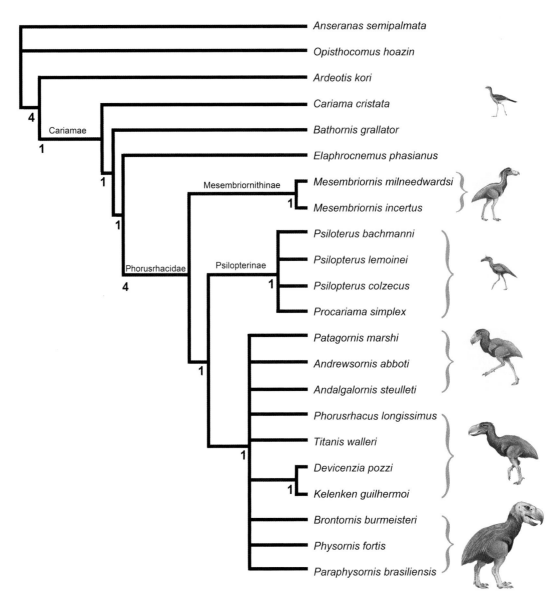

Fig. 7.1 Strict consensus cladogram resulting from the 48 most parsimonious trees from the present cladistic analysis (length: 91; CI: 0.7; RI: 0.83). Numbers in nodes express the Bremer support.

2003; Agnolin, 2006). We have included both taxa tentatively in the subfamily Psilopterinae according to the features indicated by Alvarenga & Höfling (2003).

The strict consensus of the 48 most parsimonious trees resulting from analysis of the matrix (Appendix 7B) is shown in Figure 7.1. The group traditionally named "Cariamae" is here phylogenetically defined as all clades descending from the common ancestor of *Cariama* and Phorusrhacidae. A good number of other birds from the Paleogene of North America, many of them

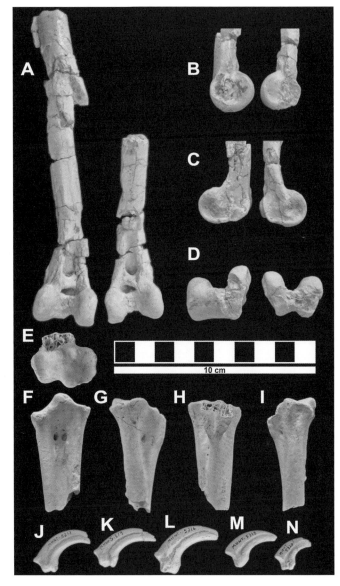

Fig. 7.2 *Paleopsilopterus itaboraiensis*: The holotype right tarsometatarsus (MNRJ-4040-V) in proximal (E), dorsal (F), lateral (G), plantar (H), and medial (I) views. Referred right tibiotasus (in the left side) and left tibiotarsus (in the right side) in ventral (A), lateral (B), medial (C), and distal (D) views. Five ungual phalanges (MHNT-5316-5320) from the same locality of the holotype are tentatively attributed to the digit III (J and K), digit II (L), digit IV (M) and digit I (N) of the same species.

attributed to the "Bathornithidae" and "Idiornithidae" or "suborder Cariamae" (Wetmore, 1944; Cracraft, 1968, 1971, 1973; Mourer-Chauviré, 1983; Mayr & Mourer-Chauviré, 2006; Mayr, 2007), are evidently close to the extant Cariamidae (Olson, 1985). In fact, a revision of these birds is necessary to define the basal cladistic position of the species of Cariamae, as well as establish a monophyletic and possibly broader Cariamidae family.

The monophyly of the phorusrhacidae

Andrews (1896, 1899) was the first to recognize the close relationship between Phorusrhacidae and the extant Cariamidae. There is a great deal of confusion, however, in terms of which genera should be included in these families. Brodkorb (1967) included *Psilopterus*, *Procariama*, and M*esembriornis* in Cariamidae, causing confusion for many subsequent authors. The extant Cariamidae are good fliers while the Phorusrhacidae form a more derived group that is wholly flightless. The phylogenetic analysis presented here firmly establishes the systematic position of these groups.

Another question regarding the monophyly of the Phorusrhacidae concerns *Brontornis*, the largest member of the group. In *Brontornis burmeisteri* the internal condyle of the tibiotarsus is medially diverted, in a way typical of Anseriformes, and this is evident in at least two specimens figured by Moreno & Mercerat (1891). Based on this feature and the morphology of an incomplete quadrate bone, Agnolin (2007) proposed recognizing *Brontornis* as a giant anseriform convergent on the Gastornithidae of the Northern Hemisphere and the Dromornithidae of Australia but coexisting with the Phorusrhacidae and ratites in the

Miocene of South America. The phylogenetic analysis presented here recovers *Brontornis* in the Phorusrhacidae, close to the other large members of this clade and relatively distant from the Anseriformes. We believe that the particular diverted medial condyle of the tibiotarsus may be an adaptation due to the excessive weight of the bird, influencing its posture and walk. Significantly, this particular character appears in *Andalgalornis* (Noriega & Agnolin, 2008, figure 6-B) and also in *Cariama*. The quadrate bone mentioned by Angolin (2007) is fragmented to the point where meaningful comparisons are impaired. An important character for *Brontornis* can be seen in the thoracic vertebra shown by Moreno & Mercerat (1891, plate VII, figures 1 and 2). There is a large *recessus pneumaticus* in the mid-centrum (character 22; Appendix 7A), which is a very clearly defined character in the Phorusrhacidae (Figure 7.3) and is well illustrated in the literature for *Psilopterus* (Sinclair & Farr, 1932, figures 7 and 9, plate XXXI) and *Titanis* (Gould & Quitmyer, 2005, figure 6-C).

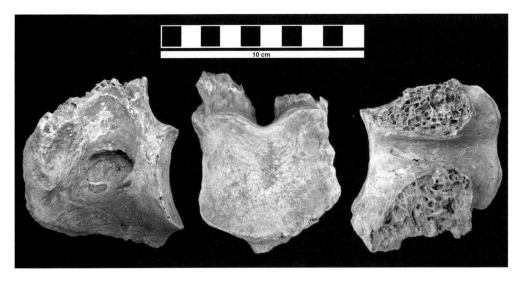

Fig. 7.3 A thoracic vertebra of *Paraphysornis brasiliensis* (DGM-1418-R) in lateral (A), cranial (B), and dorsal (C) views. Important anatomic similarities to *Brontornis burmeisteri* can be observed: the large *recessus pneumaticus* (A) in the mid-centrum (character 22) is a diagnostic feature for phorusrhacids.

Taxonomic hierarchy

Aves Linnaeus, 1758
Cariamae Fürbringer
Phorusrhacidae Ameghino, 1889

Diagnosis – The cladistic analysis offered here provides a diagnosis of Phorusrhacidae based on 13 synapomorphies, of which nine are unambiguous and two are exclusive:

upper beak tip strongly curved (character 2);
palate desmognathous (character 3);
large temporal fossa almost meet at median line (character 7);
foramen magnum oriented caudally (character 8);
processus basipterygoid present (character 10);
the pterygoid with articulation for the processus basipterygoid (character 11);
processus zygomaticus present (character 13);
acrocoracoidal process absent (character 29);
tuberculum ventrale of humerus projected proximally (character 36);
diaphysis of humerus is bowed and not in "sigma" (character 38);
processus flexorius of humerus projected distally (character 39);
trochanter majus of the femur absent or not prominent (character 49);
trochlea metatarsi II (in dorsal view) not deflected medially (character 59).

Sexual dimorphism – Alvarenga & Höfling (2003) commented on the intraspecific differences of size within the Phorusrhacidae and highlighted a variation of 33% in the size of the tarsometatarsus between two specimens of *Brontornis burmeisteri* as well as in the specimens attributed to *Psilopterus australis* by Sinclair & Farr (1932). The latter taxon also suggests important intraspecific differences in the overall size and height of the maxilla. Gould & Quitmyer (2005) summarized all the material referred to *Titanis walleri* and highlighted an important difference in the size of two quadratojugals (UF 57580 and UF 57585) and two proximal phalanges of the pedal digit III (UF 30001 and UF 171382). These differences may well be the expression of sexual dimorphism, which if confirmed would likely be female biased. Males may have been larger than females, as in the case of large flightless rails such as the Weka (*Gallirallus australis*) the Takahe (*Porphyrio mantelli*) (Taylor, 1996) and among extant cariamas (Alvarenga, personal observation), all taxa phylogentically close to Phorusrhacidae.

ANATOMY

There are many described and illustrated fossils of the Phorusrhacidae. The species of *Psilopterus*, the smallest phorusrhacid, are known from complete or almost complete skeletons and are particularly well described and illustrated (Sinclair & Farr, 1932). *Procariama simplex* is also represented by a nearly complete skeleton (FM–P14525), partially described by Alvarenga & Höfling (2003). Unfortunately the larger species are known by much less complete specimens. Among the larger phorusrhacids, *Paraphysornis brasiliensis* (Alvarenga, 1982) is the best represented with about 70% of the skeleton available for one specimen). As can be seen in Appendix 7A, a good number of anatomic characters can be determined for the phorusrhacids but in spite of the relative abundance of phorusrhacid fossils, some important anatomic questions still persist.

Cervical vertebrae

Several specimens of Phorusrhacidae show an osseous bridge from the *processus transversus* to the middle of the *corpus vertebrae*, forming large dorsal fenestrae (Mayr & Clark, 2003, character 52 and 53; Sinclair & Farr, 1932, plate XXXI; Patterson & Kraglievich, 1960, figures 4 and 5). Noriega *et al.* (2009) illustrated a cervical vertebra (certainly close to C-10) attributed to *Devincenzia pozzi*, where this character is present. In *Paraphysornis brasiliensis*, complete vertebrae such as C3 and possibly C10 or C11 do not present this character (Figure 7.4). An examination of other existing vertebral fragments also fail to confirm its presence. This character appears to be absent in *Paraphysornis* and Brontornithinae.

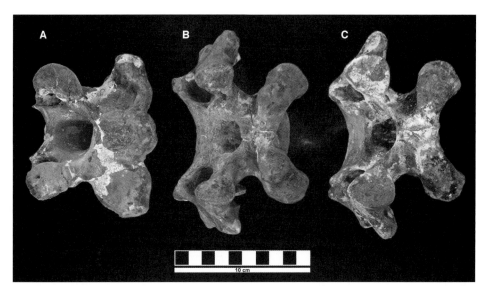

Fig. 7.4 Some complete cervical vertebrae of *Paraphysornis brasiliensis* (DGM-1418-R) in dorsal views: 3rd cervical (A), possibly the 10th cervical (B), and possibly the 11th cervical (C). The absence (at least in these vertebrae) of bridges linking the *processus transversus* to the middle of corpus of the vertebrae, forming dorsal fenestras, may be an important feature.

Thoracic vertebrae

In addition to the *recessus pneumaticus* in the middle centrum of some of the pre-synsacral thoracic vertebrae, there is another feature that relates to the *processus dorsalis* of the thoracic vertebrae. This process is very tall in all complete specimens of thoracic vertebrae (Patterson & Kraglievich, 1960, figure 6; Gould & Quitmyer, 2005, figure 6). In *Paraphysornis brasiliensis*, a fragment of *processus dorsalis* attributed to the first pre-synsacral vertebra, as well as a crest that we identified as a cranial extremity of the iliac dorsalis crest (Figure 7.5), suggest a shorter *processus dorsalis* in the thoracic vertebrae, of *Paraphysornis*. It may also be a feature of the Brontornithinae.

Uncinate process in ribs

In the illustrations and descriptions of Sinclair & Farr (1932), the ribs of the Phorusrhacidae lack uncinate processes. However, it is possible that uncinate processes were present but not fused to the ribs and were subsequently lost during fossilization. We observe that while preparing the skeletons of some birds such as Psophiidae, Aramidae and some Rallidae (*Rallus, Pardirallus*) by maceration, the uncinate processes are completely released and not fused to the ribs. If uncinate processes were present in the large phorusrhacids it seems unlikely that they were fused to the ribs (character 24; Appendix 7A).

Clavicles

Within the Phorusrhacidae, the cranial tip (*extremitas omalis claviculae*) of the coracoid is known to be fused to the clavicles only in Mesembriornithinae (Rovereto, 1914). Neither free clavicles nor a furcula are known for any other representatives of the family. In *Paraphysornis brasiliensis*, a bone fragment not originally described (Alvarenga, 1982) seems to belong to the cranial extremity of the left clavicle (Figure 7.6).

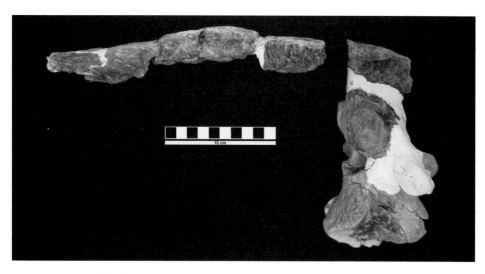

Fig. 7.5 A reconstruction of the first pre-sinsacral thoracic vertebra in lateral view, close to the remains of the cranial extremity of the dorsal iliac crest from *Paraphysornis brasiliensis* (DGM-1418R). This reconstruction suggests a short *processus dorsalis* for *Paraphysornis*.

Pubis

Andrews (1899) describes the pubis of *Patagornis marshi* as limited to its cranial portion, projecting from the floor of the acetabulum as a bar closing the *foramen obturatum*, bordering the ischium and ending at around its medium portion. Sinclair & Farr (1932) similarly described the pubis in *Psilopterus*, concluding that "the absence of the posterior pubic projection, if subsequently confirmed, should prove to be a good diagnostic character, perhaps of ordinal value." However, in a specimen of

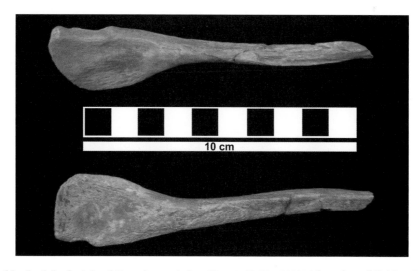

Fig. 7.6 Possibly the left clavicle of *Paraphysornis brasiliensis* (DGM-1418R) from lateral (left) and medial (right) views.

Fig. 7.7 Pelvis of *Procariama simplex* (FM- P14525) in lateral (A) and ventral (B) views. The most cranial portion of the pubis (arrow) is closed and delimits the *foramen obturatum*; the most caudal portion of the pubis is articulated with the ischiun, and is not continuous with the cranial portion.

Procariama simplex (FM-P14525), the nearly complete pelvis preserves the pubis with both the proximal and distal extremities (Figure 7.7), and a delicate medium portion adhered to the ventral surface of the ischium. Such a design is similar to that of the Accipitridae and some Falconidae. The distal extremity of the pubis may certainly have been lost or not identified in *Patagornis* and *Psilopterus* in these cases. Pelvis fragments from *Paraphysornis brasiliensis* (DGM-1014), not previously described by Alvarenga (1982), preserve the cranial portions or the ischium (Figure 7.8) and the caudal projections of the two pubes (Figure 7.9). Their morphology is very similar to that observed in *Procariama*. The pubis of Phorusrhacidae may be defined as discontinuous, with the medial filamentous portion adhering to the ventromedial surface of the ischium. The caudal extremity of the pubis articulates with the ischium (as in the Accipitridae). In spite of these conclusions, we believe that some variations, such as an open or closed obturator foramen, may occur within the Phorusrhacidae.

TAXONOMY

Alvarenga & Höfling (2003) proposed the allocation of 17 species of Phorusrhacidae among five subfamilies. Recently, Bertelli *et al.* (2007) have

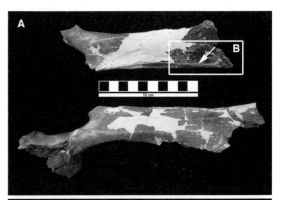

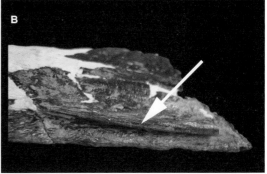

Fig. 7.8 (A) Cranial fragment of the ischium of *Paraphysornis brasiliensis* (DGM-1418-R) in ventral view. The detail of the right ischium (B) shows a pubis segment as a branch adhering to the ventral portion o the ischium (arrow).

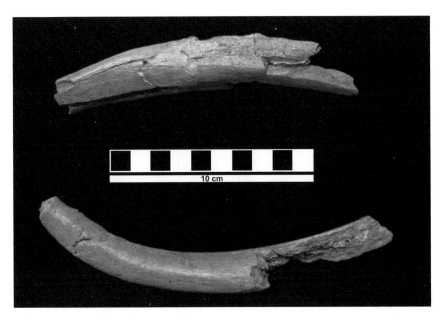

Fig. 7.9 Caudal segments of the two pubes of *Paraphysornis brasiliensis* (DGM1418-R); the conformation is similar to that seen in *Procariama simplex*.

described *Kelenken guillermoi*, an exceptionally large specimen from the Middle Miocene of Argentina as an additional genus and species. Its addition to the family brings the total to 13 genera and 18 species (Table 7.1). The updated phylogenetic analysis continues to support the five subfamilies proposed by Alvarenga & Höfling (2003) even though the cladogram *sensu stricto* does not separate the subfamilies Brontornithinae, Phorusrhacinae, and Patagornithinae. These taxa can be diagnosed by the characters described by the authors cited above. Given this, we propose the following family structure, from the more basal to the more derived groups.

Family phorusrhacidae

1 Subfamily Mesembriornithinae (Kraglievich, 1932). (Diagnosed by fusion of the coracoid to the clavicle – character 31.)
Genus *Mesembriornis* (Moreno, 1889)
M. milneedwardsi (Moreno, 1889; Figure 7.10B)
M. incertus (Rovereto, 1914)

2 Subfamily Psilopterinae (Dolgopol de Saez, 1927). (Diagnosed by the medial expansion of the articular surface of the trochea metatarsi II of the tarsometatarsus – character 59).
Genus *Psilopterus* (Moreno & Mercerat, 1891)
P. bachmanni (Moreno & Mercerat, 1891; Figure 7.10C)
P. lemoinei (Moreno & Mercerat, 1891)
P. affinis (Ameghino, 1899)
P. colzecus (Tonni & Tambussi, 1988)
Genus *Procariama* (Rovereto, 1914)
P. simplex (Rovereto, 1914)
Genus *Paleopsilopterus* (Alvarenga, 1985)
P. itaboraiensis (Alvarenga, 1985)

3 Subfamily Patagornithinae (Mercerat, 1897). (Diagnosed by Alvarenga & Höfling (2003) as medium-sized, smaller, and slimmer than the Phorusrhacinae, with a long and narrow mandibular symphysis, and a long and slender tarsometatarsi that is more than 70% of the length of the tibiotarsus.)
Genus *Patagornis* (Moreno & Mercerat, 1891)
P. marshi (Moreno & Mercerat, 1891)
Genus *Andrewsornis* (Patterson, 1941)

Fig. 7.10 Reconstructions of some phorusrhacids compared to the extant *Cariama*. (A) *Cariama cristata*; (B) *Mesembriornis milneedwardsi*; (C) *Psilopterus bachmanni*; (D) *Andalgalornis steuletti*; (E) *Phorusrhacus long-issimus*; (F) *Paraphysornis brasiliensis*; and (G) *Brontornis burmeiteri*. A man's silhouette (1.75 m) is used as scale. (Drawing by Eduardo Brettas.)

A abbotti (Patterson, 1941)
Genus *Andalgalornis* (Patterson & Kraglievich, 1960)
A steulleti (Kraglievich, 1931; Figure 7.10D)
4 Subfamily Phorusrhacinae (Ameghino, 1889). (Diagnosed by Alvarenga & Höfling, (2003) as gigantic; with a mandibular symphysis that is relatively long and narrow but shallow, and more than twice as long as the width of the base; tarsometatarsus is relatively long and slender, and is always longer than 60% of the tibiotarsus.)
Genus *Phorusrhacos* (Ameghino, 1889)
P. longissimus (Ameghino, 1899; Figure 7.10E)
Genus *Devincenzia* (Kraglievich, 1932)

D. pozzi (Kraglievich, 1931)
Genus *Kelenken* (Bertelli *et al.*, 2007)
K. guilermoi (Bertelli *et al.*, 2007)
Genus *Titanis* (Brodkorb, 1963)
T. walleri (Brodkorb, 1963)
5 Subfamily Brontornithinae (Moreno & Mercerat, 1891). (Diagnosed by Alvarenga & Höfling (2003) as gigantic; the mandibular symphysis is proportionally shorter, wider, and higher than other Phorusrhacidae; the tarsometatarsus is proportionally short, widened, and flattened dorso-ventrally; also it is possible that the condition of character 58 represents a synapomorphy to this subfamily.)
Genus *Brontornis* (Moreno & Mercerat, 1891)

B. burmeisteri (Moreno & Mercerat, 1891; Figure 7.10G)
Genus Physornis (Ameghino, 1895)
P. fortis (Ameghino, 1895)
Genus *Paraphysornis* (Alvarenga, 1993; Figure 7.10F)
P. brasiliensis (Alvarenga, 1982)

BIOGEOGRAPHY AND THE ORIGIN OF THE PHORUSRHACIDAE

The biogeographical history of the Cariamae remains unclear but several genera and species have been described from Eocene and Oligocene deposits in North America (Wetmore, 1944, 1967; Cracraft, 1968, 1971, 1973; Olson, 1985) and Europe (Mourer-Chauviré, 1983; Mayr, 2007, 2009) although the interpretation of these birds as a monophyletic group needs to be re-examined. They disappear from the fossil record of both areas in the Miocene. The Cariamidae, the sister group to the Phorusrhacidae, appear to have been almost absent from South America until the mid-Tertiary. There are only two known representatives, both from Argentina: *Chunga incerta* from the Late Miocene (Tonni, 1974) and *Cariama santacrucensis* from the Early–Middle Miocene (Noriega *et al.*, 2009).

Members of the Phorusrhacidae and also an unpublished Cariamae, closely related to the European Idiornithidae (Alvarenga, personal observation), were present in South America during the Paleocene. They must represent the South American portion of a stock of Cariamae that enjoyed an early (even Cretaceous) diversity in Europe, North America and certainly Africa. Claims supporting the presence of phorusrhacids in Europe (Mourer-Chauviré, 1983; Peters, 1987) were discussed and discarded by Alvarenga & Höfling, 2003.

It is possible that Phorusrhacidae and also Cariamidae arose in South America, but unfortunately it will be necessary to find new fossil evidence to provide direct support for reconstruction of the biogeographic history of these birds. They might have arrived in South America from Europe by traveling through Africa at a time when the southern continents were much closer together. In the Oligocene, extreme cooling of the planet and subsequent lowering of sea levels might have facilitated further movements. It is also possible that some birds made reverse movements but, later in the Tertiary, movement may have been restricted by further vicarious geographic effects.

ACKNOWLEDGMENTS

We thank Rafael Migotto, Ricardo Mendonça, and Graziella Couto-Ribeiro, from the Museu de História Natural de Taubaté, Brazil for important help in the initial phase of this paper, including on phylogenetic analysis and editing of pictures. To Gary Kaiser and Gareth Dyke for the important help in the revision on the final version of the manuscripts and also in the final treatment of the figures of this chapter. Finally, to Eduardo Brettas for the excellent artistic reconstructions of some phorusrhacids in Figure 7.10.

APPENDIX 7A: CHARACTER LIST AND CHARACTER STATES USED FOR THE PRESENT CLADISTIC ANALYSIS

Skull and mandible

1 Upper beak, maxilla and praemaxilla: wider than tall (0); taller than wide (1).
2 Upper beak, praemaxilla tip: straight or slight curved (0); strongly curved (1).
3 Palate: squizognathous (0); desmognathous (1).
4 Rostral border of antorbital fenestrae: strongly obliquous (0); obliquous (1); almost vertical (2).
5 Neurocranium: wider than tall (0); taller than wider (1).
6 Os lacrimale (in adult): not ankilosed to frontal (0); ankilosed to frontal (1).
7 Temporal fossa: small (0); large – almost meet up in median line (1).
8 Foramen magnum oriented: ventrally or ventrally-caudal (0); caudally (1).
9 Processus supraorbitales of lacrimale: short (0); caudally long (1).

10 Processus basipterigoid: absent or very small (0); present (1).

11 Os pterigoid with articulation for the processus basipterigoid: absent (0); present (1).

12 Premaxillare nasal process (in adult): not completly fused (0); completly fused (1).

13 Processus zigomaticus: present (0); absent (1).

14 Jugal bar very tall (the hight is two times or more than the wide): absent (0); tall two times (1); tall more than two times (2).

15 Mandibulae – pars symphysialis: lenght equal or bigger than one quarter of the mandibulae: absent (0); present (1).

16 Mandibulae – pars symphysialis: longer than than wide, strong and massive: absent(0); present (1).

17 Mandibulae – pars symphysialis: straight or ventrally curved (0); dorsally curved (1).

18 Fenestra caudalis mandibulae: absent (0); present (1).

19 Fenestra rostralis mandibulae:absent (0); present (1).

20 Mandibulae – processus retroarticularis: absent or small (0); present and large (1).

Vertebral column and ribs

21 Third cervical vertebrae – an osseous bridge linking the processus transversus to processus articularis (pos-zygapophysis) making a dorsal fenestrae (see Mayr & Clarke, 2003 – char. 52): absent (0); present (1).

22 Thoracicae vertebrae – a large recessus pneumaticus in the mid centrum (see Livezey & Zusi, 2006 – char. 0850): absent (0); present (1).

23 Presinsacral vertebrae form a notarium: absent (0); present (1).

24 Ribs – processus uncinati: absent or unfused to ribs (0); fused to ribs (1).

Thoracic girdle

25 Coracoid – ?pneumatic foramen directly below facies articularis scapularis (see Mayr & Clarke, 2003 –char. 66): absent (0); present (1).

26 Coracoid – foramen nervi supracoracoidei: present (0); absent (1).

27 Coracoid – pneumatic foramina in dorsal surface of extremitas sternalis: absent (0); present (1).

28 Procoracoidal process: present (0); absent (1).

29 Acrocoracoidal process; present (0); absent (1).

30 An osseous bridge linking the acrocoracoidal to the procoracoidal process: absent (0); present (1).

31 Coracoid fused to claviculae: absent (0); present (1).

32 Coracoid articular facet for the scapula: an excavated cotila (0); not a cotila (1).

33 Scapula, acromion cranially projected; absent (0); present (1).

34 Scapula corpus: curved (0); straight (1).

35 Scapula pneumatic foramen: present (0); absent (1).

Thoracic limb

36 Humerus – tuberculum ventrale projected proximally (more than caput humeri); absent (0); present (1).

37 Humerus – pneumatic foramen: absent or very small (0); large (1).

38 Humerus – diaphysis in anconal view: double curve in "sigma" (0); one curve with concavity anconal and medial (1).

39 Humerus – processus flexorius projected distally: absent (0); present (1).

40 Ulna length: equal or longer than the humerus (0); shorter than the humerus (1).

41 Carpometacarpus – distal end of metacarpale minus (see Alvarenga & Höfling, 2003, Fig. 6): same level of metacarpale majus (0); shorter than metacarpale majus (1).

42 Carpometacarpus – os metacarpale minus (shaft) (see Mayr & Clarke, 2003, character 85): almost parallel to metacarpale majus (0); bowed (1).

Pelvic girdle

43 Pelvis elongated and compressed laterally: absent (0); present (1).

44 A strong transversal crest supracetabularis ilii: absent (0); present (1).

45 Pars preacetabularis ilii: fused only in the top of spinous process of synsacral vertebrae (0); fused

in the top and lateral face of spinous process of synsacral vertebrae (1).

46 Pubis incomplete: absent (0); present (1).

Pelvic limb

47 Femur length: shorter than tarsometatarsus (0); equal or longer (1).

48 Femur length: shorter than humerus (0); equal or longer (1).

49 Femur – trochanter majus: prominent proximally (0); absent or not prominent (1).

50 Femur – fossa poplitea: shallow (90); deep (1).

51 Tibiotarsus – distal rim of condylus medialis distinctly notched (see Mayr & Clarke, 2003, char. 102): absent (0); present (1).

52 Tibiotarsus – condylus medialis medially deflected: absent (0); present (1).

53 Tarsometatarsus proportions: larger than 60% the length of tibiotarsus (0); smaller (1).

54 Tarsometatarsus strong and short: ratio of total length/width middle of diaphysis is bigger than 6 (0); the ratio is smaller than 6 (1).

55 Tarsometatarsus long and slender: the ratio of total length/width of middle of diaphysis is smaller than 12 (0); the ratio is bigger than 12 (1).

56 Tarsometatarsus – facies dorsalis excavated (an evident longitudinal sulcus): absent (0); present (1).

57 Tarsometatarsus – hypotarsus with well-developed crista/sulci (see Mayr & Clarke, 2003, character 103): absent (0); present (1).

58 Tarsometatarsus (dorsal view) – articular surface of middle trochlea – a dorsomedial expansion (see Alvarenga & Höfling, 2003, figure 8): absent (0); present (1).

59 Tarsometatarsus (dorsal view), trochlea metatarsi II: deflected medially (0); almost parallel to the trochlea III (1); articular surface extended medially (2).

60 Tarsometatarsus (distal view), trochlea metatarsi II: not deflec ted plantarly (0); deflected plantarly (1).

61 Tarsometatarsus (dorsal view), trochlea metatarsi IV, a longitudinal sulcus: present (0); absent (1).

APPENDIX 7B: CHARACTER MATRIX

Taxon	Character																														
	1	2	3	4	5	6	7	8	9	10	11	12	13	14	15	16	17	18	19	20	21	22	23	24	25	26	27	28	29	30	31
Anseranas-mipalmata	0	0	1	1	1	0	0	1	0	1	1	0	1	1	0	0	0	1	1	1	1	1	1	1	1	1	1	0	0	0	0
Ardeotis kori	0	0	0	0	0	1	0	0	0	0	0	0	0	0	0	0	1	1	1	1	1	0	0	1	0	1	0	0	0	0	0
Opisthocomus hoazin	0	0	1	1	?	0	0	0	0	0	1	0	0	0	0	0	1	1	1	1	0	1	1	1	1	?	1	0	0	1	1
Cariama cristata	0	0	0	0	0	0	0	0	1	0	0	0	0	0	0	0	0	1	0	1	1	0	0	1	0	1	1	0	1	0	0
Elaphrocnemus phasianus	0	0	0	1	0	0	0	0	1	0	0	0	1	?	?	?	?	?	?	?	?	0	?	?	0	1	0	0	0	0	0
Bathornis grallator	0	0	0	0	0	1	0	0	0	?	0	1	0	0	0	0	0	0	0	0	?	?	?	?	?	0	0	?	?	?	?
Brontornis burmeisteri	?	?	?	?	?	?	?	?	?	?	?	?	?	1	1	1	1	1	1	?	1	?	?	?	?	?	?	1	?	?	?
Physornis fortis	?	?	?	?	?	?	?	?	?	?	?	1	1	1	1	1	1	1	1	0	?	1	?	?	?	?	?	?	?	?	?
Paraphysornis brasiliensis	1	1	?	0	?	0	?	?	1	1	?	?	?	1	1	1	1	1	1	1	0	1	0	0	0	1	0	1	1	0	0
Phorusrhacos longissimus	1	1	?	1	?	?	?	?	1	?	?	?	?	1	1	1	1	1	1	0	?	?	?	?	?	?	?	?	?	?	?
Devincenzia pozzii	1	1	1	1	0	0	1	1	1	1	1	0	2	2	1	1	1	?	?	?	?	1	?	?	?	?	?	?	?	?	?
Kelenken guilhermoi	1	1	1	1	0	?	1	1	?	1	?	0	2	2	?	1	?	?	?	?	?	?	?	?	?	?	?	?	?	?	?
Titanis walleri	?	?	?	?	?	?	?	?	?	?	?	1	1	1	?	1	?	?	1	?	1	1	?	?	?	?	?	?	?	?	?
Patagornis marshi	1	1	1	1	1	0	1	1	1	1	1	0	1	1	1	0	1	1	0	0	?	0	0	0	0	0	1	0	?	0	0
Andrewsornis abbotti	1	1	1	1	?	?	1	1	?	1	?	1	?	?	1	1	1	0	?	1	?	?	?	?	?	?	?	?	?	?	?
Andalgalornis steulleti	1	1	2	1	0	1	1	1	1	1	?	0	0	0	1	1	1	1	0	0	0	0	0	0	0	?	?	0	1	0	0
Psilopterus bachmanni	1	1	0	0	0	1	1	1	1	1	0	0	0	0	1	1	1	1	0	1	1	0	0	0	0	1	1	1	0	0	0
Psilopterus lemoinei	1	1	1	1	0	1	1	1	1	1	?	0	0	0	0	?	1	1	0	1	1	1	0	0	0	1	0	1	0	0	0
Psilopterus colzecus	?	?	?	?	?	?	?	?	?	?	?	0	0	0	?	?	?	?	?	?	?	?	0	0	0	1	0	1	1	0	?
Procariama simplex	1	1	1	1	0	0	1	1	1	1	1	0	0	1	0	1	1	1	1	1	1	0	?	0	0	0	0	1	1	0	1
Mesembriornis milneedwardsi	0	?	1	1	0	0	1	1	?	1	?	0	?	?	?	?	?	?	?	0	?	?	?	?	?	?	?	1	1	0	1
Mesembriornis incertus	?	?	?	?	?	?	?	?	?	?	?	?	?	?	?	?	?	?	?	?	?	0	?	?	0	?	?	1	1	0	1

(*Continued*)

Taxon	Character																													
	32	33	34	35	36	37	38	39	40	41	42	43	44	45	46	47	48	49	50	51	52	53	54	55	56	57	58	59	60	61
Anseranas semipalmata	0	0	0	0	0	1	0	0	0	0	0	0	0	0	0	0	0	0	0	0	0	1	0	0	1	0	1	0	1	0
Ardeotis kori	0	0	1	0	1	0	1	0	0	0	0	0	0	0	0	0	0	0	0	0	0	0	0	1	1	1	0	0	0	1
Opisthocomus hoazin	0	1	0	0	1	1	0	0	0	0	0	0	0	0	0	1	0	1	0	0	0	0	0	0	0	0	0	0	0	1
Cariama cristata	1	1	0	0	1	0	0	0	0	0	1	0	0	0	0	0	0	0	0	0	1	0	0	1	0	0	0	0	0	1
Elaphrocnemus phasianus	1	?	?	0	0	0	0	1	0	0	1	0	0	?	?	0	0	1	1	0	0	0	1	1	1	0	0	0	0	0
Bathornis grallator	?	?	0	1	?	?	0	?	1	0	?	?	?	?	?	0	1	?	?	0	0	0	0	1	1	?	0	0	0	?
Brontornis burmeisteri	?	?	?	?	?	?	?	1	0	?	?	?	?	?	?	0	?	1	?	1	1	1	1	0	1	1	1	1	0	1
Physornis fortis	?	?	?	?	?	?	?	?	?	?	?	?	?	?	?	1	1	?	?	?	?	?	?	?	?	?	1	?	?	?
Paraphysornis brasiliensis	1	1	1	1	1	1	1	1	1	1	0	1	1	?	1	0	1	1	1	1	0	1	1	0	1	0	1	1	0	1
Phorusrhacos longissimus	?	?	1	?	1	1	1	?	?	?	?	?	?	?	?	0	1	1	0	1	0	0	0	1	0	0	1	0	0	1
Devincenzia pozzii	?	?	?	?	?	?	?	?	?	?	?	?	?	?	?	?	1	?	?	0	0	0	0	1	?	?	0	?	0	1
Kelenken guilhermoi	?	?	?	?	?	?	?	?	?	?	?	?	?	?	?	?	?	?	1	?	?	?	0	0	0	1	0	0	0	1
Titanis walleri	?	?	?	?	?	?	?	?	0	1	1	?	?	?	?	0	1	?	?	?	?	?	?	1	1	0	1	0	0	1
Patagornis marshi	1	1	1	1	?	1	1	1	0	1	1	1	1	1	0	1	0	1	1	0	0	0	1	1	1	0	1	0	1	1
Andrewsornis abbotti	?	?	1	1	?	?	?	?	?	?	1	1	?	?	?	?	?	?	?	?	?	?	?	1	?	?	?	?	?	?
Andalgalornis steulleti	?	?	?	1	?	?	?	?	?	?	1	1	1	1	1	?	?	?	1	?	?	?	?	1	1	?	?	?	?	1
Psilopterus bachmanni	1	1	1	1	1	1	1	1	0	1	0	1	1	1	1	0	1	1	1	0	0	0	0	0	1	0	0	2	0	1
Psilopterus lemoinei	1	1	1	1	1	1	1	1	0	1	0	1	1	1	1	0	1	1	0	0	0	0	1	1	1	0	0	2	0	1
Psilopterus colzecus	?	?	?	1	?	?	1	?	?	?	1	?	?	?	?	?	1	?	?	?	0	0	?	?	?	?	2	2	?	1
Procariama simplex	1	1	1	1	1	1	1	1	1	0	1	1	1	1	1	0	1	1	1	0	0	0	0	1	1	0	0	2	0	1
Mesembriornis milneedwardsi	1	?	?	?	1	?	1	1	1	1	?	?	?	?	0	1	1	1	1	0	0	0	0	1	1	0	0	1	1	0
Mesembriornis incertus	1	?	?	1	1	1	1	1	?	?	?	?	?	?	0	1	1	?	1	0	0	0	0	1	1	0	1	0	0	1

REFERENCES

Acosta Hospitaleche C. 2002. Nuevo registro de Hermosiornithinae (Cariamidae) del Mioceno tardio-temprano (Chasiquense) de Argentina. *Ameghiniana* **39** (2): 251–254.

Acosta Hospitaleche C, Tambussi C. 2005. Phorusrhacidae Psilopterinae (Aves) en la Formación Sarmiento de la localidad de Gran Hondonada (Eoceno Superior), Patagonia, Argentina. *Revista Española de Paleontología* **20**(2): 127–132.

Agnolin F. 2006. Posición sistemática de algumas aves fororracoideas (Ralliformes; Cariamae) argentinas. *Revista Museo Argentino Ciencias Naturales* **8**(1): 27–33.

Agnolin F. 2007. *Brontornis burmeisteri* Moreno and Mercerat, un Anseriformes (Aves) gigante del Mioceno Medio de Patagonia, Argentina. *Revista Museo Argentino Ciencias Naturales* **9**(1): 15–25.

Alfaro M, Perea D. 2004. Nuevos dados sobre la cronoestratigrafia de *Devincenzia* (Aves: Ralliformes, Cariamae) en Uruguay. *Actas del IV Congresso Uruguayo de Geologia.*

Alvarenga H. 1982. Uma gigantesca ave fóssil do Cenozóico brasileiro: *Physornis brasiliensis* sp. n. *Anais da Academia Brasileira de Ciências* **54**: 697–712.

Alvarenga H. 1983. Uma ave ratita do Paleoceno Brasileiro: Bacia Calcárea de Itaboraí. Est. Rio de Janeiro, Brasil. *Boletim do Museu Nacional, Rio de Janeiro, Geology* **41**: 1–11.

Alvarenga H. 1985. Um novo Psilopteridae (Aves: Gruiformes) dos sedimentos Terciários de Itaboraí, Rio de Janeiro, Brasil. In: Congresso Brasileiro de Paleontologia, 8. Rio de Janeiro, 1983. *NME-DNPM Série Geologia* **27**: 17–20.

Alvarenga H. 1993. Paraphysornis novo gênero para Physornis brasiliensis Alvarenga, 1982 (Aves: Phorusrhacidae). *Anais da Academia Brasileira de Ciências* **65**: 403–406.

Alvarenga H, Höfling E. 2003. Systematic revision of the Phorusrhacidae (Aves: Ralliformes). Papéis avulsos de Zoologia. *Museu de Zoologia da Universidade de São Paulo* **43**(4): 55–91.

Alvarenga H, Jones W, Rinderknecht A. 2010. The youngest record of Phorusrhacid birds (Aves: Phorusrhacidae) from the late Pleistocene of Uruguay. *Neues Jahrbuch für Geologie und Paläontologie* **256**(2): PP.

Amadon D. 1947. An estimated weight of the largest known bird. *Condor* **49**: 159–164.

Ameghino F. 1887. Enumeración sistemática de las espécies de mamíferos fósiles coleccionados por Carlos Ameghino en los terrenos Eocenos de la Patagonia austral y depositados en el Museo de La Plata. *Boletim Museo La Plata* **1**: 1–26.

Ameghino F. 1889. Contribuición al conocimiento de los mamíferos fósiles de la República Argentina. *Actas Academia Nacional Ciencias de Cordoba* **6**: 1–1028.

Ameghino F. 1891a. Mamíferos y aves fósiles Argentinos: espécies nuevas: adiciones y correciones. *Revista Argentina Historia Natural* **1**: 240–259.

Ameghino F. 1891b. Enumeración de las aves fósiles de la República Argentina. *Revista Argentina Historia Natural* **1**: 441–453.

Ameghino F. 1895. Sobre las aves fósiles de Patagonia. *Boletim Instituto Geografico Argentina* **15**: 501–602.

Ameghino F. 1898. Sinopsis geológico-paleontológica de la Argentina. *Segundo Censo de la República Argentina* **1**: 112–255.

Ameghino F. 1899. Sinopsis geológico-paleontológica. Suplemento *(Adiciones y correciones), La Plata*; 13 pp.

Ameghino F. 1901. L' age des formations sédimentaires de Patagonie. *Anales de la Sociedad Científica Argentina* **51**: 65–91.

Andors A. 1992. Reappraisal of the Eocene groundbird Diatryma (Aves Anserimorphae). *Los Angeles County Museum of Natural History, Science Series* **36**: 109–126.

Andrews C. 1896. Remarks on the Stereornithes, a group of extinct birds from Patagonia. *The Ibis* **7**: 1–12.

Andrews C. 1899. On the extinct birds of Patagonia. *Transactions of the Zoological Society of London* **15**: 55–86.

Baskin JA. 1995. The giant flightless bird *Titanis walleri* (Aves: Phorusrhacidae) from the Pleistocene coastal plain of south Texas. *Journal of Vertebrate Paleontology* **15**: 842–844.

Bertelli S, Chiappe L, Tambussi C. 2007. A new phorusrhacid (Aves: Cariamae) from the Middle Miocene of Patagonia, Argentina. *Journal of Vertebrate Paleontology* **27**(2): 409–419.

Brodkorb P. 1963. A giant flightless bird from the Pleistocene of Florida. *The Auk* **80**: 111–115.

Brodkorb P. 1967. Catalogue of fossil birds, Part III (Ralliformes, Ichthyornithiformes, Charadriiformes). *Bulletin of the Florida State Museum* **2**: 99–220.

Cabrera A. 1939. Sobre vertebrados fósiles del Plioceno de Adolfo Alsina. *Revista del Museo La Plata* **2**: 3–35.

Case JA, Woodburne M, Chaney D. 1987. A gigantic phororhacoid (?) bird from Antarctica. *Journal of Paleontology* **61**: 1280–1284.

Chandler RM. 1994. The wing of *Titanis walleri* (Aves: Phorusrhacidae) from the late Blancan of Florida.

Bulletin of Florida Museum of Natural History **36**: 175–180.

Chiappe L, Bertelli S. 2006. Skull morphology of giant terror birds. *Nature* **443**: 929.

Cracraft J. 1968. A review of the Bathornithidae (Aves, Gruiformes), with remarks on the relationships of the suborder Cariamae. *American Museum Novitates* **2326**: 1–46.

Cracraft J. 1971. Systematics and evolution of the Gruiformes (Class Aves) 2. Additional comments on the Bathornithidae, with descriptions of new species. *American Museum Novitates* **2449**: 1–14.

Cracraft J. 1973. Systematics and evolution of the Gruiformes (Class Aves) 3. Phylogeny of the suborder Grues. *Bulletin of the American Museum of Natural History* **151**(1): 1–127.

Degrange F, Tambussi C. 2008. *Psilopterus lemoinei* (Aves, Gruiformes, Phorusrhacidae, Psilopterinae) del Mioceno Inferior de Patagonia: Redescrición. *III Congreso Latinoamericano de Paleontologia de Vertebrados, Neuquén, Argentina.*

Dolgopol de Saez M. 1927. Las aves corredoras fósiles del Santacrucense. *Anales de la Sociedad Científica Argentina* **103**: 145–64.

Gould G, Quitmyier I. 2005. *Titanis walleri*: bones of contention. *Bulletin of the Florida Museum of Natural History* **45**(4): 201–229.

Hou L. 1980. (New form of the Gastornithidae from the lower Eocene of the Xichuan, Honan). *Vertebrata PalAsiatica* **18**: 111–115. (In Chinese – English abstract.)

Kraglievich L. 1931. Contribución al conocimiento de las aves fósiles de la época araucoentrerriana. *Physis* **10**: 304–315.

Kraglievich L. 1932. Una gigantesca ave fósil del Uruguay, *Devincenzia gallinali* n. gen. n. sp, tipo de una nueva familia, Devincenziidae, del Orden Stereornithes. *Anales del Museo Historia Natural de Montevideo* **3**: 323–55.

Kraglievich L. 1946. Noticia preliminar acerca de un nuevo y gigantesco Estereornito de la fauna Chapadmalense. *Anales de la Sociedad Científica Argentina* **142**: 104–121.

Livezey B, Zusi R. 2006. Higher-order phylogenetics of modern birds (Theropoda, Aves: Neornithes) based on comparative anatomy: I. Methods and characters. *Bulletin of the Carnegie Museum of Natural History* **37**: 1–544.

MacFadden B, Labs-Hochstein J, Hulbert R, Baskin J. 2006. Refined age of the late Neogene terror bird (*Titanis*) from Florida and Texas using rare earth elements. *Journal of Vertebrate Paleontology* **26** (Supplement 3): 92A.

Martin L. 1992. The status of the late Paleocene birds *Gastornis* and *Remiornis*. *Los Angeles County Museum of Natural History, Science Series* **36**: 97–108.

Matthew W, Granger W. 1917. The skeleton of *Diatryma*, a gigantic bird from the Lower Eocene of Wyoming. *Bulletin of the American Museum of Natural History* **37**: 307–326.

Mayr G. 2007. The birds from the Paleocene fissure filling of Walbeck (Germany). *Journal of Vertebrate Paleontology* **27**(2): 394–408.

Mayr G. 2009. Paleogene Fossil Birds. Heidelberg: Springer-Verlag; 262 pp.

Mayr G, Clarke J. 2003. The deep divergences of neornithine birds: a phylogenetic analysis of morphological characters. *Cladistics* **19**: 527–553.

Mayr G, Mourer-Chauviré C. 2006. Three dimensionally preserved cranial remains of *Elaphrocnemus* (Aves, Cariamae) from the Paleogene Quercy fissure fillings in France. *Neues Jahrbuch für Geology und Paläontology* **2006**: 15–27.

Mercerat A. 1897. Note sur les oiseaux fossiles de la Republique Argentine. *Anales de la Sociedad Científica Argentina* **43**: 222–240.

Moreno FP. 1889. Breve reseña de los progresos del Museo La Plata, durante el segundo semestre de 1888. *Boletin del Museo La Plata* **3**: 1–44.

Moreno FP, Mercerat A. 1891. Catálogo de los pájaros fósiles de la República Argentina conservados en el Museo de La Plata. *Anales del Museo de La Plata* **1**: 7–71.

Mourer-Chauviré C. 1983. Les Gruiformes (Aves) des Phosphorites du Quercy (France). I. Sous-ordre Cariamae (Cariamidae et Phorusrhacidae) systématique et biostratigraphie. *Paleovertebrata* **13**: 83–143.

Murray P, Vickers-Rich P. 2004. Magnificent Mirhirungs. Bloomington: Indiana University Press; 410 pp.

Noriega J, Agnolin F. 2008. El Registro paleontológico de las Aves del "Mesopotamiense" (Forrmación Ituzaingó, Mioceno tardio-Plioceno) de la Provincia de Entre Rios, Argentina. *Miscelânea INSUGEO* **17**(2): 271–290.

Noriega J, Vizcaino S, Bargo M. 2009. First record and a new species of seriema (Aves: Ralliformes: Cariamidae) from Santacucian (Early-Middle Miocene) beds of Patagonia. *Journal of Vertebrate Paleontology* **29**(2): 620–626.

Olson S. 1985. The fossil record of birds. In *Avian Biology, Volume 8*, Farner D, King J, Parkes K (eds). New York: Academic Press; 79–252.

Patterson B. 1941. A new phororhacoid bird from the Deseado formation of Patagonia. *Field Museum of Natural History, Geological Series* **8**: 49–54.

Paterson B, Kraglievich L. 1960. Sistemática y nomenclatura de las aves fororracoideas del Plioceno Argentino. *Publication del Museo Municipal Ciencias Naturales y Tradicionales de Mar del Plata* **1**: 1–51.

Peters DS. 1987. Ein "Phorusrhacidae" aus dem Mittel-Eozan von Messel (Aves: Gruiformes: Cariamae). *Documents des Laboratoires de Géologie de Lyon* **99**: 71–87.

Rovereto C. 1914. Los estratos araucanos y sus fósiles. *Anales del Museo Nacional Historia Natural de Buenos Aires* **25**: 1–247.

Sinclair W, Farr M. 1932. Aves of the Santa Cruz Beds. In: *Reports of the Princeton University Expeditions to Patagonia (1896–1899) Volume 7*, Scott W (ed.). Princeton University Press; 157–191.

Shockey B, Salas RQuispe R, Flores A, Sargis E, Acosta J, Pino A, Jarica N, Urbina M. 2006. Discovery of Deseadan fossils in the Upper Moquegua formation (Late Oligocene–?early Miocene) of southern Peru. *Journal of Vertebrate Paleontology* **26**(1): 205–208.

Swofford D. 2002. PAUP*: *Phylogenetic Analysis Using Parsimony (and Other Methods) 4.0 Beta for Macintosh*. Sunderland, MA: Sinauer Associates.

Tambussi C, Ubill M, Perea D. 1999. The youngest large carnassial bird (Phorusrhacidae, Phorusrhacinae) from South America (Pliocene–Early Pleistocene of Uruguay). *Journal of Vertebrate Paleontology* **19**: 404–406.

Taylor P. 1996. Family Rallidae (Rails, Gallinules and coots). In *Handbook of the birds of the World, Volume 3*, Hoatzin to Auks, del Hoyo J, Elliott A, Sargatal J (eds.). Barcelona: Lynx Edicions; 108–209.

Tonni E. 1974. Un Nuevo Cariámido (Aves Gruiformes) del Plioceno superior de la provincia de Buenos Aires. *Ameghiniana* **11**(4): 366–372.

Tonni E. 1980. The present state of knowledge of the cenozoic birds of Argentina. *Natural History Museum of Los Angeles County, Contributions in Science* **330**: 105–114.

Tonni E, Tambussi C. 1988. Un nuevo Psilopterinae (Aves: Ralliformes) del Mioceno tardio de la Provincia de Buenos Aires, Republica Argentina. *Ameghiniana* **25**: 155–160.

Wetmore A. 1944. A new terrestrial vulture from the upper Eocene deposits of Wyoming. *Annals of the Carnegie Museum* **30**: 57–69.

Wetmore A. 1967. Re-creating Madagascar's giant extinct bird. *National Geographic Magazine* **132**: 488–493.

Wroe S. 1998. Bills, bones and bias: did thunder birds eat meat? *Riversleigh Notes, New South Wales*, **40**: 2–4.

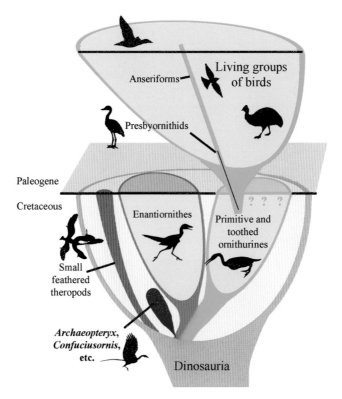

Plate 0.1   Cartoon to illustrate the basics of our current consensus regarding the pattern of the evolution of birds, relative to the Cretaceous-Paleogene (K-Pg) boundary.

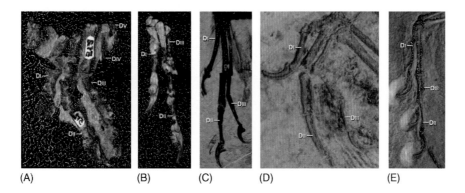

(A)   (B)   (C)   (D)   (E)

Plate 1.2   Comparison of theropod mani showing progressive reduction and loss of digits IV and V and changes in the proportions of manus elements. *Eoraptor* (A), *Guanlong* (B), *Sinornithosaurus* (C), *Archaeopteryx* (D), and *Confuciusornis* (E). Abbreviations: DI–V, digits I–V. All specimens shown at the same scale.

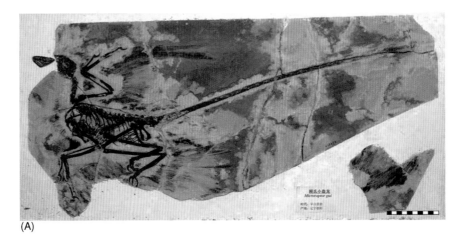

(A)

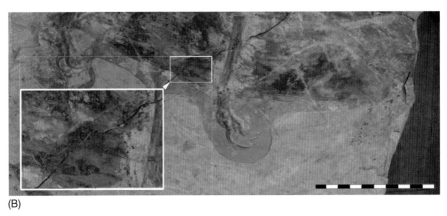

(B)

Plate 1.3 (A) Skeleton of the dromaeosaurid *Microraptor gui* from the Yixian Formation of Liaoning, China, exhibiting vaned, asymmetric feathers on both fore- and hindlimbs. (B) Detail of hindlimb primary feathers of *Pedopenna* from the Middle Jurassic of Inner Mongolia, China. *Pedopenna* is the earliest paravian fossil to exhibit vaned feathers and a four-winged body plan. The inset shows a close-up of the aligned and parallel barbs on each vane that indicate the presence of interlocking barbules, as well as the rachis. Note the large sickle claw characteristic of deinonychosaurians (= dromaeosaurs and troodontids) on digit II of the foot. Scale bars equal 5 cm. (Photographs: P. Makovicky.)

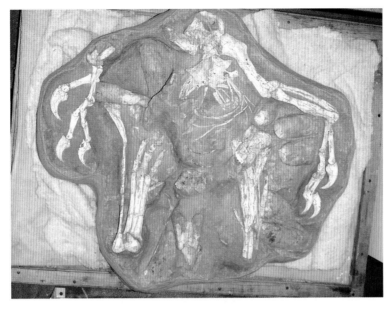

Plate 1.4 Partial skeleton of an oviraptorosaur in brooding posture on a nest of its eggs. Egg identity has been independently confirmed through embryonic remains. Specimens such as this reveal that these dinosaurs laid eggs in pairs over protracted periods (diachronous laying), and brooded them with direct contact indicative of synchronous hatching. Such associations of eggs and sexually mature individuals are now known from multiple nonavian maniraptoran taxa. (Photograph L. Zanno.)

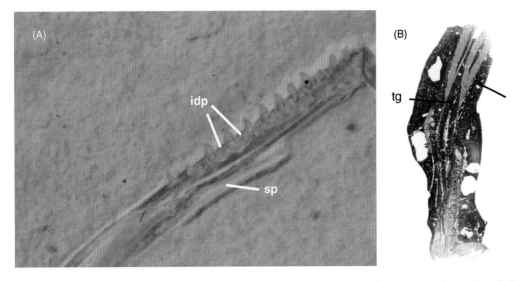

Plate 1.5 (A) Lower jaws of the Munich specimen of *Archaeopteryx* revealing the presence of interdental plates. (B) Cross-section of the dentary of *Allosaurus* revealing continuous histological ultrastructure between the bone below the alveoli and the interdental plates and demonstrating that the latter are not separate ossifications. Abbreviations: idp, interdental plates; sp, splenial; tg, germinating tooth. Specimens not to scale. (Photographs P. Makovicky.)

Plate 8.6 Reconstruction of the early Tertiary pseudo-toothed bird *Dasornis toliapica*, based on fossils from Paleocene–Eocene strata of Morocco and from Eocene strata of England.

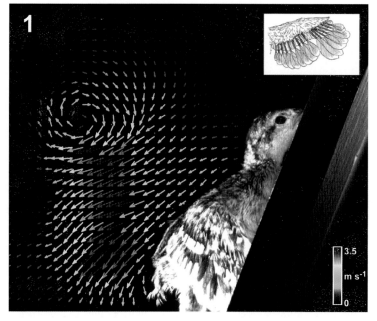

Plate 10.6 Airflow in the wake of a flightless chukar partridge (*Alectoris chukar*) chick engaged in wing-assisted incline running (WAIR). Velocity in the flow field was revealed using particle imagery velocimetry (PIV); the wake reveals evidence of lift production in a manner similar to juvenile and adult birds that are capable of flight. At day 8 of development, the chick has symmetrical remiges (inset, upper right). (Adapted from Tobalske & Dial, 2007.)

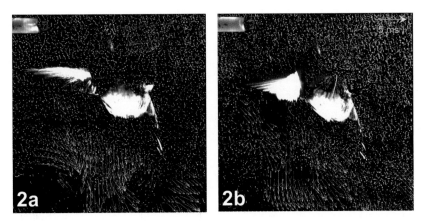

Plate 10.15 Hummingbird wing presentation and flow field in the wake at mid-downstroke (a) and mid-upstroke (b). (a) A red line is drawn above the dorsal surface of the wing to highlight the camber of the wing. (b) During upstroke, the proximal part of the wing (red line) is not as supinated as the distal portion (yellow line). The vector scale is at top right. (From Warrick *et al.*, 2005).

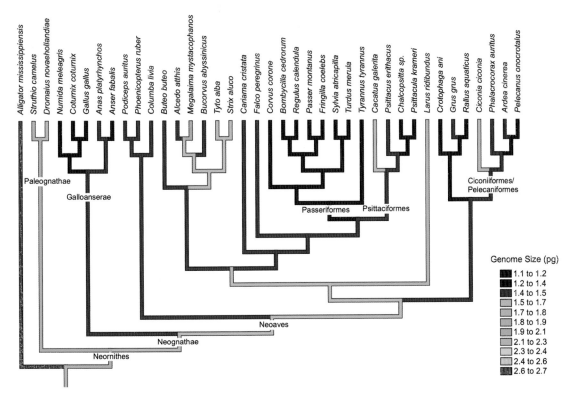

Plate 13.2 Diagram of genome size variation within modern birds (Neornithes). Data were obtained from the animal genome size database (Gregory, 2007) and mapped onto a pruned version of the most recent phylogenetic framework for Neornithes (Hackett *et al.*, 2008) used squared-change parsimony (Maddison & Maddison, 2008).

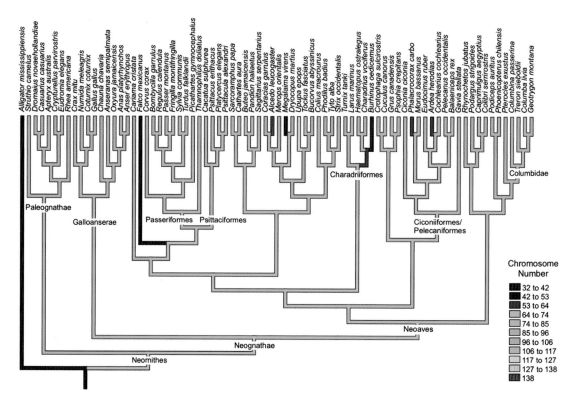

Plate 13.3 Diagram of karyotype variation within modern birds (Neornithes). Conspecific karyotype data used for Asian barbets (*Megalaima*) (Kaul & Ansari, 1981), owls (*Strix*) (Takagi & Sasaki, 1974), and cormorants (*Phalacrocorax*) (Ebied *et al.*, 2005). Other karyotype data were obtained from the literature (Benirschke, 1977; Waldrigues & Ferrari, 1982; Christidis, 1990; Qingsong *et al.*, 1995; Nishida *et al.*, 2008). Karyotype was mapped onto a pruned version of the most recent phylogenetic framework for Neornithes (Hackett *et al.*, 2008) used squared-change parsimony (Maddison & Maddison, 2008).

Plate 15.11 Global occurrence of Procellariiformes (light blue and light green) and Alcidae (light green only). Breeding areas are shown in darker tones. Occasionally, members of the Procellariiformes may wander into an uncolored area but it is not a regular part of their habitat. The large figure eight (red), in the Pacific Ocean, indicates the approximate track of the annual migration by the sooty shearwater (*Puffinus griseus*).

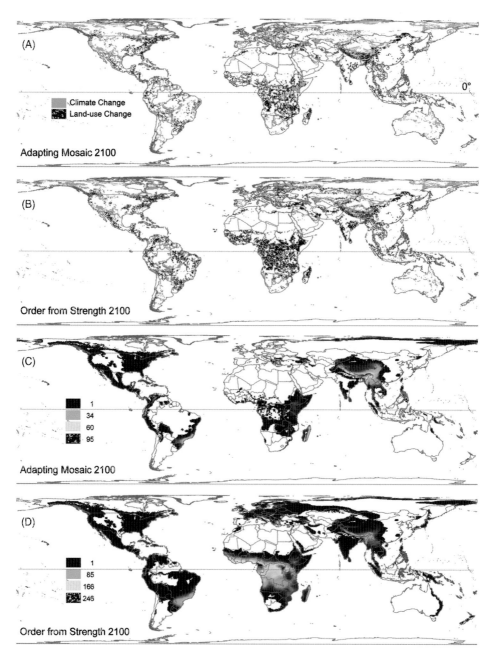

Plate 16.3 Projected impacts of global change on geographic patterns of avian species richness. Land cover conversion in 2100 due to climate and land-use based on projections of (A) the environmentally proactive Adapting Mosaic and (B) the environmentally reactive Order from Strength models. Species richness of birds with projected range declines of ≥ 50% on a 0.5° grid for (C) the Adapting Mosaic and (D) Order from Strength models. (From Jetz *et al.*, 2007.)

8 The Pseudo-toothed Birds (Aves, Odontopterygiformes) and their Bearing on the Early Evolution of Modern Birds

ESTELLE BOURDON

American Museum of Natural History, New York, USA

The pseudo-toothed birds or false-toothed birds (Odontopterygiformes, Pelagornithidae) are an extinct group of large seabirds with a huge bill bearing spiny osseous processes along the tomia (Figure 8.1A–F). Those tooth-like projections were hollow outgrowths of the mandibula and maxilla and lacked enamel, dentine, or cementum (Howard, 1957). They were well suited for holding soft or slippery prey (Zusi & Warheit, 1992). False-toothed birds most likely grasped prey such as soft-skinned fish and squid near the water surface while they were in flight or swimming (Zusi & Warheit, 1992). Bowing of the mandible to a great degree admitted large prey into the throat (Zusi & Warheit, 1992).

Their skeleton was highly pneumatized with elongated, straight, and slender wingbones (Figure 8.1G). The architecture of their forelimb indicates that they were pelagic soaring seabirds that must have filled a niche similar to that of albatrosses (Olson, 1985). Very long wings and short legs may have required peculiar takeoff and landing conditions. The wingspan of the smallest representatives of these birds was slightly smaller than that of a northern gannet (around 1.6 m: Bourdon, 2006a; Bourdon *et al.*, 2010),

and the largest were truly gigantic, reaching a wingspan of 5.5–6 m (Olson, 1985), which is two times larger than that of extant albatrosses. The morphology of the proximal humerus is highly distinctive (Figure 8.2A–D): the caput, tuberculum dorsale, and tuberculum ventrale are in virtually the same proximodistal plane, and the caput is an oblong diagonal (*sic.*) in proximal view (Olson, 1985). This would have restricted the rotary movement, typical of birds that use flapping flight, of the caput humeri in the cavitas glenoidalis (Olson, 1985). The structure of the bony labyrinth and cerebellum of the Paleogene pseudo-toothed bird *Dasornis toliapica* (Owen 1873) tends to confirm that this bird was adapted as a gliding snatch feeder, and was probably not a particularly aerobatic flyer (Milner & Walsh, 2009), which is congruent with previous paleoecological inferences (Olson, 1985; Zusi & Warheit, 1992).

Since Lartet (1857) described the first pelagornithid bone from the Miocene of France, numerous discoveries of these enigmatic birds from various Tertiary localities have punctuated the history of avian paleontology. Pseudo-toothed birds are now known to have had a worldwide

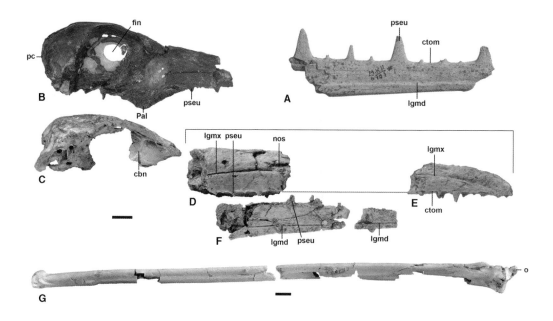

Fig. 8.1   (A) LACM22444, *Osteodontornis* from the Miocene of North America, fragment of right mandibula in lateral view. (B–G) *Dasornis toliapica* from the Lower Paleogene of England and Morocco; (B) BMNH 44096, cranium in right lateral view; (C) OCP.DEK/GE 1076, cranium in right lateral view; (D) OCP.DEK/GE 1185, fragment of proximal part of maxilla in right lateral view; (E) D1-0027E, distal end of maxilla in right lateral view; (F) OCP.DEK/GE 1166, fragment of right mandibula in lateral view; (G) MHNL 20-149215, left ulna in cranial view. cbn, caudal border of nasal cavity; ctom, crista tomialis; fin, fonticulus interorbitalis; lgmd, lateral groove of mandibula; lgmx, lateral groove of maxilla; nos, apertura nasi ossea; o, olecranon; Pal, os palatinum; pc, prominentia cerebellaris; pseu, pseudo-tooth. Scale bars equal 10 mm.

distribution and wandered the seas for approximately 55.5 Myr, from the late Paleocene to the latest Pliocene. The extinction of these highly specialized soarers could be linked to the climatic cooling that marked the Pliocene–Pleistocene transition (Lisiecki & Raymo, 2007).

In Eurasia, remains are known from the upper Paleocene (Harrison, 1985), lower Eocene (Harrison & Walker, 1976; Mayr, 2008) and possibly lower Oligocene (Harrison & Walker, 1979) strata of England; middle Eocene strata of Belgium (Mayr & Smith, 2010); Miocene strata of France (Lartet, 1857) and Portugal (Mayr, 2008); upper Paleocene strata of Kazakhstan and Eocene strata of Uzbekistan (Averianov *et al.*, 1991); Oligocene strata of Azerbaijan (Aslanova & Burchak-Abramovich, 1999); Oligocene (Hasegawa *et al.*, 1986; Okazaki, 1989), Miocene

(Ono, 1989; Matsuoka *et al.*, 1998), and Pliocene strata (Ono *et al.*, 1985) of Japan.

In Africa, pseudo-toothed birds were first described from the middle Eocene strata of Nigeria (Andrews, 1916). Some abundant pseudo-toothed bird remains have been found in the upper Paleocene–lower Eocene phosphates of Morocco (Gheerbrant *et al.*, 2003; Bourdon, 2005, 2006a,b; Bourdon *et al.*, 2010). A few undescribed specimens have also been recovered from the middle Eocene phosphates of Togo (Bourdon, 2006a). Recently, Mourer-Chauviré & Geraads (2008) described some fossils from the uppermost Pliocene strata of Morocco. These fossils represent the most recent remains of pseudo-toothed birds so far known.

Odontopterygiformes are also abundantly represented in the New World, with fossils

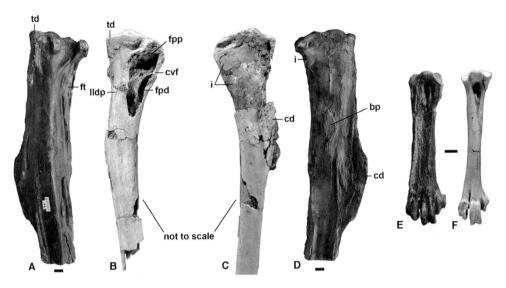

Fig. 8.2 Comparison of *Dasornis* from the Lower Paleogene of Morocco and *Pelagornis/Osteodontornis* from the Miocene of North America. (A–D) proximal parts of left humeri: (A) cf. *Pelagornis* USNM 335794, caudal view; (B) *Dasornis toliapica* reconstructed after OCP.DEK/GE 1229 and MHNL 20-149229, caudal view; (C) *D. toliapica* reconstructed after OCP.DEK/GE 1116, OCP.DEK/GE 1229 and MHNL 20-149229, cranial view; (D) cf. *Pelagornis* USNM 335794, cranial view. (E–F) right tarsometatarsi in dorsal view: (E) *Osteodontornis* LACM 128424; (F) *Dasornis emuinus* reversed OCP.DEK/GE 1106. bp, bicipital prominence; cd, crista deltopectoralis; cvf, crus ventrale fossae; fpd, distal foramen pneumaticum; fpp, proximal foramen pneumaticum; ft, fossa tricipitalis; i, intumescentia humeri; lldp, proximal part of linea musculi latissimi dorsi; td, tuberculum dorsale. Scale bars equal 10 mm.

described from the Eocene (Goedert, 1989; Olson, 1999), Oligocene (Shufeldt, 1916; Wetmore, 1917; Hopson, 1964; Olson, 1985) and Miocene strata (Howard, 1957; Howard & White, 1962; Howard, 1978; Olson, 1984, 1985; Rasmussen, 1998; Olson & Rasmussen, 2001; Stidham, 2004) of North America; Oligocene strata of Canada (Cope, 1894; Olson, 1985); middle Eocene (González-Barba *et al.*, 2002) and upper Miocene strata (González-Barba *et al.*, 2004) of Central America; Miocene and Pliocene strata of South America (Cheneval, 1993; Walsh & Hume, 2001; Rincón & Stucchi, 2003; Chavez *et al.*, 2007), including undescribed material from Miocene strata of Peru (GoGeometry, 2009).

Finally, some remains were also recovered from middle Eocene (Stilwell *et al.*, 1998) to upper Eocene strata of Antarctica (Tonni, 1980; Tonni & Tambussi, 1985; Tambussi & Hospitaleche, 2007) and from Miocene (Howard &

Warter, 1969; Scarlett, 1972) to Pliocene (McKee, 1985) strata of New Zealand.

Some authors have regarded the pseudo-toothed birds as being "intermediate" between the Procellariiformes and the Pelecaniformes, and assigned them to a distinct order, the Odontopterygiformes (Howard, 1957; Harrison & Walker, 1976). In a number of recent studies, however, the pseudo-toothed birds have been assigned to the family Pelagornithidae within the traditional order Pelecaniformes (e.g. McKee, 1985; Olson, 1985; Goedert, 1989; Averianov *et al.*, 1991; Cheneval, 1993; Rasmussen, 1998; Walsh & Hume, 2001; González-Barba *et al.*, 2002; Warheit, 2002; Rincón & Stucchi, 2003; Chavez *et al.*, 2007). A recent phylogenetic study including part of the pseudo-toothed birds from the Paleogene phosphates of Morocco has shown that the pseudo-toothed birds are sister to the Anseriformes (Bourdon, 2005). Given that waterfowl and pseudo-toothed birds differ in many respects, it is justified to assign the pseudo-toothed birds to the higher taxon Odontopterygiformes, as initially proposed by Howard (1957) and Harrison & Walker (1976).

ODONTOPTERYGIFORM DIVERSITY

Taxonomists who described odontopterygiform remains often proposed new names for incomplete and/or badly preserved specimens (e.g. Harrison & Walker, 1976; Aslanova & Burchak-Abramovich, 1982; Harrison, 1985; Averianov *et al.*, 1991). Harrison & Walker (1976) encouraged this over split taxonomy by proposing no less than four families inside the Odontopterygiformes, without convincing justification (see Olson, 1985; Mayr, 2008).

My examination of the available material led me to conclude that two morphotypes can be distinguished within the pseudo-toothed birds. In a previous phylogenetic study, Bourdon (2005) showed that *Odontopteryx* Owen 1873 and *Argillornis* Owen 1878 were part of the same clade, which was sister to *Pelagornis* Lartet 1857. Based on a new skull of *Dasornis* Owen 1870 from the lower Eocene strata of England, Mayr (2008) proposed that *Argillornis* is a junior synonym of *Dasornis*. The abundant material from the early Tertiary of Morocco permits further simplification of the taxonomy of the pseudo-toothed birds (Bourdon *et al.*, submitted): first, *Macrodontopteryx oweni* Harrison and Walker 1976 and *Neptuniavis minor* Harrison and Walker 1977 are synonymized with *Odontopteryx toliapica* Owen 1873, and second, *Odontopteryx* and *Macrodontopteryx* Harrison and Walker 1976 are junior synonyms of *Dasornis*. Pseudo-toothed birds from the Paleocene–Eocene phosphates of Morocco thus comprise the giant *Dasornis emuinus* (Bowerbank 1854) and the albatross-sized *Dasornis toliapica* (Owen 1873) (Bourdon *et al.*, 2010), both species also occurring in the lower Eocene London Clay. A third, gannet-sized new species from the lower Paleogene of Morocco is assigned to *Dasornis* and represents the smallest

pseudo-toothed bird ever known (Bourdon *et al.*, 2010). In the current state of knowledge, middle Eocene fossils from Belgium tentatively referred to *Dasornis emuinus* and *Macrodontopteryx oweni* by Mayr & Smith (2010) cannot be assigned confidently to *Dasornis* (Bourdon *et al.*, 2010). In sum, the taxon *Dasornis* comprises pseudo-toothed birds from the Upper Paleocene–Lower Eocene of Morocco (Bourdon, 2005, 2006a; Bourdon *et al.*, 2010) and Lower Eocene of England (Harrison & Walker, 1976; Mayr, 2008).

While comparing more recent pseudo-toothed birds with *Dasornis*, I found that they were no differences between the genera *Pelagornis* and *Osteodontornis* Howard 1957. *Osteodontornis* is most probably a junior synonym of *Pelagornis*. This synonymy cannot be established without a thorough systematic revision of the Odontopterygiformes. However, I regard all the *Pelagornis* and *Osteodontornis* specimens as pertaining to a single taxonomic entity, which corresponds to the *Pelagornis* morphotype. So far, the *Pelagornis/ Osteodontornis* material is known from the upper Eocene to Miocene strata of North America (Howard, 1957, 1978; Howard & White, 1962; Hopson, 1964; Olson, 1984, 1985; Goedert, 1989; Olson & Rasmussen, 2001; Stidham, 2004), Miocene and Pliocene strata of South America (Chavez *et al.*, 2007); Miocene strata of Europe (Lartet, 1857; Mayr *et al.*, 2008) and Pliocene strata of North Africa (Mourer-Chauviré & Geraads, 2008). Numerous specimens also most probably belong to *Pelagornis*: Miocene to Pliocene strata of South America (Cheneval, 1993; Walsh & Hume, 2001; Rincón & Stucchi, 2003) including a well preserved but still undescribed skull from the Miocene of Peru (GoGeometry, 2009); Miocene to Pliocene strata of Japan (Ono *et al.*, 1985; Ono, 1989); and Miocene strata of New Zealand (Scarlett, 1972).

Aside from these two morphotypes, odontopterygiform specimens were given a variety of different names, such as *Cyphornis magnus* Cope 1894, *Gigantornis eaglesomei* Andrews 1916, *Palaeochenoides miocaenus* Shufeldt

1916, *Tympanonesiotes wetmorei* Hopson 1964, *Neodontornis stirtoni* (Howard and Warter 1969), *Caspiodontornis kobystanicus* Aslanova and Burchak-Abramovich 1982 and no less than four different species assigned to the genus *Pseudodontornis* Lambrecht 1930 (Spulski, 1910; Harrison & Walker, 1976; Harrison, 1985; Averianov *et al.*, 1991). Whether part or all of these names are actually synonymous with either *Dasornis* or *Pelagornis* must await a thorough taxonomic revision of the group. At present, the number of species assigned to the Odontopterygiformes remains uncertain, but it is obvious that the number of names proposed in the literature exceeds the number of taxa.

The morphology of the humerus in *Dasornis* differs strikingly from that of *Pelagornis* (Figure 8.2A–D). In the *Pelagornis* morphotype, the morphological peculiarities of the humerus that are supposedly related to gliding flight are more pronounced than in *Dasornis*: the caput humeri is caudocranially wider; the tuberculum dorsale and tuberculum ventrale are more prominent and in more proximal position; the crista deltopectoralis is somewhat square in outline and located far distally from the caput. The sternum of *Pelagornis* (Mayr *et al.*, 2008) exhibits a more derived morphology than a sternum putatively assigned to *Dasornis* (Mayr & Smith, 2010): it is strongly vaulted and the carina sterni shows a marked cranial projection for the articulation with the furcula. These features probably reduced loadings linked to gigantic size and sustained gliding flight (Mayr *et al.*, 2008). Moreover, *Pelagornis* had relatively shorter and stouter legs than *Dasornis* (Figure 8.2E and F), and was possibly more clumsy on land. Altogether, these differences between the two morphotypes suggest that *Pelagornis* was more narrowly specialized for soaring flight than *Dasornis*. Pseudo-toothed birds pertaining to the *Dasornis* morphotype were more generalists and could probably use flapping flight, even if limited. In contrast, *Pelagornis* was most likely unable of sustained flapping flight, and had to rely almost entirely on winds to provide lift (Olson, 1985).

The following institutional abbreviations are used in this chapter: AMNH, American Museum of Natural History, New York, USA; BMNH and NHM, Natural History Museum, London, United Kingdom; D1, Rhinopolis Association, Gannat, France; LACM, Natural History Museum of Los Angeles County, Los Angeles, USA; MHNL, Muséum d'Histoire Naturelle de Lyon, France; MNHN, Muséum National d'Histoire Naturelle, Paris, France; MNHN-LAC, Muséum National d'Histoire Naturelle, Laboratoire d'Anatomie Comparée; OCP.DEK/GE, Office Chérifien des Phosphates, Direction des Exploitations de Khouribga, Service Géologie, Maroc; USNM, National Museum of Natural History, Washington, USA.

PHYLOGENETIC ANALYSIS

The initial character–taxon matrix including the Odontopterygiformes provided in Bourdon [2005, 2006a] was revised and expanded for the purpose of this study. The present matrix comprises 21 taxa and 128 osteological characters. A list of characters included in the analysis is provided in Appendix 8A, and the character–taxon matrix is shown in Appendix 8B. Outgroup taxa now comprise three Mesozoic nonneornithine Ornithurae, *Apsaravis* (Clarke & Norell, 2002), *Hesperornis* (Marsh, 1880; Witmer & Martin, 1987; Bühler *et al.*, 1988; Witmer, 1990; Elzanowski, 1991) and *Ichthyornis* (Marsh, 1880; Clarke, 2004). Odontopterygiform taxa included in the ingroup are *Dasornis* and *Pelagornis/Osteodontornis* (see above). Anatomical terminology of character descriptions follows Baumel *et al.* (1993), and Livezey & Zusi (2006) unless stated otherwise.

The parsimony analysis was performed using PAUP*4b10 (Swofford, 1998) and Winclada (Nixon, 1999). The 13 multistate characters (1, 48, 54, 55, 65, 88, 98, 100, 110, 112, 113, 114 and 123) were treated as ordered. The branch-and-bound search option was used for the PAUP* program. Node support was assessed using Bremer-support indices (Bremer, 1994), calculated by searching

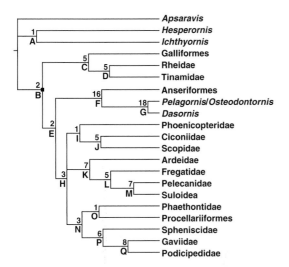

Fig. 8.3 Single most parsimonious tree showing the phylogenetic position of Odontopterygiformes within modern birds. Length (L), 207; consistency index (CI), 0.7; retention index (RI), 0.79. Numbers above branches correspond to Bremer-support indices. List of unambiguous synapomorphies (homoplastic ones marked with an asterisk): node A: 2(0)*; node B: 3(1), 4(1), 5(1), 6(1); node C: 7(1), 8(1), 9(1), 10(1), 11(1), 12(1), 13(1), 14(1), 15 (1), 16(1), 17(1)*, 18(1), 19(1), 20(1), 21(1), 23(1), 24(1)*, 25 (1), 26(1); node D: 27(1), 28(1), 29(1), 30(1), 34(0)*, 54(0)*; node E: 46(1)*, 49(1); node F: 1(2), 48(2)*, 54(2), 55(2), 56(1), 57(1), 58(1), 59(1), 61(1), 62(1), 63(1), 64(1), 65(1), 66(1), 67 (1), 68(1), 69(1), 71(1); node G: 1(3), 54(3), 65(2), 72(1), 73 (1), 74(1), 75(1), 76(1), 77(1), 78(1)*, 79(1), 80(1)*, 81(1), 82 (1)*, 83(1), 84(1), 85(1), 86(1), 87(1); node H: 54(0)*, 88(1), 89(1); node I: 123(1), 124(1); node J: 123(2), 125(1), 126(1), 127(1), 128(1); node K: 88(2), 109(1)*, 110(1), 111(1), 112 (1), 113(1), 114(1); node L: 110(2), 112(2), 113(2), 114(2), 115(1), 116(1); node M: 113(3), 117(1)*, 118(1), 119(1), 120 (1), 121(1), 122(1); node N: 90(1), 91(1), 92(1); node O: 78 (1)*, 93(1)*, 94(1), 95(1); node P: 4(0)*, 98(2), 99(1), 100(1), 101(1), 102(1); node Q: 98(3), 100(2), 103(1)*, 104(1)*, 105 (1), 106(1), 107(1), 108(1).

suboptimal trees up to 18 extra steps with the branch-and-bound search option of PAUP*. Cladistic analysis of the character–taxon matrix in Appendix 8B resulted in one most parsimonious tree (Length = 207; CI = 0.7; RI = 0.79), which is shown in Figure 8.3. Only major results are provided here. Readers are referred to the caption of

Figure 8.3 for a complete list of unambiguous synapomorphies. In the consensus tree, Neornithine birds split into a well-supported clade (Galliformes plus Palaeognathae) on one side and a poorly supported clade comprising all remaining modern birds on the other side. The latter clade includes the strongly supported Odontoanserae (Anseriformes plus Odontopterygiformes) and a loosely supported clade comprising all other neoavian groups included in the study.

The monophyly of the Odontoanserae (Odontopterygiformes plus Anseriformes) is based on 18 characters (Figures 8.4 and 8.5), one of which is homoplastic: origin of the musculus extensor carpi ulnaris on the humerus forming two distinct concavities surrounded by a prominent ridge of even width (character 48; Bourdon, 2005, figure 3). Seventeen uniquely derived characters support the Odontoanserae clade: condylus occipitalis pedonculate, large and bilobate (character 54; Figure 8.4A); processus rostropterygoideus present, lipped, with basal support (character 55; Figure 8.4A; Bourdon, 2005, figure 3); processus paroccipitalis strongly protruding caudoventrally, caudally convex, with wide lateral side for origin of musculus depressor mandibulae; the processus is continuous with stout processus lateralis parasphenoidalis and ala parasphenoidalis, so that the cavitas tympanica is deeply recessed (character 56; Figure 8.4A and B); processus postorbitalis stout and projecting obliquely rostrally beneath orbita (character 57; Figure 8.4A and B); lamina parasphenoidalis caudorostrally narrow and triangular in shape with tuberculum basilare at caudal corner and curved caudal border close to but higher than condylus occipitalis; fossa subcondylaris deep (character 58; Figure 8.4A); cranium with impressio musculi adductoris mandibulae externus pars coronoidea in medial position (character 59; Figure 8.4A; Bourdon, 2005, figure 3); os meseth-moidale well developed rostrally, showing deep depression for concha caudalis in caudal half and conspicuous oblique column bordering this depression rostroventrally (character 61; Figure 8.4A; Bourdon, 2005, figure 3); lateral border of os lacrimale above processus orbitalis

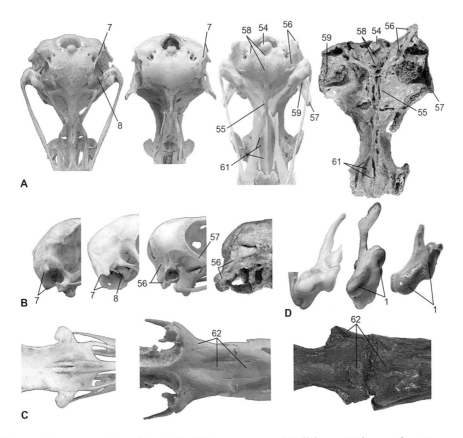

Fig. 8.4   Illustrated synapomorphies of the clades Odontoanserae and Galliformes–Palaeognathae. Taxa are enumerated from left to right. (A) Crania in ventral view: *Nothoprocta ornata* AMNH 6500, *Lophura erythrophthalma* AMNH 4820, *Cairina moschata* AMNH 11024 and *Dasornis toliapica* OCP.DEK/GE 1044. (B) Crania in right lateral view: *N. ornata* AMNH 6500, *L. erythrophthalma* AMNH 4820, *C. moschata* AMNH 11024 and *D. toliapica* OCP.DEK/GE 1076. (C) Crania in dorsal view: *L. erythrophthalma* AMNH 4820, *Chloephaga* sp. MNHN-LAC 1884-848 and *D. toliapica* BMNH 44096. (D) Right ossa quadrati in ventral view: *Tetrao urogallus* AMNH 12662, *Anseranas semipalmata* MNHN-LAC 2004-151 and *D. toliapica* BMNH 44096. Not to scale.

forming an elongated, vertical and slightly concave facet that is roughly parallel to the long axis of the cranium; surface between ossa lacrimales concave from side to side; apex of maxilla a flat to slightly concave triangular surface just rostral to abrupt zona flexoria craniofacialis (character 62; Figure 8.4C); processus mandibularis of os quadratum with 2 condylae, the condylus medialis being distinctly rostral to the condylus lateralis (character 1; Figure 8.4D); extremitas distalis humeri: condylus ventralis round, distally prominent, with well-marked caudal border distinctly

cranial to fossa olecrani and in the same plane as condylus dorsalis; wide and deep incisura intercondylaris (character 63; Figure 8.5B; Bourdon, 2005, figure 1e); extremitas proximalis ulnae wider than it is deep, with strongly convex facies caudodorsalis showing impressio musculi scapulotricipitalis in distodorsal position; flat cotyla dorsalis with straight cranial border and pointed extremity; depression for attachment of meniscus radioulnaris poorly developed (character 64; Figure 8.5C); extremitas distalis ulnae forming an isosceles triangle with condylus dorsalis,

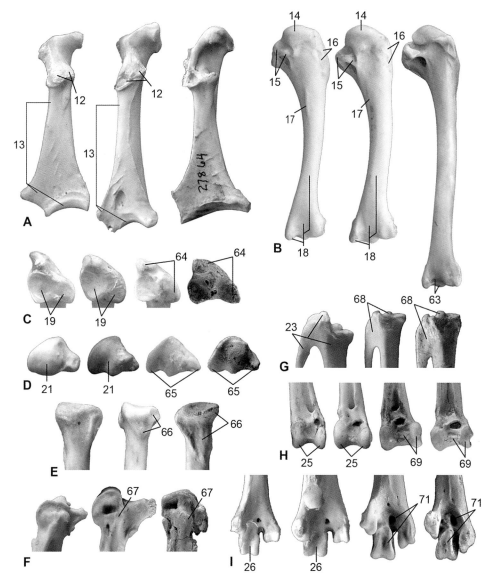

Fig. 8.5 Illustrated synapomorphies of the clades Odontoanserae and Galliformes–Palaeognathae. Taxa are enumerated from left to right. (A) left ossa coracoidei in dorsal view: *Rhynchotus rufescens* AMNH 3533, *Tetrao urogallus* AMNH 12662 and *Anas platyrhynchos* AMNH 27864. (B) right humeri in caudal view: *R. rufescens* AMNH 3533, *Lophura erythrophthalma* AMNH 4820 and reversed *A. platyrhynchos* AMNH 21525. (C) Proximal ends of left ulnae in proximal view: *Tinamus solitarius* AMNH 21983, *L. erythrophthalma* AMNH 4820, *A. platythynchos* AMNH 27864 and *Dasornis toliapica* BMNH 44096. (D) Distal ends of right ulnae in distal view: *R. rufescens* AMNH 3533, *L. erythrophthalma* AMNH 4820, *A. platyrhynchos* AMNH 27864 and *D. toliapica* OCP.DEK/GE 1198. (E) Proximal ends of left radii in caudal view: *T. urogallus* AMNH 12662, *A. platyrhynchos* AMNH 27864 and *Dasornis emuinus* OCP. DEK/GE 1224. (F) Proximal ends of left carpometacarpi in ventral view: *Argusianus argus* MNHN-LAC 1972-86, *Chloephaga* sp. MNHN-LAC 1884-848 and *D. toliapica* OCP.DEK/GE 1152. (G) Distal parts of right carpometacarpi in

condylus ventralis and tuberculum carpale of same width and smooth aspect; base of tuberculum carpale prominent, thick and oblique (character 65; Figure 8.5D); radius, ventral border of cotyla humeralis convex, forming distinct overhang and continuous with caudal edge of tuberculum bicipitale; surface dorsal to the latter and distal to facies articularis ulnaris flat and triangular (character 66; Figure 8.5E); processus pisiformis of carpometacarpus thick and prominent, with proximal border reaching trochlea carpalis; rostral border extending far distally and roughly paralleling the long axis of corpus (character 67; Figure 8.5F); carpometacarpus, facies ventralis: long symphysis metacarpalis distalis with os metacarpale minus close and nearly parallel to os metacarpale majus, which shows curved median ridge; high and well-defined caudal protuberance on facies articularis digitalis major (character 68; Figure 8.5G); tibiotarsus with wide incisura intercondylaris extending onto proximal part of condylus medialis; the latter protrudes more rostrally and is narrower than condylus lateralis (character 69; Figure 8.5H); tarsometatarsus, plantar side of trochlea metatarsi III prominent, elongated with pointed extremity, slightly oblique; foramen vasculare distale in low position with a recessed opening (character 71; Figure 8.5I).

The Galliformes are grouped with the Palaeognathae (represented here by two taxa, namely Tinamidae and Rheidae) on the basis of 19 synapomorphies, two of which are homoplastic: humerus, linea musculi latissimi dorsi either in median position between margo dorsalis and margo ventralis or closer to margo ventralis (character 17; Figure 8.5B); proximal part of facies caudalis of corpus tibiotarsi flat (character 24). Seventeen strict synapomorphies support the monophyly of the clade Galliformes +

Palaeognathae: processus paroccipitalis, a thick hemispherical flange completely enclosing the middle ear region caudally (character 7; Figure 8.4A and B); ala parasphenoidalis greatly thickened, cancellous, with evenly curved border (character 8; Figure 8.4A and B); apertura nasi ossea of large size, bounded caudally by very thin processus maxillaris of os nasale (character 9); sternum showing greatly elongated and stout processus craniolaterales with rostrolateral orientation (character 10); facies articularis humeralis scapulae strongly protruding, caudocranially short, roughly round in shape, and facing dorsally (character 11); os coracoideum: indistinct facies articularis scapularis and short blunt processus procoracoideus located either at the level of or slightly omal to the former (character 12; Figure 8.5A); os coracoideum, corpus coracoidei straight, slender, elongated, mediolaterally compressed and ventrally convex; facies articularis sternalis thick and narrow (character 13; Figure 8.5A); caput humeri caudocranially thick and rectangular in shape (character 14; Figure 8.5B); humerus showing prominent ridge joining crista dorsale fossae and caput humeri, and continuous with edge of blunt tuberculum ventrale (character 15; Figure 8.5B); humerus, caudal surface of crista deltopectoralis convex and bearing proximodistally elongated impressio musculi supracoracoidei (character 16; Figure 8.5B); corpus humeri sigmoid, widening and flattening towards extremitas distalis, which shows smooth facies dorsalis; well developed processus flexorius that protrudes mediodistally (character 18; Figure 8.5B); extremitas proximalis ulnae, cotylae forming single flat surface and crista intercotylaris absent (character 19; Figure 8.5C); ulna strongly bowed and dorsoventrally compressed (character 20); extremitas distalis ulnae with

---

Fig. 8.5 *(Continued)* ventral view: *A. argus* MNHN-LAC 1972-86, *Chloephaga* sp. MNHN-LAC 1884-848 and *D. toliapica* OCP.DEK/GE 1102. (H) Distal ends of right tibiotarsi in rostral view: *R. rufescens* AMNH 3533, *L. erythrophthalma* AMNH 4820, *Chloephaga* sp. MNHN-LAC 1884-848 and *D. toliapica* OCP.DEK/GE 1152. (I) Distal ends of right tarsometatarsi in plantar view: *R. rufescens* AMNH 3533, *L. erythrophthalma* AMNH 4820, *C. moschata* AMNH 11024 and *D. toliapica* reversed OCP.DEK/GE 1146. Not to scale.

poorly defined sulcus intercondylaris plus condylus ventralis (character 21; Figure 8.5D); carpometacarpus, os metacarpale majus dorsally flat with
sulcus tendinosus in extreme rostral position; os
metacarpale minus curved, caudocranially flattened and longer than os metacarpale majus, so
that facies articularis digitalis minor markedly
distal to facies articularis digitalis major (character
23; Figure 8.5G); extremitas distalis tibiotarsi:
condylae wide, parallel to each other, caudocranially short and proximodistally high; condylus
medialis protruding more rostrally than round
condylus lateralis (character 25; Figure 8.5H); tarsometatarsus, trochlea metatarsi III widening
towards extremity and elongated at the base; in
plantar aspect, the trochlea is short and distinctly
asymmetrical, its lateral part extending further
proximally than its medial one (character 26;
Figure 8.5I).

ODONTOPTERYGIFORMES AND EARLY
NEORNITHINE EVOLUTION

The present study reinforces the hypothesis that
pseudo-toothed birds and ducks are each other's
closest relatives (Bourdon, 2005). Interestingly,
a recent study of the brain and endocranium
of Lower Eocene birds from England provides
some support for this hypothesis (Milner &
Walsh, 2009): in contrast to Procellariiformes
and most 'Pelecaniformes', both Anseriformes
and *Dasornis toliapica* have the carotid rami
enclosed into bony tunnels and exhibit a long
anastomosis intercarotica. False-toothed birds
are part of large clade that mainly includes birds
adapted to aquatic life. Most species included in
the Anseranatidae-Anatidae clade are freshwater
birds and a few species live in the coastal zone (del
Hoyo *et al.*, 1992). A few anseriforms depart from
the ecology of "typical" ducks, such as the screamers that live in flooded open areas (del Hoyo
et al., 1992), and the extinct Dromornithidae,
large flightless birds of the Tertiary and Pleistocene of Australia (Murray & Vickers-Rich, 2004).
Though Odontopterygiformes are fully aquatic
birds, their habits are very distinct from those of

their close relatives. The life reconstruction
of *Dasornis* proposed here is based mainly on
albatross morphology (Figure 8.6) because the
elongated and straight forelimb bones of the
pseudo-toothed birds are reminiscent of the wing
proportions of extant albatrosses (Diomedeidae).
Osteodontornis orri (Feduccia, 1999, p. 192) is
represented with a naked gular pouch and an elongated, S-shaped neck based on "pelecaniform"
morphology, especially that of the Pelecanidae.
A naked gular pouch has been shown to be synapomorphic for the Steganopodes (Cracraft, 1985;
Mayr, 2003) and a kinked neck is found in

E.Bourdon

Fig. 8.6 Reconstruction of the early Tertiary pseudo-
toothed bird *Dasornis toliapica*, based on fossils from
Paleocene–Eocene strata of Morocco and from Eocene
strata of England. [This figure appears in color as Plate 8.6.]

Steganopodes (Pelecanidae, Phalacrocoracidae, Anhingidae) and in their closest relatives (Figure 8.3), the Ardeidae. I have chosen not to represent these features in my own reconstruction, first because the false-toothed birds are sister to the Anseriformes, and second because the few illustrated pelagornithid cervical vertebrae appear massive and not caudocranially elongated as in pelicans (Howard, 1957, figures 2, 3; Olson & Rasmussen, 2001, plate 11h; Chavez *et al.*, 2007, figure 4).

In recent decades, the vast majority of studies, both molecular and morphological, have shown that the first divergence within modern birds separates the Palaeognathae from the Neognathae, and that the latter taxon splits into Galloanserae (Galliformes plus Anseriformes) and Neoaves (all remaining neognathous birds; e.g. Groth & Barrowclough, 1999; van Tuinen *et al.*, 2000; Cracraft & Clarke, 2001; Mayr & Clarke, 2003; Chubb, 2004; Cracraft *et al.*, 2004; Fain & Houde, 2004; Simon *et al.*, 2004; Harshman, 2007; Hugall *et al.*, 2007; Livezey & Zusi, 2007; Hackett *et al.*, 2008). In spite of the fact that classical osteological characters supporting the monophyly of the Neognathae (characters 31–45) and Galloanserae (characters 50–55) were included in the present analysis (Livezey, 1986, 1997; Cracraft, 1988; Ericson, 1997; Cracraft & Clarke, 2001; Clarke & Norell, 2002; Mayr & Clarke, 2003; Livezey & Zusi, 2006), the Galliformes are grouped with the Palaeognathae, which are represented by two taxa in this study: Tinamidae and Rheidae. This result not only refutes the hypothesis of Galloanserae monophyly, but also contradicts the Paleognathae/Neognathae dichotomy, which has been established since the dawn of the Twentieth Century (Pycraft, 1900).

An alliance between Anseriformes and Galliformes came from the idea that the screamers (Anhimidae) bear some resemblances with gallinaceous birds (Huxley, 1867; Garrod, 1873, 1874, 1876; Seebohm, 1889; Shufeldt, 1901; Mayr & Amadon, 1951). The weakness of this hypothesis is that the osteological synapomorphies of the Galloanserae are only cranial. Surprisingly, not a single convincing postcranial

character of the Galloanserae has ever been proposed. I have been unable to distinguish the character states defined in the character 2073 of Livezey & Zusi (2006), which deals with the shape of the cristae cnemiales tibiotarsi. Olson & Feduccia (1980b) stated: "The most current hypothesis is that the Anseriformes are related to the Galliformes...the morphological basis for this rests on the alleged similarities in the pterygoid–parasphenoid articulation and the retroarticular process of the mandible in the two groups. Were it not for this and the superficially fowl-like appearance of the bill of screamers, it is extremely doubtful that a relationship between ducks and galliforms would ever have been entertained, for all other aspects of their morphology are so utterly different."

Although basal divergences of modern birds are now regarded as fully resolved, a small minority of authors have continued to debate the pairing of Galliformes with Anseriformes in the Galloanserae (Olson & Feduccia, 1980a,b; Ericson, 1996; Hope, 2002). Some cladistic analyses based on morphology (Ericson, 1997; Clarke & Chiappe, 2001; Ericson *et al.*, 2001; Bourdon, 2005) or DNA sequences (Ericson *et al.*, 2001) have supported an initial split of the Neognathae into Galliformes and all remaining Neognathae. Furthermore, it is of interest to note that various molecular studies have challenged the Palaeognathae–Neognathae dichotomy in the past two decades. Galloanserae are grouped with the Palaeognathae in the large DNA–DNA hybridization study by Sibley & Ahlquist (1990, figure 353). Nonetheless, this result comes from an unrooted tree obtained by average linkage clustering of DNA–DNA hybridization distances. Some phylogenetic studies based on partially complete to complete mitochondrial DNA genomes have proposed that the closest relatives of the Palaeognathae are either the Galloanserae (Mindell *et al.*, 1997, 1999; Haring *et al.*, 2001) or the Galliformes (Härlid *et al.*, 1998; Härlid & Arnason, 1999). They suggest that the paleognathous characteristics have been secondarily acquired from the corresponding morphological features characterizing neornithine birds. It appears, however, that these mtDNA studies are based on a limited taxon sampling and suffer from an outgroup problem (García-Moreno *et al.*, 2003). In fact, a study based on four mitochondrial genes, with a larger taxon sampling and both turtles and crocodiles used as outgroups, supports the traditional Paleognathae–Neognathae dichotomy (García-Moreno *et al.*, 2003).

Formerly, many taxonomists used morphology (e.g. osteology, pterylosis) as evidence for a close relationship between gallinaceous birds and paleognathous birds, especially the tinamous (see Sibley & Ahlquist (1990) for a review). The clade Galliformes–Palaeognathae proposed here (Figure 8.3) is based on a new set of osteological characters that were found while revising the previous phylogeny by Bourdon (2005). None of these features occur in the three nonneornithine Ornithurae used for outgroup comparison. In the present context, similarities between Galliformes and Palaeognathae are reinterpreted as derived within modern birds. These comprise cranial as well as postcranial characters including features of the pectoral girdle, wing and hindlimb.

At present, the earliest representatives of the pseudo-toothed birds are late Paleocene in age (Harrison, 1985; Averianov *et al.*, 1991; Bourdon, 2005, 2006a,b). The phylogenetic placement of the late Maastrichtian *Vegavis* from Antarctica (Clarke *et al.*, 2005) provides the first reliable record of a modern bird in the Cretaceous and the earliest certain record for the Anseriformes. This discovery refutes a proposed early Tertiary origin of modern birds (Feduccia, 1995, 2003) and implies that the age of diversification of the Odontoanserae is older than Late Cretaceous. The tree proposed here may constitute the first step towards an alternative view of the history of modern birds. Here I suggest that the Neognathae are paraphyletic, and that the oldest divergence within Neornithes might not be the currently recognized one. Whether the Galliformes–Palaeognathae clade is either set apart from all remaining modern birds or nested within the Neornithes remains to be solved, as the clade comprising all neornithine birds except Galliformes and Palaeognathae is not well supported in the present study. Moreover, further

work is needed to assess the phylogenetic position of the Odontoanserae within modern birds. The next step would be to include representatives of all order-level neornithine groups, including key Cretaceous–Paleogene fossils. This would permit the assessment of the number of diversification events that punctuated the history of modern birds prior to the Late Cretaceous, and the evaluation of their degree of survivorship at the K–P boundary.

ACKNOWLEDGMENTS

Some of the photographs were taken by D. Serrette and P. Loubry (MNHN, Paris). J. L. Cracraft provided useful remarks on a late draft of this manuscript. The LACM, MHNL, NHM, OCP, Rhinopolis Association, and USNM kindly provided access to their collections. This work was supported by the Collège de France (Paris) and the AMNH (F. M. Chapman Memorial Fund, Department of Ornithology). The editors and one anonymous referee made useful comments on the manuscript.

APPENDIX 8A

1 Os quadratum, processus mandibularis: not as follows (0); with 3 condylae (1) (Livezey, 1997, character 51, modified); 2 condylae, and condylus medialis distinctly rostral to condylus lateralis (2) (Livezey, 1997, character 52, modified); condylus medialis with protruding caudolateral corner (3) (Bourdon, 2005, character 25).

2 Mandibula, os dentale, ossified symphysis mandibularis: no (0); yes (1) (Mayr & Clarke, 2003, character 42).

3 Humerus, crista deltopectoralis with cranial deflection: no (0); yes (1) (Clarke & Norell, 2002, character 112). This is coded nonapplicable in *Hesperornis*.

4 Humerus with well marked fossa pneumotricipitalis bearing foramen pneumaticum: no (0); yes (1) (see Clarke & Norell, 2002, character 118; Mayr & Clarke, 2003, character 77).

5 Tibiotarsus, ossified pons supratendineus: no (0); yes (1) (Clarke & Norell, 2002, character 180; Mayr & Clarke, 2003, character 100).

6 Tarsometatarsus, corpus tarsometatarsi, facies plantaris distinctly angular-convex, with well marked cristae et sulci: no (0); yes (1) (Livezey & Zusi, 2006, character 2294).

7 Os exoccipitale, processus paroccipitalis a thick hemispherical flange completely enclosing the middle ear region caudally: no (0); yes (1).

8 Ala parasphenoidalis greatly thickened, cancellous, with evenly curved border: no (0); yes (1).

9 Apertura nasi ossea of large size, bounded caudally by very thin processus maxillaris of os nasale: no (0); yes (1). The processus maxillaris of os nasale is secondarily thickened in most Galliformes.

10 Sternum, greatly elongated and stout processus craniolaterales with rostrolateral orientation: no (0); yes (1).

11 Scapula, facies articularis humeralis strongly protruding, caudocranially short, roughly round in shape, and facing dorsally: no (0); yes (1). This is coded nonapplicable for Rheidae.

12 Os coracoideum: indistinct facies articularis scapularis and short blunt processus procoracoideus located either at the level of or slightly omal to the former: no (0); yes (1) (modified from Ericson, 1997, p. 451). This is coded nonapplicable for Rheidae.

13 Os coracoideum, corpus coracoidei straight, slender, elongated, mediolaterally compressed and ventrally convex; facies articularis sternalis thick and narrow: no (0); yes (1). This is coded nonapplicable for Rheidae.

14 Humerus, caput humeri caudocranially thick and rectangular in shape: no (0); yes (1).

15 Humerus: prominent ridge joining crista dorsale fossae and caput humeri, continuous with edge of blunt tuberculum ventrale: absent (0); present (1). This is coded nonapplicable for Rheidae, in which the caput humeri and the tuberculum ventrale are fused into one single structure.

16 Humerus, caudal surface of crista deltopectoralis convex and bearing proximodistally elongated impressio musculi supracoracoideus: no (0); yes (1) (Ericson, 1997, p. 455, modified). In Rheidae, the impressio is displaced distally; this is regarded as an autapomorphic feature. This is coded nonapplicable in *Hesperornis*.

17 Humerus, linea musculi latissimi dorsi: closer to margo dorsalis (0); in median position between margo dorsalis and margo ventralis or closer to margo ventralis (1) (modified from Bourdon, 2005, character 11).

18 Humerus, corpus humeri sigmoid, widening and flattening towards extremitas distalis, which shows smooth facies dorsalis (no marked sulci); well developed processus flexorius that protrudes mediodistally: no (0); yes (1).

19 Ulna, extremitas proximalis: cotyla dorsalis and cotyla ventralis forming single flat surface; crista intercotylaris absent: no (0); yes (1).

20 Ulna strongly bowed and dorsoventrally compressed: no (0); yes (1).

21 Ulna, extremitas distalis, poorly defined sulcus intercondylaris and condylus ventralis (modified from Ericson, 1997, 458): no (0); yes (1).

22 Radius, distal part of corpus strongly twisted dorsally: no (0); yes (1).

23 Carpometacarpus: os metacarpale majus dorsally flat with sulcus tendinosus in extreme rostral position; os metacarpale minus curved, caudocranially flattened and longer than os metacarpale majus, so that facies articularis digitalis minor markedly distal to facies articularis digitalis major: no (0); yes (1). Due to their highly apomorphic condition, the Rheidae are coded nonapplicable for this character.

24 Tibiotarsus, proximal part of facies caudalis of corpus: convex (0); flat (1) (modified from Bourdon, 2005, character 16). This is coded variable (nonapplicable) in the Suloidea.

25 Tibiotarsus, extremitas distalis: condylae wide, parallel to each other, caudocranially short and proximodistally high; condylus medialis protruding more rostrally than round

condylus lateralis: no (0); yes (1). In Rheidae, the condylae are continuous with each other and truncated proximally; this is regarded as an autapomorphic feature.

26 Tarsometatarsus, trochlea metatarsi III: widening towards extremity and with elongated base; in plantar aspect, the trochlea is short and distinctly asymmetrical, its lateral part extending further proximally than its medial one: no (0); yes (1). In Rheidae, the two plantar edges of trochlea metatarsi III are equal in length; this is regarded as an autapomorphic feature.

27 Os exoccipitale, nervus vagus and nervus glossopharyngeus exiting via single common foramen: no (0); yes (1) (Pycraft, 1900, p. 173).

28 Basis rostri parasphenoidalis showing paired elongated processus basipterygoidei: no (0); yes (1) (e.g. Pycraft, 1901).

29 Fossa temporalis, impressio musculi adductor mandibulae externus, pars coronoidea (Zusi and Livezey, 2000) indistinct and area of origin of musculus pseudotemporalis superficialis laterally facing and greatly enlarged: no (0); yes (1).

30 Os quadratum, processus oticus, capitula continuous with each other and forming a single, lateromedially wide surface: no (0); yes (1).

31 Tuba auditiva opening close to mid-line: no (0); yes (1) (Cracraft, 1988; Cracraft & Clarke, 2001, character 12; Clarke & Norell, 2002, character 27; Mayr & Clarke, 2003, character 29; Livezey & Zusi, 2006, character 126).

32 Sutura frontoparietalis: not obliterated in adults (0); obliterated in adults (1) (Clarke & Norell, 2002, character 51; Mayr & Clarke, 2003, character 32; Livezey & Zusi, 2006, character 213).

33 Os palatinum meets its counterpart at midline, contacts rostrum parasphenoidale plus os premaxillare, and has only simple, primarily dorsoventral articulatio pterygo-palatina (Cracraft, 1988; Ericson, 1997, p. 441; Cracraft & Clarke, 2001, characters 6–8; Clarke & Norell, 2002, characters 15–16; Mayr & Clarke, 2003, character 22; Livezey & Zusi, 2006, characters 579 and 601).

34 Os quadratum, facies articularis pterygoidea condylar, forming well-projected tubercle: no (0); yes (1) (Clarke & Norell, 2002, character 32; Livezey & Zusi, 2006, characters 523 and 600).

35 Os quadratum, processus oticus, incisura between cotylae squamosum et oticum (double headed cranial articulation): no (0); yes (1) (Cracraft, 1988; Cracraft & Clarke, 2001, character 18; Clarke & Norell, 2002, character 36).

36 Costa vertebralis and processus uncinatus completely synostosed in adults: no (0); yes (1) (Livezey & Zusi, 2006, character 1096).

37 Sternum, corpus sterni, facies muscularis sterni, lineae intermusculares: absent (0); present (1) (Clarke & Norell, 2002, character 77; Livezey & Zusi, 2006, character 1106).

38 Humerus, distinct fossa musculi brachialis: absent (0); present (1) (see Mayr & Clarke, 2003, character 79). This is coded nonapplicable in *Hesperornis*.

39 Humerus, well-developed sulcus scapulotricipitalis: absent (0); present (1) (see Clarke & Norell, 2002, character 127; Mayr & Clarke, 2003, character 81; Livezey & Zusi, 2006, character 1488). This is coded non applicable in *Hesperornis*.

40 Ilium forming the major part of tuberculum preacetabulare: no (0); yes (1) (Livezey & Zusi, 2006, character 1809).

41 Fenestra ilioischiadica (synchondrosis ilioischiadica caudalis): absent (0); present (1) (Clarke & Norell, 2002, character 154; Mayr & Clarke, 2003, character 94; Livezey & Zusi, 2006, characters 1789 and 1953).

42 Tibiotarsus, facies articulares medialis et lateralis well delimited mutually by area interarticularis and fossae retrocristales: no (0); yes (1) (Livezey & Zusi, 2006, character 2068).

43 Tibiotarsus, caput tibiotarsi, facies articularis fibularis present as a short ridge extending distal to margo capitis: no (0); yes (1) (see Livezey & Zusi, 2006, character 2108).

44 Tibiotarsus, os tibiale forming only condylus lateralis of extremitas distalis tibiotarsi: no (0); yes (1) (Livezey & Zusi, 2006, character 2209).

45 Tarsometatarsus, hypotarsus with well developed cristae and sulci: no (0); yes (1) (modified from Clarke & Norell, 2002, character 192; Mayr & Clarke, 2003, character 103; Livezey & Zusi, 2006, character 2217).

46 Humerus, tuberculum ventrale: poorly developed (0); well developed (1) (Ericson, 1997, p. 455).

47 Humerus, processus flexorius, origin of musculus flexor carpi ulnaris: caudal scar absent or indistinct (0); caudal scar a well-marked concavity (1) (see Ericson, 1997, p. 457). This is coded nonapplicable in *Hesperornis* and Rheidae.

48 Humerus, origin of musculus extensor carpi ulnaris: poorly marked and shallow (0); forming two distinct concavities (1); forming two distinct concavities surrounded by prominent ridge of even width (2) (Bourdon, 2005, character 13). This is coded nonapplicable in *Hesperornis*.

49 Ulna, cotyla dorsalis and cotyla ventralis in the same plane, with crista intercotylaris absent to poorly developed (0); cotylae not in the same plane and distinctly separated by well-developed crista intercotylaris (1) (Bourdon, 2005, character 14; see also Livezey & Zusi, 2006, character 1497). This is coded nonapplicable in *Hesperornis*.

50 Basis cranii externa with deep fossa parabasalis and strongly convex lamina parasphenoidalis meeting rostrum parasphenoidale at very acute angle: no (0); yes (1) (Mayr & Clarke, 2003, character 26; Livezey & Zusi, 2006, character 117).

51 Mandibula, long slender and dorsally oriented processus medialis: absent (0); present (1) (Cracraft & Clarke, 2001, character 41; Mayr & Clarke, 2003, character 45).

52 Mandibula, processus retroarticularis long, curving, strongly compressed lateromedially: absent (0); present (1) (Cracraft & Clarke, 2001, character 42; Mayr & Clarke, 2003, character 44). This is coded present in Phoenicopteridae.

53 Os quadratum, processus oticus, well developed eminentia articularis (Lowe, 1926): absent (0); present (1) (Livezey, 1997, character 49, modified; Mayr & Clarke, 2003, character 35).

54 Condylus occipitalis: not as follows (0); strongly bilobate (1) (modified from Livezey, 1997, character 1); pedonculate, large and bilobate (2); very large (3) (modified from Bourdon, 2005, character 48). In Galliformes, the condylus occipitalis is sessile and smaller than in Odontoanserae. Galliformes are thus assigned state 1.

55 Processus rostropterygoideus (Weber, 1996): absent (0); present as an inconspicuous surface without basal support (1); present, lipped, with basal support (2) (Livezey, 1986, character 20).

56 Os exoccipitale, processus paroccipitalis: strongly protruding caudoventrally, caudally convex, with wide lateral side for origin of musculus depressor mandibulae; the processus is continuous with stout processus lateralis parasphenoidalis and ala parasphenoidalis, so that the cavitas tympanica is deeply recessed: no (0); yes (1) (modified from Bourdon, 2005, character 26). In Anhimidae, the processus is rostrally curved; this is regarded as an autapomorphic feature.

57 Processus postorbitalis stout and projecting obliquely rostrally beneath orbita: no (0); yes (1) (Murray and Vickers-Rich, 2004, p. 152, modified).

58 Lamina parasphenoidalis caudorostrally narrow and triangular in shape with tuberculum basilare at caudal corner and distinct curved caudal border that is close to but higher than condylus occipitalis; fossa subcondylaris deep: no (0); yes (1) (modified from Bourdon, 2005, character 49).

59 Impressio musculi adductoris mandibulae externus, pars coronoidea (Weber, 1996) in medial position: no (0); yes (1) (see Bourdon, 2005, character 27).

60 Sutura lacrimofrontalis and/or sutura lacrimonasalis obliterated in adults: no (0); yes (1) (see Cracraft, 1968; Harrison & Walker, 1976; Olson, 1985, fig. 10; Ericson, 1997, p. 440). This is coded nonapplicable for Anseriformes given that the condition is variable in this

taxon. Ciconiidae are assigned state 0 given this is the most common condition found in this taxon.

61 Os mesethmoidale well developed rostrally, showing deep depression for concha caudalis in caudal half and conspicuous oblique column bordering this depression rostroventrally: no (0); yes (1) (Bourdon, 2005, character 28).

62 Cranium and maxilla: not as follows (0); lateral border of os lacrimale above processus orbitalis an elongated, vertical and slightly concave facet roughly parallel to long axis of cranium; surface between ossa lacrimales concave from side to side; apex of maxilla a flat to slightly concave triangular surface just rostral to abrupt zona flexoria craniofacialis (1) (Bourdon, 2005, character 29, modified). In Anhimidae, the roof of the orbit is greatly widened and swollen, extending beyond the lateral border of os lacrimale. This is regarded as an autapomorphic feature.

63 Humerus, extremitas distalis: condylus ventralis round, distally prominent, with well-marked caudal border distinctly cranial to fossa olecrani and in the same plane as condylus dorsalis; incisura intercondylaris wide and deep: no (0); yes (1) (Bourdon, 2005, character 30, modified). In Anhimidae, the incisura intercondylaris is narrower and shallower and the condylus ventralis is dorsoventrally elongated so that its ventral border is less distinct. This is regarded as an autapomorphic feature.

64 Ulna, extremitas proximalis wider than it is deep, with strongly convex facies caudodorsalis showing impressio musculi scapulotricipitalis in distodorsal position; flat cotyla dorsalis with straight cranial border and pointed extremity; depression for attachment of meniscus radioulnaris poorly developed: no (0); yes (1) (see Bourdon, 2005, character 31).

65 Ulna, extremitas distalis: not as follows (0); forming an isosceles triangle with condylus dorsalis, condylus ventralis and tuberculum carpale of same width and smooth aspect; base of tuberculum carpale prominent, thick and oblique (1) (modified from Bourdon, 2005, character 32); base of tuberculum carpale continuing far proximally (2). In Anhimidae, the extremitas distalis ulnae is dorsoventrally compressed and the caudal end of condylus ventralis is pointed. This is regarded as an autapomorphic feature.

66 Radius, ventral border of cotyla humeralis convex, forming distinct overhang and continuous with caudal edge of tuberculum bicipitale; surface dorsal to the latter and distal to facies articularis ulnaris flat and triangular: no (0); yes (1) (modified from Bourdon, 2005, character 33).

67 Carpometacarpus, thick and prominent processus pisiformis with proximal border reaching trochlea carpalis; rostral border extending far distally and roughly paralleling long axis of corpus: no (0); yes (1) (Bourdon, 2005, character 34).

68 Carpometacarpus, facies ventralis: long symphysis metacarpalis distalis with os metacarpale minus close and nearly parallel to os metacarpale majus; the latter shows median ridge that curves caudally at distal extremity; high and well-defined caudal protuberance on facies articularis digitalis major: no (0); yes (1) (Bourdon, 2005, character 35).

69 Tibiotarsus, wide incisura intercondylaris extending onto proximal part of condylus medialis; the latter protrudes more rostrally and is narrower than condylus lateralis: no (0); yes (1) (Bourdon, 2005, character 36).

70 Tarsometatarsus, crista medialis hypotarsi in line with plantaromedial corner of cotyla medialis, which forms a proximally protruding process: no (0); yes (1) (Bourdon, 2005, character 37, modified). In Anhimidae, the process is poorly developed. This is regarded as an autapomorphic feature.

71 Tarsometatarsus, facies plantaris: trochlea metatarsi III prominent, proximodistally elongated with pointed extremity, slightly oblique; low foramen vasculare distale with recessed opening: no (0); yes (1) (Bourdon, 2005, character 38).

72 Cotyla quadratica squamosi and cotyla quadratica otici rostrally continuous: no (0); yes (1) (Bourdon, 2005, character 50).

73 Maxilla, deep groove extending rostral and caudal to minute apertura nasi ossea, which is in distal position: absent (0); present (1) (Bourdon, 2005, character 53).

74 Cristae tomiales: pseudo-teeth absent (0); pseudo-teeth present (1) (Bourdon, 2005, character 52).

75 Humerus: caput tapering ventrally and showing dorsal prominence on facies cranialis that extends distally; tuberculum dorsale dorsoventrally wide, distinctly separated from caput, and continuing far distally as a smooth convexity: no (0); yes (1) (Bourdon, 2005, character 54).

76 Humerus, short sulcus transversus perpendicular to long axis of corpus, sharply defined, evenly tapering toward pointed ventral extremity, and facing fully cranially: no (0), yes (1) (Bourdon, 2005, character 55).

77 Humerus: distal part of linea musculi latissimi dorsi thick, well defined, close to margo dorsalis and continuing far distally to crista deltopectoralis, which extends far distally on corpus; impressio musculi pectoralis low with evenly curved ventral border: no (0); yes (1) (Bourdon, 2005, character 56, modified).

78 Corpus humeri straight: no (0); yes (1) (Mayr, 2003, character 30).

79 Humerus, epicondylus dorsalis a very prominent, proximodistally short convexity: no (0); yes (1) (Bourdon, 2005, character 58).

80 Corpus ulnae straight: no (0); yes (1) (Bourdon, 2005, character 59).

81 Radius, sharp crest extending from caput radii to a point slightly distal to tuberculum bicipitale: no (0); yes (1) (modified from Bourdon, 2005, character 33).

82 Carpometacarpus, os metacarpale alulare greatly elongated and parallel to os metacarpale majus: no (0), yes (1).

83 Carpometacarpus, proximal synostosis of os metacarpale minus and os metacarpale majus proximodistally elongated; distal to synostosis, os metacarpale minus nearly straight with transverse section oval to cylindrical: no (0), yes (1).

84 Femur, distinct triangular facet at mid height of corpus femoris, located on facies caudalis and close to facies medialis: absent (0); present (1) (Bourdon, 2005, character 60).

85 Femur, prominent tuberculum musculi gastrocnemialis lateralis: absent (0); present (1) (Bourdon, 2005, character 61).

86 Tibiotarsus, condylus medialis perpendicular to long axis of corpus, strongly protruding rostrally, showing convex proximal border and lacking prominent crista trochleae: no (0), yes (1) (Bourdon, 2005, character 62).

87 Tarsometatarsus, trochlea metatarsi II: extending less far distally than trochlea metatarsi IV, projecting plantarly, and having smooth semicircular external edge higher than internal one and continuous with proximally pointed process: no (0), yes (1) (see Bourdon, 2005, character 63).

88 Temporal region: not as follows (0); with well-developed processus zygomaticus dissociated from os quadratum; large fossa temporalis mainly occupied by musculus adductor mandibulae externus, pars coronoidea (Zusi & Livezey, 2000) (1) (modified from Cracraft & Clarke, 2001, character 10); processus zygomaticus continuous with high crest separating musculus adductor mandibulae externus pars coronoidea rostrally and musculus adductor mandibulae externus pars articularis caudally (2) (Bourdon, 2005, character 66).

89 Os palatinum: not as follows (0); showing well-developed pars lateralis plus variably developed crista ventralis, and forming broad articulatio palatorostralis (1) (modified from Mayr & Clarke, 2003, characters 15–16; Bourdon, 2005, character 3).

90 Ulna, extremitas proximalis: cotyla ventralis caudocranially elongated; crista intercotylaris sharp; deep incisura radialis bounded by sharp processus dorsalis extending far distally; olecranon low and blunt, craniocaudally elongated and dorsoventrally compressed, in dorsal position (modified from

Bourdon, 2005, character 82). Peculiar conditions found in Pelecanoididae and Hydrobatidae are regarded as autapomorphic.

91 Ulna, extremitas distalis dorsoventrally flattened with sharp, dorsodistally protruding condylus dorsalis: no (0); yes (1) (modified from Bourdon, 2005, character 74).

92 Tarsometatarsus, trochleae metatarsi with sharp borders and deep wide median groove continuing far proximally on facies dorsalis; foramen vasculare distale in high position: no (0); yes (1) (modified from Bourdon, 2005, character 77). In Podicipedidae, the grooved aspect is present only on plantar facies, the dorsal facies is smooth, and the foramen vasculare distale is displaced distally; this is regarded as an autapomorphic feature.

93 Humerus, condylus ventralis undercut, proximodistally narrow: no (0); yes (1) (Bourdon, 2005, character 73, modified).

94 Radius, extremitas distalis: cranial border of sulcus tendinosus bearing prominent convexity and small tubercle distal to the latter: no (0); yes (1) (see Bourdon, 2005, character 75).

95 Carpometacarpus, facies articularis digitalis major with conspicuous, hemispherical and dorsoventrally flattened process at rostral corner: absent (0): present (1) (Bourdon, 2005, character 76).

96 Fossa glandulae nasalis: absent (0); present (1) (Mayr, 2003, character 17; Mayr & Clarke, 2003, character 25). In Podicipedidae, this structure is present but secondarily displaced towards the outer rim of os frontale. This taxon is thus coded state 1.

97 Os coracoideum, processus acrocoracoideus dorsoventrally short with facies articularis clavicularis forming medial overhang and projecting ventrally; this processus is located on a long, dorsoventrally narrow peduncle: no (0), yes (1).

98 Tibiotarsus, cristae cnemiales: not as follows (0); with proximodistally extensive cranial prominence and lateral concavity of crista cnemialis cranialis (1) (see Mayr & Clarke, 2003, character 99; Livezey &

Zusi, 2006, character 2076); thick and raised proximally into a sharp point; sulcus intercnemialis facing rostrally and having rough aspect (2) (Bourdon, 2005, character 93); very strongly raised proximally with crista cnemialis cranialis flaring far distally on corpus (3) (Cracraft, 1988, p. 350).

99 Cranium: not as follows (0); crista nuchalis transversa forming flange (supraoccipital crest) on either side of prominentia cerebellaris (Bourdon, 2005, character 90, modified); the base of this flange is pierced by the medial foramen rami occipitalis arteriae ophthalmicae externae that is in ventral position (Bourdon et al., 2005, character 18, modified); surface above cotyla quadratica squamosi horizontal and noticeably convex (Cracraft, 1985, character 48) (1). These features are obviously correlated to each other, and are grouped here in one single character.

100 Os quadratum: not as follows (0); corpus mediolaterally compressed; condylus medialis with strongly protruding and very sharp caudal and medial edges (1); processus orbitalis very long, slender, mediolaterally compressed (Bourdon et al., 2005, character 41) (2).

101 Humerus, proximal part of corpus ventrally curved and crista deltopectoralis poorly developed, proximodistally elongated and cranially bent: no (0); yes (1).

102 Carpometacarpus dorsoventrally flattened; straight os metacarpale minus parallel to os metacarpale majus and narrow spatium intermetacarpale: no (0); yes (1) (Bourdon, 2005, character 92).

103 Fossa hypophysialis: not as follows (0); caudorostrally elongated (1) (Bourdon, 2005, character 109).

104 Fossae temporales extending to mid-line of cranium: absent (0); present (1) (Cracraft, 1985, character 47). This is coded nonapplicable (variable) for Suloidea.

105 Humerus, incisura capitis narrow, nearly perpendicular to long axis of corpus, and bounded proximally by sharp border of caput: no (0); yes (1) (Bourdon, 2005, character 95).

106 Ulna, cotyla ventralis sharply defined caudally, projecting far ventrally to olecranon and far cranially to cotyla dorsalis; the latter is caudocranially narrow and much smaller than the former: no (0); yes (1) (Bourdon, 2005, character 96).

107 Femur, facies articularis antitrochanterica strongly compressed caudorostrally; corpus strongly curved with distinct depression proximal to sulcus patellaris; condylus fibularis (Howard, 1929) enlarged and twisted laterally: no (0); yes (1) (Cracraft, 1988, p. 350).

108 Tarsometatarsus, corpus strongly compressed lateromedially; lateral ridge of sulcus extensorius raised as thin ridge and twisted medially; trochlea metatarsi II extending less far distally than trochlea metatarsi IV, projecting plantarly, strongly compressed lateromedially and twisted medially; trochlea metatarsi III distinctly asymmetrical in plantar aspect: no (0); yes (1) (Cracraft, 1988, p. 350; Bourdon, 2005, character 98).

109 Well-marked impressio glandulae nasalis situated within roof of orbit, in rostromedial position: absent (0); present (1) (Bourdon, 2005, character 105; Bourdon *et al.*, 2005, character 39). This structure is very small and shallow in Fregatidae and Anhingidae. This is coded absent in Phaethontidae.

110 Maxilla and cavitas nasalis: not as follows (0); maxilla elongated with well-defined groove distal to apertura nasi ossea, the proximal end of which is in proximal position, very close to zona flexoria craniofacialis; median part of maxilla wide and strongly convex, distinctly separated from lateral part (1); tiny apertura nasi ossea (2) (Bourdon *et al.*, 2005, character 2, modified).

111 Mandibula: smooth ridge medial and slightly ventral to crista tomialis, extending from symphysis to angulus: no (0); yes (1). Ardeidae are assigned state 1 given this is the most common condition found in this taxon (absent in *Cochlearius*). The ridge is poorly developed, but present, in Pelecanidae.

112 Os quadratum: not as follows (0); condylus caudalis in high position; deep fossa between the three condylae (1); corpus sharply defined laterally and caudally flat; condylus lateralis forming long peduncle and cotyla quadratojugalis in far ventral position (2); os quadratum wrung and laterally flattened, so that peduncle of condylus lateralis shorter, cotyla quadratojugalis higher and fossa between the three condylae shallow (3).

113 Tibiotarsus: not as follows (0); incisura intercondylaris very wide and perpendicular to inner side of the condylae; condylus medialis with sharp lateral margin, truncated aspect (1); corpus caudorostrally flattened in distal part; sulcus extensorius wide with low borders; condylus medialis with wide flattened rostral border (2); sulcus extensorius in lateral position with distal end of canalis extensorius distinctly medial to proximal one; condylus lateralis rounded in shape, very shortened caudocranially (3) (Bourdon, 2005, character 103, modified).

114 Tarsometatarsus: not as follows (0); asymmetrical, with corpus widening only towards trochlea metatarsi II; the latter projects as far or further distally than trochlea metatarsi IV; trochlea metatarsi III deeply grooved with facies dorsalis strongly protruding dorsally and extending close to foramen vasculare distale (1); short stout corpus with strong widening towards medially deflected trochlea metatarsi II (2) (Bourdon, 2005, character 104, modified).

115 Scapula, acromion greatly elongated, slender, dorsoventrally narrow, evenly curved and dorsally oriented: no (0); yes (1). This is coded nonapplicable for Rheidae.

116 Os coracoideum, facies articularis scapularis and processus procoracoideus located on dorsally protruding support: no (0); yes (1) (Bourdon, 2005, character 102). This is coded nonapplicable for Rheidae.

117 Vena occipitalis externa: piercing occipital plate to exit brain cavity (0); exiting brain cavity via foramen magnum (1) (Bourdon *et al.*, 2005, character 19).

118 Rostrum parasphenoidale very narrow: no (0); yes (1) (Bourdon *et al.*, 2005, character 37).

119 Os palatinum, pars lateralis elongated, fused with its counterpart over a great length and continued by prominent crista ventralis; processus maxillopalatinus straight, forming flat bar distant from its counterpart: no (0); yes (1) (modified from Mayr, 2003, character 10; Mayr & Clarke, 2003, character 11).

120 Sternum, carina sterni located in rostral part of sternum, strongly projecting rostrally and showing nearly straight ventral edge; apex carinae with facies articularis furculae: no (0); yes (1). Pelecanidae are coded state 1 (a facies articularis furculae is present in juveniles pelicans).

121 Os coracoideum: not as follows (0); facies articularis humeralis strongly protruding laterally, neck long with lateral side concave, and overhanging lateral border of processus acrocoracoideus (1) (Bourdon, 2005, character 106). This is coded nonapplicable for Rheidae.

122 Tarsometatarsus, crista medialis hypotarsi prominent, forming distinct flat surface: no (0); yes (1) (Bourdon, 2005, character 107).

123 Os quadratum, processus mandibularis: not as follows (0); condylus caudalis continuous with condylus lateralis and bearing caudomedial prominence; condylus medialis very wide (1); condylus caudalis medially deflected and condylus medialis slightly pedicellate (2). In *Anastomus*, the processus mandibularis is strongly compressed lateromedially. This is regarded as an autapomorphic condition.

124 Tibiotarsus, incisura intercondylaris very deep, extending far medially and undercutting pons supratendineus; proximal inner border of condylus medialis very sharp and oblique, abruptly interrupted by incisura intercondylaris; distal end of canalis extensorius in medial position: no (0); yes (1).

125 Elongated maxilla with concave ventral surface, sharp apex and sides strongly sloping ventrally: no (0); yes (1).

126 Humerus, linea musculi latissimi dorsi thick, elongated, evenly curved, either at mid-width of corpus or closer to margo ventralis and continuous with prominent crus dorsale fossae: no (0); yes (1).

127 Ulna, extremitas distalis: not as follows (0); overhangs corpus at the level of depressio radialis; condylus dorsalis short and strongly protruding; caudal edge of condylus ventralis wide and sharp; tuberculum carpale short, abruptly ending proximally (1).

128 Tarsometatarsus showing very high eminentia intercotylaris with medial buttress and convex lateral edge: no (0); yes (1).

APPENDIX 8B: CHARACTER–TAXON MATRIX USED FOR PHYLOGENETIC ANALYSIS

Taxa	Characters

```
                      0000000001111111111222222222233333333334444444444555555555566666
                      1234567890123456789012345678901234567890123456789012345678901234

Apsaravis             ?10000?????000?00000?0??00??????????00?0???00000??00?????????00
Hesperornis           00?000000000000?00????000000000000000?00??00??00???000010000000000?
Ichthyornis           000000??00000000000000000??00??1000000000000111000001?00?000000
Rheidae               1110011111???1?1111110?11111110000000000000?00000000000000000
Tinamidae             111110111111111111111111111111110000000000000000000000000010000
Galliformes           011111111111111111111111111100001111111111111110000111110000000000
Anseriformes          21111100000000000000000000000001111111111111111121111112211111?1111
Pelagornis/Osteodon.  30111100000000000000000?00?00010?11?111?????1112100003?111111111
Dasornis              3?1111000??000000000000000000010111??11??11?111210??032111111111
Phaethontidae         111111000000000000000010000001111111111111111111000000000000000
Procellariiformes     111111000000000000000001111111111111111111110000000000000000
Spheniscidae          11101100000000?0000000000001111111?1111111011100010000000000000
Gaviidae              111010000000000000000000011111111111111111110000000000000000
Podicipedidae         11101000000000000000000011111111111111111000010000000000000
Phoenicopteridae      111111000000000000000000001111111111111111111001000000000000000
Ciconiidae            11111100000001000000000001111111111111111111100000000000000
Scopidae              11111100000001000000000001111111111111111111100000000000000
Ardeidae              11111100000001000000000001111111111111111111100000000000000
Fregatidae            11111100000000000000000001111111111111111111100000000000000
Pelecanidae           10111100000000000000000001111111111111111121000000000010000
Suloidea              11111100000000000000000?0000001111111111111111111100000000000000
```

```
                      000000000000000000000000000000000001111111111111111111111111111
                      666667777777777888888888899999999990000000000111111111222222222
                      567890123456789012345678901234567890123456789012345678901234 5678

Apsaravis             ?00?000?00000010000?00??0?000??0???00??0000????0000??000?0?0?0
Hesperornis           ????00000000000???0000000?00??1000000??10?000000000000000000?0
Ichthyornis           00000000000000000000000?000001000000?100000000000000??0000000
Rheidae               0000000000000000000000000000000?00000000000000000??0000?0000000
Tinamidae             0000000000000000000000000000000000000000000000000000000000000
Galliformes           0000000000000000000000000000000000000000000000000000000000000
Anseriformes          1111111000000000000000000000000000000000000000000000000000000
Pelagornis/Osteodon.  211?1011111111111111??0000??000?000?00000100000000???0?0000000
Dasornis              2111111111111111111111000000000000000000000?0000?00000000
Phaethontidae         000000000000010000000001111111000000000000000000000000000000
Procellariiformes     000000000000101000000011111111110000000001000000000000000000
Spheniscidae          00000000000000?000000011???00011211110000000000000000000000000
Gaviidae              000000000000010000011110001131211111110000000000000000000
Podicipedidae         00000000000000111100001131211111110000000000000000000
Phoenicopteridae      0000000000000000001100000000000000000000000000000110000
Ciconiidae            000000000000000001100000000000000000000000000000211111
Scopidae              0000000000000000011000000000000000000000000100000211111
Ardeidae              0000000000000000002100000000001100001111110000000000000
Fregatidae            00000000000010000000021000100000000000012122211000000000000
Pelecanidae           0000000000000000000021000000000000000001213321111111111000000
Suloidea              00000000000000000000021000000000000001?0000121232111111111000000
```

REFERENCES

Andrews CW. 1916. Note on the sternum of a large carinate bird from the (?) Eocene of Southern Nigeria. *Proceedings of the Zoological Society of London* **1916**: 519–524.

Aslanova SM, Burchak-Abramovich NI. 1982. The first and unique find of the fossil of Perekishkul toothed bird in the terrirory of USSR and in the Asiatic continent. *Izvestiâ Akademii nauk SSR. Seriâ biologiceskaâ* **8**: 406–412.

Aslanova SM, Burchak-Abramovich NI. 1999. A detailed description of *Caspiodontornis kobystanicus* from the Oligocene of the Caspian seashore. *Acta Zoologica Cracoviensia* **42**: 423–433.

Averianov AO, Panteleyev AV, Potapova OR, Nessov LA. 1991. Bony-toothed birds (Aves: Pelecaniformes: Odontopterygia) from the Late Paleocene and Eocene of the Western margin of Ancient Asia. *Proceedings of the Zoological Institute, USSR Academy of Sciences* **239**: 3–12.

Baumel JJ, King AS, Breazile JE, Evans HE, Vanden Berge JC. 1993. *Handbook of Avian Anatomy: Nomina Anatomica Avium*, 2nd edn. In A. Raymond and J. Paynter (eds). Nuttall Ornithological Club, Cambridge, 779 pp.

Bourdon E. 2005. Osteological evidence for sister group relationship between pseudo-toothed birds (Aves: Odontopterygiformes) and waterfowls (Anseriformes). *Naturwissenschaften* **92**: 586–591.

Bourdon E. 2006a. *L'avifaune du Paléogène des phosphates du Maroc et du Togo: diversité, systématique et apports à la connaissance de la diversification des oiseaux modernes (Neornithes).* Unpublished PhD thesis, Muséum National d'Histoire Naturelle, Paris, 330 pp.

Bourdon E. 2006b. A new avifauna from the Early Tertiary of the Ouled Abdoun Basin, Morocco: contribution to higher-level phylogenetics of modern birds (Neornithes). *Journal of Vertebrate Paleontology* **26**: 44A.

Bourdon E, Amaghzaz M, Bouya B. 2010. Pseudo-toothed birds (Aves, Odontopterygiformes) from the Early Tertiary of Morocco. *American Museum Novitates* **3704**: 1–71.

Bremer K. 1994. Branch support and tree stability. *Cladistics* **10**: 295–304.

Bühler P, Martin LD, Witmer LM. 1988. Cranial kinesis in the Late Cretaceous birds *Hesperornis* and *Parahesperornis*. *The Auk* **105**: 111–122.

Chavez M, Stucchi M, Urbina M. 2007. El registro de Pelagornithidae (Aves: Pelecaniformes) y la avifauna neógena del Pacífico sudeste. *Bulletin de l'Institut Français d'Études Andines* **36**: 175–197.

Cheneval J. 1993. L'Avifaune Mio-Pliocène de la Formation Pisco (Pérou). Etude Préliminaire. *Documents des Laboratoires de Géologie de Lyon* **125**: 85–95.

Chubb AL. 2004. New nuclear evidence for the oldest divergence among neognath birds: the phylogenetic utility of ZENK. *Molecular Phylogenetics and Evolution* **30**: 140–151.

Clarke JA. 2004. Morphology, phylogenetic taxonomy, and systematics of *Ichthyornis* and *Apatornis* (Avialae: Ornithurae). *Bulletin of the American Museum of Natural History* **286**: 1–179.

Clarke JA, Chiappe LM. 2001. A new carinate bird from the Late Cretaceous of Patagonia (Argentina). *American Museum Novitates* **3323**: 1–23.

Clarke JA, Norell MA. 2002. The morphology and phylogenetic position of *Apsaravis ukhaana* from the Late Cretaceous of Mongolia. *American Museum Novitates* **3387**: 1–46.

Clarke JA, Tambussi CP, Noriega JI, Erickson GM, Ketcham RA. 2005. Definitive fossil evidence for the extant avian radiation in the Cretaceous. *Nature* **433**: 305–308.

Cope ED. 1894. On *Cyphornis*, an extinct genus of birds. *Journal of the National Academy of Science of Philadelphia* **9**: 449–452.

Cracraft J. 1968. The lacrimal-ectethmoid complex in birds: a single character analysis. *American Midland Naturalist* **80**: 316–359.

Cracraft J. 1985. Monophyly and phylogenetic relationships of the Pelecaniformes: a numerical cladistic analysis. *The Auk* **102**: 834–853.

Cracraft J. 1988. The major clades of birds. In *The Phylogeny and Classification of the Tetrapods, Volume 1: Amphibians, Reptiles, Birds*, Benton MJ (ed.). Oxford: Clarendon Press; 339–361.

Cracraft J, Clarke J. 2001. The basal clades of modern birds. In *New Perspectives on the Origin and Early Evolution of Birds*, Gauthier J, Gall LF (eds). New Haven: Peabody Museum of Natural History; 143–156.

Cracraft J, Barker FK, Braun MJ, Harshman J, Dyke GJ, Feinstein J, Stanley S, Cibois A, Schikler P, Beresford P, García-Morena J, Sorenson MD, Yuri T, Mindell DP. 2004. Phylogenetic relationships among modern birds (Neornithes): toward an avian tree of life. In *Assembling the Tree of Life*, Cracraft J, Donoghue MJ (eds). New York: Oxford University Press; 468–489.

Del Hoyo J, Elliott A, Sargatal J. 1992. *Handbook of the Birds of the World, Volume 1, Ostrich to Ducks*. Barcelona: Lynx Edicions; 696 pp.

Elzanowski A. 1991. New observations on the skull of *Hesperornis*, with reconstructions of the bony palate and otic region. *Postilla* **207**: 1–20.

Ericson PGP. 1996. The skeletal evidence for a sister-group relationship of anseriform and galliform birds – a critical evaluation. *Journal of Avian Biology* **27**: 195–202.

Ericson PGP. 1997. Systematic relationships of the Palaeogene family Presbyornithidae (Aves: Anseriformes). *Zoological Journal of the Linnean Society* **121**: 429–483.

Ericson PGP, Parsons TJ, Johansson US. 2001. Morphological and molecular support for nonmonophyly of the Galloanseres. In *New Perspectives on the Origin and Early Evolution of Birds*, Gauthier J, Gall LF (eds). New Haven: Peabody Museum of Natural History; 157–168.

Fain MG, Houde P. 2004. Parallel radiations in the primary clades of birds. *Evolution* **58**: 2558–2573.

Feduccia A. 1995. Explosive evolution in Tertiary birds and mammals. *Science* **267**: 637–638.

Feduccia A. 1999. *The Origin and Evolution of Birds*, 2nd edn. New Haven: Yale University Press; 466 pp.

Feduccia A. 2003. "Big bang" for Tertiary birds? *Trends in Ecology and Evolution* **18**: 172–176.

García-Moreno J, Sorenson MD, Mindell DP. 2003. Congruent avian phylogenies inferred from mitochondrial and nuclear DNA sequences. *Journal of Molecular Evolution* **57**: 27–37.

Garrod AH. 1873. On certain muscles of the thigh of birds and on their value in classification. Part I. *Proceedings of the Zoological Society of London* **1873**: 626–644.

Garrod AH. 1876. On the anatomy of *Chauna derbiana*, and on the systematic position of the screamers. *Proceedings of the Zoological Society of London* **1876**: 189–200.

Gheerbrant E, Sudre J, Cappetta H, Mourer-Chauviré C, Bourdon E, Iarochène M, Amaghzaz M, Bouya B. 2003. Les localités à mammifères des carrières de Grand Daoui, bassin des Ouled Abdoun, Maroc, Yprésien: premier état des lieux. *Bulletin de la Société Géologique de France* **174**: 279–293.

Goedert JL. 1989. Giant Late Eocene marine birds (Pelecaniformes: Pelagornithidae) from Northwestern Oregon. *Journal of Paleontology* **63**: 939–944.

GoGeometry. 2009. Fossil skull of giant toothy seabird found in Ocucaje, Ica, Peru (http://www.gogeometry.com/incas/peru_fossil_skull_giant_bird.html).

González-Barba G, Scwennicke T, Goedert JL, Barnes LG. 2002. Earliest Pacific Basin record of the Pelagor-nithidae (Aves: Pelecaniformes). *Journal of Vertebrate Paleontology* **22**: 722–725.

González-Barba G, Goedert JL, Scwennicke T. 2004. First record of *Osteodontornis* (Aves: Pelagornithidae) from Mexico. *Abstracts of the 6th International Meeting of the Society of Avian Paleontology and Evolution*, Quillan, 28 September–3 October.

Groth JG, Barrowclough GF. 1999. Basal divergences in birds and the phylogenetic utility of the nuclear RAG–1 gene. *Molecular Phylogenetics and Evolution* **12**: 115–123.

Hackett SJ, Kimball RT, Reddy S, Bowie RCK, Braun EL, Braun MJ, Chojnowski JL, Cox WA, Han K-L, Harshman J, Huddleston CJ, Marks BD, Miglia KJ, Moore WS, Sheldon FH, Steadman DW, Witt CC, Yuri T. 2008. A phylogenomic study of birds reveals their evolutionary history. *Science* **320**: 1763–1768.

Haring E, Kruckenhauser L, Gamauf A, Riesing MJ, Pinsker W. 2001. The complete sequence of the mitochondrial genome of *Buteo buteo* (Aves, Accipitridae) indicates an early split in the phylogeny of raptors. *Molecular Biology and Evolution* **18**: 1892–1904.

Härlid A, Arnason U, 1999. Analyses of mitochondrial DNA nest ratite birds within the Neognathae: supporting a neotenous origin of ratite morphological characters. *Proceedings of the Royal Society of London Series B (Biological Sciences)* **266**: 305–309.

Härlid A, Janke A, Arnason U. 1998. The complete mitochondrial genome of *Rhea americana* and early avian divergences. *Journal of Molecular Evolution* **46**: 669–679.

Harrison CJO. 1985. A bony-toothed bird (Odontopterygiformes) from the Palaeocene of England. *Tertiary Research* **7**: 23–25.

Harrison CJO, Walker CA. 1976. A review of the bony-toothed birds (Odontopterygiformes): with description of some new species. *Tertiary Research Special Paper* **2**: 1–62.

Harrison CJO, Walker CA. 1979. Birds of the British Lower Oligocene. *Tertiary Research Special Paper* **5**: 29–43.

Harshman J. 2007. Classification and phylogeny of birds. In *Reproductive Biology and Phylogeny of Birds*, Jamieson BGM (ed.). Enfield: Science Publishers; 1–35.

Hasegawa Y, Ono K, Koda Y. 1986. Odontopterygid bird from the Oligocene Iwaki Group. *Abstracts of the Annual Meeting of the Palaeontological Society of Japan*; 24.

Hope S. 2002. The Mesozoic radiation of Neornithes. In *Mesozoic Birds: Above the Heads of Dinosaurs*, Chiappe LM, Witmer LM (eds). Berkeley: University of California Press; 339–388.

Hopson JA. 1964. *Pseudodontornis* and other large marine birds from the Miocene of South Carolina. *Postilla* **83**: 1–19.

Howard H. 1929. The avifauna of Emeryville Shellmound. *University of California Publications in Zoology* **32**: 301–394.

Howard H. 1957. A gigantic "toothed" marine bird from the Miocene of California. *Bulletin of the Department of Geology of the Santa Barbara Museum of Natural History* **1**: 1–23.

Howard H. 1978. Late Miocene marine birds from Orange County, California. *Natural History Museum of Los Angeles County, Contributions in Science* **290**: 1–25.

Howard H, Warter SL. 1969. A new species of bony-toothed bird (Family Pseudodontornithidae) from the Tertiary of New Zealand. *Records of the Canterbury Museum* **8**: 345–357.

Howard H, White JA. 1962. A second record of *Osteodontornis*, Miocene "toothed" bird. *Natural History Museum of Los Angeles County, Contributions in Science* **52**: 1–12.

Hugall AF, Foster R, Lee MSY. 2007. Calibration choice, rate smoothing, and the pattern of tetrapod diversification according to the long nuclear gene RAG–1. *Systematic Biology* **56**: 543–563.

Huxley TH. 1867. On the classification of birds; and on the taxonomic value of the modifications of certain of the cranial bones observable in that class. *Proceedings of the Zoological Society of London* **1867**: 415–472.

Lartet E. 1857. Note sur un humérus fossile d'oiseau, attribué à un très grand palmipède de la section des Longipennes. *Compte Rendu Hebdomadaire des Séances de l'Académie des Sciences* **44**: 736–741.

Lisiecki LE, Raymo ME. 2007. Plio-Pleistocene climate evolution: trends and transitions in glacial cycle dynamics. *Quaternary Science Reviews* **26**: 56–69.

Livezey BC. 1986. A phylogenetic analysis of recent anseriform genera using morphological characters. *The Auk* **103**: 737–754.

Livezey BC. 1997. A phylogenetic analysis of basal Anseriformes, the fossil *Presbyornis*, and the interordinal relationships of waterfowl. *Zoological Journal of the Linnean Society* **121**: 361–428.

Livezey BC, Zusi RL. 2006. Higher-order phylogeny of modern birds (Theropoda, Aves: Neornithes) based on comparative anatomy. I. Methods and characters. *Bulletin of the Carnegie Museum of Natural History* **37**: 1–556.

Livezey BC, Zusi RL. 2007. Higher-order phylogeny of modern birds (Theropoda, Aves: Neornithes) based on comparative anatomy. II. Analysis and discussion. *Zoological Journal of the Linnean Society* **149**: 1–95.

Lowe P. 1926. More notes on the quadrate as a factor in avian classification. *The Ibis* **2**: 152–189.

Marsh OC. 1880. Odontornithes: a monograph on the extinct toothed birds of North America. *Reports of the United States Geological Exploration of the Fortieth Parallel* **7**: 1–201.

Matsuoka H, Sakakura F, Ohe F. 1998. A Miocene pseudododontorn (Pelecaniformes: Pelagornithidae) from the Ichishi Group of Misato, Mie Prefecture, Central Japan. *Paleontological Research* **2**: 246–252.

Mayr E, Amadon D. 1951. A classification of recent birds. *American Museum Novitates* **1496**: 1–42.

Mayr G. 2003. The phylogenetic affinities of the shoebill (*Balaeniceps rex*). *Journal für Ornithologie* **144**: 157–175.

Mayr G. 2008. A skull of the giant bony-toothed bird *Dasornis* (Aves: Pelagornithidae) from the Lower Eocene of the Isle of Sheppey. *Palaeontology* **51**: 1107–1116.

Mayr G, Clarke J. 2003. The deep divergences of neornithine birds: a phylogenetic analysis of morphological characters. *Cladistics* **19**: 527–553.

Mayr G, Hazevoet CJ, Dantas P, Cachao M. 2008. A sternum of a very large bony-toothed bird (Pelagornithidae) from the Miocene of Portugal. *Journal of Vertebrate Paleontology* **28**: 762–769.

Mayr G, Smith T. 2010. Bony-toothed birds (Aves: Pelagornithidae) from the Middle Eocene of Belgium. *Palaeontology* **53**: 365–376.

McKee JWA. 1985. A pseudodontorn (Pelecaniformes: Pelagornithidae) from the Middle Pliocene of Hawera, Taranaki, New Zealand. *New Zealand Journal of Zoology* **12**: 181–184.

Milner AC, Walsh SA. 2009. Avian brain evolution: new data from Palaeogene birds (Lower Eocene) from England. *Zoological Journal of the Linnean Society* **155**: 198–219.

Mindell DP, Sorenson MD, Huddleston CJ, Miranda HCJ, Knight A, Sawchuk SJ, Yuri T. 1997. Phylogenetic relationships among and within select avian orders based on mitochondrial DNA. In *Avian Molecular Evolution and Systematics*, Mindell DP (ed.). San Diego: Academic Press; 213–247.

Mindell DP, Sorenson MD, Dimcheff DE, Hasegawa JC, Yuri AT. 1999. Interordinal relationships of birds and other reptiles based on whole mitochondrial genomes. *Systematic Biology* **48**: 138–152.

Mourer-Chauviré, Geraads CD 2008. The Struthionidae and Pelagornithidae (Aves: Struthioniformes,

Odontopterygiformes) from the late Pliocene of Ahl al Oughlam, Morocco. *Oryctos* **7**: 169–194.

Murray PF, Vickers-Rich P. 2004. *Magnificent Mihirungs. The Colossal Flightless Birds of the Australian Dreamtime*. Bloomington: Indiana University Press; 410 pp.

Nixon KC. 1999. *WinClada 0.9.9 (Beta)*. Ithaca, New York. [Published by the author.]

Okazaki Y. 1989. An occurence of fossil bony-toothed bird (Odontopterygiformes) from the Ashiya Group (Oligocene), Japan. *Bulletin of the Kitakyushu Museum of Natural History* **9**: 123–126.

Olson SL. 1984. A brief synopsis of the fossil birds from the Pamunkey River and other Tertiary marine deposits in Virginia. In *Stratigraphy and paleontology of the outcropping Tertiary beds in the Pamunkey River region, central Virginia Coastal Plain – Guidebook for Atlantic Coastal Plain Geological Association 1984 Field Trip*, Ward LW, Krafft K (eds). Norfolk, VA: Atlantic Coastal Plain Geological Association; 217–223.

Olson SL. 1985. The fossil record of birds. In *Avian Biology, Volume 8*, Farner D, King J, Parkes K (eds). New York: Academic Press; 79–256.

Olson SL. 1999. Early Eocene birds from Eastern North America: a faunule from the Nanjemoy Formation of Virginia. In *Early Eocene Vertebrates and Plants from the Fisher/Sullivan site (Nanjemoy Formation) Stafford County, Virginia*, Weems RE, Grimsley GJ (eds). Virginia Division of Mineral Resources; Publication **152**, 123–132.

Olson SL, Feduccia A. 1980a. Relationships and evolution of flamingos (Aves: Phoenicopteridae). *Smithsonian Contributions to Zoology* **316**: 1–73.

Olson SL, Feduccia A. 1980b. *Presbyornis* and the origin of the Anseriformes (Aves: Charadriomorphae). *Smithsonian Contributions to Zoology* **323**: 1–24.

Olson SL, Rasmussen PC. 2001. Miocene and Pliocene birds from the Lee Creek Mine, North Carolina. *Smithsonian Contributions to Paleobiology* **90**: 233–365.

Ono K. 1989. A bony-toothed bird from the Middle Miocene, Chichibu Basin, Japan. *Bulletin of the National Science Museum of Tokyo, Series C* **15**: 33–38.

Ono K, Hasegawa Y, Kawakami T. 1985. Part IV. First record of the Pliocene bony-toothed bird (Odontopterygiformes) from Japan. *Bulletin of the Iwate Prefectural Museum* **3**: 155–157.

Pycraft WP. 1900. On the morphology and phylogeny of the Palaeognathae (Ratitae and Crypturi) and Neognathae (Carinatae). *Transactions of the Zoological Society of London* **15**: 149–290.

Pycraft WP. 1901. Some points in the morphology of the palate of the Neognathae. *Linnean Society Journal of Zoology* **28**: 343–357.

Rasmussen PC. 1998. Early Miocene avifauna from the Pollack Farm Site, Delaware. In *Geology and Paleontology of the Lower Miocene Pollack Farm Fossil Site, Delaware*, Benson RN (ed.). Delaware Geological Survey; Special Publication, 149–151.

Rincón AD, Stucchi M. 2003. Primer registro de la familia Pelagornithidae (Aves: Pelecaniformes) para Venezuela. *Boletín de la Sociedad Venezolana de Espeleología* **37**: 27–30.

Scarlett RJ. 1972. Bone of a presumed Odontopterygian bird from the Miocene of New Zealand. *Journal of Geology and Geophysics* **15**: 269–274.

Seebohm H. 1889. An attempt to diagnose the suborders of the ancient ardeino-anserine assemblage of birds by the aid of osteological characters alone. *The Ibis* **1**: 92–104.

Shufeldt RW. 1901. On the osteology and systematic position of the screamers (*Palamedea: Chauna*). *American Naturalist* **35**: 455–461.

Shufeldt RW. 1916. New extinct bird from South Carolina. *Geological Magazine* **6**: 343–347.

Sibley CG, Ahlquist JE. 1990. *Phylogeny and Classification of Birds: A Study in Molecular Evolution*. New Haven: Yale University Press.

Simon J, Laurent S, Grolleau G, Thoraval P, Soubieux D, Rasschaert D. 2004. Evolution of preproinsulin gene in birds. *Molecular Phylogenetics and Evolution* **30**: 755–766.

Spulski B. 1910. *Odontopteryx longirostris* n. sp. *Zeitschrift der Deutschen Geologischen Gesellschaft* **62**: 507–524.

Stidham TA. 2004. New skull material of *Osteodontornis orri* (Aves: Pelagornithidae) from the Miocene of California. *PaleoBios* **24**: 7–12.

Stilwell JD Jones CM, Levy RH, Harwood DM. 1998. First fossil bird from East Antarctica. *Antarctic Journal* **23**: 12–16.

Swofford DL. 1998. *PAUP*: Phylogenetic Analysis Using Parsimony (and Other Methods) 4.0 Beta for Macintosh*. Sunderland, MA: Sinauer Associates.

Tambussi CP, Acosta Hospitaleche C. 2007. Antarctic birds (Neornithes) during the Cretaceous-Eocene times. *Revista de la Asociación Geológica Argentina* **62**: 604–617.

Tonni EP. 1980. Un pseudodontornítido (Pelecaniformes, Odontopterygia) de gran tamaño, del Terciario de Antártida. *Ameghiniana* **17**: 273–276.

Tonni EP, Tambussi CP. 1985. Nuevos restos de Odontopterygia (Aves: Pelecaniformes) del Terciario temprano de Antártida. *Ameghiniana* **21**: 121–124.

Van Tuinen M, Sibley CG, Hedges SB. 2000. The early history of modern birds inferred from DNA sequences of nuclear and mitochondrial ribosomal genes. *Molecular Biology and Evolution* **17**: 451–457.

Walsh SA, Hume JP. 2001. A new Neogene marine avian assemblage from North-Central Chile. *Journal of Vertebrate Paleontology* **21**: 484–491.

Warheit KI. 2002. The seabird fossil record and the role of paleontology in understanding seabird community structure. In *Biology of Marine Birds*, Schreiber EA, Burger J (eds). Boca Raton: CRC Press; 17–55.

Weber E. 1996. Das Skelet-Muskel-System des Kieferapparates von *Aepypodius arfakianus* (Salvadori, 1877) (Aves, Megapodiidae). *Courier Forschungsinstitut Senckenberg* **189**: 1–132.

Wetmore A. 1917. The relationships of the fossil bird *Palaeochenoides miocaenus*. *Journal of Geology* **15**: 555–557.

Witmer LM. 1990. The craniofacial air sac system of Mesozoic birds (Aves). *Zoological Journal of the Linnean Society* **100**: 327–378.

Witmer LM, Martin LD. 1987. The primitive features of the avian palate, with special reference to Mesozoic birds. *Documents des Laboratoires de Géologie de Lyon* **99**: 21–40.

Zusi RL, Warheit KI. 1992. On the evolution of intraramal mandibular joints in pseudodontorns (Aves: Odontopterygia). In *Papers in Avian Paleontology Honoring Pierce Brodkorb*, Campbell KE (ed.). Los Angeles: Natural History Museum of Los Angeles County, Science Series **36**, 351–360.

Zusi RL, Livezey BC. 2000. Homology and phylogenetic implications of some enigmatic cranial features in galliform and anseriform birds. *Annals of the Carnegie Museum* **69**: 157–193.

9   Phylogeny and Diversification of Modern Passerines

F. KEITH BARKER

Department of Ecology, Evolution and Behavior; and, Bell Museum of Natural History,
University of Minnesota, St. Paul, USA

"...We know little about the inter-relations of over 60 families in the suborder Oscines (song birds), an assemblage comprising nearly half of all bird species. The situation...is most serious as it affects the ornithological contribution to the history of world faunas. This contribution can only be a misleading one so long as the arrangement of these families into a phylogenetic tree remains unaccomplished. With few exceptions, the relationships of groups occupying widely-separated faunal regions may only be guessed at now." (Beecher, 1953)

Passerines (Aves: Order Passeriformes) represent over half of modern avian species diversity. This skewed diversity pattern is even more striking given the relative morphological uniformity of the group. Indeed, Hans Gadow, the noted avian anatomist, complained that "In talking of these 'families' we are apt to forget, or rather we never appreciate, the solemn fact that, strictly speaking, all the Oscines [– approximately 80% of all passerines –] together are of the rank of one family only!" (Gadow, 1892). Despite their relative uniformity, these birds exhibit an astonishing distributional breadth and trophic diversity: passerine birds are found on every continent except Antarctica (and by inference they used to occur there, as well), feed on everything from arthropods to nectar to carrion, and are generally the numerically dominant form in terrestrial avian communities. Moreover, two classic cases of adaptive radiation involve passerine birds: the Darwin's finches of the Galapagos, and the Hawaiian honeycreepers. The apparent contrasts in species and morphological diversity have brought a great deal of attention to bear on the group, and understanding the mechanisms that have generated this pattern has been a major point of emphasis for ornithologists, ecologists, and evolutionary biologists. Hypotheses explaining passerine diversity include the effects of body size, the impact of vocal learning, nesting behaviors, sexual selection, or various combinations of these and additional factors (Raikow, 1986, Fitzpatrick, 1988, Kochmer & Wagner, 1988, Baptista & Trail, 1992, Collias, 1997, Raikow & Bledsoe, 2000). However, proper interpretation of this diversity is intimately tied to understanding the phylogeny of the group – both its monophyly and relationships to other avian orders, and relationships within the order. This is because diversity patterns can only be understood comparatively – the question "why are there so many passerine birds?" (e.g. Raikow, 1986) is meaningless in itself. The real question is: "are there really more passerine birds than expected relative to other bird groups?" (Nee et al., 1992, Raikow & Bledsoe, 2000), and more specifically, "are there really more of some sorts of passerine birds than others?" (Ricklefs, 2003). If not, then the inordinate attention paid to understanding their diversity might not be merited.

Living Dinosaurs: The Evolutionary History of Modern Birds, First Edition. Edited by Gareth Dyke and Gary Kaiser.
© 2011 John Wiley & Sons, Ltd. Published 2011 by John Wiley & Sons, Ltd.

MONOPHYLY, RELATIONSHIPS AND INORDINATE DIVERSITY OF PASSERINE BIRDS

An apparently simple way of interpreting passerine diversity is by comparison with the sister group (or groups) of the order (Raikow & Bledsoe, 2000). Under a simple null model of equal per-lineage diversification rates, the expectation of sister lineage diversity is quite straightforward to derive (Slowinski & Guyer, 1989), and thus the probability of particular sister lineage disparities in diversity can be calculated. This presupposes: (i) monophyly of the group; and (ii) knowledge of its sister taxon. Unfortunately, progress on both of these points has been remarkably slow.

Monophyly of the Passeriformes has rarely been questioned (but see Feduccia, 1977), and a variety of characters have been cited as uniting the group (reviewed in Raikow, 1982). However, character support for this monophyly was not evaluated cladistically until the important work of Raikow (1982), who was the first to outline the surprisingly few (in fact, five) morphological synapomorphies of the group. Unfortunately, the relationships of passerines to other birds has remained an area of controversy. Traditionally, passerines have been associated with a loosely defined group of perching birds called the "higher land birds" (e.g. "Division 9" of Cracraft, 1981), which also includes the Coliiformes, Coraciiformes, Piciformes, and Trogoniformes. However, the question of passerine relationships is really the same as the question of avian ordinal phylogeny – which has a long history beyond the scope of this review (e.g. Livezy & Zusi, 2001, 2007; Cracraft et al., 2004). I will focus on two perspectives on passerine relationships – those afforded by the DNA–DNA hybridization-based phylogenetic hypothesis of Sibley & Ahlquist (1990), and a more recent hypothesis derived from sequences of multiple nuclear genes (Hackett et al., 2008).

The use of DNA–DNA hybridization in phylogenetics had its acme in the 1980s to early 1990s, with application primarily in mammalian and avian systematics (Sibley & Ahlquist, 1984;

Kirsch et al., 1997; Sheldon & Bledsoe, 1993). In particular, Charles G. Sibley and his student and colleague Jon Ahlquist were extremely prolific in their generation of data bearing on avian phylogeny, culminating in the 1990 publication of their *Phylogeny and Classification of Birds*. This publication has proven to be a landmark in avian systematics and in our understanding of avian evolution. On the one hand, their phylogenetic hypotheses stimulated a wave of additional molecular phylogenetic studies, this time based on DNA sequence data. On the other hand, their phylogenetic hypothesis not only provided a new classification for birds, but an explicit relative and absolute temporal framework for interpreting avian evolution, which has been exploited to that end up to the present day (Healy & Guilford, 1990; Briskie & Montgomerie, 1992; Promislow et al., 1992; Cotgreave & Harvey, 1994; Höglund & Sillén-Tullberg, 1994; Barraclough et al., 1995; Ricklefs et al., 1996; Owens & Bennett, 1997; Poiani & Pagel, 1997; Westneat & Sherman, 1997; Hallgrimsson & Maiorana, 2000; von Euler, 2001; Ricklefs, 2003; Sol et al., 2005; Sol & Price, 2008; Swanson & Garland, 2009).

The DNA–DNA hybridization data recovered monophyly of the Passeriformes, as predicted by the available morphological data. In addition, these data provided the first explicit evidence-based phylogenetic hypothesis for the sister-group of passerines – a large unnamed assemblage including the traditional orders Columbiformes, Gruiformes, Charadriiformes, Pelecaniformes, Podicipediformes, Falconiformes, Gaviiformes, Sphenisciformes and Procellariiformes (Sibley & Ahlquist, 1990). Using this sister group relationship, Raikow & Bledsoe (2000) calculated the probability of a group as large as the Passeriformes (5712 species) being sister to an assemblage of the size proposed by Sibley and Ahlquist (1536 species), to be 0.42, using the methods of Slowinski & Guyer (1989). Thus, on the face of it the DNA–DNA hybridization hypothesis suggested that passerine birds are not particularly special with regard to phylogeny – that is, given that the order was rooted deeply enough in the avian tree, its diversity was no greater than might be expected by the operation of uniform speciation and extinction processes.

Recent data bearing on passerine ordinal relationships prompt reevaluation of this comparison. Two recent studies using largely independent data sets (Ericson *et al.*, 2006; Hackett *et al.*, 2008) have pointed to a monophyletic grouping of passerines, Psittaciformes (parrots), Falconidae (falcons), and Cariamidae (seriemas). Aside from monophyly of passerines and the other families, the first study leaves relationships within this clade unresolved, whereas the larger study of Hackett *et al.* yielded some support for a grouping of passerines with parrots. This is an extremely intriguing result insofar as these are two of three avian orders (the other being the Apodiformes) where vocal learning is known to occur, suggesting a potentially ancient origin for this behavior. Recalculation of Slowinski and Guyer's asymmetry statistic for passerines versus their putative sister group the parrots yields a much lower *p*-value (0.12), a number that rises to 0.14 if we conservatively consider falcons, parrots, and the seriema as the sister group to passerines. While not statistically significant in themselves, these numbers suggest that diversity of passerines is more unusual than indicated by the Sibley and Ahlquist's ordinal phylogeny. By contrast, comparison of the diversity of passerines and parrots to that of falcons (or falcons and seriems both) yields a probability of 0.02, well below the standard significance threshold (although we must bear the multiple comparisons in mind). This leads to the fascinating possibility that vocal learning – shared by oscine passerines, a few suboscines (Kroodsma, 2004; Saranathan *et al.*, 2007), and parrots (Pepperberg, 2004), might have played an important role in the diversification of this group, as has been suggested previously for passerines alone (Vermeij, 1988; Baptista & Trail, 1992).

As noted above, Sibley and Ahlquist's data provided not only an estimate of phylogeny, but a temporal scale for avian diversification. This is extremely useful, because the temporal scale of passerine diversification relative to nonpasserines is of great importance – knowing the relative ages of avian lineages allows comparison of passerine

diversity to that of multiple lineages of equal age, rather than just to its sister taxon. Analyses of this type have been performed (Nee *et al.*, 1992; Nee, 2004), yielding the conclusion that while passerines as a whole may not be excessively diverse, the oscine passerines (Sibley and Ahlquist's suborder Passeri) do depart from a null assumption of equal rates (Figure 9.1A). However, this conclusion is necessarily hedged by many considerations, including the assumption of rate constancy and the validity of the phylogenetic hypothesis, both of which have been called into question on the merits of the data themselves and by collection of additional data from other sources (Cracraft, 1992; Lanyon, 1992b; Harshman, 1994; Fain & Houde, 2004; Ericson *et al.*, 2006; Hackett *et al.*, 2008).

The publication of a new well-sampled (both in terms of taxa and characters) phylogenetic hypothesis for major lineages of birds (Hackett *et al.*, 2008) yields the opportunity to re-evaluate patterns of avian diversity. Fortunately, all that is necessary is a relative-time-scaled ultrametric tree, so that the number of contemporaneous lineages at any given time (e.g. at the origin of passerines) can be evaluated, and compared to extant diversity patterns. I used nonparametric rate smoothing (Sanderson, 1997) in order to derive a relative-time tree for the taxa included in the Hackett *et al.* sample, then assigned extant avian species diversity to the major lineages following taxonomy (Dickinson, 2003). Based on this crude analysis, some 39 avian lineages are of at least the same age as the Passeriformes (approximately 73 Ma, when scaling the root of the avian tree to 119 Ma, following van Tuinen, 2009). If we apportion the approximately 9718 species of birds to these 39 lineages, we obtain a histogram as shown in Figure 9.1B.

The pattern in this analysis is – if anything – even more startling than that uncovered in the original Nee *et al.* analysis of avian diversity (Figure 9.1A), although the contrast between passerines and nonpasserines is exaggerated here by the use of species rather than sublineages at the second time horizon. This pattern should be interpreted with some caution, given that it is based on

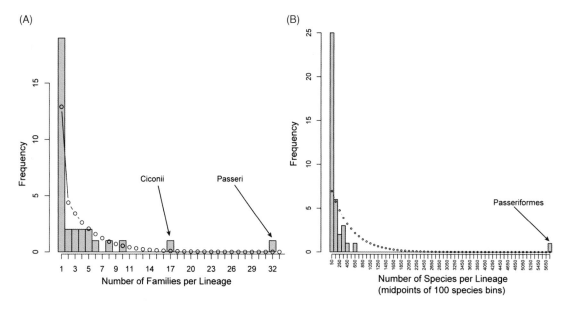

Fig. 9.1   Nonuniformity of avian lineage diversities. (A) The number of avian families ($n_2 = 137$) descending from lineages coexisting at $\Delta T_{50}H = 18$ ($n_1 = 32$), according to the DNA hybridization data (Sibley & Ahlquist, 1990). The expected distribution is geometric or "broken stick" (Nee *et al.*, 1992; shown by points and lines). The Passeri are the oscine passerines, and the Ciconii a suborder of Sibley and Ahlquist's order Ciconiiformes (see text). (B) Distribution of the number of avian species descending from lineages coexisting at the time of origin of the Passeriformes, as evaluated by nonparametric rate smoothing (Sanderson, 1997) of Hackett *et al.*'s (2008) hypothesis of avian relationships (expected geometric distribution shown by points and lines).

a very cursory analysis of Hackett *et al.*'s data, which did not incorporate multiple internal fossil constraints. Also, some apparently contemporaneous clades may not have had their earliest splits sampled (e.g. the earliest split within parrots is not included here; de Kloet & de Kloet, 2005, Wright *et al.*, 2008), leading to a slight undersampling of ancestral lineages. However, if the relative timing of avian cladogenesis uncovered in this analysis is even approximately correct, then this analysis reinforces the notion that passerine birds exhibit notable diversity relative to other avian lineages. Moreover, it counters the suggestion given above regarding a possible causal role for vocal learning in the diversity of passerines plus parrots – clearly whatever burst of diversification is responsible for the pattern in Figure 9.1 has occurred within the passerines in particular, rather than some more inclusive group. Unfortunately, because of the

nested hierarchical nature of phylogenetic relationships, asymmetries of diversity can be hard to isolate to a given level (Purvis *et al.*, 1995) – more detailed analyses of diversity patterns within passerines are necessary to elucidate this pattern.

RELATIONSHIPS AND DIVERSITY PATTERNS WITHIN PASSERIFORMES – DNA–DNA HYBRIDIZATION DATA

Analysis of passerine diversity patterns is critically dependent on understanding phylogenetic relationships within the group. Unfortunately, work on relationships within passerine birds has languished because of their relative uniformity. A few major divisions were apparent in the earliest days classification. In particular, examination of the syrinx (the avian vocal apparatus) proved informative (reviewed in Ames, 1971), allowing recognition of the so-called Oscine passerines (with relatively complex intrinsic syringeal musculature), the tracheophones (a less complex but specialized group), and the suboscine passerines (with the simplest, essentially plesiomorphic syringeal form). However, most of the attention of early systematists was focused on delimiting family groups. As noted above, defining family and higher-level groups with morphology alone proved challenging, and even now, only a handful of passerine families have definite morphological synapomorphies (e.g. the swallows – Hirundinidae; larks – Alaudidae; broadbills – Eurylaimidae). Worse, since much of the variation in passerine form was related to trophic adaptations (e.g. bill shape, hindlimb proportions), and classification preceded the advent of cladistic theory, many groupings have turned out to be para- or polyphyletic when evaluated with modern methods or other character systems (e.g. molecular data).

No comprehensive phylogenetic hypothesis for passerine birds above the family level existed prior to the work of Sibley and Ahlquist. A number of single character or character complex studies yielded many important insights (Wallace, 1874; Miller, 1924; Beecher, 1953; Tordoff, 1954a,b; Bock, 1960, 1962; Ames, 1971, 1975; Feduccia, 1974, 1975a,b; Henly *et al.*, 1978; Raikow, 1982, 1987), but no general synthesis. Importantly, Feduccia (1974; 1975a, b; 1977; 1979) and Raikow (Raikow, 1982, 1987) worked on the monophyly of passerines and on basal relationships among major passerine groups, and in addition, Raikow and collaborators (1976, 1977, 1978, 1980) developed significant character-based analysis of the relationships of a number of passerine families. In terms of molecular data, a number of studies were published prior to Sibley and Ahlquist, some of which had bearing on higher-level passerine relationships (Smith & Zimmerman, 1976; Barrowclough & Corbin, 1978; Avise *et al.*, 1980a,b, 1982; Lanyon, 1985; Marten & Johnson, 1986; Johnson *et al.*, 1988; Johnson & Marten, 1989), though primarily at the family level and below. However, the vast majority of higher-level molecular work on passerines was

published by Sibley and coauthors on the long run up to the *magnum opus* (Ahlquist *et al.*, 1984; Sibley, 1970, 1974, Sibley & Ahlquist, 1982a,b,c, 1984, 1985a,b,c, Sibley *et al.*, 1982, 1984).

Sibley's work supported some traditional notions of passerine relationships, contradicted others, but more generally established for the first time a frame of reference for the relationships of the majority of passerine groups, whose affinities had for most of ornithological history remained enigmatic. Notably, the DNA–DNA hybridization data appeared to strongly support the traditional notion of a split between suboscine and oscine passerines. By contrast, the data contradicted vague and speculative notions such as the existence of a monophyletic assemblage of Old World insectivorous oscines, the Muscicapidae (Hartert, 1910; Mayr & Amadon, 1951). Fairly early on, Sibley noted that molecular differences (first egg white protein electrophoretic patterns, Sibley, 1976; later DNA–DNA hybridization distances, Sibley & Ahlquist, 1985a,b) between Australian and other putatively closely related passerines were significantly larger than expected. This led him to the conclusion – perhaps somewhat ahead of the data (see below) – that Australian oscine passerines represented a separate evolutionary radiation from oscines (Sibley & Ahlquist, 1985a,b) . Thus, one of the major conclusions of the DNA–DNA hybridization work was a basal split within oscines between what he termed the parvorders Corvida and Passerida. The Corvida, although worldwide in distribution, were considered to be ancestrally Australasian, and the Passerida likewise widespread but conversely northern in origin. On the one hand, the Corvida included traditionally crow-like birds (e.g. ravens and jays – Corvidae; shrikes – Laniidae), as well as many Australian endemic groups (e.g. bowerbirds – Ptilonorhynchidae; birds of paradise – Paradisaeidae; lyrebirds – Menuridae). On the other hand, the Passerida included many primarily Holarctic groups (nuthatches – Sittidae; finches – Fringillidae; thrushes – Turdinae). More or less surprisingly, this split separated a number of traditionally associated groups, including the

Petroicidae (dawn robins) and Muscicapidae (thrushes, chats, and flycatchers), the Neosittini (sittellas) and Sittidae (nuthatches), the Pomatostomidae (Australian babblers), and Timaliini (true babblers). This novel view emphasized the importance of the Southern Hemisphere in avian – and in particular passerine – diversification (Olson, 1988; MacLean, 1990; Christidis & Schodde, 1991; Vickers-Rich, 1991; Boles, 1997; Cracraft, 2001). In combination with the ecological similarities observed between lineages in these two groups – as reflected in traditional notions of relationship – this view suggested two parallel radiations of oscine passerines, one primarily "northern" (ancestrally Eurasian or possibly African) and one primarily Australasian.

As discussed for birds as a whole above, in addition to providing a new perspective on passerine relationships, Sibley and Ahlquist's work placed these relationships in a temporal context. Perhaps unfortunately (see below), this temporal component was explicitly incorporated into classification, as Sibley attempted to apply ranks to monophyletic groups according to the degree of genetic distance (and thus of age, under a strict clock). Thus, family rank was given to clades separated at 9–11 $\Delta T_{50}H$ (a measure of genetic distance), subfamilies at 7–9, tribes at 4.5–7, and so on (Sibley & Ahlquist, 1990, p. 254). Just as the temporal and phylogenetic framework provided by these data has strongly influenced our understanding of passerine diversity relative to other bird lineages – the same is true for our understanding of diversity patterns within passerines.

In a series of papers starting in 2003, Robert Ricklefs analyzed passerine diversity using taxa recognized by Sibley & Ahlquist (1990), and the classification derived from that work (Sibley & Monroe, 1990). All of the analyses presented in these papers are founded on Sibley's ranking of taxa more or less according to their genetic divergence (see above). For instance, family-level taxa on average should have originated at the same time, or more accurately should all have coexisted as independent undifferentiated lineages at some time in the past (i.e. a line drawn through a phylogeny of the group would pick out each fam-

ily as a single lineage, as in the analyses shown in Figure 9.1). At the level of tribes, all tribes as well as families and subfamilies without lineages differentiated at the tribal level of divergence could likewise be treated as equal-age contemporaries. Ricklefs (2003) analyzed passerine family and tribal diversity under expectations of the geometric distribution, much as shown for birds as a whole above. However, Ricklefs presented the data in a slightly different form, plotting the natural logarithm of clade diversity ranks versus clade diversities (Figure 9.2). These figures identify a few taxa with significantly more species than expected under a constant-rates diversification model (e.g. the Fringillidae), but perhaps more notably they show a large number of taxa with significantly *fewer* species than expected. In fact, Ricklefs concluded from the contrasting results from the family and tribal level analyses (Figure 9.2A versus 2B), as well as from the lack of phylogenetic correlation of lineage diversity patterns, that whatever factors may have contributed to the excessive family-level diversity were transient, and probably related to biogeographic effects (e.g. invasion of South America) rather than intrinsic lineage characteristics (i.e. key innovations). Instead, Ricklefs focused on explaining the large number of species-poor lineages, concluding that these groups were either ecologically or geographically peripheral. Notably, 14 out of 32 tribe-to-family level clades with fewer than five species are Australian in distribution, which Ricklefs suggested might reflect the biogeographic history of passerines (see below) and the current restriction of rainforest habitats in Australia.

Subsequently, Ricklefs (2004, 2005) proceeded to analyze the diversity of these same tribe-to-family level clades in terms of morphological variation. For each of these groups, he measured eight external morphological characters (total length, wing, tail, tarsus, middle toe, and the length, width, and breadth of the culmen) for between 1 (monotypic lineages only) and 83 (Thraupini) exemplars (median of 6; three clades were unsampled). Multiple regression analyses of these data indicated a correlation of species diversity with

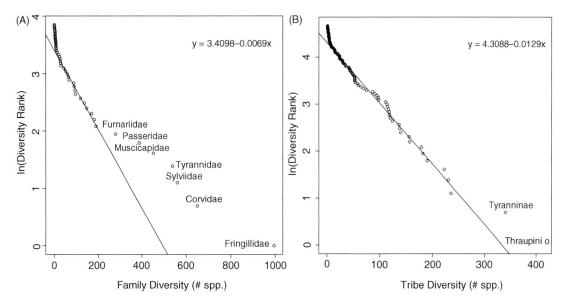

Fig. 9.2 Diversity patterns of taxa within Passeriformes, based on the classification of Sibley & Monroe (1990), following Ricklefs (2003). Shown is the natural log of taxon diversity rank (e.g. 1 = most diverse, 2 = next most, etc.) as a function of taxon diversity, for (A) passerine families, and (B) passerine tribe-to-family level clades. The lines are fitted to the approximately linear sections of each scatterplot, and yield estimates of the natural log of the expected number of lineages (α), and the parameter of a geometric distribution of clade size (β; see Ricklefs (2003) for derivation), for a uniform diversification process. Points above the fitted line are either less (upper left) or more (lower right) species-rich than expected.

morphological diversity, independent of clade age, suggesting an effect of cladogenesis on trait evolution (Ricklefs, 2004). Although this result could support a straightforward model of punctuated equilibrium, Ricklefs argued for speciation events "setting the stage" for divergent selection to operate, possibly by differences in novel geographic areas being occupied, by differing sets of divergent competitors being encountered, or by setting up competition between newly speciated and ecologically exchangeable sister taxa. Ricklefs (2005) also used these data to address the question of ecological marginality in small passerine clades (see above), finding that many of these clades (e.g. Bombycillidae) appear at the periphery of morphospace as defined by the characters measured. In combination with diversity patterns, this suggests that these lineages have had extremely low extinction and speciation rates for long periods of time, due to their ecological specialization.

A potentially key problem with these analyses is the use of named clades from Sibley and Ahlquist's work. Theoretically, tribe-level clades are monophyletic lineages differentiated anywhere from $\Delta T_{50}H$ 4.5–7°C, which lack of precision in itself is a source of error. In practice, only some lineages in this range were recognized taxonomically by Sibley and Ahlquist, and in fact quite a large number of lineages were not. This is illustrated in Figure 9.3, which shows the stem age and age of first within-lineage split for all of the passerine tribe-to-family-level clades included in the "Tapestry." The geometric expectation of extant species diversity for the 106 clades analyzed by Ricklefs requires that those clades co-occurred with extant diversity of one at some point in the past. As shown in the figure, the maximum number of co-occurring undifferentiated lineages in this sample is 62, based on published divergence data in Sibley & Ahlquist (1990). Thus, the sample

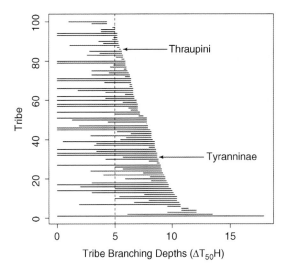

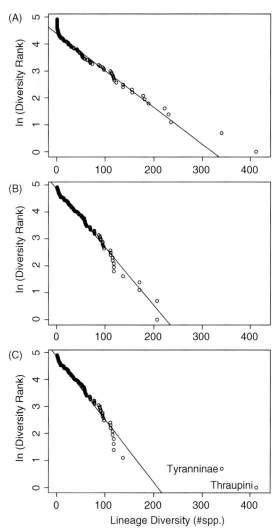

Fig. 9.3 Stem and basal intragroup divergence distances for 99 tribe-to-family level lineages of passerines, based on Sibley & Ahlquist (1990). This is a subset of 106 lineages studied by Ricklefs (2003) for which distance information is actually represented in the "Tapestry." Solid lines indicate the stem distance (right), and distance of the first sampled split within that lineage (left: lineages with only a single representative extend to zero); tribes are sorted in order of stem divergence. If all groups coexisted as single lineages at some time in the past (the basic assumption required for a geometric expectation of descendant – i.e. contemporary – diversity levels), there should be a point where a vertical line would intersect all of these lines. The maximum number of contemporary lineages possible for these data is 62 (dashed line, $\Delta T_{50}H = 5$).

Fig. 9.4 Hypothetical diversification patterns within passerines, dividing lineages contemporaneous at $\Delta T_{50}H$ into (A) a single species sister to the remaining species, (B) two equal parts, and (C) into two equal parts, excepting the Tyranninae and Thraupini. The latter choice is arbitrary, in order to illustrate the potential (currently unknown) impact of including heterogeneous lineages in a diversification analysis of this type.

of 106 taxa includes a significant number of lineages with more species than they properly should have for true equivalence, as well as some lineages that are significantly younger than others included in the sample. This variation could lead to significant biases in terms of sub-lineage diversities, depending on the symmetry of diversity patterns within clades. At two extremes, if those unnamed subclades tend to be species-poor divergent lineages within each tribe, then the pattern of Figure 9.2 would be essentially unchanged, whereas if the unnamed splits evenly divide tribal diversity, an even more uniform

process would be inferred than that found for tribes (Figure 9.4). On the other hand, if many of the splits are uniform, but some are asymmetrical, then the diversity of particular groups might be even more emphasized. These thought experiments serve to emphasize that uncertainty both in relative divergence times and in phylogenetic relationships may have serious implications for our analyses of passerine diversity using these methods.

RELATIONSHIPS AND DIVERSITY PATTERNS WITHIN PASSERIFORMES – DNA SEQUENCE DATA

Nearly coincident with publication of Sibley and Ahlquist's work, tests of their hypotheses based on analyses of other types of data began to be published (Baverstock, 1991; Christidis & Schodde, 1991; Christidis *et al.*, 1993). Notably, Christidis & Schodde (1991) found using allozyme data that members of the Passerida nested within a grade of Corvidan lineages, and a particularly large genetic distance between *Menura* (the Australian lyrebird) and other oscines (although its phylogenetic placement depended on the analysis), making the first cracks in the newly minted hypothesis. However, the community quickly moved from protein-based techniques toward the genetic material itself. The advent of polymerase chain reaction (PCR) and DNA sequencing, and latterly of efficient high-throughput sequencing technologies, has led to significant progress in inferring relationships within passerines. This work stands fair to clarify our understanding of passerine history and the processes that have contributed to the outstanding diversity of the order. While a comprehensive treatment of this work is beyond the scope of this review, consideration of even the broadest results is informative.

Perhaps the earliest DNA sequence analysis of passerine relationships was that of Edwards *et al.* (1991). Although plagued by a minor laboratory error (Edwards & Arctander, 1997), these data nevertheless established the utility of mitochondrial DNA (mtDNA) sequences in inferring passerine relationships, notably recovering monophyly of oscine and suboscine birds, and (laboratory error corrected) monophyly of the

Passerida. Subsequent studies through the 1990s (Lanyon, 1992a, 1994; Helm-Bychowski & Cracraft, 1993; Christidis *et al.*, 1996a,b ; Sheldon & Gill, 1996; Burns, 1997; Cibois *et al.*, 1999; Helbig & Seibold, 1999; Johnson & Lanyon, 1999; Lanyon & Omland, 1999; Omland *et al.*, 1999; Sheldon *et al.*, 1999; Honda & Yamagishi, 2000) bolstered this utility, especially at middle to lower taxonomic levels (i.e. families to species). At higher levels, early analyses of mtDNA data variation among bird orders yielded some apparent artifacts, including recovery of passerines as sister to all other birds (Mindell *et al.*, 1999; Johnson, 2001). These results appear to have be attributable to small character and/or taxon samples, rapid evolution, and model inadequacy (Braun & Kimball, 2002; Slack *et al.*, 2007). The late 1990s saw the advent of nuclear gene DNA sequences in avian phylogenetics (Cooper & Penny, 1997; Barrowclough & Groth, 1999; Lovette & Bermingham, 2000; Prychitko & Moore, 2000), although there had been previous work based on amino acid sequencing (Laskowski & Fitch, 1989). These studies proved pivotal in shifting the emphasis in higher-level phylogenetics from rapidly evolving mtDNA genes to more slowly evolving nuclear genes. Subsequent sequencing of the chicken (*Gallus gallus*) and zebra finch (*Taeniopygia guttata*) genomes has made amplification and sequencing of such loci in passerines increasingly easy (Backström *et al.* 2008; Kimball *et al.*, 2009), such that most studies now include both mitochondrial and nuclear DNA data, or data from multiple nuclear loci.

The most comprehensive studies of higher-level passerine relationships to date have been those conducted by myself and colleagues (Barker *et al.*, 2002), by Ericson and colleagues (Ericson *et al.*, 2000, 2002a,b, 2006; Ericson & Johansson, 2003; Irestedt *et al.*, 2001, 2002; Norman *et al.*, 2009), and by Hackett *et al.* (2008). These studies, in addition to a multitude of lower-level studies focused on specific passerine groups, have made a great deal of progress in testing hypotheses first forwarded by the DNA hybridization data, as well as in fleshing out our understanding of passerine relationships at all hierarchical levels. Briefly, many aspects of the DNA hybridization tree are borne out by analysis of sequence data, and many are flatly contradicted (Figure 9.5). More importantly, the degree to which the many conflicts that have been found are significant in understanding passerine diversification varies from trivially to profoundly. Some of the most critical conflicts are discussed below.

Sequence data unequivocally support the paraphyly of suboscine passerines (Figure 9.5). This not too surprising when it is considered that prior to Sibley's DNA hybridization work, which recovered suboscine monophyly, relationships of suboscines had been controversial on morphological grounds. Specifically, the suboscine/oscine split is based on the presence versus absence of specialized intrinsic syringeal musculature in oscines (in addition to a critical contribution of learning to vocal development). Absence of this musculature is not synapomorphic, as it is shared with many nonpasserine groups (Ames, 1971). In fact, Feduccia (1975b) showed that one group of suboscines, the Acanthisittidae (the four species of New Zealand Wrens), differ from all other suboscines in the form of the stapes, which it shares with oscine passerines. However, the same form is found in many other nonpasserines, and therefore likely to be symplesiomorphic. For a brief time, Feduccia actually entertained the notion that suboscines (exclusive of the Acanthisittidae) might be more closely related to some Coraciiform birds that share a similar stapedial morphology, than to oscine passerines (Feduccia, 1977), a notion he later rejected (Feduccia, 1979). Analysis of hindlimb musculature of major passerine lineages also indicated uncertainty in the relationships of early branching lineages (Raikow, 1987). Raikow recovered the Acanthisittidae as sister to a monophyletic oscine passerines, with the two of them sister to the remaining oscines – this arrangement is perfectly congruent with both the syringeal and stapedial data. However, this relationship was only supported by a single derived character state, and Raikow did not claim any significant conflict with the DNA hybridization-based phylogeny on that basis. Sequence data now

unequivocally place Acanthisittidae as sister to all other passerine birds (Barker *et al.*, 2002, 2004; Ericson *et al.*, 2002; Hackett *et al.*, 2008), consistent with both syringeal and stapedial morphology, but in conflict with Raikow's single hindlimb muscle character.

The second major conflict between the results of DNA sequence analyses and the DNA hybridization tree regards the basal split within oscine passerines. Recall that Sibley and Ahlquist inferred the parallel radiation of crow-like birds (Corvida) primarily in Australasia, and of all other oscines (Passerida) outside of that region. This relationship has been unequivocally rejected by analyses of sequence data (Figure 9.5; Barker *et al.*, 2002, 2004; Ericson *et al.*, 2002a,b ; Hackett *et al.*, 2008). Specifically, a clade more or less corresponding to the Passerida (with some emendation), rather than being sister to all corvoid lineages, appears to be *nested well within* the corvoid radiation (Figure 9.5). In fact, as many as the first five (and possibly more) lineage splits within oscine passerines appear to have occurred within Australasia, judging by the current distributions of these groups (Barker *et al.*, 2002, 2004). The seemingly inescapable conclusion is that the oscines had their origin on the Australian continental plate – the only alternative would be the previous existence of a widespread oscine lineage that invaded Australia multiple times and then went extinct with the exception of the Passerida.

Taken together, these primary results suggest that major lineages of passerines had their origin on specific Gondwanan fragments (Figure 9.6). That is, the sister to all other passerines is distributed in New Zealand, a Gondwanan component that rifted approximately 80 Ma (Weissel *et al.*, 1977; Mayes *et al.*, 1990; Storey *et al.*, 1999). The suboscines of the New World – one of two major lineages of this group, which is monophyletic once Acanthisittidae have been removed – are distributed primarily in South America, with a few lineages (most notably the Tyrannidae or flycatchers) making it into Central and North America. Finally, the phylogenetic relationships of oscine passerines, which are now worldwide in distribution, trace their history

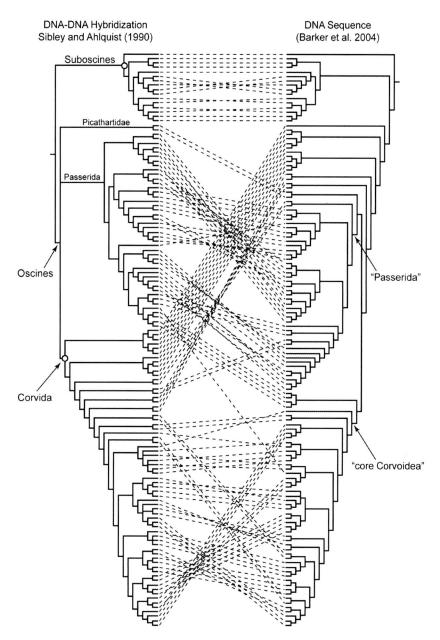

Fig. 9.5 Direct comparison of passerine relationships in the "Tapestry" (Sibley & Ahlquist, 1990), with the most comprehensive sequence analysis of passerine relationships to date (Barker *et al.*, 2004). Both trees have been pruned to include only shared genera, which are connected by dashed lines between their locations on each tree. This representation is not the minimal "tanglegram," and thus the frequency of crossing connections may exaggerate conflict between the trees. Open circles on the DNA hybridization tree indicate two important conflicts between these hypotheses of relationship.

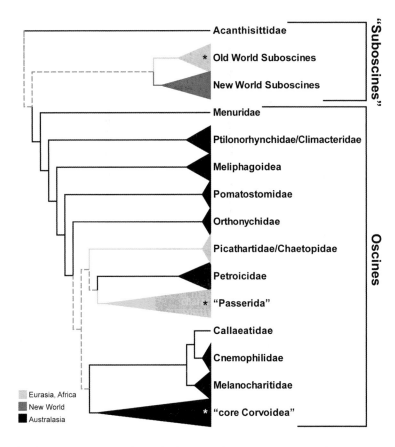

Fig. 9.6   A phylogenetic hypothesis for major passerine groups, based on sequences of RAG1 and RAG2 (Barker et al., 2004). This tree summarizes analysis of these genes from 144 passerine species, rooted using two nonpasserine outgroups. The inferred geographic origins of each group are represented by coloring of terminals and branches (black = Australia, New Guinea, and New Zealand; light gray = Old World exclusive of Australasia; and dark gray = New World; note that the Old World Suboscines, Passerida, and "core Corvoidea" are all distributed in multiple areas, and their shading reflects inference of ancestral areas).

back to Australia. The only lineage that is not endemic to a single Gondwanan landmass is the Old World suboscines, which are broadly distributed through Africa and Asia, into northern Australia, with no clear center of diversity. It is tempting to suggest that this lineage owes its distribution to an Indian origin (e.g. Cracraft, 2001; Ericson et al., 2003) but the distributional and phylogenetic data render this completely speculative.

A Gondwanan origin for extant passerines immediately raises the question of time scale. That is, this only makes sense if the origin of passerines pre-dates Gondwanan breakup, or at least the final severing of subaerial connections between New Zealand, Antarctica, Australia, and South America. Unfortunately, the question of the timing of bird divergences in general and of passerine divergences in particular is a vexed one,

involving interpretation of both molecular and fossil data. Although a comprehensive review of these issues is beyond the scope of this chapter, I will discuss the specifics as they pertain to passerine origins and diversification. Sibley & Ahlquist (1990) interpreted DNA hybridization distances as directly proportional to time since divergence, reckoning some 4.7 Myr per degree $\Delta T_{50}H$ with some allowance for the effects of generation time. This rate calibration would suggest an origin of stem passerines at 102 Ma, with the split beween oscines and suboscines (including Acanthisittidae) approximately 93 Ma. Subsequently, many molecular studies with a variety of taxon samples and calibration sets have yielded a range of estimates for these splits (Cooper & Penny, 1997; van Tuinen & Hedges, 2001; Ericson et al., 2006; Pereira & Baker, 2006; Slack et al., 2006; Brown et al., 2007; Brown et al., 2008). In general, these studies agree in placing the origin of passerines in the late Cretaceous (~100–65 Ma), with the earliest splits among the extant members of the order no later than ~60 Ma (Ericson *et al.*, 2006).

By contrast with the dates suggested by molecular data, the earliest passerine fossils found to date are from the Eocene (~55 Ma). These remains, found in Australia (Boles, 1995), are fragmentary and of uncertain taxonomic affinity, and therefore relatively uninformative regarding the timescale of passerine evolution. The earliest identifiable passerine remains in Australia date to the Miocene, and in some cases can be attributed to extant oscine families (Boles, 1995, 1997). The majority of the early passerine fossil record derives from Oligocene and Miocene deposits in northern Europe, primarily France and Germany (Mayr & Manegold, 2004; Manegold *et al.*, 2004). Intriguingly, many of the fossils that have been recovered appear to fall outside of the crown Passeriformes (Eupasseres), possibly representing an earlier radiation of passerines that has been replaced by modern members of the order (Manegold *et al.*, 2004, Mayr & Manegold, 2004). It is unclear whether crown-group passerines occurred in Europe prior to the Mid-Miocene (Manegold *et al.*, 2004). In sum, the passerine record provides few usable constraints on our interpretation of modern passerine diversity.

In evolutionary terms, what are the implications of accepting a Gondwanan origin for crown Passeriformes, with early divergences on the order of 80 Ma? One of the primary consequences is that this reorients our interpretation of adaptive diversification in the group. As described above, Sibley and Ahlquist's hypothesis implied parallel adaptive radiations (*sensu lato*) of oscines within and outside of Australia. The new perspective derived from DNA sequence data implies that these adaptive radiations did not occur separately in space, but rather in time. That is, subsequent to their origin in Australia, oscines experienced an initial adaptive radiation there, while the continent was still dominated by mesic forest rather than woodland and scrub habitats (Tedford *et al.*, 1975, Behrensmayer *et al.*, 1992). As

Australia began its northward journey after its final split with Antarctica, one lineage of oscines (the Passerida) successfully dispersed out of Australia, where it underwent a second adaptive radiation, essentially replicating the ecological diversity found within the Australian passerine fauna. Analysis of the timing of this dispersal, based on the assumption that the Acanthisittidae split from other passerines concurrent with New Zealand's rifting from Antarctica at 82 Ma, indicates that it occurred in the Eocene, ~45 Ma (Barker *et al.*, 2004).

There is some question as to the cradle of passeridan diversification. Barker *et al.* (2002, 2004) suggested that the ancestor of the Passerida likely dispersed across the narrowing gap between Australia and Southeast Asia, as the former began its headlong crash into the latter. In this scenario, the lineage would have subsequently spread from Eurasia into Africa and the New World, diversifying as it expanded its distribution. However, details of phylogenetic relationships within the Passerida and its closest relatives have suggested an alternative interpretation. Sequence data agree that the closest relatives of the Passerida include the Australian Petroicidae (dawn robins), and a clade of African endemics including the families Picathartidae (rockfowl) and Chaetopidae (rockjumpers). However, there is conflict among data sets, with some data supporting the Petroicidae as the sister group to the Passerida (e.g. Barker *et al.*, 2002, 2004), and others supporting the African clade (Ericson & Johansson, 2003; Irestedt & Ohlson, 2008). Assuming that the African clade is truly the sister taxon to the Passerida (based primarily on a shared single codon insertion in the c-*myc* gene), and performing a dispersal-vicariance analysis on a subsampled supertree of oscine passerines, Jønsson & Fjeldsa (2006) suggested that the Passerida have an African origin. This hypothesis was apparently reinforced by the discovery of the early branching position of an African lineage of Passerida (the genus *Hyliota*; Fuchs *et al.*, 2006), and by the discovery that a southeast Asian endemic, *Eupetes macrocerus*, was a member of the "African" Picathartidae/Chaetopidae clade, specifically sister to *Chaetops* (Jønsson *et al.*, 2007). This hypothesis suggests that the ancestor of the Passerida, rather than dispersing from Australia to Asia, flew directly from Australia to Africa, a distance of over 5000 km, possibly aided by the existence of exposure of the Kerguelen Plateau or ridge-system-associated islands in the southern Indian Ocean (Jønsson *et al.*, 2007).

While trans-Indian Ocean dispersal at the origin of the Passerida is an intriguing possibility, the data on which the hypothesis is based are tenuous at best. For example, the nesting of *Eupetes* within the African Picathartidae–Chaetopidae clade suggests an African-to-Asian polarity – however, the existence of a single extinct Asian lineage would render this optimi-

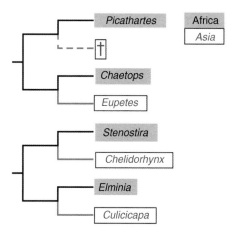

Fig. 9.7 Near biogeographic parallelism in two clades of passerines: (A) the rockfowl, rockjumpers, and rail-babbler (Picathartidae, Chaetopidae, and Eupetidae), and (B) the Stenostiridae or fairy flycatchers (Jønsson *et al.*, 2007; Fuchs *et al.*, 2009). Each African group, with the exception of *Picathartes*, has a Southeast Asian sister taxon. Absence of one Asian lineage in the former favors an African origin in the former, whereas presence of Asian sister taxa in the latter renders their ancestral distribution ambiguous. Since both of these lineages are deep within the oscine tree (the former possibly sister to Passerida, and the latter an early branch within the Old World Sylvioid assemblage – one of three or four major groups of Passerida), their distributions have a strong influence on inferences of ancestral areas for the Passerida.

zation ambiguous (Figure 9.7). Precisely this pattern is shown by another passerine group, the Stenostiridae or fairy flycatchers (Fuchs *et al.*, 2009). In this family, a deeply divergent Asian–African disjunction is repeated in parallel, with a single southern African endemic (*Stenostira*) sister to a single Southeast Asian endemic (*Chelidorhynx*, formerly *Rhipidura hypoxantha*), and a small African clade (*Elminia*) sister to two southeast Asian species (*Culicicapa*), with each of these pairs sister to one another (Figure 9.7). Furthermore, although the basal optimization of the Passerida exclusive of the Picathartidae–Chaetopidae–*Eupetes* clade appears African (Jønsson & Fjeldsa, 2006), this is critically dependent upon relationships among major lineages of the group, which are not as yet well resolved (Barker *et al.*, 2002, 2004; Ericson & Johansson, 2003; Johansson *et al.*, 2008). As a cautionary note regarding taxon sampling, the novel phylogenetic placement of *Chelidorhynx* in the Stenostiridae (Fuchs *et al.*, 2009) renders the basal distribution of the "Sylvioidea", one of these major groups of Passerida, ambiguous rather than African as asserted in Jønsson & Fjeldsa (2006). We can only speculate regarding the impact of the assumptions going into the construction of the oscine supertree. As it stands, I am forced to conclude that the evidence for an African origin of Passerida is weak at best. Conversely, there is little in favor of the Australia-to-Asian dispersal hypothesis save a certain edge in plausibility. Additional work on resolving basal relationships in the Passerida, as well as fleshing out the phylogenies of specific critical groups (e.g. the Alaudidae and Nectariniidae) should go far to resolving this issue.

Consideration of the biogeography of oscine diversification leads to some interesting questions regarding among-lineage patterns of diversity. There is no question that dispersal out of Australasia triggered a prodigious increase in diversification within the Passerida. Whether the sister group of Passerida is the Petroicidae or the African rockfowl and allies, the asymmetry in species number between these groups is highly significant ($p = 0.002$ or 0.024, respectively, using the test of Slowinski & Guyer, 1989). However, the Passerida are not the only oscine lineage that successfully dispersed out of Australasia. Specifically, many corvoid lineages have also dispersed and met with varying success throughout the globe. However, it is notable that the diversity of these corvoid groups neither individually nor in sum (ca. 400 species) comes close to that of the Passerida (ca. 3500 species). Given that the Passerida appear to have dispersed out of Australasia significantly earlier than the earliest corvoid lineage (a clade containing the New World Vireonidae plus the Asian "babblers" *Erpornis* and *Pteruthius*; Barker *et al.*, 2004; Reddy & Cracraft, 2007), it may be that the time differential of dispersal is a sufficient explanation. However, this proves difficult to assess. Notably, the maximum dispersal time difference between Passerida and the vireos is 8 Myr (±3.6 Myr, bootstrap standard error) as estimated by recombination activating genes (RAG) sequences with a vicariant calibration for divergence of the Acanthisittidae. A crude estimate of the diversification rate of Passerida is ln(diversity)/(clade age), which yields an estimate of 0.18 species/Myr. With exponential growth, this minimum differential of 8 Myr would only be expected to yield ca. 4 species at that rate of diversification. However, several caveats must be borne in mind. First, it is quite likely that most corvoid dispersals occurred much later than the vireonid dispersal – more extensive taxon sampling and time-calibrated analyses will be necessary to assess this, and significant progress is being made in this regard (e.g. Moyle *et al.*, 2006; Fuchs *et al.*, 2007; Pasquet *et al.*, 2007; Dumbacher *et al.*, 2008; Jønsson *et al.*, 2008). Second, the crude estimate of diversification rate given above is almost certainly an underestimate, because it is only a valid estimator under a pure birth model, which is unlikely for a group of this age. Notably, empirical studies of multiple avian lineages show evidence of decreased diversification over time (Phillimore & Price, 2008; Rabosky & Lovette, 2008a;), which may be due to density-dependent speciation (Rabosky & Lovette, 2008a,b). Thus, at this point it is difficult to reject the notion that differences in dispersal times contribute to this

asymmetry. If so, then early occupancy of Southeast Asia (either directly or via dispersal from an African center of origin), might be a sufficient explanation for the subsequent species-poorness of corvoid lineage. If not, this would suggest that intrinsic lineage characteristics, either of Passerida or of corvoid birds, might be important in explaining this asymmetry. In this regard, it is interesting to note that cooperative breeding, a social system common among corvoids, has been suggested to limit diversification in passerines (Cockburn, 2003; but see Ricklefs, 2005).

ONE OF SIX?

Clearly, our developing understanding of passerine phylogeny is providing new perspectives on passerine biogeography and diversification. Just as clearly, a great deal of work remains to be done. In the last decade and more, a wide array of new methods have been developed for assessing diversity patterns in phylogenetic data (Nee *et al.*, 1992; Purvis *et al.*, 1995; Pybus & Harvey, 2000; Paradis, 2005; Rabosky, 2006; Bokma, 2008). Most of these methods are critically dependent on well- or completely sampled phylogenies, as well as some timescale, be it relative or absolute. Application of molecular phylogenetic methods to increasingly large sequence data sets gathered from passerines are rapidly bringing such diversification analyses within reach for many passerine groups. Even now, application of simpler methods such as the lineage/sublineage comparisons of Nee *et al.* (1992; see also Ricklefs, 2003) to sequence-based passerine phylogenies is possible. Figure 9.8 summarizes such an analysis, as applied to the oscine passerine portion of the time-scaled passerine phylogeny of Barker *et al.* (2004; the suboscines were too poorly sampled in that phylogeny for a reliable comparison). This analysis of 67 oscine lineages is similar to the results obtained by Ricklefs (2003) for 47 family-level clades. Notably, the Fringillidae (*sensu* Sibley & Monroe, 1990, comprising the finches and New World nine-primaried oscines) is highlighted in both analyses as significantly more diverse than expected.

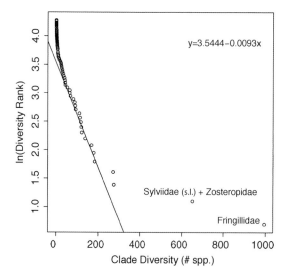

y=3.5444−0.0093x

Sylviidae (s.l.) + Zosteropidae

Fringillidae

ln(Diversity Rank)

Clade Diversity (# spp.)

Fig. 9.8 Diversity patterns of taxa within oscine passerines (following Ricklefs, 2003), based on clades co-occuring 26 Ma, as estimated from the phylogeny and timescale of Barker *et al.* (2004). Shown is the natural log of taxon diversity rank (e.g. 1 = most diverse, 2 = next most, etc.) as a function of taxon diversity (see Figure 9.2 for explanation).

Likewise, both analyses highlight the unusually large number of species-poor clades in the group. In the sequence-based analysis, the Sylviidae (the Old World warblers, babblers, and white eyes – which turn out to derive from within a single genus of babblers [Cibois, 2003; Moyle *et al.*, 2009]) also appear to be much more diverse than expected under a uniform process – to a much greater degree than in the DNA hybridization-based analysis. Unfortunately, current sampling precludes more detailed analyses (e.g. equivalent to Figure 9.2B), but given the uncertainties and potential biases inherent in previous work (e.g. Figures 9.3 and 9.4), it is likely that additional discrepant patterns will be uncovered. Of course, increased sampling will also make much more detailed and rigorous analyses of passerine biogeography possible, which will to improve our understanding of diversification in this group.

A recent analysis of vertebrate diversity identified six lineages significantly more species rich and three lineages significantly less rich than expected under a process of uniform diversification. One of these exceptional species-rich lineages was the Neoaves (nonratite, nonGalloanserae birds). As discussed above, previous analyses of avian diversity based on DNA hybridization data and current reanalyses based on DNA sequence data both agree in identifying the Passeriformes (or a significant subset thereof) as the most significant outlier in terms of avian diversity patterns. Although this result merits rigorous study of the sequence data in the context of multiple fossil calibrations, its persistence across multiple data sets and analyses is suggestive. If so, then the passerine birds are likely to represent one of the six most notable radiations of vertebrates remaining on the planet (the others being Euteleost, Ostariophysan, and Percomorph fishes, Boroeutherian mammals, and nonGekkonid squamate reptiles; Alfaro *et al.*, 2009). As such, future focus on establishing detailed knowledge of their phylogenetic relationships and relative timescale of divergences, in combination with careful analyses of their biogeography, morphological diversity, and ecology cannot but be instructive.

REFERENCES

Ahlquist JE, Sheldon FH, Sibley CG. 1984. The relationships of the Bornean bristlehead (*Pityriasis gymnocephala*) and the black-collared thrush (*Clamydochaera jefferyi*). *Journal of Ornithology* **125**: 129–140.

Ames PL. 1971. The morphology of the syrinx in passerine birds. *Bulletin of the Peabody Museum of Natural History* **37**: 194.

Ames PL. 1975. The application of syringeal morphology to the classification of the Old World insect eaters (Muscicapidae). *Bonner Zoologische Beitrage* **26**: 107–134.

Avise JC, Patton JC, Aquadro CF. 1980a. Evolutionary genetics of birds. I. Relationships among North American thrushes and allies. *The Auk* **97**: 135–147.

Avise JC, Patton JC, Aquadro CF. 1980b. Evolutionary genetics of birds. II. Conservative protein evolution in North American sparrows and relatives. *Systematic Zoology* **29**: 323–334.

Avise JC, Aquadro CF, Patton JL. 1982. Evolutionary genetics of birds. V. Genetic distances within Mimidae (mimic thrushes) and Vireonidae (vireos). *Biochemical Genetics* **20**: 95–104.

Backström N, Fagerberg S, Ellegren H. 2008. Genomics of natural bird populations: a gene-based set of reference markers evenly spread across the avian genome. *Molecular Ecology* **17**: 964–980.

Baptista LF, Trail PW. 1992. The role of song in the evolution of passerine diversity. *Systematic Biology* **41**: 242–247.

Barker FK, Barrowclough GF, Groth JG. 2002. A phylogenetic hypothesis for passerine birds: taxonomic and biogeographic implications of an analysis of nuclear DNA sequence data. *Proceedings of the Royal Society of London Series B (Biological Sciences)* **269**: 295–308.

Barker FK, Cibois A, Schikler PA, Feinstein J, Cracraft J. 2004. Phylogeny and diversification of the largest avian radiation. *Proceedings of the National Academy of Sciences, USA* **101**: 11040–11045.

Barraclough TG, Harvey PH, Nee S. 1995. Sexual selection and taxonomic diversity in passerine birds. *Proceedings of the Royal Society of London Series B (Biological Sciences)* **259**: 211–215.

Barrowclough GF, Corbin KW. 1978. Genetic variation and differentiation in the Parulidae. *The Auk* **95**: 691–702.

Baverstock PR, Schodde R, Christidis L, Kreig M, Sheedy C. 1991. Microcomplement fixation: preliminary results from the Australasian avifauna. In *Acta XX Congressus Internationalis Ornithologici, Christchurch, 1990*, Bell B.D. *et al.* (eds). Wellington: New Zealand Ornithological Trust Board.

Beecher WJ. 1953. A phylogeny of the oscines. *The Auk* **70**: 270–333.

Behrensmayer AK, Damuth JD, DiMichele WA, Potts R, Sues H-D, Wing SL. (eds). 1992. *Terrestrial Ecosystems Through Time: Evolutionary Paleoecology of Terrestrial Plants and Animals.* Chicago: University of Chicago Press.

Bock WJ. 1960. The palatine process of the premaxilla in the Passeres. *Bulletin of the Museum of Comparative Zoology, Harvard University* **122**: 359–488.

Bock WJ. 1962. The pneumatic fossa of the humerus in the Passeres. *The Auk* **79**: 425–443.

Bokma F. 2008. Detection of "punctuated equilibrium" by bayesian estimation of speciation and extinction rates, ancestral character states, and rates of anagenetic and cladogenetic evolution on a molecular phylogeny. *Evolution* **62**: 2718–2726.

Boles WE. 1995. The world's oldest songbird. *Nature* **374**: 21–22.

Boles WE. 1997. Fossil songbirds (Passeriformes) from the early Eocene of Australia. *Emu* **97**(1): 43–50.

Braun EL, Kimball RT. 2002. Examining basal avian divergences with mitochondrial sequences: model complexity, taxon sampling, and sequence length. *Systematic Biology* **51**: 614–625.

Briskie JV, Montgomerie R. 1992. Sperm size and sperm competition in birds. *Proceedings of the Royal Society of London Series B (Biological Sciences)* **247**: 89–95.

Burns KJ. 1997. Molecular systematics of tanagers (Thraupinae): Evolution and biogeography of a diverse radiation of Neotropical birds. *Molecular Phylogenetics and Evolution* **8**: 334–348.

Christidis L, Schodde R. 1991. Relationships of Australo-Papuan songbirds – protein evidence. *Ibis* **133**: 277–285.

Christidis L, Schodde R, Robinson NA. 1993. Affinities of the aberrant Australo-Papuan honeyeaters, *Toxorhamphus, Oedistoma, Timeliopsis* and *Ephthianura*: protein evidence. *Australian Journal of Zoology* **41**: 423–432.

Christidis L, Norman JA, Scott IAW, Westerman M. 1996a. Molecular perspectives on the phylogenetic affinities of lyrebirds (Menuridae) and treecreepers (Climacteridae). *Australian Journal of Zoology* **44**: 215–222.

Christidis L, Leeton PR, Westerman M. 1996b. Were bowerbirds part of the New Zealand fauna? *Proceedings of the National Academy of Sciences, USA* **93**: 3898–3901.

Cibois A. 2003. Mitochondrial DNA phylogeny of babblers (Timaliidae). *The Auk* **120**(1): 35–54.

Cibois A, Pasquet E, Schulenberg TS. 1999. Molecular systematics of the Malagasy babblers (Passeriformes: Timaliidae) and warblers (Passeriformes: Sylviidae), based on cytochrome b and 16S rRNA sequences. *Molecular Phylogenetics and Evolution* **13**: 583–595.

Cockburn, A. 2003. "Cooperative breeding in passerines: does sociality inhibit speciation?" *Proceedings of the Royal Society of London Series B-Biological Sciences* **270**(1530): 2207–2214.

Collias NE. 1997. On the origin and evolution of nest building by passerine birds. *Condor* **99**: 253–270.

Cooper A., Penny D. 1997. Mass survival of birds across the Cretaceous–Tertiary boundary: molecular evidence. *Science* **275**(5303):1109–1113.

Cotgreave P, Harvey PH. 1994. Associations among biogeography, phylogeny and bird species diversity. *Biodiversity Letters* **2**: 46–55.

Cracraft J. 1981. Toward a phylogenetic classification of the recent birds of the world (Class Aves). *The Auk* **98** (4): 681–714.

Cracraft J. 1992. Review: Phylogeny and Classification of Birds. *Molecular Biology and Evolution* **9**: 182–186.

Cracraft J. 2001. Avian evolution, Gondwana biogeography and the Cretaceous-Tertiary mass extinction event. *Proceedings of the Royal Society of London Series B (Biological Sciences)* **268**: 459–469.

Cracraft J, Barker FK, Braun MJ, Harshman J, Dyke GJ, Feinstein J, Stanley S, Cibois A, Schikler P, Beresford P, García-Morena J, Sorenson MD, Yuri T, Mindell DP. 2004. Phylogenetic relationships among modern birds (Neornithes): toward an avian tree of life. In Assembling the Tree of Life, Cracraft J, Donoghue MJ (eds). New York: Oxford University Press; 468–489.

De Kloet RS, de Kloet SR. 2005. The evolution of the spindlin gene in birds: Sequence analysis of an intron of the spindlin W and Z gene reveals four major divisions of the Psittaciformes. *Molecular Phylogenetics and Evolution* **36**: 706–721.

Dickinson EC. (ed) 2003. *The Howard and Moore Complete Checklist of the Birds of the World*. Princeton, NJ: Princeton University Press.

Dumbacher JP, Deiner K, Thompson L, Fleischer RC. 2008. Phylogeny of the avian genus Pitohui and the evolution of toxicity in birds. *Molecular Phylogenetics and Evolution* **49**: 774–781.

Edwards SV, Arctander P. 1997. Congruence and phylogenetic re-analysis of perching bird cytochrome b sequences. *Molecular Phylogenetics and Evolution* **7**: 266–271.

Edwards SV, Arctander P, Wilson AC. 1991. Mitochondrial resolution of a deep branch in the genealogical tree for perching birds. *Proceedings of the Royal Society of London Series B (Biological Sciences)* **243**: 99–107.

Ericson PGP, Johansson US. 2003. Phylogeny of Passerida (Aves: Passeriformes) based on nuclear and mitochondrial sequence data. *Molecular Phylogenetics and Evolution* **29**: 126–138.

Ericson PGP, Johansson US, Parsons TJ. 2000. Major divisions in oscines revealed by insertions in the nuclear gene c-*myc*: a novel gene in avian phylogenetics. *The Auk* **117**: 1069–1078.

Ericson PGP, Christidis L, Cooper A, Irestedt M, Jackson J, Johansson US, Norman JA. 2002a. A Gondwanan origin of passerine birds supported by DNA sequences of the endemic New Zealand wrens. *Proceedings of the Royal Society of London Series B (Biological Sciences)* **269**: 235–241.

Ericson PGP, Christidis L, Irestedt M, Norman JA. 2002b. Systematic affinities of the lyrebirds (Passeriformes: Menura), with a novel classification of the major groups of passerine birds. *Molecular Phylogenetics and Evolution* **25**: 53–62.

Ericson PGP, Irestedt M, Johansson US. 2003. Evolution, biogeography, and patterns of diversification in passerine birds. *Journal of Avian Biology* **34**: 3–15.

Ericson PGP, Anderson CL, Britton T, Elzanowski A, Johansson US, Källersjö M, Ohlson JI, Parsons TJ, Zuccon D, Mayr G. 2006. Diversification of Neoaves: integration of molecular sequence data and fossils. *Biology Letters* **2**: 543–547.

Fain MG, Houde P. 2004. Parallel radiations in the primary clades of birds. *Evolution* **58**: 2558–2573.

Feduccia A. 1974. Morphology of the bony stapes in New and Old World suboscines: New evidence for common ancestry. *The Auk* **91**: 427–429.

Feduccia A. 1975a. Morphology of the bony stapes (columella) in the Passeriformes: Evolutionary implications. *University of Kansas Museum of Natural History, Miscellaneous Publication* **63**: 34.

Feduccia A. 1975b. Morphology of the bony stapes in the Menuridae and Acanthisittidae: Evidence for oscine affinities. *Wilson Bulletin* **87**: 418–420.

Feduccia A. 1977. A model for the evolution of perching birds. *Systematic Zoology* **26**: 19–31.

Feduccia A. 1979. Comments on the phylogeny of perching birds. *Proceedings of the Biological Society of Washington* **92**: 689–696.

Fitzpatrick JW. 1988. Why so many passerine birds? A response to Raikow. *Systematic Zoology* **37**: 71–76.

Fuchs J, Cruaud C, Couloux A, Pasquet E. 2007. Complex biogeographic history of cuckoo-shrikes and allies (Passeriformes: Campephagidae) revealed by mitochondrial and nuclear sequence data. *Molecular Phylogenetics and Evolution* **44**: 138–153.

Fuchsa J, Pasquetc, Arnaud Couloux E, Fjeldså J, Bowiea RCK. 2009. A new Indo-Malayan member of the Stenostiridae (Aves: Passeriformes) revealed by multilocus sequence data: Biogeographical implications for a morphologically diverse clade of flycatchers. *Molecular Phylogenetics and Evolution* **53**(2): 384–393.

Gadow H. 1892. On the classification of birds. *Proceedings of the Zoological Society of London* **1892**: 229–256.

Hackett SJ, Kimball RT, Reddy S, Bowie RCK, Braun EL, Braun MJ, Chojnowski JL, Cox WA, Han K-L, Harshman J, Huddleston CJ, Marks BD, Miglia KJ, Moore

WS, Sheldon FH, Steadman DW, Witt CC, Yuri T. 2008. A phylogenomic study of birds reveals their evolutionary history. *Science* **320**: 1763–1768.

Hallgrimsson B, Maiorana V. 2000. Variability and size in mammals and birds. *Biological Journal of the Linnean Society* **70**: 571–595.

Harshman J. 1994. Reweaving the tapestry: What can we learn from Sibley and Ahlquist (1990)? *The Auk* **111**: 377–388.

Hartert E. 1910. *Die Vögel der paläarktischen Fauna, Volume 1.* Berlin: R. Friedländer.

Healy S, Guilford T. 1990. Olfactory-bulb size and nocturnality in birds. *Evolution* **44**: 339–346.

Helbig AJ, Seibold I. 1999. Molecular phylogeny of Palearctic-African *Acrocephalus* and *Hippolais* warblers (Aves: Sylviidae). *Molecular Phylogenetics and Evolution* **11**: 246–260.

Helm-Bychowski K, Cracraft J. 1993. Recovering phylogenetic signal from DNA sequences: relationships within the corvine assemblage (class aves) as inferred from complete sequences of the mitochondrial DNA cytochrome-b gene. *Molecular and Biological Evolution* **10**(6): 1196–1214.

Henley C, Feduccia A, Costello DP. 1978. Oscine spermatozoa: a light- and electron-microscopy study. *Condor* **80**: 41–48.

Höglund J, Sillén-Tullberg B, 1994. Does lekking promote the evolution of male-biased size dimorphism in birds? On the use of comparative approaches. *American Naturalist* **144**: 881–889.

Honda M, Yamagishi S. 2000. A molecular perspective on oscine phylogeny, with special reference to interfamilial relationships. *Japanese Journal of Ornithology* **49**: 175–184.

Irestedt M, Ohlson JI. 2008. The division of the major songbird radiation into Passerida and 'core Corvoidea' (Aves: Passeriformes) – the species tree vs. gene trees. *Zoologica Scripta* **37**: 305–313.

Irestedt M, Johansson US, Parsons TJ, Ericson PGP. 2001. Phylogeny of major lineages of suboscines (Passeriformes) analysed by nuclear DNA sequence data. *Journal of Avian Biology* **32**: 15–25.

Irestedt M, Fjeldså J, Johansson US, Ericson PGP. 2002. Systematic relationships and biogeography of the tracheophone suboscines (Aves: Passeriformes). *Molecular Phylogenetics and Evolution* **23**: 499–512.

Johanssona US, Fjeldså J, Bowie RCK. 2008. Phylogenetic relationships within Passerida (Aves: Passeriformes): a review and a new molecular phylogeny based on three nuclear intron markers. *Molecular Phylogenetics and Evolution* **48**(3): 858–876.

Johnson KP. 2001. Taxon sampling and the phylogenetic position of Passeriformes: Evidence from 916 avian cytochrome *b* sequences. *Systematic Biology* **50**: 128–136.

Johnson KP, Lanyon SM. 1999. Molecular systematics of the grackles and allies, and the effect of additional sequence (cyt *b* and ND2). *The Auk* **116**: 759–768.

Johnson NK, Zink RM, Marten JA. 1988. Genetic-evidence for relationships in the avian family vireonidae. *Condor* **90**: 428–445.

Johnson NK, Marten JA, Ralph CJ. 1989. Genetic evidence for the origin and relationships of Hawaiian honeycreepers (Aves: Fringillidae). *Condor* **91**: 379–396.

Jønsson KA, Fjeldså J. 2006. A phylogenetic supertree of oscine passerine birds (Aves: Passeri). *Zoologica Scripta* **35**: 149–186.

Jønsson KA, Fjeldsa J, Ericson PGP, Irestedt M. 2007. Systematic placement of an enigmatic Southeast Asian taxon Eupetes macrocerus and implications for the biogeography of a main songbird radiation, the Passerida. *Biology Letters* **3**: 323–326.

Jønsson KA, Irestedt M, Fuchs J, Ericson PGP, Christidis L, Bowie RCK, Norman JA, Pasquet E, Fjeldsa J. 2008. Explosive avian radiations and multi-directional dispersal across Wallacea: Evidence from the Campephagidae and other Crown Corvida (Aves). *Molecular Phylogenetics and Evolution* **47**: 221–236.

Kimball RT, Braun EL, Barker FK, Bowie RCK, Braun MJ, Chojnowski JL, Hackett SJ, Han KL, Harshman J, Heimer-Torres V, Holznagel W, Huddleston CJ, Marks BD, Miglia KJ, Moore WS, Reddy S, Sheldon FH, Smith JV, Witt CC. Yuri T. 2009. A well-tested set of primers to amplify regions spread across the avian genome. *Molecular Phylogenetics and Evolution* **50**: 654–660.

Kirsch JAW, Lapointe FJ. Springer MS. 1997. DNA-hybridisation studies of marsupials and their implications for metatherian classification. *Australian Journal of Zoology* **45**: 211–280.

Kochmer JP, Wagner RH. 1988. Why are there so many kinds of passerine birds? Because they are small. A reply to Raikow. *Systematic Zoology* **37**: 68–69.

Kroodsma DE. 2004. The diversity and plasticity of birdsong. In *Nature's Music: The Science of Birdsong*, Marler P, Slabbekoorn H (eds). San Diego, CA: Elsevier, Academic Press; 108–131.

Lanyon SM. 1985. Molecular perspective on higher-level relationships in the Tyrannoidea (Aves). *Systematic Zoology* **34**: 404–418.

Lanyon SM. 1992a. Interspecific brood parasitism in blackbirds (Icterinae): a phylogenetic perspective. *Science* **255**: 77–79.

Lanyon SM. 1992b. Review: Phylogeny and classification of birds: A study in molecular evolution. *Condor* **94**: 304–307.

Lanyon SM. 1994. Polyphyly of the blackbird genus *Agelaius* and the importance of assumptions of monophyly in comparative studies. *Evolution* **48**: 679–693.

Lanyon SM, Omland KE. 1999. A molecular phylogeny of the blackbirds (Icteridae): five lineages revealed by cytochrome-b sequence data. *The Auk* **116**(3): 629–639.

Laskowski, M, Jr. Fitch W. 1989. Evolution of avian ovomucoids and of birds. In *The Hierarchy of Life*, Fernholm B, Bremer K, Jornvall H (eds). Amsterdam: Elsevier; 371–387.

Livezy BC, Zusi RL. 2001. Higher-order phylogenetics of modern Aves based on comparative anatomy. *Netherlands Journal of Zoology* **51**: 179–205.

Livezy BC, Zusi RL. 2007. Higher-order phylogeny of modern birds (Theropoda, Aves: Neornithes) based on comparative anatomy. II. Analysis and discussion. *Zoological Journal of the Linnean Society* **149**: 1–95.

Lovette IJ, Bermingham E. 2000. c-*mos* variation in songbirds: molecular evolution, phylogenetic implications, and comparisons with mitochondrial differentiation. *Molecular Biology and Evolution* **17**: 1569–1577.

Maclean GL. 1990. Evolution of the passerines in the Southern Hemisphere. MORE DETAILS Pages 1–11.

Manegold A, Mayr G, Mourer-Chauviré C. 2004. Miocene songbirds and the composition of the European passeriform avifauna. *Auk* **121**: 1155–1160.

Marten JA, Johnson NK. 1986. Genetic relationships of North American cardueline finches. *Condor* **88**: 409–420.

Mayes CL, Lawver LA, Sandwell DT. 1990. Tectonic history and new isochron chart of the South Pacific. *Journal of Geophysical Research* **95**: 8543–8567.

Mayr E, Amadon D. 1951. A classification of recent birds. Pages 10227 In Am. Mus. Novit. MORE DETAILS.

Mayr G, Manegold A. 2004. The oldest European fossil songbird from the early Oligocene of Germany. *Naturwissenschaften* **91**(4): 173–177.

Miller WD. 1924. Variations in the structure of the aftershaft and their taxonomic value. *American Museum Novitates* **140**: 1–7.

Mindell DP, Sorenson MD, Dimcheff DE, Hasegawa MC, Ast JC, Yuri T. 1999. Interordinal relationships of birds and other reptiles based on whole mitochondrial genomes *Systematic Biology* **48**(1): 138–152.

Moyle RG, Cracraft J, Lakim M, Nais J, Sheldon FH. 2006. Reconsideration of the phylogenetic relationships of the enigmatic Bornean Bristlehead (Pityriasis gymnocephala). *Molecular Phylogenetics and Evolution* **39**: 893–898.

Moyle RG, Filardi CE, Smith CE, Diamond J. 2009. Explosive Pleistocene diversification and hemispheric expansion of a "great speciator." *Proceedings of the National Academy of Sciences, USA* **106**(6): 1863–1868.

Nee S. 2004. Extinct meets extant: simple models in paleontology and molecular phylogenetics. *Paleobiology* **30**: 172–178.

Nee S, Mooers AØ, Harvey PH. 1992. Tempo and mode of evolution revealed from molecular phylogenies. *Proceedings of the National Academy of Sciences, USA* **89**: 8322–8326.

Norman JA, Ericson PGP, Jonsson KA, Fjeldsa J, Christidis L. 2009. A multi-gene phylogeny reveals novel relationships for aberrant genera of Australo-Papuan core Corvoidea and polyphyly of the Pachycephalidae and Psophodidae (Aves: Passeriformes). *Molecular Phylogenetics and Evolution* **52**: 488–497.

Olson SL. 1988. Aspects of global avifaunal dynamics during the Cenozoic. Pages 2023–2029 In *Acta XIX Congressus Internationalis Ornithologici, Ottawa, 1986*, Ouellet H (ed.). University of Ottawa Press, Ottawa: National Museum of Natural Sciences.

Omland KE, Lanyon SM, Fritz SJ. 1999. A molecular phylogeny of the New World Orioles (*Icterus*): The importance of dense taxon sampling. *Molecular Phylogenetics and Evolution* **12**: 224–239.

Owens IP, Bennett FPM. 1997. Variation in mating system among birds: ecological basis revealed by hierarchical comparative analysis of mate desertion. *Proceedings of the Royal Society of London Series B (Biological Sciences)* **264**: 1103–1110.

Paradis E. 2005. Statistical analysis of diversification with species traits. *Evolution* **59**: 1–12.

Pasquet E, Pons JM, Fuchs J, Cruaud C, Bretagnolle V. 2007. Evolutionary history and biogeography of the drongos (Dicruridae), a tropical Old World clade of corvoid passerines. *Molecular Phylogenetics and Evolution* **45**: 158–167.

Pepperberg IM. 2004. Grey parrots: learning and using speech. In *Nature's Music: The Science of Birdsong*, Marler P, Slabbekoorn H (eds). San Diego, CA: Elsevier, Academic Press; 363–373.

Phillimore A.B., Price T.D., 2008. Density-dependent cladogenesis in birds. *PLoS Biology* **6**(3): e71. doi:10.1371/journal.pbio.0060071

Poiani A, Pagel M. 1997. Evolution of avian cooperative breeding: comparative tests of the nest predation hypothesis. *Evolution* **51**: 226–240.

Promislow DEL, Montgomerie R, Martin TE. 1992. Mortality costs of sexual dimorphism in birds. *Proceedings of the Royal Society of London Series B (Biological Sciences)* **250**: 143–150.

Purvis A, Nee S, Harvey PH. 1995. Macroevolutionary inferences from primate phylogeny. *Proceedings of the Royal Society of London Series B (Biological Sciences)* **260**: 329–333.

Pybus OG, Harvey PH. 2000. Testing macro-evolutionary models using incomplete molecular phylogenies. *Proceedings of the Royal Society of London Series B (Biological Sciences)* **267**: 2267–2272.

Rabosky DL. 2006. Likelihood methods for detecting temporal shifts in diversification rates. *Evolution* **60**: 1152–1164.

Rabosky DL, Lovette IJ. 2008a. Density-dependent diversification in North American wood warblers. *Proceedings of Biological Sciences* **275**(1649): 2363–2371.

Rabosky DL, Lovette IJ. 2008b. Explosive evolutionary radiations: decreasing speciation or increasing extinction through time? *Evolution* **62**: 1866–1875.

Raikow RJ. 1976. Pelvic appendage myology of the Hawaiian honeycreepers (Drepanididae). *The Auk* **93**: 774–792.

Raikow RJ. 1977. Pectoral appendage myology of the Hawaiian honeycreepers (Drepanididae). *The Auk* **94**: 331–342.

Raikow RJ. 1978. Appendicular myology and relationships of the New World, nine-primaried oscines (Aves: Passeriformes). *Bulletin of the Carnegie Museum of Natural History* **7**: 1–43.

Raikow RJ. 1982. Monophyly of the Passeriformes: test of a phylogenetic hypothesis. *The Auk* **99**: 431–445.

Raikow RJ. 1986. Why are there so many kinds of passerine birds? *Systematic Zoology* **35**: 255–259.

Raikow RJ. 1987. Hindlimb myology and evolution of the Old World suboscine passerine birds (Acanthisittidae, Pittidae, Philepittidae, Eurylaimidae). *Ornithological Monographs* **41**: 1–81.

Raikow RJ, Bledsoe AH. 2000. Phylogeny and evolution of the passerine birds. *BioScience* **50**: 487–499.

Raikow RJ, Polumbo PJ, Borecky SR. 1980. Appendicular myology and relationships of the shrikes (Aves: Passeriformes: Laniidae). *Annals of the Carnegie Museum, Carnegie Museum of Natural History* **49**: 131–152.

Reddy S, Cracraft J. 2007. Old world Shrike-babblers (*Pteruthius*) belong with New World Vireos (Vireonidae). *Molecular Phylogenetics and Evolution* **44**: 1352–1357.

Ricklefs RE. 2003. Global diversification rates of passerine birds. *Proceedings of the Royal Society of London Series B (Biological Sciences)* **270**: 2285–2291.

Ricklefs RE. 2004. Cladogenesis and morphological diversification in passerine birds. *Nature* **430**: 338–341.

Ricklefs RE. 2005. Small clades at the periphery of passerine morphological space. *American Naturalist* **165**: 43–659.

Ricklefs RE, Konarzewski M, Daan S. 1996. The relationship between basal metabolic rate and daily energy expenditure in birds and mammals. *American Naturalist* **147**: 1047–1071.

Sanderson MJ. 1997. A nonparametric approach to estimating divergence times in the absence of rate constancy. *Molecular Biology and Evolution* **14**: 1218–1231.

Saranathan V, Hamilton D, Powell GVN, Kroodsma DE, Prum RO. 2007. Genetic evidence supports song learning in the three-wattled bellbird Procnias tricarunculata (Cotingidae). *Molecular Ecology* **16**: 3689–3702.

Sheldon FH, Bledsoe AH. 1993. Avian molecular systematics, 1970s to 1990s. *Annual Review of Ecology and Systematics* **24**: 243–278.

Sheldon FH, Gill FB. 1996. A reconsideration of songbird phylogeny, with emphasis on the evolution of titmice and their sylvioid relatives. *Systematic Biology* **45**: 473–495.

Sheldon FH, Whittingham LA, Winkler DW. 1999. A comparison of cytochrome *b* and DNA hybridization data bearing on the phylogeny of swallows (Aves: Hirundinidae). *Molecular Phylogenetics and Evolution* **11**: 320–331.

Sibley, CG. 1970. A comparative study of the egg-white proteins of passerine birds. *Bulletin of the Peabody Museum of Natural History* **32**: 1–131.

Sibley CG. 1974. The relationships of the lyrebirds. *Emu* **74**: 65–79.

Sibley CG. 1976. Protein evidence of the relationships of some Australian passerine birds. Pages 557–570 In *Proceedings of the 16th International Ornithological Congress*, Frith H, Calaby JH (eds.). Canberra: Australian Academy of Sciences.

Sibley CG, Ahlquist JE. 1982a. The relationships of the Australo-Papuan scrub-robins *Drymodes* as indicated by DNA–DNA hybridization. *Emu* **82**: 173–176.

Sibley CG, Ahlquist JE. 1982b. The relationships of the vireos (Vireoninae) as indicated by DNA-DNA hybridization. *Wilson Bulletin* **94**: 114–128.

Sibley CG, Ahlquist JE. 1982c. The relationships of the Wrentit (*Chamaea fasciata*) as indicated by DNA–DNA hybridization. *Condor* **84**: 40–44.

Sibley CG, Ahlquist JE. 1984. The relationships of the starlings (Sturnidae: Sturnini) and the mockingbirds (Sturnidae: Miminae). *The Auk* **101**: 230–243.

Sibley CG, Ahlquist JE. 1985a. The phylogeny and classification of the Australo-Papuan passerine birds. Emu **85**: 1–14.

Sibley CG, Ahlquist JE. 1985b. The phylogeny and classification of the New World suboscine birds (Passeriformes: Oligomyodi: Tyrannides). In *Neotropical Ornithology*, Buckley PA, Foster M, Morton E, Ridgely R, Smith N (eds.). Washington: American Ornithologists' Union; 396–428.

Sibley CG, Ahlquist JE. 1985c. The phylogeny and classification of the passerine birds, based on comparisons of the genetic material, DNA. In *Procedings of the 18th International Ornithological Congress*, Ilyichev VD, Gavrilov VM (eds). Moscow: Nauka; 83–121.

Sibley CG, Ahlquist JE. 1990. *Phylogeny and Classification of Birds: A Study in Molecular Evolution*. New Haven: Yale University Press.

Sibley CG, Monroe BL. 1990. *Distribution and Taxonomy of Birds of the World*. New Haven: Yale University Press.

Sibley CG, Williams GR, Ahlquist JE. 1982. The relationships of the New Zealand wrens (Acanthisittidae) as indicated by DNA–DNA hybridization. *Notornis* **29**: 113–130.

Sibley CG, Lanyon SM, Ahlquist JE. 1984. The relationships of the Sharpbill (*Oxyruncus cristatus*). *Condor* **86**: 48–52.

Slack KE, Delsuc F, McLenachan PA, Arnason U, Penny D. 2007. Resolving the root of the avian mitogenomic tree by breaking up long branches. *Molecular Phylogenetics and Evolution* **42**: 1–13.

Slowinski JB, Guyer C. 1989. Testing the stochasticity of patterns of organismal diversity: An improved null model. *American Naturalist* **134**: 907–921.

Smith JK, Zimmerman EG. 1976. Biochemical genetics and evolution of North American blackbirds, family Icteridae. *Comparative Biochemistry and Physiology Part B: Comparative Biochemistry* **53**: 319–324.

Sol D, Price TD. 2008. Brain size and the diversification of body size in birds. *American Naturalist* **172**: 170–177.

Sol D, Duncan RP, Blackburn TM. Cassey P, Lefebvre L. 2005. Big brains, enhanced cognition, and response of birds to novel environments. *Proceedings of the National Academy of Sciences of the United States of America* **102**: 5460–5465.

Storey BC, Leat PT, Weaver SD, Pankhurst RJ, Bradshaw JD, Kelley S. 1999. Mantle plumes and Antarctica-New Zealand rifting: evidence from Mid-Cretaceous mafic dykes. *Journal of the Geological Society of London* **156**: 659–671.

Swanson DL, Garland T. 2009. The evolution of high summit metabolism and cold tolerance in birds and its impact on present-day distributions. *Evolution* **63**: 184–194.

Tedford RH, Banks MR, Kemp NR, McDougall I, Sutherland FL. 1975. Recognition of oldest known fossil marsupials from australia. *Nature* **255**: 141–142.

Tordoff HB. 1954a. Relationships of the New World nine-primaried oscines. *The Auk* **71**: 273–284.

Tordoff HB. 1954b. A systematic study of the avian family Fringillidae based on the structure of the skull. *Miscellaneous Publications of the Museum of Zoology, University of Michigan* **81**.

Van Tuinen M. 2009. Birds (Aves). In *The Timetree of Life*, Hedges SB, Kumar S (eds). Oxford: Oxford University Press; 409–411.

Vermeij GJ. 1988. The evolutionary success of passerines: A question of semantics? *Systematic Zoology* **37**: 69–71.

Vickers-Rich P. 1991. The Mesozoic and Tertiary history of birds on the Australian plate. Pages 722–808 In *Vertebrate Palaeontology of Australasia*, Vickers-Rich P, Baird RF, Monaghan J, Rich TH. (eds). Melbourne: Thomas Nelson.

Von Euler F. 2001. Selective extinction and rapid loss of evolutionary history in the bird fauna. *Proceedings of the Royal Society of London Series B (Biological Sciences)* **268**: 127–130.

Wallace AR. 1874. On the arrangement of the families constituting the order *Passeres*. *Bulletin of the British Ornithological Club* **22**: 406–416.

Weissel JK, Hayes DE, Herron EM. 1977. Plate tectonics synthesis: the displacements between Australia, New Zealand and Antarctica since the late Cretaceous. *Marine Geology* **25**: 231–277.

Westneat DF, Sherman PW. 1997. Density and extra-pair fertilizations in birds: a comparative analysis. *Behavioral Ecology and Sociobiology* **41**: 205–215.

Wright TF, Schirtzinger EE, Matsumoto T, Eberhard JR, Graves GR, Sanchez JJ, Capelli S, Mueller H, Scharpegge J, Chambers GK, Fleischer RC. 2008. A multilocus molecular phylogeny of the parrots (Psittaciformes): Support for a Gondwanan origin during the Cretaceous. *Molecular Biology and Evolution* **25**: 2141–2156.

Part 3 The Evolution of Key Avian Attributes

10 Morphological and Behavioral Correlates of Flapping Flight

BRET W. TOBALSKE,[1] DOUGLAS R. WARRICK,[2]
BRANDON E. JACKSON,[1] AND KENNETH P. DIAL[1]

[1]University of Montana, USA
[2]Oregon State University, USA

Active flight, characterized in birds by wing flapping, requires greater power output than swimming, walking or running (Schmidt-Nielsen, 1972; Harrison & Roberts, 2000). The power required for flight varies as a function of flight speed approximately according to a U-shaped curve, with more power required for hovering and fast flight than for flight at intermediate speeds (Pennycuick, 1975; Rayner, 1985; Tobalske *et al.*, 2003; Askew & Ellerby, 2007; Tobalske, 2007). Metabolic rates during flight are up to 30 times greater than basal metabolic rate (Nudds & Bryant, 2000). We begin this chapter by exploring the anatomy of the muscles that generate this power output and the skeletal elements that provide support for these muscles. In separate sections we then examine how the functional morphology of the flight apparatus affects flight performance. In each case, we observe that wing morphology and body size are key elements governing flight performance. We begin with the ontogeny of flight ability in precocial birds and use this model system to describe a novel, testable model for the origin of flight. We then turn to a style of flight that requires high power output: vertical escape after take-off. Next, we examine intermittent flight styles that offer energetic savings relative to continuous flapping. We move to maneuvering, an area that clearly needs new data and a modern synthesis since much of what is predicted about the ability to maneuver is based upon fixed-wing aerodynamics, pertinent only to gliding, and the highly flexible, morphing bird wing is scarcely ever fixed in shape, even during glides. Finally, we turn to hovering, the ultimate exertion of control during flight.

FUNCTIONAL MORPHOLOGY OF THE WING

There are a variety of features of the wings of birds that are associated with the production of high power output. The primary flight muscles include the major downstroke muscle, the pectoralis, and the major upstroke muscle, the supracoracoideus (Figure 10.1). Empirical studies of the function of these muscles using *in vivo* electromyography, sonomicrometry, and bone strain measurements as well as *in vitro* ergometry all indicate that these two muscles are generally designed to produce relatively high force per unit cross-sectional area (stress) while undergoing a relatively large length change (strain) during contraction (Dial *et al.*, 1997; Biewener *et al.*, 1998; Hedrick *et al.*, 2003; Tobalske *et al.*, 2003; Askew & Ellerby, 2007; Figure 10.2). While the pectoralis is comprised exclusively of fast-twitch fibers in most flying birds, some soaring birds have a deep anterior

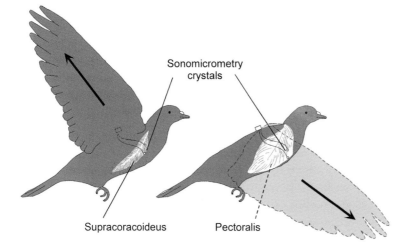

Fig. 10.1 The primary flight muscles in bird flight are the supracoracoideus (SUPRA) and pectoralis (PECT). These muscles function to decelerate and accelerate the wing, and these functions have been revealed *in vivo* using sonomicrometry transducers to measure changes in muscle length, and electromyography to measure neuromuscular activation. (From Tobalske & Biewener, 2008.)

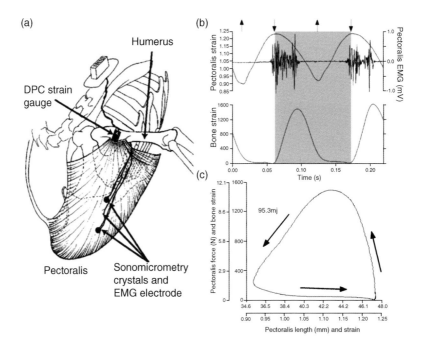

Fig. 10.2 In flying birds, the primary flight muscles appear to be designed to maximize the output work and power rather than isometric force. (A) This conclusion has emerged from *in vivo* measures of mechanical work that are obtained using sonomicrometry, electromyography, and strain-gauge measurements on the deltopectoral crest (DPC) of the humerus. These data reveal length change, neuromuscular activation, and force development in the muscle (B). Plotting muscle force as a function of muscle length produces a work loop (C); the area inside the work loop is the net work-output by the muscle (From Hedrick *et al.*, 2003; Tobalske *et al.*, 2003.)

portion of their pectoralis that consists of slow fibers and is thought to be a specialization for maintaining economical isometric contractions (Rosser & George, 1986a,b; Rosser *et al.* 1994; Meyers & Stakebake, 2005).

During flapping, the pectoralis muscle decelerates the wing at the end of upstroke and reaccelerates it at the beginning of downstroke (Dial, 1992a). The peak force observed in the muscle occurs at the middle of downstroke (Figure 10.2a), and the muscle typically changes between 20 and 42% of its resting length during contraction (Figure 10.2b). The large stress and strain in the muscle are evident in work loops obtained from *in vivo* measurements (Figure 10.2c). The area inside the work loop is a measure of work output by the muscle, and the rate of accomplishing this work, a function of wingbeat frequency, is the power output by the muscle.

Similar length change (muscle strain) and even higher force per unit area (muscle stress) are exhibited by the primary upstroke muscle, the supracoracoideus (Figure 10.1). This muscle decelerates the wing at the end of downstroke and reaccelerates it at the beginning of upstroke. A key function of the supracoracoideus is to accomplish long-axis rotation (supination) of the wing during the transition between downstroke and upstroke (Poore *et al.* 1997; Tobalske & Biewener, 2008). The supracoracoideus features a long tendon that inserts dorsally on the proximal humerus via a foramen triosseum that is bordered by the coracoids, furcula, and scapula (Baumel *et al.*, 1993). The tendon elastically stores and releases energy put into by the supracoracoideus, and this process may contribute up to 60% of the net work of the muscle (Tobalske & Biewener, 2008). The furcula may also function to elastically store and release energy (Jenkins *et al.*, 1988).

It may be that the pectoralis is the minimum muscle required for level flapping flight in birds, as experiments have shown that birds can fly without use of their supracoracoideus (Sokoloff *et al.*, 2001) or the distal muscles of the forearm and wrist (Dial, 1992b). However, future study should seek to clarify the relative contribution of other muscles of the wing to power output.

Consider, for example, the scapulohumeralis caudalis. This is the third largest muscle of the wing, it inserts ventrally on the humerus, and the timing of its activation suggests that it is involved in wing pronation and depression at the start of downstroke (Dial, 1992a). Based upon patterns of neural recruitment, the intensity of electromyography signals, it is thought that the distal muscles of the wing are primarily used to alter wing shape to permit a bird to engage in different modes of flight or maneuver. A four-bar linkage system made up of the humerus, radius, ulna, and proximal metacarpus is hypothesized to automatically flex and extend the distal wing when proximal muscles such as the pectoralis are activated (Dial, 1992b).

Skeletal elements provide surface areas for the origins and insertions of the wing muscles, and, acting as levers, they transmit muscle forces to the air. Key features of the skeleton that support powered flight include the proportionally massive keel of the sternum, the enlarged deltopectoral crest (DPC) of the humerus, and the strut-like, stout coracoids (Figures 10.2a and Figures 10.3). The keel provides the origin for the pectoralis and supracoracoideus, and the deltopectoral crest provides the insertion for the pectoralis.

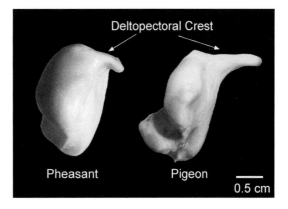

Fig. 10.3 The ventral side of the deltopectoral crest of the humerus (seen in medial view) is the insertion site for the primary downstroke muscle, the pectoralis. There is considerable diversity of shape in the deltopectoral crest as is evident in this comparison of bones from a ring-necked pheasant (*Phasianus colchicus*) and a rock dove (*Columba livia*). (From Tobalske & Dial, 2000.)

Proportionally large areas for muscle attachment presumably lessen the risk of detachment by holding tendon stress (force per unit area) below the point of failure, although this safety factor has not been studied explicitly for the pectoral girdle of birds. Among species, there is considerable diversity in DPC size and shape (Figure 10.3). Since this is the point of force transmission from the pectoralis to the rest of the wing, with clear implications for the majority of the lift produced by the wing, the functional significance of the diversity in DPC shape deserves study. The coracoids are oriented and shaped to resist compression of the thorax during contraction of the pectoralis and supracoracoideus (Pennycuick, 1968; Baier *et al.*, 2007). The furcula shows variation in form that is consistent with different uses of the wing, and flapping fliers exhibit less variation in shape compared with, for example, soaring birds or subaqueous flappers (Hui, 2002). As the furcula can contribute to elastic energy storage (Jenkins *et al.*, 1988), comparative mechanical analysis of the furcula is also warranted.

ONTOGENY AS A MODEL FOR THE EVOLUTION OF FLAPPING FLIGHT

Few subjects in science ignite such polarizing discussions as the origin and evolution of avian powered flight. Until recently, the vast literature on the subject remained firmly entrenched within two camps referred to as the ground-up (cursorial) and tree-down (arboreal) proponents (for review see Witmer, 2002). Cursorial hypotheses contend that the ancestors of birds ran bipedally using their long and slender theropod hind limbs, while their clawed and feathered forelimbs functioned to grab prey. Flapping the forelimbs in order to generate aerodynamic power and sustain powered flight came later. An extant model for this behavior is not apparent. The arboreal hypothesis suggests proto-birds quadrupedally climbed trees or other elevated structures to gain potential energy and then glided downwards (as observed in extant flying squirrels, e.g. *Glaucomys*; Bishop, 2006). The putative sequence of steps between gliding and powered

flight is not fully resolved. Dudley *et al.* (2007) maintain that small motions of the appendages permit an animal to control the direction of descent during a glide, thus offering a precursor to flapping. Likewise, small-amplitude flapping motions may contribute to stability in flying squirrels (Bishop, 2006). Significantly, though, no extant gliders have been observed to actively flap their webbed appendages or fins (e.g. flying fish, Exocoetidae, Davenport, 1994) in an effort to produce thrust and extend their glide distance. An alternative, hypothesis-based approach to the origin of avian flight, explored by Garner *et al.* (1999) gave rise to a "pouncing predator" model, which satisfies several major phylogenetic assumptions. Nonetheless, an extant analog of the pouncing predatory model has also not been identified, so it is not presently possible to empirically test the functional morphology – mechanics and physiology – of the model.

Where can we find extant analogs to the origin of powered flight in birds? Where else can one find an incipient avian wing but on a baby bird? Before juvenile birds can fly, they readily use their wings in a form of escape behavior known as wing-assisted incline running (WAIR) that consists of flapping the wings during climbing (Dial, 2003; Dial *et al.*, 2006). This escape behavior may be used by ground-dwelling species such as the Galliformes when they have access to a sloped terrain (cliff, boulder, tree, etc.), and is common among nestlings of a diverse array of bird species (Dial *et al.*, 2008b). If partially developed wings in precocial birds are reasonably analogous to the incipient wings that the presumed ancestors of modern birds possessed, then the ontogeny of WAIR in extant species offers a novel, testable biomechanical model for the origin of powered flight in birds (Bundle & Dial, 2003; Dial, 2003; Dial *et al.*, 2006). This model assumes development in external wing morphology is representative of transitional adaptive stages (Bock, 1965) that led to the complex structure of the extant avian wing. An obvious limitation of the model is uncertainty in how extant avian neuromuscular control and contractile behavior as well as external wing motions compare with ancestral forms.

During ontogeny in chukar partridge (*Alectoris chukar*), feathers are structurally symmetrical (i.e. equal feather surface on either side of rachis) from day 6 through to day 14 (Figure 10.4). Potentially analogous feather symmetry is apparent in theropod fossils hypothesized to represent ancestors of extant birds (Quiang *et al.*, 1998; Xu *et al.*, 1999). In chukars, wing surface area increases in a near-linear fashion with age during the first 30 days and is asymptotic by day 45. The growth in body mass is such that wing loading (weight per unit surface area of the combined wings) remains relatively constant throughout their normal growth phase, with the lowest wing-loading values recorded during the first 30 days of development.

In adult chukar, there is no significant variation in the patterns of wing motion used during WAIR, descending, and level flight (Dial *et al.*, 2008b; Figure 10.5). Likewise, developing birds move their incipient wings, and, later, their fully developed wings, through a stereotypic kinematic pathway so that they may flap–run over obstacles, control descending flight and ultimately perform level flapping flight (Figure 10.5). As the same basic wing motion can allow an adult bird to accomplish disparate modes of locomotion, and baby birds use this basic pattern of wing motion before they can fly, Dial *et al.* (2008b) proposed an "ontogenetic–transitional wing hypothesis" that the transitional stages leading to the evolution of avian flight correspond both behaviorally and morphologically to the transitional stages observed in ontogenetic forms.

To reveal the aerodynamics of incipient wings during WAIR, Tobalske & Dial (2007) used particle image velocimetry (PIV) and measured flow dynamics in the wake of these animals as they engaged in WAIR and ascending flight (Tobalske & Dial 2007; Figure 10.6). The ontogeny of lift production was evaluated using three age classes:

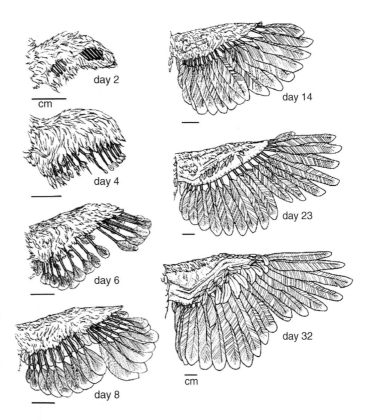

day 2

cm

day 4

day 6

day 8

day 14

day 23

day 32

cm

Fig. 10.4 Wing and feather development for the chukar partridge (*Alectoris chukar*) during ontogeny. By day 8, flapping the wings provides aerodynamic force that enhances the ascending and descending performance of the chicks. (From Dial *et al.*, 2006.)

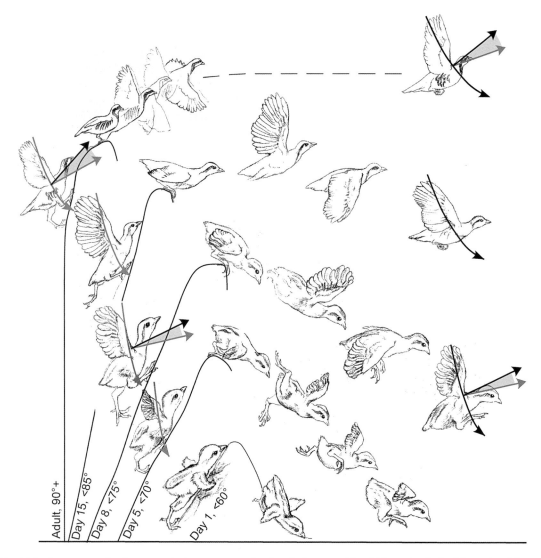

Fig. 10.5 Locomotor development during ontogeny in the chukar partridge (*Alectoris chukar*) from hatching to adulthood. The sequence of transitional stages during development in an extant species may be relevant to understanding the origin and evolution of extinct forms. Stroke curves represent the trajectory of the wing during wing-assisted incline running (WAIR) (grey) and flight (black). Vectors indicate average lift during WAIR (grey) and the estimated lift (black) during slow level flight and descent. (From Dial *et al.*, 2008b.)

baby birds incapable of flight (5–8 days post-hatching) and volant juveniles (25–28 days) and adults (45+ days). All three age classes of birds, including baby birds with partially emerged, symmetrical wing feathers (Figure 10.6), generate circulation

with their wings and share a wake structure that consists of discrete vortex rings shed once per downstroke. Unlike during flight when the wings produce lift to support body weight and match drag, during WAIR, lift from the wings

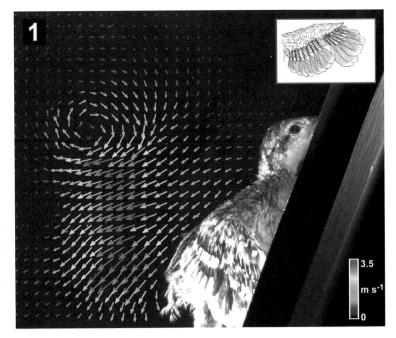

Fig. 10.6 Airflow in the wake of a flightless chukar partridge (*Alectoris chuk*ar) chick engaged in wing-assisted incline running. Velocity in the flow field was revealed using particle imagery velocimetry (PIV); the wake reveals evidence of lift production in a manner similar to juvenile and adult birds that are capable of flight. At day 8 of development, the chick has symmetrical remiges (inset, upper right). (Adapted from Tobalske & Dial, 2007.) [This figure appears in color as Plate 10.6.]

accelerates the body toward the surface of the substrate being climbed, thereby increasing friction and aiding the feet in gaining purchase. These data show that partially developed wings, not yet capable of flight, can produce useful lift during WAIR.

These aerodynamic experiments show that factors besides external wing morphology may be functioning as primary constraints upon the onset of flight ability during development (Tobalske & Dial, 2007). Potential variables that should be tested include neuromuscular control and power output of the muscles moving the wings. Nonetheless, the aerodynamics of WAIR in baby chukar provides new insight into how an ancestral incipient wing that was not capable of supporting flight may have been an exaptation (Gould & Vrba, 1982) originally used solely for WAIR.

TAKE-OFF AND ESCAPE FLIGHT

When flying to escape a predator, or voluntarily initiating flight from the ground, take-off and the gain in potential energy that occurs during flight

demand more power than most other forms of flight (Pennycuick, 1975; Rayner, 1979a,b, 1985; Ellington, 1991). Although some of the mechanical power from the flight muscles is used to overcome profile (pressure and skin-friction) drag on the wings, the majority of power output during take-off and vertical flight is used to induce a massive downward velocity to the air. Induced power is the product of this induced velocity multiplied by body mass and any net vertical or horizontal acceleration (including gravity) (Askew *et al.*, 2001). More broadly, flight speeds at take-off are relatively slow, and induced power output is modeled as being greatest at low speed, decreasing exponentially with increasing air velocity over the wing.

Since few birds are capable sustaining flight at zero velocity (see *Hovering*, below), and acceleration requires even more muscle power than hovering, take-off imposes induced-power demands beyond the capabilities of most avian wings. Birds therefore depend on their legs to provide assistance. The contribution of the legs to the velocity of the bird at the end of take-off, defined as the end of the first downstroke after the feet have left the ground varies from 59% (rufous hummingbird, *Selasphorus rufus*; Tobalske *et al.*, 2004) to 90% (blue-breasted quail, *Coturnix chinensis*; Earls, 2000). Peak jumping forces can be as low as 1–3 times body mass during voluntary take-off (Heppner & Anderson, 1985; Bonser & Rayner, 1996) and reach about 4–5 times body weight in escape flight in the passerines (Passeriformes; Earls, 2000; Jackson, unpublished data) and 7.8 times body weight in blue-breasted quail (Earls, 2000).

The legs are only in contact with the ground for a fraction of a second, so the wings must eventually take over, and short wings are better suited for rapid take-off. Power output is a function of work per wingbeat (Figure 10.2b) divided by the duration of the wingbeat. Thus, everything else being equal, the higher the wingbeat frequency, the more induced power a bird can produce, and the quicker it can accelerate vertically. Within a group of similarly shaped birds (Tobalske & Dial, 2000), wingbeat frequency during take-off decreases with increasing body mass (m), approximately proportional to the cube-root of mass ($m^{-1.3}$). Comparing species of different wing shapes but similar mass, however, it appears that wingbeat frequency is inversely related to wing length (Pennycuick, 1996). Consider, for example, a species such as an albatross (Diomedeidae) with long and pointed (high aspect ratio) wings. Although the wing shape is thought to be extremely efficient for gliding, the birds have a difficult time getting off the ground, and usually have to run into prevailing winds before taking off (Pennycuick, 1975). Comparatively, a gallinaceous bird of similar mass such as the wild turkey (*Meleagris gallpavo*), which has short and rounded wings, is ideally suited for high-acceleration take-off (Tobalske & Dial, 2000; Askew *et al.*, 2001).

During take-off and vertical flight, birds must use muscle power to do work to raise their center of mass against gravity and to accelerate. The amount of mechanical power produced by the muscles in relation to body-mass (i.e. mass-specific power) therefore largely determines the actual flight performance. Extant flying birds range in mass from a 2 g bee hummingbird (*Mellisuga helenae*) to a 14 kg mute swan (*Cygnus olor*;

Dunning, 1993). While it is readily observed that a swan is not capable of hovering at a flower or even taking-off vertically like a hummingbird, the underlying mechanism, the mass-specific power available for flight relative to the amount required, is not fully understood. Bird species in general scale isometrically (Greenewalt, 1962), meaning that muscle masses are the same proportion of body mass, and wing-lengths are the same proportion of body length. Scaling theory would, therefore, predict that available mass-specific power should scale proportional to wingbeat frequency ($m^{-1/3}$; Hill, 1950; Pennycuick, 1975; Ellington, 1991). According to this line of reasoning, since large species tend to have lower wingbeat frequencies, their muscles produce less mass-specific power, which translates into lower take-off performance compared to smaller species. This could account for the observed trend of decreasing take-off performance with increasing size in birds if the mass-specific power required for flight is independent of body mass (Figure 10.7; Tobalske & Dial, 2000; Dial *et al.*, 2008a). On the other hand, aerodynamic modeling suggests that the mass-specific power output during take-off actually increases with body mass (Askew *et al.*, 2001). Consistent with the notion that mass-specific power is not limiting flight performance in larger birds, proportional load-lifting ability increases with increasing body mass (Marden, 1994) and larger hummingbirds exhibit greater ability to climb with added load or support their weight in reduced-density air compared with smaller hummingbirds (Chai & Millard, 1997; Altshuler *et al.*, 2004). For hummingbirds, nonisometric scaling of muscle morphology or physiology may compensate for the impact of body mass (Chai & Millard, 1997). While variation in relative muscle mass, muscle morphology, and fiber physiology, and wing shape and size all could explain some of the variation in take-off performance, body-mass is likely a fundamental determinant of burst flight performance during take-off.

Some bird species experience significant fluctuations in body mass due to migratory fat loading or egg production. Since their body mass increases but muscle mass and wing size typically do not,

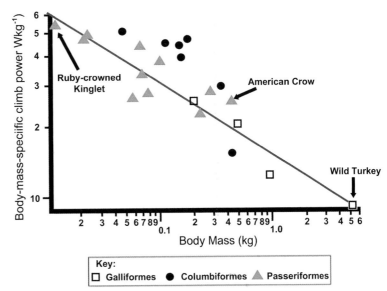

Fig. 10.7 Vertical escape-flight performance in three orders of birds spanning three orders of magnitude of body mass: Galliformes (squares), Columbiformes (circles), and Passeriformes (triangles). Whole-body (external) mass-specific power output is proportional to mass raised to the -0.3 power ($m^{-0.3}$). (From Dial et al., 2008a.)

individuals of these species may experience reduced take-off performance (Hedenström & Alerstam, 1992; Witter et al., 1994; Kullberg et al., 1998). The trade-off between fat-loading to decrease risk of starvation vs. foraging minimally to maintain take-off and predator-escape performance is a rich area of study. However, not all species demonstrate reduced take-off performance with fat-loading, some reduce acceleration or velocity, while others reduce only the angle of ascent. The factors that drive the variation in strategy, and the ecological and evolutionary implications of this trade-off, are mostly unknown.

INTERMITTENT FLIGHT

The vast majority of small and medium sized birds use one of two forms of intermittent flight during which they regularly interrupt flapping phases to hold their wings either in a flexed-wing "bound" posture, during which the wings are held tightly against the body, or in an extended-wing "glide" (Rayner, 1985; Tobalske, 2001; Figure 10.8). Some species, such as the budgerigar (*Melopsittacus undulatus*) use bounds, glides, and partial-bounds

during which the wings are partially extended (Tobalske & Dial, 1994). These flight styles are characterized by undulating flight paths as the birds gain altitude using flapping and lose altitude during the fixed-wing pauses (Figure 10.8). The flight style of the black-billed magpie (*Pica*

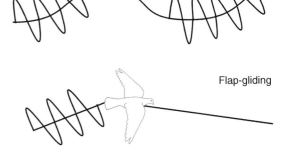

Fig. 10.8 Intermittent flight features regular, brief pauses in between flapping phases. A bound occurs if the bird flexes its wings against its body and a glide occurs when the birds holds its wings extended. Some species exhibit intermediate wing postures (partial bounds or glides).

hudsonia) is a novel form of intermittent flight, which consists of regular variation in wingbeat frequency and amplitude during flapping phases as well as intermittent bounds and glides (Tobalske *et al.*, 1997). Flap-bounding is readily observed during foraging and migratory flights in many small passerines (Passeriformes; Danielson, 1988) and woodpeckers (*Picidae*; Tobalske, 1996). Flap-gliding flight is exhibited during flight in a diverse array of birds including swallows (Hirundinidae; Bruderer *et al.* 2001), swifts (Apodidae), accipiters (Accipitridae), wood pigeons (*Columba palumbus*), and northern harriers (*Circus cyaneus*).

Intermittent flight appears to be a strategy for saving energy by reducing the average power required for flight in comparison with that required for continuously flapping. Mathematical models developed from aerodynamic theory indicate that flap-bounding can be an attractive strategy when flying relatively fast (Rayner, 1985; Ward-Smith, 1984a), while flap-gliding may offer greater advantages at slower speeds (Ward-Smith, 1984b; Rayner, 1985). The production of lift by the body and tail may help extend the range of aerodynamically attractive speeds for flap-bounding to include maximum range speed, the speed predicted to be optimal for sustained cruising flight (Rayner, 1985; Tobalske *et al.*, 1999). Measurements of body acceleration and wake dynamics in live birds as well as force measurements on prepared specimens all indicate that birds can support 10–15% of their body weight even with their wings fully flexed in a bound posture (Csicsáky, 1977; Tobalske *et al.*, 1999, 2009). The contribution of "turn-out" phases during which the wings are extended after a bound may allow flap-bounding to offer an advantage over a broad range of speeds (DeJong, 1983). Likewise, variation in flight speed and thrust can result in predicted energetic advantages for both flap-bounding and flap-gliding over a wide range of speeds (Rayner *et al.*, 2001). Kinematics reveal that variation in flight speed is typical of intermittent flight (Tobalske, 1995; Tobalske *et al.*, 1999), and correlations between body motion and muscle activity suggest that thrust likely varies as well (Tobalske & Dial,

1994; Tobalske, 1995; Tobalske *et al.*, 2005; Askew & Ellerby, 2007).

Activity in the major flight muscles decreases during intermittent pauses compared with during flapping phases (Meyers, 1993; Tobalske & Dial, 1994; Tobalske, 1995, 2001; Tobalske *et al.* 2005; Askew & Ellerby, 2007; Figure 10.9). During intermittent glides, the pectoralis exhibits an isometric contraction and the supracoracoideus is inactive, whereas during bounds, both muscles are inactive (Tobalske, 2001). Sonomicrometry reveals that the pectoralis does not change length during intermittent pauses (Tobalske *et al.*, 2005; Askew & Ellerby, 2007).

There are prominent effects of body size and wing shape upon the performance of intermittent flight. Small birds with rounded, low-aspect ratio wings such as the zebra finch (*Taeniopygia guttata*; 13 g; aspect ratio, $AR = 4.2$) appear to only use intermittent bounds (Tobalske *et al.*, 1999, 2005). In contrast, species of about the same body mass but with more pointed, high-aspect-ratio wings such as the barn swallow (*Hirundo rustica*; 20 g; $AR = 6.2$) and house martin (*Delichon urbica*; 17 g, $AR = 6.5$), use both bounds (or partial bounds) and glides (Bruderer *et al.*, 2001). Regardless of aspect ratio, species of intermediate mass between 34 g budgerigars ($AR = 7.1$) and 150 g black-billed magpie ($AR = 4.1$) use both forms of intermittent flight. Above 300 g, birds do not appear to be able to use intermittent bounds so, for example, the rock dove, *Columba livia* only uses gliding during pauses in wing flapping (Tobalske & Dial, 1996).

What limits the upper-size range for the ability to bound? The largest species observed to regularly bound is the pileated woodpecker (*Dryocopus pileatus*; 270 g; Tobalske, 1996, 2001). As described above for whole-body power output during take-off and vertical flight performance, there is an observed decline in the performance of bounds as body size goes up. The percentage of time spent flapping increases with increasing body mass among passerines engaged in migratory flight (Danielson, 1988) and woodpeckers engaged in foraging flight (Tobalske, 1996, 2001; Figure 10.10). The scaling is proportional to mass raised

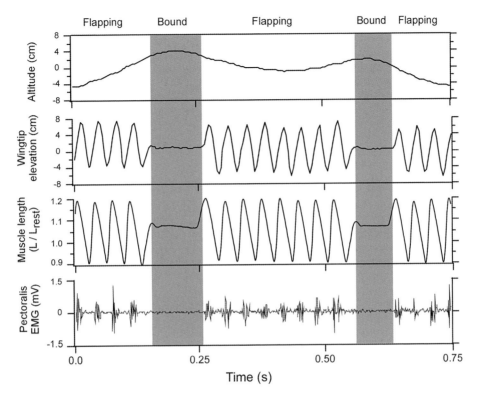

Fig. 10.9 Patterns of wing and body motion and muscle contractile behavior during flap-bounding flight in a zebra finch (*Taniopygia guttata*). The bird gains altitude during the latter half of flapping phases and loses altitude during the latter half of bounds. The pectoralis is inactive during bounds: there is no neuromuscular activity as measured using electromyography, and there is no change in muscle length as measured using sonomicrometry. (From Tobalske *et al.*, 2005.)

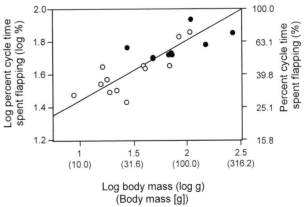

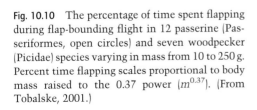

Fig. 10.10 The percentage of time spent flapping during flap-bounding flight in 12 passerine (Passeriformes, open circles) and seven woodpecker (Picidae) species varying in mass from 10 to 250 g. Percent time flapping scales proportional to body mass raised to the 0.37 power ($m^{0.37}$). (From Tobalske, 2001.)

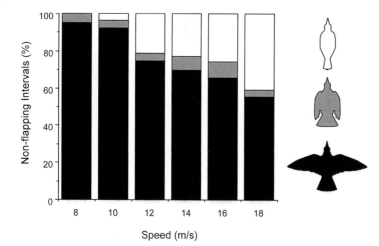

Fig. 10.11 Flight speed affects the type of wing posture assumed during intermittent pauses in some species. These data, from European starlings (*Sturnus vulgaris*), show the percentage of bounds (white), partial bounds (gray), and glides (black) among all nonflapping phases exhibited at a given speed. As speed increased, the percentage of glides decreased while the percentage of bounds increased. (From Tobalske, 1995.)

to the 0.37 power $(m^{0.37})$. A potential explanation for this is that the sustainable mass-specific power available from the flight muscles is proportional to wingbeat frequency, and, therefore, decreases as a function of increasing body mass (Hill, 1950; Pennycuick, 1975). Consistent with such a hypothesis, wingbeat frequency scales proportional to $m^{-0.37}$ in flap-bounding birds (Tobalske, 1995) Alternatively, the lift per unit power output may decrease with increasing body mass (Marden, 1994). The aerodynamic mechanisms responsible for this decrease in relative lift production need to be measured empirically.

Flight speed has significant effects upon intermittent flight behavior as it appears to influence the percentage of time spent flapping as well as the nonflapping postures adopted during intermittent pauses. In zebra finch, a species that only flap-bounds, there is a decrease in time spent flapping from 89% during brief hovering episodes to 55% during fast forward flight $(14\,\mathrm{m\,s^{-1}})$. In the budgerigar and European starling, species that use both intermittent bounds and glides, the percentage of time spent flapping varies according to an upwardly concave, "U-shaped" curve (Tobalske, 2001). Similarly, mean effective wingbeat frequency varies as a U-shaped curve in barn swallows and house martins (Bruderer *et al.*, 2001). Among the species that use both bounds and

glides, there is a tendency to flap-glide at slow speeds and flap-bound during faster flight (Tobalske & Dial, 1994; Tobalske, 1995; Bruderer *et al.* 2001; Figure 10.11). However, recent research did not reveal the same trend to switch from the use of bounds to the use of glides as flight speed increased in rose-colored starlings (*Sturnus roseus*; Engel *et al.*, 2006).

MANEUVERING

The high velocities – and hence, kinetic energy – characteristic of flight must place a selective premium on control. Clearly, the broad utility of avian flight would not have been realized without development of effective stability and maneuvering (Thomas & Taylor, 2001; Taylor & Thomas, 2002; Warrick *et al.*, 2002).

During gliding (e.g. Pennycuick, 1971), a maneuvering bird can be described by well-understood aircraft dynamics: turns are effected by creating a bilateral force asymmetry, imparting a rolling moment about the long axis of the body to establish a bank angle, thus redirecting the lift force to provide a centripetal force. Maneuverability in this case has been defined by radius of turn (Norberg & Rayner, 1987); with a fixed-wing assumption, the radius of turn will be determined by wing loading (mass × wing area^{-1}; Pennycuick, 1971). Even given the assumption of fixed-wings, maneuverability has considerable explanatory power; variation in wingloading and maneuverability have been used to cogently describe differences in habitat use in bats (Aldridge, 1987; Norberg & Rayner, 1987), and foraging behavior and prey selection in swallows (Warrick, 1998). Further, they have the desirable feature of being among the few flight performance parameters that can be inferred when the fossil record provides reliable estimates of body mass and wing area (e.g. Pennycuick, 1988).

To create force asymmetries to produce roll, a bird can manipulate one or more lift variables: wing surface area, angle of attack, or wing speed. In fast gliding flight, a bird can merely increase the angle of attack by supinating a wing, while simultaneously pronating to decrease angle of attack on the other. The wings can be used to create moments around the other two body axes as well. By moving the wings' center of lift forward or aft of the bird's center of mass, birds create pitching rotation to change whole-body angle (Thomas & Taylor, 2001). Likewise, any asymmetry in area or angle of attack will produce not only differential lift but also differential drag, causing yawing rotations.

Agility, the ability to create angular velocities in rolling, pitching, or yawing movements, has been distinguished from maneuverability (Norberg & Rayner, 1987) as a meaningful performance criterion of its own for some ecotypes (e.g. coursing insectivorous birds such as swallows (Hirundinidae; Warrick, 1998). However, as function of the strength of forces available relative to the inertia of a body around its three rotational axes, most small birds are intrinsically agile. Viewed another way, relative to their terrestrial ancestors, birds are intrinsically unstable – perhaps a result of selection for a compact and therefore robust body able to withstand the rigors of high frequency, periodic support (Taylor & Thomas, 2002).

While the tail during low-speed flight seems to be restricted to acting as a lifting device (Gatesy & Dial, 1996; Thomas, 1996a,b; Berg & Biewener, 2008), at high speed the avian tail can function in pitch and yaw control, both to augment maneuvering performance and stabilize level flight (Thomas & Taylor, 2001). Unilaterally depressing the tail creates a laterally directed force and yawing moment (Hummel, 1992; Thomas, 1993) away from the depressed side of the tail. Functioning much like the rudder of an airplane, this force can thus be used for countering the so-called adverse yaw that is created during wing asymmetries, when the higher lift wing must also create more drag, yawing the animal in a direction opposite to its intended direction of flight (Warrick, 1998). Empirical (Hummel, 1992) and theoretical (Thomas, 1993) studies show that the forces created by the tail are small relative to those created by the wings. This may make the tail even more useful as a stabilizing device; its ineffectiveness allows coarse motor control to produce fine-scale aerodynamic force.

For a complete understanding of maneuverability in birds, a fixed-wing assumption is inadequate. But abandoning it introduces a staggering level of complexity; not surprisingly, no single functional pathway for the control of maneuvering during flapping flight has yet been identified. However, studies of the dynamics of maneuvering flight illustrate both the central role of the pectoral architecture and the importance of the intrinsic wing muscles in controlling slow, flapping flight.

Assuming an aerodynamically inactive – or simply less active – upstroke (Rayner, 1979a; Tobalske, 2000), maneuvering in slow flight will be to some degree a saltatory affair. That is, when aerodynamic force production ceases, the centripetal force ceases, and the bird will move in a straight line until the next downstroke. Nevertheless, the smallest radius turn – a radius of zero – is available only to a flapping bird: in a hover, no centripetal force is required, and the bird simply rotates around its center mass and heads off in a new direction. More generally, a bird able to maintain high incident air velocity over its wing through flapping, while the velocity of the body is low, will produce turns of small radius.

As in gliding flight, birds may modulate aerodynamic force by varying surface area and angle of attack, but, during flapping, they may also vary

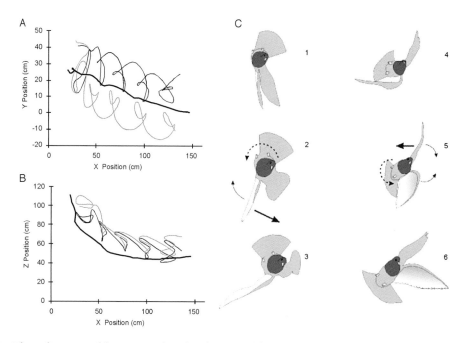

Fig. 10.12 Three dimensional kinematics (A, B) and wing and body orientation (C) of a rock dove (Columba livia) maneuvering after being held inverted and then dropped. Wingtip (right = grey; black = left) and body (bold black) kinematics of a pigeon, held inverted and dropped. With one asymmetrical wingbeat, the pigeon rights itself; with two further wingbeats (200 ms) it has arrested its descent, and flies to a perch. (D.R.Warrick, unpublished data.)

downstroke velocity. The kinematics of pigeons (Warrick & Dial 1998; Figure 10.12) and parrots (Psittaciformes; Hedrick & Biewener, 2007a; Hedrick *et al.*, 2007) show that birds use asymmetries in downstroke velocity – and to some degree in pigeons, upstroke – to create roll and yaw during slow flight. Both were shown to produce these asymmetries in the first half of downstroke, and reverse the asymmetry in the second, thus halting the rolling momentum before the upstroke. Warrick & Dial (1998) assumed that the velocity asymmetries observed were used to create aerodynamic force asymmetries, but Hedrick & Biewener (2007a,b) and Hedrick *et al.* (2007) showed that birds may also take advantage of the body rotations resulting from asymmetric wing movement. This inertial reorientation was shown to be particularly important in changing body angle within a wingbeat, and allows for an immediate, transient, and easily reversed bank angle. As these

studies illustrate, the ability to produce transient bank within a wingbeat, with no net change in bank, gives the maneuvering bird an opportunity to move stepwise through its environment.

Patterns of wing motion (Warrick & Dial 1998; Hedrick & Biewener 2007a,b; Hedrick *et al.*, 2007; Figure 10.12) and measurements of the force experienced by the wings (Warrick *et al.*, 1998; Figure 10.13) show that birds frequently produce a series of asymmetries, with higher force on the outside wing, rather than simply creating a bank, holding that bank and flying symmetrically through the turn. While there has been no rigorous examination of the advantages of this maneuvering strategy, the higher success of pigeons exhibited this pattern in negotiating an obstacle course (Warrick *et al.*, 1998), and the proficiency of both these phylogenetically distant species in creating these incremental maneuvers, suggests that slow maneuvering flight is a tightly controlled

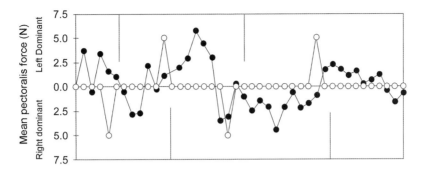

Fig. 10.13 Peak force asymmetries (black points) at mid-downstroke measured at the deltopectoral crest of a pigeon, superimposed on the obstacle course flown it was flying. The open points would be the expected asymmetry pattern if the birds simply established a bank with one force asymmetry and then flew around the barriers. The black vertical lines indicate the position of the barriers around which the birds maneuvered. (From Warrick *et al.*, 1998)

behavior. Whether reorienting through asymmetry in aerodynamic force production or inertia, these studies illustrate the intrinsic instability of birds (roll accelerations greater than 20,000° s^{-2}; Warrick & Dial, 1998; Hedrick & Biewener, 2007a,b; Hedrick *et al.*, 2007). However, a theoretical examination of stability in flapping flight suggests that symmetrical flapping itself does little to destabilize the bird (Taylor & Thomas, 2002).

While we currently lack a complete description of the muscular control of these maneuvering events, electromyogram (EMG) studies of rose-breasted cockatoos (*Eolophus roseicapillus*; Hedrick & Biewener, 2007b) and denervation studies of pigeons (Dial, 1992a) suggest that distal muscles may function to modulate the activity of the pectoralis through pronation, supination, or flexion. While Hedrick & Biewener (2007a) found no muscle activity asymmetries in these intrinsic wing muscles consistently associated with particular maneuvering kinematics, these muscles did display more asymmetry during maneuvering than during level flight, suggesting a complex synergism. In contrast, the asymmetry in recruitment of the pectoralis was consistently correlated with maneuvering kinematics. Thus it appears that, as a primary provider of both aerodynamic power and flapping wing inertia, the timing and force production of the pectoralis is critical, and

may be the "key innovation" (Liem, 1973; also see Raikow, 1986) in the evolution of control of low-speed maneuvering flight.

HOVERING

As we have reinforced in this chapter, flight is an energetic affair, and control is the purpose of maneuvering; thus, no discussion of flight can be complete without exploring how birds control their kinetic energy by flying slowly. In this sense, the ultimate flight maneuver is one that requires no maneuvering at all: the hover.

True hovering – the ability to fly with incident airspeed of zero over the body of the bird is an option probably available to all small and medium-sized birds (Pennycuick, 1975; Ellington, 1991). Even if for only one or two seconds, the flexibility it provides a bird in safely moving through its environment – particularly during landing – may be profoundly important (e.g. Green & Cheng, 1998). Sustained hovering, using aerobic metabolism for indefinite time intervals (Lasiewski, 1963), is a different matter, seemingly confined to hummingbirds (Trochilidate). Hovering in still air is a particularly demanding flight style in terms of power requirements because the bird is solely responsible for inducing a large downward velocity into the air to support its weight. These induced high velocities require high power output from the flight muscles. In contrast, in forward flight, or during hovering with a headwind in birds such as kingfishers (Coraciiformes), incoming air (wind) contributes to the production of lift and the induced velocities required to support body weight are, therefore, less (Pennycuick, 1975; Rayner, 1979a,b, 1985).

The hovering ability of hummingbirds is related to their small body size (Altshuler & Dudley, 2002): most species have body masses between 2 and 8 g, and the giant hummingbird (*Patagona gigas*), unusually large for the family, is only 20 g (Dunning, 1993). Hummingbirds also exhibit a range of morphological and physiological specializations that are well suited for sustaining high power output during hovering (Altshuler & Dudley, 2002). For example, they have pectoralis and supracoracoideus muscles with relatively small-diameter fibers, high mitochondrial density, and high capillary density (Suarez *et al.*, 1991; Mathieu-Costello *et al.*, 1992). These attributes allow their muscles to sustain the highest mass-specific metabolic rates known for vertebrate skeletal muscle (Suarez *et al.*, 1991). Their primary flight muscles make up a relatively large proportion of their body mass (ca. 25%; Greenewalt, 1962; Wells, 1993). Their wing dimensions also exhibit positive allometry, meaning that wing length and area increase at a greater rate with increasing body mass than one would expect based on an assumption of geometric similarity and observed trends in other clades of birds (Greenewalt, 1962).

The first descriptions of the wing motions of hummingbirds illustrated a wingbeat dramatically different from all other birds (Stolpe & Zimmer, 1939; Greenewalt, 1962; Figure 10.14), which quickly set them apart, likened them to insects (Weis-Fogh, 1972; Wells, 1993), and eventually led to two aerodynamic classifications of avian hovering: symmetrical, and asymmetrical (Norberg, 1990). Aerodynamic symmetry of the two half strokes was thought to be a prerequisite for sustained, aerobic hovering, and the general similarities between hummingbird and hovering insect

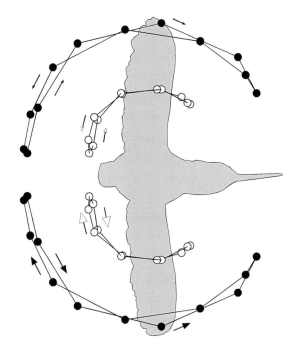

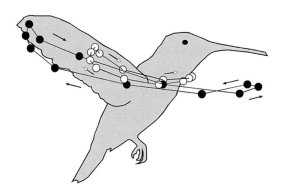

Fig. 10.14 Wing motion during hovering in a hovering rufous hummingbird (*Selasphorus rufus*). Black circles indicate position of wingtips, and white circles indicate position of wrists. Circles and arrows indicate sequential position and local direction of movement. (From Tobalske *et al.*, 2007.)

kinematics suggested a remarkable convergence in form and function in these long divergent (500+ Myr) taxa (Weis-Fogh, 1972). Attractive though this suggestion was, direct measurements

Fig. 10.15 Hummingbird wing presentation and flow field in the wake at mid-downstroke (a) and mid-upstroke (b). (a) A red line is drawn above the dorsal surface of the wing to highlight the camber of the wing. (b) During upstroke, the proximal part of the wing (red line) is not as supinated as the distal portion (yellow line). The vector scale is at top right. (From Warrick *et al.*, 2005.) [This figure appears in color as Plate 10.15.]

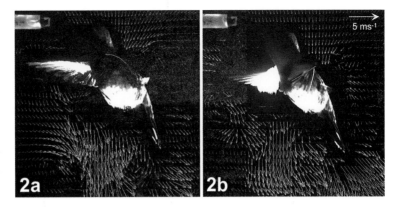

of airflow in the wake of hummingbirds show that the majority of weight support (75%) is provided by downstroke (Warrick *et al.*, 2005), and subtle asymmetries between downstroke and upstroke in wing velocity, area, camber, and long-axis twist (Figure 10.15) result in a two- to three-fold disparity in the lift production. Previously, it was unclear why the supracoracoideus to pectoralis mass ratio is approximately 0.5 (Wells, 1993), but relatively lower force production during upstroke helps account for this.

Although it does not produce an equal amount of force as downstroke, it is, nevertheless, upstroke that appears to be unique in hummingbirds. During upstroke, they leave their wings extended and markedly supinated. Their wingtips trace a path through the air that resembles a "figure-8" in lateral view (Figure 10.14). A dorsal view reveals that the tips and wrists trace approximately the same path during both halves of the wingbeat. In contrast, other species flex their wings to some extent during upstrokes of slow flight and hovering. Birds with rounded wings tend to adduct their entire wing during upstroke, while birds with pointed wings tend to adduct only their wrists and supinate their hand wing (Figure 10.16). There are exceptions to this general pattern. For example, Galliform birds, with rounded wings, supinate their handwing during upstroke of take-off flight (Tobalske & Dial, 2000). There is some argument that a supinated, extended handwing can produce useful lift, drag, or inertial forces,

potentially representing a precursor to the hummingbird-style wingbeat, but such functions have not yet been clearly revealed (Tobalske, 2000, 2007; Tobalske & Biewener, 2008). The ability to supinate the handwing has been attributed to wrist anatomy in the mallard duck (*Anas platyrhynchos*; Vasquez, 1992); intriguingly, Vasquez (1992) observed that the relevant hummingbird wrist anatomy was different from that of the duck.

The ability to hover permits hummingbirds to exploit nectar, a concentrated source of glucose, as a food source, and recent study demonstrates that glucose oxidation in hummingbirds requires less oxygen compared with fatty-acid oxidation (Welch *et al.*, 2007). Other small nectivorous species routinely hover for brief intervals (≤ 15 s). These include two passerine groups: sunbirds (Nectariniidae; Hambly *et al.*, 2004; Köhler *et al.*, 2006) and honeyeaters Meliphagidae (Collins & Clow, 1978). Unfortunately, quantitative descriptions are lacking for wing kinematics and other details of flight styles in sunbirds and honeyeaters; such data would likely improve understanding about the relative specialization of hummingbirds and the processes that led to the independent evolution of hovering ability in what are hypothesized to be relatively distantly related clades (Sibley & Ahlquist, 1990; Livezey & Zusi, 2007).

Metabolic data for hovering sunbirds and honeyeaters reveal that hovering is more costly in terms of energy than slow or fast forward flight

Fig. 10.16 Wing-tip reversal (supination) during upstroke of slow flight in a rock dove (*Columba livia*).

(Hambly *et al.*, 2004). In contrast, data from hummingbirds suggests that there is no significant increase in metabolic power between hovering and forward flight up to speeds of $7\,\mathrm{m\,s^{-1}}$ (Berger, 1985; Ellington, 1991). This suggests that hummingbirds are uniquely efficient at hovering such that costs vary according to a "J-shaped" curve with flight speed rather than a "U-shaped" curve that may be observed when other species are flown over a wide range of speeds (Bundle *et al.*, 2007). One proposed mechanism that could account for higher efficiency in hummingbirds is elastic energy storage in the flight muscles during deceleration of the wing at the end of each half stroke (Wells, 1993)

Given their unique wingbeat patterns (Figure 10.15) and high mass-specific metabolism, what ultimately limits hovering performance in hummingbirds? This question has been explored in laboratory experiments in which air density and the partial pressure of oxygen are varied within a sealed chamber (Chai & Dudley, 1995, 1996; Altshuler *et al.* 2001; Altshuler & Dudley 2003), and also with measurements of hovering performance in the field along elevational gradients in mountains (Altshuler *et al.*, 2001, 2004; Atshuler

& Dudley, 2003). These studies indicate that variation in air density is a more significant constraint than oxygen availability even though low partial pressures of oxygen can make it impossible for hummingbirds to sustain hovering (Altshuler *et al.*, 2001). As air density decreases, hummingbirds compensate by increasing wingbeat amplitude but not wingbeat frequency, and when wingbeat amplitude reaches 180° they can no longer hover (Chai & Dudley, 1995). Populations living at higher altitudes compensate for low density by having relatively longer wings (Altshuler *et al.*, 2004). Increasing oxygen availability does not improve performance at low air densities (Altshuler *et al.*, 2001).

ACKNOWLEDGMENTS

We thank the many collaborators who have assisted us with our empirical studies of bird flight. Much of our research has been supported by the National Science Foundation, most recently via IOS-0923606 to BWT and DRW and IOS-0919799 to KPD and BWT.

REFERENCES

Aldridge HDJN. 1987. Turning flight of bats. *Journal of Experimental Biology* **128**: 419–425.

Altshuler DL, Dudley R. 2002. The ecological and evolutionary interface of hummingbird flight physiology. *Journal of Experimental Biology* **205**: 2325– 2336.

Altshuler DL, Dudley R. 2003. Kinematics of hovering hummingbird flight along simulated and natural elevational gradients. *Journal of Experimental Biology* **206**: 3139–3147.

Altshuler DL, Chai P, Chen JSP. 2001. Hovering performance of hummingbirds in hyperoxic gas mixtures. *Journal of Experimental Biology* **204**: 2021–2027.

Altshuler DL, Dudley R, McGuire JA. 2004. Resolution of a paradox: hummingbird flight at high elevation does not come without a cost. *Proceedings of the National Academy of Science USA* **101**: 17731–17736.

Askew GN, Ellerby DJ. 2007. The mechanical power requirements of avian flight. *Biology Letters* **3**: 445–448.

Askew GN, Marsh RL, Ellington CP. 2001. The mechanical power output of the flight muscles of blue-breasted quail (*Coturnix chinensis*) during take-off. *Journal of Experimental Biology* **204**: 3601–3619.

Baier DB, Gatesy SM, Jenkins, FA. Jr 2007. A critical ligamentous mechanism in the evolution of avian flight. *Nature* **445**: 307–310.

Baumel JJ, King AS, Breazile JE, Evans HE (ed.). 1993. *Handbook of Avian Anatomy: Nomina Anatomica Avium*. Cambridge, MA: Nuttall Ornithological Club Publication 23.

Berg AM, Biewener AA. 2008. Kinematics and power requirements of ascending and descending flight in the pigeon (*Columba livia*). *Journal of Experimental Biology* **211**: 1120–1130.

Berger M. 1985. Sauerstoffverbrauch von Kolibris (*Colibri coruscans* und *C. thalassinus*) beim Horizontalflug. In *BIONA* Nachtigall W (ed.). Stuttgart: G. Fischer; Report 3 307–314.

Biewener AA, Corning WR, Tobalske BW. 1998. *In vivo* pectoralis muscle force-length behavior during level flight in pigeons (*Columba livia*). *Journal of Experimental Biology* **201**: 3293–3307.

Bishop KL. 2006. The relationship between 3D kinematics and gliding performance in the southern flying squirrel, *Glaucomys volans*. *Journal of Experimental Biology* **209**: 689–701.

Bock WJ. 1965. The role of adaptive mechanisms in the origin of the higher levels of organization. *Systematic Zoology* **14**: 272–287.

Bonser RHC, Rayner JM. V. 1996. Measuring leg thrust forces in the common starling. *Journal of Experimental Biology* **199**: 435–439.

Bruderer L, Liechti F, Bilo D. 2001. Flexibility in flight behaviour of barn swallows (*Hirundo rustica*) and house martins (*Delichon urbica*) tested in a wind tunnel. *Journal of Experimental Biology* **204**: 1473–1484.

Bundle MW, Dial KP. 2003. Mechanics of wing-assisted incline running (WAIR). *Journal of Experimental Biology* **206**: 4553–4564.

Bundle MW, Hansen KS, Dial KP. 2007. Does the metabolic rate-flight speed relationship vary among geometrically similar birds of different mass? *Journal of Experimental Biology* **210**: 1075–1083.

Chai P, Dudley R. 1995. Limits to vertebrate locomotor energetics suggested by hummingbirds hovering in heliox. *Nature* **377**: 722–725.

Chai P, Dudley R. 1996. Limits to flight energetics of hummingbirds hovering in hypodense and hypoxic gas mixtures. *Journal of Experimental Biology* **199**: 2285–2295.

Chai P, Millard D. 1997. Flight and size constraints: hovering performance of large hummingbirds under maximal loading. *Journal of Experimental Biology* **200**: 2757–2763.

Collins BG, Clow H. 1978. Feeding behaviour and energetics of the western spinebill, *Acanthorhynchus superciliosis* (Aves: Meliphagidae). *Australian Journal of Zoology* **26**: 269–277.

Csicsáky MJ. 1977. Body-gliding in the zebra finch. *Fortschritte der Zoologie. Neue Folge* **24**: 275–286.

Danielson R. 1988. Parametre for fritflyvende småfugles flugt. *Dansk Ornithologisk Forenings Tidsskrift* **82**: 59–60.

Davenport J. 2004. How and why do flying fish fly? *Reviews in Fish Biology and Fisheries* **4**: 184–214.

DeJong MJ. 1983. *Bounding flight in birds*. PhD thesis, University of Wisconsin, Madison.

Dial KP. 1992a. Activity patterns of the wing muscles of the pigeon (Columba livia) during different modes of flight. *Journal of Experimental Zoology* **262**: 357–373.

Dial KP. 1992b. Avian forelimb muscles and nonsteady flight: can birds fly without using the muscles of their wings? *The Auk* **109**: 874–885.

Dial KP. 2003. Wing-assisted incline running and the evolution of flight. *Science* **299**: 402–404.

Dial KP, Biewener AA, Tobalske BW, Warrick DR. 1997. Mechanical power output of bird flight. *Nature* **390**: 67–70.

Dial KP, Randall RJ, Dial TR. 2006. What use is half a wing in the ecology and evolution of birds? *BioScience* **56**: 437–455.

Dial KP, Greene E, Irschick DJ. 2008a. Allometry of behavior. *Trends in Ecology and Evolution* **23**: 394–401.

Dial KP, Jackson BE, Segre P. 2008b. A fundamental avian wing-stroke provides a new perspective on the evolution of flight. *Nature* **451**: 985–989.

Dudley R, Byrnes G, Yanoviak SP, Borrell B, Brown RM, McGuire JA. 2007. Gliding and the functional origins of flight: Biomechanical novelty or necessity? *Annual Review of Ecology, Evolution, and Systematics* **38**: 179–201.

Dunning JB. Jr (ed.). 1993. *CRC Handbook of Avian Body Masses.* Boca Raton, FL: CRC Press.

Earls KD. 2000. Kinematics and mechanics of ground take-off in the starling *Sturnis vulgaris* and the quail *Coturnix coturnix. Journal of Experimental Biology* **203**: 725–739.

Ellington CP. 1991. Limitations on animal flight performance. *Journal of Experimental Biology* **160**: 71–91.

Engel S, Biebach H, Visser GH. 2006. Metabolic costs of avian flight in relation to flight velocity: a study in rose coloured starlings (*Sturnus roseus*, Linnaeus). *Journal Of Comparative Physiology B: Biochemical, Systemic, And Environmental Physiology* **176**: 415–427.

Garner J, Taylor G, Thomas A. 1999. On the origins of birds: the sequence of character acquisition in the evolution of avian flight. *Proceedings Biological Sciences* **266**: 1259–1266.

Gatesy SM, Dial KP. 1996. Locomotor modules and the evolution of avian flight. *Evolution* **50**: 331–340.

Gould SJ, Vrba ES. 1982. Exaptation; a missing term in the *Science* of form. *Paleobiology* **8**: 4–15.

Green PR, Cheng P. 1998. Variation in kinematics and dynamics of the landing flights of pigeons on a novel perch. *Journal of Experimental Biology* **201**: 3309–3316.

Greenewalt CH. 1962. Dimensional relationships for flying animals. *Smithsonian Miscellaneous Collection* **144**: 1–46.

Hambly C, Pinshow B, Wiersma P, Verhulst S, Piertney SB, Harper EJ, Speakman JR. 2004. Comparison of the cost of short flights in a nectarivorous and a non-nectarivorous bird. *Journal of Experimental Biology* **207**: 3959–3968.

Harrison JF, Roberts SP. 2000. Flight respiration and energetics. *Annual Review of Physiology* **62**: 179–205.

Hedenström A, Alerstam T. 1992. Climbing performance of migratory birds as a basis for estimating limits for fuel carrying capacity and muscle work. *Journal of Experimental Biology* **164**: 19–38.

Hedrick TL, Biewener AA. 2007a. Experimental study of low speed turning flight in cockatoos and cockatiels. *45th AIAA Aerospace Sciences Meeting and Exhibit*, 8–11 January, Reno, Nevada; AIAA Paper 2007-0044.

Hedrick TL, Biewener AA. 2007b. Low speed maneuvering flight of the rose- breasted cockatoo (*Eolophus roseicapillus*). I. Kinematic and neuromuscular control of turning. *Journal of Experimental Biology* **210**: 1897–1911.

Hedrick TL, Tobalske BW, Biewener AA. 2003. How cockatiels (*Nymphicus hollandicus*) modulate pectoralis power output across flight speeds. *Journal of Experimental Biology* **206**: 1363–1378.

Hedrick TL, Usherwood JR, Biewener AA. 2007. Low speed maneuvering flight of the rose-breasted cockatoo (*Eolophus roseicapillus*). II. Inertial and aerodynamic reorientation. *Journal of Experimental Biology* **210**: 1912–1924.

Heppner FH, Anderson JGT. 1985. Leg thrust in flight take-off in the pigeon. *Journal of Experimental Biology* **114**: 285–288.

Hill AV. 1950. The dimensions of animals and their muscular dynamics. *Science in Progress* **38**: 209–230.

Hui CA. 2002. Avian furcula morphology may indicate relationships of flight requirements among birds. *Journal of Morphology* **251**: 284–293.

Hummel D. 1992. Aerodynamic investigations on tail effects in birds. *Zeitschrift für Flugwissenschaften und Weltraumforschung* **16**: 159–168.

Jenkins FA. Jr, Dial KP, Goslow GE. Jr 1988. A cineradiographic analysis of bird flight: the wishbone is a spring. *Science* **241**: 1495–1498.

Kohler A, Verburgt L, Nicolson SW. 2006. Short-term energy regulation of whitebellied sunbirds (*Nectarinia talatala*): effects of food concentration on feeding frequency and duration. *Journal of Experimental Biology* **209**: 2880–2887.

Kullberg C, Jakobsson S, Fransson T. 1998. Predator-induced take-off strategy in great tits (Parus major). *Proceedings of the Royal Society of London Series B (Biological Sciences)* **265**: 1659–1664.

Lasiewski RC. 1963. Oxygen consumption of torpid, resting, active and flying hummingbirds. *Physiological Zoology* **36**: 122–140.

Liem KF. 1973. Evolutionary strategies and morphological innovations: cichlid pharyngeal jaws. *Systematic Zoology* **22**: 425–441.

Livezey BC, Zusi RL 2007. Higher-order phylogeny of modern birds (Theropoda, Aves: Neornithes) based on comparative anatomy: II. – Analysis and discussion. Zoological Journal of the Linnean Society **149**: 1–94.

Marden JH. 1994. From damselflies to pterosaurs; how burst and sustainable flight performance scale with size. *American Journal of Physics* **266**: R1077–R1084.

Mathieu-Costello O, Suarez RK, Hochachka PW. 1992. Capillary-to-fiber geometry and mitochondrial density in hummingbird flight muscle. *Respiratory Physiology* **89**: 113–132.

Meyers RA. 1993. Gliding flight in the American kestrel (*Falco sparverius*): an electromyographic study. *Journal of Morphology* **215**: 213–224.

Meyers RA, Stakebake EF. 2005. Anatomy and histochemistry of spread-wing posture in birds. 3. Immunohistochemistry of flight muscles and the "shoulder lock" in albatrosses. *Journal of Morphology* **263**: 12–29.

Norberg UM. 1990. *Vertebrate Flight: Mechanics, Physiology, Morphology, Ecology and Evolution.* New York: Springer-Verlag.

Norberg UM, Rayner JMV. 1987. Ecological morphology and flight in bats (Mammalia; Chiroptera): Wing adaptations, flight performance, foraging strategy and echo-location. *Proceedings of the Royal Society of London Series B (Biological Sciences)* **316**: 335–427.

Nudds RL, Bryant DM. 2000. The energetic cost of short flights in birds. *Journal of Experimental Biology* **203**: 1561–1572.

Pennycuick CJ. 1968. Power requirements for horizontal flight in the pigeon Columba livia. *Journal of Experimental Biology* **49**: 527–555.

Pennycuick CJ. 1971. Gliding flight of the white-backed vulture Gyps africanus. *Journal of Experimental Biology* **55**: 13–38.

Pennycuick CJ. 1975. Mechanics of flight. In *Avian Biology, Volume 5,* Farner DS. King JR (ed.). New York: Academic Press; 1–75.

Pennycuick CJ. 1988. On the reconstruction of pterosaurs and their manner of flight, with notes on vortex wakes. *Biological Reviews* **63**: 299–331.

Pennycuick CJ. 1996. Wingbeat frequency of birds in steady cruising flight: new data and improved preditions. *Journal of Experimental Biology* **199**: 1613–1618.

Poore SO, Ashcroft A, Sanchez-Haiman A, Goslow GE. Jr 1997. The contractile properties of the M. supracoracoideus in the pigeon and starling: a case for long-axis rotation of the humerus. *Journal of Experimental Biology* **200**: 2987–3002.

Quiang J, Currie PJ, Norell MA, Shu-An J. 1998. Two feathered dinosaurs from northeastern China. *Nature* **393**: 753–761.

Raikow RJ. 1986. Why are there so many kinds of passerine birds? *Systematic Zoology* **35**: 255–259.

Rayner JMV. 1979a. A vortex theory of animal flight. Part I. The vortex wake of a hovering animal. *Journal of Fluid Mechanics* **91**: 697–730.

Rayner JMV. 1979b. A vortex theory of animal flight. Part II. The forward flight of birds. *Journal of Fluid Mechanics* **91**: 731–763.

Rayner JMV. 1985. Bounding and undulating flight in birds. *Journal of Theoretical Biology* **117**: 47–77.

Rayner JMV, Viscardi PW, Ward S, Speakman JR, 2001. Aerodynamics and energetics of intermittent flight in birds. *American Zoologist* **41**: 188–204.

Rosser BWC, George JC. 1986a. The avian pectoralis: histochemical characterization and distribution of muscle fiber types. *Canadian Journal of Zoology* **64**: 1174–1185.

Rosser BWC, George JC. 1986b. Slow muscle fibers in the pectoralis of the Turkey Vulture (*Cathartes aura*): An adaptation for soaring flight. *Zoologischer Anzeiger* **217**: 252–258.

Rosser BWC, Waldbillig DM, Wick M, Bandman E. 1994. Muscle fiber types in the pectoralis of the white pelican, a soaring bird. *Acta Zoologica* **75**: 329–336.

Schmidt-Nielsen K. 1972. Locomotion: energy cost of swimming, running and flying. *Science* **177**: 222–228.

Sibley CG, Ahlquist JE. 1990. *Phylogeny and Classification of Birds – A Study in Molecular Evolution.* New Haven: Yale University Press.

Sokoloff AJ, Gray-Chickering J, Harry JD, Poore SO, Goslow Jr GE. 2001. The function of the supracoracoideus muscle during takeoff in the European starling (Sternus vulgaris): Maxheinz Sy revisited. In *New Perspectives on the Origin and Early Evolution of Birds: Proceedings of the International Symposium in Honor of John H. Ostrom,* Gauthier J, Gall LF (eds.). New Haven: Peabody Museum of Natural History: 319–332.

Stolpe VM, Zimmer K. 1939. Der Schwirrflug des Kolibri im Zeitlupenfilm. *Journal of Ornithology* **87**: 136–155.

Suarez RK, Lighton JRB, Brown GS, Mathieu-Costello O. 1991. Mitochondrial respiration in hummingbird flight muscles. *Proceedings of the National Academy of Science USA* **88**: 4870–4873.

Taylor GK, Thomas ALR. 2002. Animal flight dynamics. II. Longitudinal stability in flapping flight. *Journal of Theoretical Biology* **214**: 351–370.

Thomas ALR. 1993. On the aerodynamics of birds' tails. *Proceedings of the Royal Society of London Series B (Biological Sciences)* **340**: 361–380.

Thomas ALR. 1996a. The flight of birds that have wings and a tail: variable geometry expands the envelope of flight performance. *Journal of Experimental Biology* **183**, 237–245.

Thomas ALR. 1996b. Why do birds have tails? The tail as a drag reducing flap, and trim control. *Journal of Experimental Biology* **183**, 247–253.

Thomas ALR, Taylor GK. 2001. Animal flight dynamics. I. Stability in gliding fight. *Journal of Theoretical Biology* **212**: 399–424.

Tobalske BW. 1995. Neuromuscular control and kinematics of intermittent flight in European starlings (*Sturnus vulgaris*). *Journal of Experimental Biology* **198**: 1259–1273.

Tobalske BW. 1996. Scaling of muscle composition, wing morphology, and intermittent flight behavior in woodpeckers. *The Auk* **113**: 151–177.

Tobalske BW. 2000. Biomechanics and physiology of gait selection in flying birds. *Physiological and Biochemical Zoology* **73**: 736–750.

Tobalske BW. 2001. Morphology, velocity, and intermittent flight in birds. *American Zoology* **41**: 177–187.

Tobalske BW. 2007. Biomechanics of bird flight. *Journal of Experimental Biology* **210**: 3135–3146.

Tobalske BW, Biewener AA. 2008. Contractile properties of the pigeon supracoracoideus during different modes of flight. *Journal of Experimental Biology* **211**: 170–179.

Tobalske BW, Dial KP. 1994. Neuromuscular control and kinematics of intermittent flight in budgerigars (Melopsittacus undulatus). *Journal of Experimental Biology* **187**: 1–18.

Tobalske BW, Dial KP. 1996. Flight kinematics of black-billed magpies and pigeons over a wide range of speeds. *Journal of Experimental Biology* **199**: 263–280.

Tobalske BW, Dial KP. 2000. Effects of body size on take-off flight performance in the Phasianidae (Aves). *Journal of Experimental Biology* **203**: 3319–3332.

Tobalske BW, Dial KP. 2007. Aerodynamics of wing-assisted incline running in birds. *Journal of Experimental Biology* **210**: 1742–1751.

Tobalske BW, Olson NE, Dial KP. 1997. Flight of the black-billed magpie: variation in wing kinematics, neuromuscular control, and muscle composition. *Journal of Experimental Zoology* **279**: 313–329.

Tobalske BW, Peacock WL, Dial KP. 1999. Kinematics of flap-bounding flight in the zebra finch over a wide range of speeds. *Journal of Experimental Biology* **202**: 1725–1739.

Tobalske BW, Hedrick TL, Dial KP, Biewener AA. 2003. Comparative power curves in bird flight. *Nature* **421**: 363–366.

Tobalske BW, Altshuler DL, Powers DL. 2004. Take-off mechanics in hummingbirds (Trochilidae). *Journal of Experimental Biology* **207**: 1345–1352.

Tobalske BW, Puccinelli LA, Sheridan DC. 2005. Contractile activity of the pectoralis in the zebra finch according to mode and velocity of flap-bounding flight. *Journal of Experimental Biology* **208**: 2895–2901.

Tobalske BW, Hearn JWD, Warrick DR. 2009. Aerodynamics of intermittent bounds in flying birds. *Experiments in Fluids* **46**: 963–973.

Tobalske, BW, Warrick, DR, Clark, CJ, Powers, DR, Hedrick, TL, Hyder, GA ,Biewener, AA. 2007. Three-dimensional kinematics of hummingbird flight. *Journal of Experimental Biology.* **210**: 2368–2382.

Vasquez RJ. 1992. Functional osteology of the avian wrist and the evolution of flapping flight. *Journal of Morphology* **211**: 259–268.

Ward-Smith AJ. 1984a. Analysis of the aerodynamic performance of birds during bounding flight. *Mathematical Biosciences* **8**: 137–147.

Ward-Smith A.J. 1984b. Aerodynamic and energetic considerations relating to undulating and bounding flight in birds. *Journal of Theoretical Biology* **111**: 407–417.

Warrick DR. 1998. The turning and linear manervering performance of birds: the cost of efficiency for coursing insectivores. *Canadian Journal of Zoology* **76**: 1063–1079.

Warrick DR, Dial KP. 1998. Kinematic, aerodynamic and anatomical mechanisms in the slow, maneuvering flight of pigeons. *Journal of Experimental Biology* **201**: 655–672.

Warrick DR, Dial KP, Biewener AA. 1998. Asymmetrical force production in the slow maneuvering flight of birds. *The Auk* **115**: 916–928.

Warrick DR, Bundle MW, Dial KP. 2002. Bird maneuvering flight: blurred bodies, clear heads. *Integrative and Comparative Biology* **42**: 141–148.

Warrick DR, Tobalske BW, Powers DP. 2005. Aerodynamics of the hovering hummingbird. *Nature* **435**: 1094–1097.

Weis-Fogh T. 1972. Energetics of hovering flight in hummingbirds and in *Drosophila*. *Journal of Experimental Biology* **56**: 79–104.

Welch KC, Altshuler DL, Suarez RK. 2007. Oxygen consumption rates in hovering hummingbirds reflect substrate-dependent differences in P/O ratios: carbohydrate as a "premium fuel". *Journal of Experimental Biology* **210**: 310–316.

Wells D. 1993. Muscle performance in hovering hummingbirds. *Journal of Experimental Biology* **178**: 39–57.

Witmer LM. 2002. The debate on avian ancestry: phylogeny, function, and fossils. In *Mesozoic Birds: Above the Heads of Dinosaurs*, Chiappe LM, Witmer LM (eds.). Berkeley: University of California Press; 3–30.

Witter MS, Cuthill IC, Bonser RHC. 1994. Experimental investigations of mass-dependent predation risk in the European starling, *Sturnus vulgaris*. *Animal Behaviour* **48**: 201–222.

Xu X, Tang Z, Wang X. 1999. A therizinosauroid dinosaur with integumentary structures from China. *Nature* **399**: 350–354.

11 Evolution of the Avian Brain and Senses

STIG WALSH[1] AND ANGELA MILNER[2]

[1]National Museums Scotland, Edinburgh, Scotland
[2]The Natural History Museum, London, UK

The popular view of birds as small-brained and dim-witted poorer relations of mammals is gradually being overturned. A growing volume of experimental as well as anecdotal evidence is demonstrating a previously unsuspected level of cognitive ability in many avian clades (e.g. Nicolakakis & Lefebvre, 2000; Timmermans et al., 2000; Sol et al., 2002, 2005; Lefebvre et al., 2004; Hunt & Gray, 2007), while improvements in histological and in vivo brain imaging techniques indicate a degree of structural and functional complexity comparable with that of many mammalian groups (Eccles, 1992; Dubbeldam, 1998; Medina & Reiner, 2000; Reiner et al., 2004; Striedter, 2005). There traditionally has been a temptation among researchers to assign mammalian homologies to avian brain structures (Figure 11.1), but the validity of such assignments is generally questionable (Pearson, 1972; Breazile & Kuenzel, 1993; Dubbeldam, 1998; Reiner et al., 2004). It seems more likely that most apparently similar advanced structures of the avian and mammalian central nervous system arose in parallel, since the last common ancestor of birds and mammals presumably lived in the Carboniferous Period, possibly as much as 330 Ma (Benton & Donoghue, 2007). In fact, histological studies have emphasized that avian and mammalian cognition has been achieved via different evolutionary routes (Rehkämper & Zilles, 1991;

Reiner et al., 2004), making birds a useful comparison group for studying brain evolution in hominids and, ultimately, our own species.

Despite a tendency towards conservatism in the avian skeleton, bird brains are highly variable in size and shape (Dubbeldam, 1998). However, when corrected for body mass, the overall brain volume of all living birds is significantly greater than that of typical diapsid reptiles (Jerison, 1973). With some exceptions (e.g. Pelecaniformes) brain shape appears to be relatively consistent within clades (Stingelin, 1957) suggesting that the form of the brain may sometimes be phylogenetically informative. This variation is affected by a variety of factors including the shape of the skull and position of the brain within the skull (Dubbeldam, 1989). However, brain shape is to some extent a correlate of the relative size of specific regions such as the telencephalon (forebrain), cerebellum or optic lobes of the mesencephalon. The size or expansion of these regions relative to each other and the brain as a whole is a result of increased neuronal packing, which in turn occurs because of a greater emphasis on functions normally performed by those regions in certain taxa (Striedter, 2005). In other words, like the musculature of an athlete, brain shape reflects some aspect of what a bird "does" and is thus connected with its cognitive and sensory abilities, and possibly even its ecology and behavior

Living Dinosaurs: The Evolutionary History of Modern Birds, First Edition. Edited by Gareth Dyke and Gary Kaiser.
© 2011 John Wiley & Sons, Ltd. Published 2011 by John Wiley & Sons, Ltd.

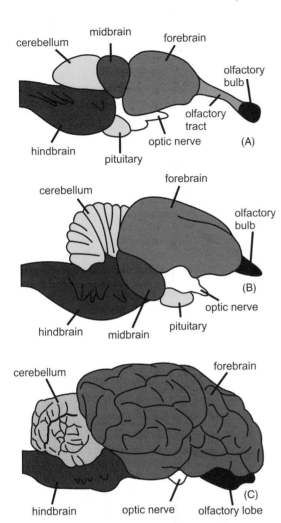

Fig. 11.1 Comparison of generalized brain morphology in (A) reptiles (*Crocodylus* sp.), (B) birds (*Columbia livia*) and (C) mammals (*Ovis aries*). Note the relative enlargement of the forebrain (telencephalon) in birds compared with reptiles. In mammals the enlargement is more extreme, and can include complete enclosure of the mid-brain as well as fissuring to increase surface area. Not drawn to scale.

(Iwaniuk *et al.*, 2004a; Iwaniuk & Hurd, 2005). The relative importance of specific senses in a given species indicated by brain region size, can be corroborated to some extent by the relative size and development of nerve bundles connecting the brain and sense organs (e.g. Witmer & Ridgely, 2008).

Examination of the avian brain can therefore clearly provide a wealth of information about a given species. Perhaps more importantly in the context of this volume, features of the central nervous system of living birds presumably coevolved with more obvious avian specializations such as flight, and are therefore an important aspect of the evolution of living birds as a whole. What follows is by no means an in-depth treatment of avian neurology; detailed descriptions of neuronal structures (e.g. nuclei, cellular sections of specific brain regions) and their function and ontogenetic development are available elsewhere (e.g. Pearson, 1972; Dubbeldam, 1998; Reiner *et al.*, 2004), and the subject is itself sufficiently complex to fill several volumes. Instead, we aim to provide the reader with a basic overview of the senses and gross morphology of the brain of extant birds set in an evolutionary context. There are good reasons for concentrating on morphology. Brain region size and shape are generally the only characteristics that can be observed in primitive bird fossils, and as such are relevant to discussion of avian neurology in an evolutionary sense. While recent recognition of avian–mammalian neurostructural homologies provide new insight into the relative timing of the appearance of some structures (e.g. Reiner *et al.*, 2004) within Archosauria, fossils remain the only direct evidence of brain evolution within Aves and specific avian clades at known points in time.

THE AVIAN BRAIN: AN ANATOMICAL PRIMER

The avian brain shares many features with the brains of living reptiles (Figure 11.1a and b) but, as mentioned above, is notably larger when viewed as a percentage component of overall body size. Much of this relative increase in neural mass can be accounted for by selective expansion of the telencephalon and cerebellum, and to some extent the optic lobes of the mesencephalon (Jerison, 1973; Dubbeldam, 1998). The main axis of the brain of most species deviates from the main axis of the spinal cord, resulting in a noticeably flexed form in lateral view (Figure 11.1b). Most of this flexing occurs within the region of the mesencephalon (Pearson, 1972), with a strong angle often developed between the long axis of the telencephalon and that of the brain as a whole (Dubbeldam, 1989). The angle of intersection between the main axis of the brain and the bill also differs strongly, with extremes being found in *Phalacrocorax* (Figure 11.2a) where the brain is oriented nearly parallel to the main axis of the bill, and *Gallinago* (Figure 11.2b) where the brain axis is oriented almost vertically in the skull and possesses a large angle between bill and main brain axis (Portmann & Stingelin, 1961).

The difficulty in homologizing structures of the avian brain with those of reptiles and mammals has led to a potentially confusing variety of anatomical terms. For accessibility to a wider audience we have used the more commonly encountered anglicized terms for gross anatomical structures, with the nomenclature of Reiner *et al.* (2004) and Dubbeldam (1998) for consideration of structures at the cellular level.

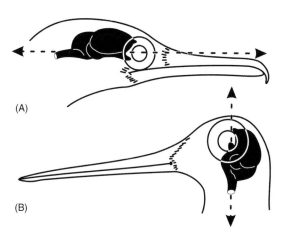

Fig. 11.2 Morphological extremes in avian bill to main brain axis intersection, with (A) the cormorant (*Phalacrocorax* sp.) demonstrating low angular deviation, and (B) the common snipe (*Gallinago gallinago*) unusually high deviation. (Modified after Portmann & Stingelin 1961.)

Forebrain

From a functional point of view, the forebrain is largely involved in higher-level processing of sensory information, cognition, and memory. Traditionally the forebrain is divided into two main structural regions, the telencephalon and the diencephalon.

Telencephalon

In dorsal view (Figure 11.3b and d–g) the brain of most birds exhibits something of an "ace of spades" outline due to the strong lateral and caudal expansion of the telencephalic hemispheres. A notable feature of the avian brain is that the telencephalon is expanded to the extent that it occludes the rounded hemispheres of the mesencephalon in dorsal view; in reptiles those hemispheres are fully visible (e.g. crocodiles, Figure 11.1a). At the cellular level, the avian telencephalon is composed of stacked layers of differentiated cells, a comprehensive review of which can be found in Reiner *et al.* (2004). The hippocampus, not visible externally as a gross morphological feature, is situated in a caudal position on the telencephalon and is implicated in memory functions and spatial awareness, including navigation (Gagliardo *et al.*, 1999; Streidter, 2005). Relative development of this region has been well studied in connection with food storage behavior in food caching species (e.g. Krebs *et al.*, 1996; Clayton, 1998).

Unlike the fissured telencephalic hemispheres characteristic of many mammals (Figure 11.1c), the dorsal surface of each hemisphere in birds is mostly smooth, but bears a distinct prominence, the eminentia sagittalis. This dorsally projected swelling, widely known as the "wulst" (meaning "bulge" in German), is in many species delimited from the rest of the telencephalon by a shallow groove called the vallecula. All modern species possess a wulst, but its shape, size, and position are variable. Stingelin (1957) distinguished two main types based on its position on the telencephalon. In Type A the wulst is positioned rostrally, while Type B is caudally positioned but includes intermediate forms in which the feature is situated more centrally on the telencephalon (Figure 11.3d–g). In common with its function in visual cognition, the wulst is largest in species with strong visual specializations, such as Strigiformes (Iwaniuk & Wylie, 2006).

Paired olfactory bulbs (olfactory lobes) are normally visible at the rostral extremity of the telencephalon (Figure 11.3a and c). The relative size of these structures is variable within bird species, but in all cases the lobes represent a far smaller proportion of the overall brain size than they do in reptiles, particularly in crocodiles and alligators (Figure 11.1a; Pearson, 1972). In general, seabirds (e.g. *Diomedea* spp.) and carrion-feeding birds such as New World vultures (e.g. *Cathartes aura*) possess relatively large olfactory bulbs, although those of the kiwi (*Apteryx* spp.) are particularly large. By comparison, the olfactory bulbs of Psittacidae and Passeridae are tiny with respect to the rest of the brain. The olfactory lobes terminate rostrally in paired olfactory nerve bundles (cranial nerve (CN) I).

Diencephalon

Only the pineal and the pituitary (hypophysis) glands of the diencephalon are easily observable

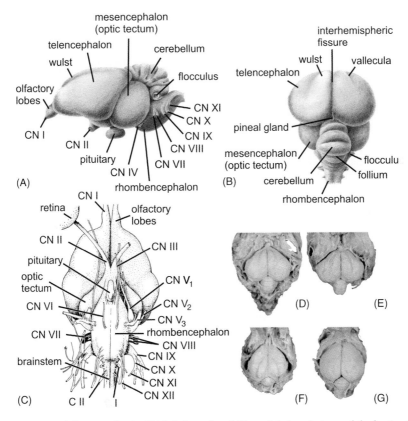

Fig. 11.3 Gross anatomy of the avian brain. (A) Left lateral and (B) caudal-dorsal views of the brain of *Columba livia*. (C) *Anser* sp. in ventral view. (Modified after Portmann & Stingelin, 1961.) (D–G) Variability of wulst position and development illustrated through dorsal views of prepared skulls revealing alcohol-fixed brains of (D) *Larus canus* (rostral position), (E) *Larus argentatus* (rostral-central position), (F) *Anas* sp. (caudal position) and (G) *Sturnus vulgaris* (wulst occupying most of dorsal telencephalon).

on the external surface of the brain. The pineal is normally relatively small and visible on the dorso-caudal surface of the brain between the rostral-most extent of the cerebellum and the caudalmost margins of the telencephalic hemispheres (Figure 11.3b). The pituitary is a rounded structure of variable form (Wingstrand, 1951) that projects ventrally from between the caudal region of the optic chiasma and rostral region of the rhomben-cephalon (Figure 11.3a and c).

The optic nerves (CN II) exit the rostral portion of the diencephalon and their form is variable. For instance, in the macaw (*Ara* sp.) the two nerves more or less bifurcate prior to exiting the brain, whereas in the tropicbird (*Phaethon* sp.) there is a relatively long optic tract before the two nerves enter the orbits. As in mammals, the optic nerves cross each other in the optic chiasma, projecting to various centers in the diencephalon and mesen-cephalon (Martin *et al.*, 2007b). Birds possess two major visual neural pathways. The tectofugal pathway connects retinal nerve fiber input to the ectostriatal core of the telencephalon via the superficial laminae of the optic tectum and nucleus rotundus thalami of the diencephalon (Benowitz & Karten, 2004). The thalamofugal pathway projects to the visual wulst via the thal-amus (Güntürkün *et al.*, 1993). In species with laterally directed eyes, the tectofugal pathway is dominant (Cook, 2000).

Mid-brain

The mid-brain (mesencephalon) of birds is broadly like that of most vertebrates in that it is composed of a tegmentum (primarily concerned with general motor- and ocularmotor control) and a tectum (further separated into the optic tectum and torus semicircularis involved in visual and audi-tory stimuli integration and routing to the dien-cephalon). The occulomotor nerve (CN III) originates in this region, projecting rostrally into the orbit (Figure 11.3c). The origin of the trochlear nerve (CN IV) is approximately in the ventro-lateral region of the mesencephalon.

The visible regions of the mesencephalon com-prise laterally projecting semihemispheres

normally referred to as "optic lobes" or the "optic tectum". Since only a portion of the structure is actually involved with integration of visual sti-muli, some authors (e.g. Breazile & Kuenzel, 1993; Dubbeldam, 1998) have adopted the term tectum mesencephali to avoid confusion with avian–mammalian homologies. The optic tectum acts as a motor reflex pathway for the occulomotor nerves (Jones *et al.*, 2007). During the early devel-opment of the embryonic brain the optic tectum hemispheres occupy a dorsal position as they do in reptiles. However, as the telencephalic hemi-spheres expand caudally the lobes are displaced laterally (Huber, 1949), such that the tegmentum becomes entirely hidden. There is some variation in the shape of the external contact of the optic tectum and the telencephalon, and the position of the lobes relative to the brain stem. In both cases, this variation is a result of the relative size of the telencephalon (Dubbeldam, 1998). The semi-circular torus occupies a caudal position, although the inner ear labyrinth of birds is mostly situated on the lateral and caudal surfaces of the cerebellum.

HINDBRAIN

The hindbrain (rhombencephalon) is composed primarily of the cerebellum and medulla, and is mostly involved in motor control.

Cerebellum

The cerebellum is situated caudal of the two tel-encephalic hemispheres, and dorsal of the medulla. The cerebellum is large in birds (Dubbeldam, 1998), although once again its size relative to other brain regions varies greatly between species. For example, in many seabirds (e.g. *Phaethon, Larus*; see Figure 11.3d and e) the cerebellum is relatively large, but tiny relative to the telencephalon in others such as Psittaciformes (e.g. *Ara*; Walsh, personal observation). The shape of the cerebellum in dorsal view is also variable between taxa; it may be relatively broad or narrow, and may possess parallel sides or taper rostrally or caudally (Kuenzi, 1918).

Unlike the telencephalon, the cerebellum is not smooth but bears 11 primary folia (numbered caudally I–X, with IX divided into IXab and cd), with secondary folia also developed (Larsell, 1967; Iwaniuk *et al.*, 2007). A greater degree of foliation is present in Psittaciformes, Corvidae, and seabirds (Iwaniuk *et al.*, 2006), and the development of each folium is also subject to phylogenetic and behavior-related variation across species (Iwaniuk *et al.*, 2007). The avian cerebellum possesses a distinct flocculus, which projects from its lateral walls through the arch of the anterior semicircular canal of the inner ear labyrinth. The flocculus is large and prominent in many seabirds (Milner & Walsh, 2009), although it may be difficult to distinguish, as it is in passerines.

Rhombencephalon

The avian rhombencephalon (medulla) is an elongate structure situated ventral to the cerebellum. It grades caudally into the spinal chord, often making its caudalmost extent difficult to determine, although this is normally taken to be rostral of cervical nerve I (Pearson, 1972). The rhombencephalon comprises autonomic centers that control heartbeat, respiration, and digestion, and also acts as a bridge between higher brain regions and input from the peripheral nervous system (Dubbeldam, 1998).

Seven cranial nerves (numbered caudally from CN V to XII) exit the rhombencephalon along the lateral and ventral surfaces (Figure 11.3c). The trigeminal (CN V) normally splits into two branches close to its origin on the lateral surface of the medulla. CN V_1 extends rostrally to relay sensory impulses from the eye, while the second branch subdivides into the maxilliary CN V_2 and mandibular CN V_3 that conduct mechanoreceptor information, particularly from the beak. Since object manipulation is mainly achieved using the beak in nearly all bird species, this nerve in birds is large. Cranial nerve VI, the abducens, exits the medulla ventrally and carries motor impulses to the lateral rectus and bulbar retractor muscles of the eye. The facial nerve (CN VII) serves both motor and sensory functions, including gustatory

information (Gentle & Clarke, 1985). Birds possess virtually no facial musculature, and correspondingly the avian CN VII is more poorly developed than in mammals. The vestibulocochlear nerve (CN VIII) is involved in only sensory functions, conducting auditory signals from the cochlear duct, and vestibular information from the saccule and semicircular canals. Cranial nerve IX (glossopharyngeal) relays sensory (particularly taste) and motor impulses, and generally exits the lateral surface of the medulla with the vagus nerve. The vagus itself (CN X) performs central roles in conducting sensory and motor impulses that regulate autonomic functions, principally in the heart, digestive tract, and lungs. The accessory nerve (CN XI) forms a further branch off the main CN IX/X branch, and conveys mainly motor impulses to muscles of the neck. The hypoglossal nerve (CN XII) is the final cranial nerve, carrying mostly motor impulses to the tongue and throat.

AVIAN SENSES

Vision

The vast majority of avian species rely on sight as the dominant sense. The avian eye is the largest relative to body size of all terrestrial vertebrates (e.g. Martin, 1985), and often represents around 50% of the cranial volume compared with 5% in humans (Jones *et al.*, 2007). This large size results in less mobility of the eye in its orbit than in mammals, and birds need to direct their field of view more by head turning than mammals do (Land, 1999). The structure of the avian eye is, however, largely similar to that of mammals (Figure 11.4), although its shape is not spherical; a ring of bony ossicles embedded within the sclera, the sclerotic ring, allows the lens and cornea to protrude ahead of the body of the eye and outside the orbit (Hall, 2008). The body of the eye itself also varies between being flattened (distance between the cornea and retina less than the maximum depth) in most birds, to tubular (distance between cornea and retina greater than the maximum depth) in species with the greatest acuity

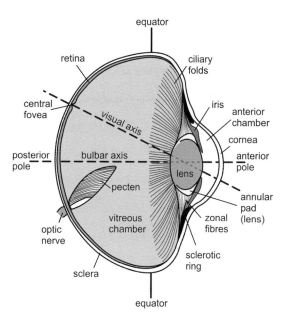

Fig. 11.4 Simplified diagrammatic representation of an avian eye, sectioned horizontally.

(Young, 1992). Similar relationships have been found between the diameter of the cornea and eye length between nocturnal (relatively large corneal diameters compared with eye length) and diurnal (relatively small corneal diameters compared with eye length) species (Hall & Ross, 2007).

The lens differs from that of mammals in that it possesses an annular pad around its core. The pad is separated from the core by a fluid-filled chamber, and this arrangement is believed to facilitate rapid focusing (accommodation) through effective transmission of force from the ciliary muscle to the lens core (Jones *et al.*, 2007). A great deal of accommodation also occurs through changes in the curvature of the cornea (Jones *et al.*, 2007).

Unlike mammals, the surface of the retina possesses no blood vessels that would otherwise interfere with light reception. Instead, a folded vascular structure called the pecten oculi protrudes from the retinal surface, providing nutrients which diffuse through the vitreous humour (Kiama *et al.*, 2006; Jones *et al.*, 2007). The photoreceptor cells of the retina comprise cone

cells, which form the basis of color vision, and rod cells, which possess a far greater sensitivity to light. However, because rod receptors are far larger than cones, fewer rods can fit onto a retinal surface of a given size. Consequently there is a trade-off between the greater visual acuity and color discrimination of the cone-dominated eyes of diurnal birds, and the greater light sensitivity of rod-dominated nocturnally adapted species. The nocturnal cave-dwelling oilbird (*Steatornis caripensis*) has few cone cells, but has maximized light sensitivity by greatly increasing rod receptor packing through reducing rod cell size, and placing the rods in a banked arrangement (Martin *et al.*, 2004).

Many avian species possess areas of the retina that have concentrations of cone receptors, which serve to enhance acuity at those points. Some of these concentrations include pits (foveae) where acuity is at its greatest (Pumphrey, 1948). Unlike human vision, in which a single fovea is situated along the central visual axis, birds may possess multiple foveae that enhance acuity at various angles. The arrangement and morphology of these centers of acuity appear to relate to the behavioral ecology of particular species (e.g. Land, 1999; Tucker, 2000; Jones *et al.*, 2007). For instance, Tucker (2000) showed that the different acuity of the deep and shallow foveae of raptors allow these species greatest acuity at different ranges but at different angles. Although the position of the foveae in raptors varies depending on the position of the eyes in the head, the longest range fovea is more laterally directed and used on the long approach to prey. A spiral flight path allows raptors to use this long range fovea, while retaining an aerodynamic head-forwards posture before the short range fovea comes into use as the prey comes within the binocular forward visual field (Tucker, 2000). The single fovea of owls is probably related to their wider binocular field and activity in low light conditions (Jones *et al.*, 2007). Foveae have also been implicated in the avian detection of polarized light (Kreithen & Keeton, 1974).

Birds are known to possess tetrachromatic color vision, and in contrast to the three-cone trichromatic system (red, blue, and green) of humans, many birds can perceive light in the red (ca. 650 nm), green (ca. 510 nm), blue (ca. 470 nm) and violet to ultraviolet (ca. 355–426 nm; Ödeen and Håstad, 2003) wavelengths. The ability of birds to detect ultraviolet (UV) wavelengths has long been known (Bennett & Cuthill, 1994), and several hypotheses for its purpose have been proposed. These include functions in orientation (Bennett & Cuthill, 1994), prey detection (Honkavaara *et al.*, 2002; Håstad *et al.*, 2005) and intraspecific signaling (e.g. Andersson & Amundsen, 1997; see Cuthill *et al.* (2000) for a review). How birds actually see color is difficult to test, but the presence of these four chromatic pigment cone types suggests that they are capable of differentiating differences in hue that humans cannot detect (Ödeen and Håstad, 2003).

Avian cones often contain droplets of carotenoid oil, which are generally red, orange, yellow, green, or colorless in diurnal species, and pale yellow to colorless in nocturnal taxa (Partridge, 1989; Young, 1992). These droplets act to low-pass filter light wavelengths before they reach the photo-pigment of the cone, possibly providing an enhanced discrimination of hue (Vorobyev *et al.*, 1998; Jones *et al.*, 2007). Combinations of such filters appear to have specific functions related to the ecology of the species in question (Bowmaker & Martin, 1985; Partridge, 1989; Hart, 2001; Hart & Vorobyev, 2005). Double cones also occur in the avian retina, as they do in all vertebrates other than placental mammals (Sillman, 1973). Double cones may have a role to play in spatial orientation through detection of polarized light (Delius *et al.*, 1976).

Avian visual fields are mostly a function of the position of the eyes in the skull, but the closeness of the eyes to each other, either side of the inter-orbital septum, places restrictions on the direction of the primary visual axis (Martin, 2009). Several taxa such as owls and nightjars appear to possess eyes that face forwards like those of primates, but in fact some lateral direction occurs in all species (Martin, 2009). However, binocular vision in which the visual field of both eyes overlaps is present in all species, although the field of overlap in the straight ahead sector may be as little as

5°–10° in some species such as *Scolopax rusticola* and *Anas platyrhynchos* (Martin *et al.*, 2007a). Such species are unlikely to use vision to guide the beak since the beak tip would be invisible to them (Martin *et al.*, 2007a). Nonetheless, overall flying ability does not appear to be significantly compromised by possession of a narrow binocular field, suggesting binocular vision plays a relatively small role in controlling flight in birds (Martin, 2009). In prey species a near panoramic field of view would seem to be more advantageous. The degree of binocular overlap in owls is particularly broad at around 50°, and these species are widely considered to possess genuine visual depth perception (stereopsis; Iwaniuk & Wylie, 2006). However, Martin (2009) has suggested stereopsis in owls may have come about because the large size of the eyes and outer ears prevents a more lateral position of the eyes.

Hearing

Like reptiles, birds lack a functional external ear composed of soft tissue. However, although the external auditory meatus is naked in Casuari-formes, Struthioniformes and Falconiformes, most species possess some form of modified feathers in this region (Pearson, 1972). In the case of Strigiformes, particularly the nocturnal barn owl (*Tyto alba*), these feathers form sound channeling structures that help the bird to localize prey effectively by sound alone (Figure 11.5a; Koch & Wagner, 2002). In eared owls (*Asio* spp.) and other strigiforms parts of the auditory meatus can also be closed voluntarily, allowing accurate localization of sounds occurring behind the bird (Pearson, 1972; Norberg, 1978). Asymmetry of the outer ear also helps to locate prey in the vertical plane through higher frequency sounds, and appears to have arisen independently in the five owl lineages (Figure 11.5b; Norberg, 2002).

The auditory meatus in living birds is normally short and slightly curved. As in most reptiles, sound waves transmitted along the passage are transmitted to the oval window of the cochlea via the tympanic membrane and a single middle ear ossicle, the columella, the morphology of

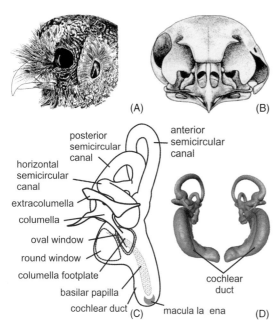

Fig. 11.5 Avian inner and outer ear hearing apparatus. (A) Facial disc feathers of *Strix virgata*, showing directional folding of feathers. (B) Rostral view of *Aegolius funereus* illustrating hearing-related cranial asymmetry. (C) Simplified diagram of a generalized inner ear labyrinth including middle ear ossicles in position; (D) Paired inner ear labyrinths segmented from a μCT scan of the skull of *Tyto alba*, illustrating the extreme (for birds) curvature of the cochlear ducts, which almost meet at the midline. (A and B after Norberg, 2002; (C) modified after Manley 1990.)

which is often taxonomically diagnostic (e.g. Feduccia, 1975). The sensory organ, the basilar papilla (analogous to the Organ of Corti in mammals; Manley, 1990) is situated in the cochlear duct (Figure 11.5c) and detects vibrations transmitted by the columella through stimulation of hair cells organized along its length. These hair cells range in number from a few thousand to more than 10,000 in some owls, and may be of up to four different types (Manley, 1990). The basilar papilla is generally wider and shorter than in most mammals, with the longest basilar papillae being found in owls (Gleich *et al.*, 2005). Also, unlike the strongly curved, often spiral form

of the mammalian cochlear duct, the avian cochlear duct is generally slightly curved or straight (Manley, 1990).

The frequency sensitivity of most birds falls within the range of 1000–5000 Hz with the greatest sensitivity between 2000 and 3000 Hz (Dooling, 2002). However, there is a relatively large range of variation in birds. For instance, pigeons (*Columba livia*) possess an unusual auditory sensitivity to very low frequency sounds as low as 0.05 Hz (Kreithen & Quine, 1979), and guinea fowl (Theurich *et al.*, 1984) and possibly grouse (Moss & Lockie, 1979) can also detect infrasound. The hearing threshold is lower at high frequencies in Passeriformes than in non Passeriformes, with the latter showing lower hearing thresholds at low frequencies than Passeriformes (Dooling, 2002). The overall hearing range of *Dromaius novaehollandiae* falls between 80 and 3500 Hz (Manley *et al.*, 1997), while that of *Tyto alba* spans 500–9500 Hz (Köppl & Gleich, 2007). Mammalian sensitivity is generally higher; a guinea pig can detect 50–5000 Hz, and a cat 40–8500 Hz (Fay, 1988). One reason for this difference is that the columella (the stapes, inherited from the reptilian ancestor of birds) is less effective at transmitting a range of frequencies than the mammalian three-bone chain, the malleus, incus, and stapes (Manley, 1990). Another is that the avian basilar papilla is generally shorter than the mammalian Organ of Corti, and is consequently capable of detecting a narrower range of frequencies. There may, nonetheless, be a limit to the maximum possible size for the avian cochlear duct. *Tyto alba* possesses one of the longest cochlear ducts of any living bird, and although the duct comes close to completing a half turn in the skull, further lengthening is not possible as the distal ends of the duct almost meet at the midline (Figure 11.5d). We have observed a similar condition in the diminutive skull of the goldcrest (*Regulus regulus*), and we suspect that *Tyto* and *Regulus* occupy an extreme position in cochlear duct length relative to skull size. It is therefore possible that skull size and architecture are factors that limit the potential length of the cochlear duct in birds.

Hearing is of course a fundamental factor in one of the most striking of avian attributes: vocal communication. Not surprisingly, the range of hearing sensitivity of a given species comfortably overlaps that of its vocalization frequency range, and in fact the vocalization frequencies of most birds fall within the lower half of their hearing sensitivity (Konishi, 1970). The complexity and speed of frequency changes in the songs of many birds (particularly Passeriformes) are too great for human perception (Borror & Reese, 1956), and it is clear that many avian species have an ability to resolve such vocalizations at remarkably fine temporal resolutions. This capacity may also be important for the ability of some taxa such as the oilbird (*Steatornis caripensis*) to echo-locate.

Olfaction and taste

The olfactory system of birds is similar to most other vertebrates: the mucosal olfactory epithelium houses olfactory receptors, which detect airborne chemical stimuli and relay this information to the olfactory lobes of the telencephalon via the olfactory nerve. However, the olfactory lobes of most birds are small, and until the latter half of the 20th century birds were thought to possess little or no olfactory capability (Stager, 1967). Nonetheless, the role of scent in locating food (Stager, 1964; Grubb, 1972; Smith & Palsek, 1986; Verheyden & Jouventin, 1994), navigation (Waldvogel, 1989; Papi, 1990), and possibly reproduction (Jones & Gentle, 1984; Lambrechts & Hossaert-McKey, 2006; Balthazart & Taziaux, 2009) is now well known, and olfactory sensitivity thresholds have also been estimated for many species (Roper, 1999). Consequently, scent must remain an important sensory modality in most birds.

Kiwis (*Apteryx* spp.) possess what is probably the best-developed sense of smell, and this is echoed by the large size of the olfactory lobes and large and complex olfactory chamber in these species. Unlike most birds the external nostrils of *Apteryx* are positioned at the tip of the bill, allowing the bird to employ olfaction as one strategy (see below) in the search for invertebrate prey by probing in the forest leaf mould (Roper, 1999; Cunningham

et al., 2007). The olfactory apparatus of Procellariiformes is also well developed, and various species are able to detect and distinguish between food sources from a distance (Verheyden & Jouventin, 1994; Nevitt, 1999) and find their way back to their nesting sites in the dark (Bonadonna et al., 2001). The smaller New World vultures such as the turkey vulture (*Cathartes aura*: Cathartidae) are well known for the development of their olfactory apparatus and their ability to locate carrion from considerable distances (Smith & Palsek, 1986). This sensitivity has been utilized by workers searching for leaks in long-distance gas pipelines; compounds that mimic decaying carcasses are introduced to the gas line and the resulting congregation of vultures indicates the position of the leak. Interestingly, even species with a supposedly poor sense of smell (e.g. pigeons) appear to be able to detect and use geographically-specific atmospheric trace gasses for the purposes of navigation (Walraff, 2004a,b).

Taste receptors mostly occur within the mandible and upper jaw, and the tongue (Berkhoudt, 1985) and send sensory information to the brain by way of the facial nerve (CN VII) and the lingual branch of the glossopharyngeal nerve (CN IX) (Berkhoudt, 1985; Ganchrow et al., 1986). Although birds generally have fewer taste receptors than most mammals and reptiles, their number among species is variable. There is some suggestion that the apparent tolerance of some species to strong acid and alkaline flavors permits otherwise unpalatable items such as unripe fruit to be exploited (Mason & Clark, 2000), and the comparatively poor sense of taste of most bird species suggests that this sense was not of prime importance during avian evolutionary history. Nonetheless, some species seem to have a very well-developed sense of taste. Sandpipers (Scolopacidae) can detect where worms have been in sand (van Heezik et al., 1983), and hummingbirds (Trochilidae) can determine the concentration of sugars in solution (Hainsworth & Wolf, 1976). Taste is apparently important for avoiding poisonous prey species (Skelhorn & Rowe, 2006), and has been observed to mediate food selection in passerines based on preferences for salt, fructose content, and certain amino acids (Espaillat & Mason, 1990).

Touch and balance

Four kinds of mechanoreceptors are present in birds. Herbst corpuscles resemble the Pacinian corpuscles of mammals, and detect rapid mechanical deformation (vibration). These receptors are the most abundant and are found in the dermis across the entire body, but are also present in large numbers in the beak, in joint capsules and in feather follicles, indicating an importance for proprioreception in flight (Gottschaldt, 1985). Grandry corpuscles are found only in the bills of aquatic birds, where they are most numerous at the tip (Cunningham *et al.*, 2007). In sandpipers Herbst and Grandry corpuscles occur in large numbers in pits around the bill tip, allowing these species to locate invertebrate prey within the substrate through minute vibrations (Piersma *et al.*, 1998; Nebel *et al.*, 2005). Kiwis possess a similar sensory pad with Herbst and Grandry-like corpusules, suggesting that they may also be able to localize prey using a similar vibrotactile sensory mechanism (Cunningham *et al.*, 2007). Merkel corpuscles (slowly adapting pressure sensors) are found on the tongue and in the beak of land birds (Necker, 2000). Ruffini endings (stretch receptors) have not been widely reported in birds outside a few anseriform and galliform species (Necker, 2000). Muscle spindles (detecting changes in muscle length) are of course important for proprioreception and are abundant throughout the body (Maier, 1992). In addition to unmyelinated nerve endings that transmit pain stimuli, thermoreceptors are present as free nerve endings within the body and in the skin. Their response is differentiated depending on increases and decreases in temperature, with cold receptors being more numerous than warm receptors (Necker, 2000).

In addition to integumentary and muscular proprioreceptors, balance and spatial awareness for flight and normal posture are monitored by the semicircular canals of the inner ear. These three canals send information to the brain about angular accelerations of the head as it is turned,

tilted or thrust forward, via sensory hair cells anchored to the walls of the canals, which are excited as the endolymphatic fluid moves relative to the canal wall during motion. This vestibular information is processed primarily in the cerebellum, and particularly in the floccular region of the cerebellum. The cerebellar flocculus is involved in processing of the vestibuloocular reflex, which serves to maintain stable vision of unmoving targets through eye stabilization as the head moves, and the optokinetic reflex, which stabilizes the eyes on moving scenes. The semicircular canals are arranged approximately orthogonal to each other, so are able to measure rotations along the axial, saggital and transverse planes. Although the anterior semicircular canal is always largest in birds, there is considerable variation in the arc length, direction of arc apex, canal diameter, and cross-sectional shape. These differences are presumably functionally important, and their purpose has been the focus of considerable interest (Hopkins, 1906; Hadžiselimović & Savković, 1964; Ten Kate *et al.*, 1970; Muller, 1999; Sipla, 2007). Specializations of the lumbosacral region of the vertebral column have also been suggested to be involved in balance during walking. Here, canals in the enlarged glycogen body may function in a similar way to the semicircular canals of the inner ear (Necker, 2006).

Magnetoreception

In addition to using scent traces (see above) and orientation from the Sun and stars, many birds are apparently able to navigate using Earth's magnetic field as both as a map (determining location) and compass (direction finding; see Beason (2005) for a review). Homing pigeons have recently been found to possess the magnetic iron compounds magnetite and maghaemite in three areas under the skin of the upper bill, and associated with the trigeminal nerve (Fleissner *et al.*, 2007). Minute changes in the orientation of these particles as the bird changes its orientation within the magnetic field are thought to be detectable by strain sensors. This system could provide a three-axis magnetometer sensitive to small changes in the strength and direction of the magnetic field, that could be used for direction finding and determining position within the magnetic field.

Neurons within the optic tectum of some species have been found to react to changes in the magnetic field, but only in the presence of light (Beason, 2005). This suggests that a magnetic input to the visual system is being directed by stimulation of the retinal photoreceptors or, more likely, that a light dependent magnetoreceptor is present in the visual system (Beason, 2005; Maeda *et al.*, 2008). Although no receptor in the avian eye has been shown to be sensitive to the weak magnetic field of the Earth, Maeda *et al.* (2008) have at least demonstrated the feasibility of such a compound. If this mechanism is correct, birds could potentially be able to "see" the lines of the magnetic field independent of any trigeminal magnetite–maghaemite sensor they may possess. It seems unlikely that birds are restricted to the use of one system, which in order to work would in any case also require input and integration of other more generalized sensory modalities such as vision, vestibular sense (Beason, 2005), and possibly hearing (Hagstrum, 2001).

FOSSIL EVIDENCE FOR AVIAN BRAIN EVOLUTION

The endocranial cavity of birds and mammals represents a reasonably accurate approximation of the shape and size of the brain that it houses (Iwaniuk & Nelson, 2002; Striedter, 2005). Consequently, the morphology and volume of the brain in fossil birds can be studied where sediment has filled the endocranium after death and subsequently been lithified. Such endocasts are nonetheless exceedingly rare (Milner & Walsh, 2009), and those that do survive are very often damaged by mechanical processes to the point that only gross morphology may remain. A potentially more serious problem is that the loss of surrounding skull material generally makes even higher level taxonomic identification of isolated endocasts extremely difficult. Both factors severely limit the information such natural endocasts may pro-

vide through comparative morphological studies. Some information can also be gained through examination of the endocranial surface, though only in cases where the skull has been damaged. Current approaches avoid all of these problems by using micro X-ray computer tomography (µCT) to examine the internal structures of skulls noninvasively, offering the possibility to visualize interactively the internal space as a positive three-dimensional object, or "virtual" endocast (Witmer *et al.*, 2008). This approach is greatly increasing the number of taxa for which brain morphology is known, allowing a better assessment of trends in avian brain evolution through time.

The oldest and best studied endocast is the specimen partially exposed on the London specimen of *Archaeopteryx lithographica* (NHMUK 37001). Examination of the incomplete morphology of this specimen led early workers (e.g. Edinger, 1926; de Beer, 1954) to regard the brain of *Archaeopteryx* as more reptilian than bird-like, in that the visible half of the telencephalon does not appear to be significantly expanded laterally or caudally (Figure 11.6c and d). Jerison (1968) reconstructed the endocast as more bird-like, but this view was not confirmed until Domínguez *et al.* (2004) used µCT analysis on the specimen to investigate brain and inner ear development, demonstrating beyond doubt that the brain of this basal avialan was more bird- than reptile-like in shape and overall volume. *Archaeopteryx* shows no evidence of a wulst, indicating that this structure had not yet appeared in early avialans.

Brain morphology is known for very few other Mesozoic bird species. A composite endocast of the Late Triassic *Protoavis texensis* was described by Chatterjee (1991) as an early bird, but the avian affinities of this taxon are disputed (Dingus & Rowe, 1998; Feduccia, 1999). The endocast as reconstructed (Figure 11.6a) does appear to be bird-like, although the telencephalon and cerebellum are not as expanded as in the reconstruction of *Archaeopteryx* made by Domínguez *et al.* (2004). Chatterjee identified a rostrally situated wulst on his reconstruction, but its form and position make it more likely that this structure represents dorsally expanded olfactory lobes (Figure 11.6a and b).

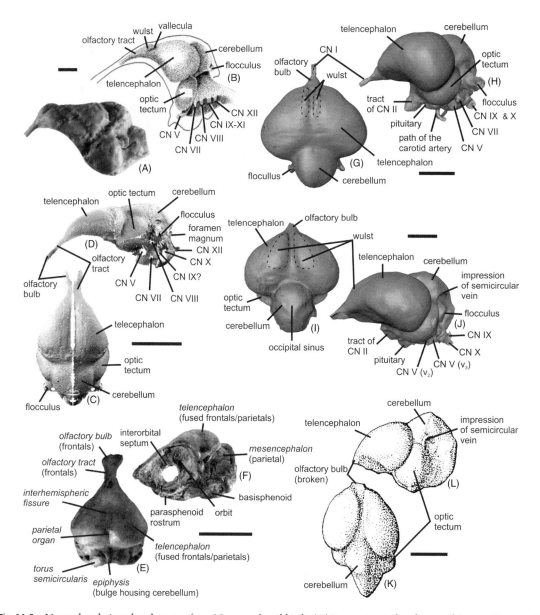

Fig. 11.6   Natural and virtual endocasts of pre-Neogene fossil birds. (A) Reconstructed endocast of *Protoavis texensis* (Late Triassic) in left lateral view, with (B) original author's interpretation of morphology (modified after Chatterjee, 1991); virtual endocast of the holotype of *Archaeopteryx lithographica* (Late Jurassic) in (C) dorsal and (D) left lateral views (modified after Domínguez *et al.* 2004); *Cerebavis cenomanica* (early Late Cretaceous) in (E) dorsal and (F) left lateral views (original interpretation of Kurochkin *et al.* (2007) labeled in italics, with our revised interpretation given in normal font; modified after Kurochkin *et al.*, 2007); digital endocasts of *Odontopteryx toliapica* in (G) dorsal and (H) left lateral views, and *Prophaethon shrubsolei* in (I) dorsal and (J) left lateral views; line drawings of "*Numenius*" *gypsorum* in (K) dorsal and (L) left lateral views (modified after Dechaseaux, 1970). (Images of *Protoavis texensis* (Chatterjee, 1991, figure 16a and e, p. 296) and *Cerebavis cenomanica* (Kurochkin *et al.*, 2007, figure 2a and b, p. 310) reproduced with permission of Royal Society Publishing. Images of *Archaeopteryx lithographica* ©Nature Publishing.)

Fragmentary endocasts of the Late Cretaceous toothed birds *Ichthyornis victor* (Ichthyornithes) and *Hesperornis regalis* (Hesperonithes) were reconstructed by Marsh (1880) as being distinctly reptilian in having enlarged, elongate olfactory bulbs and a small telencephalon that did not overstep the optic tectum of the mesencephalon. However, this material was later shown by Edinger (1951) to be too incomplete to provide much evidence of the evolutionary grade at this time, and the true condition in these taxa is not fully resolved. Elzanowski & Galton (1991) provided a detailed description of the inside of the braincase of *Enaliornis barretti* (Hesperornithes), based on three specimens from the Early Cretaceous of England. They regarded the brain of *Enaliornis* as rather primitive, with a relatively unexpanded telencephalon. However, examination of these specimens was only possible due to damage in each, and the morphology of the rostralmost portion of the brain of *Enaliornis* remains unknown. No wulst is apparent on the preserved regions of the telencephalon, but its presence on the missing rostral portions is not precluded.

Most recently a potentially important new specimen from the Late Cretaceous of Russia was described as a "fossil brain," probably from an enantiornithine (Kurochkin *et al.*, 2006, 2007). *Cerabavis cenomanica* (Figure 11.6e and f) was erected on this specimen and regarded as displaying a mix of characters typical of modern birds (enlarged cerebellum, relatively smaller mesencephalon) and more primitive taxa (enlarged olfactory bulbs). However, selected μCT slice images published in Kurochkin *et al.* (2006) show clearly that fossilized bone material fully surrounds the specimen, and bony internal structures can be identified within the object that are continuous with the outer bone. There can be little doubt that the specimen is an abraded skull rather than an endocast. The published μCT slices do indicate that the external shape of the skull closely conforms to the shape of the endocranium, demonstrating that the brain possessed a short but laterally and dorsally expanded telencephalon, a large mesencephalon and small cerebellum. Unfortunately, the taxon was erected without

a clear, unambiguous diagnosis, and must presently be regarded as a *nomen dubium* until a thorough redescription is undertaken. Assuming "*Cerebavis*" is indeed avian, this specimen is likely to be important for future studies of avian paleoneurology, particularly with respect to whether the telencephalon of this taxon shows evidence of a wulst. However, the preservation of the specimen makes its assignment to a higher taxon problematic. Nonetheless, the dermal skull roof of "*Cerebavis*" appears to be fully fused, a condition found in extant adult neognaths but not Enantiornithes, suggesting it is unlikely to be referable to that clade.

More endocasts are known from Cenozoic deposits than from Mesozoic strata. A relatively complete natural endocast of "*Numenius gypsorum*" (Figure 11.6k and l) from the late Eocene of the Paris Basin was described by Dechaseaux (1970). This specimen is clearly modern in shape, although previous authors (Dechaseaux, 1970; Jerison, 1973) did not detect the presence of a wulst on the dorsal telencephalon. However, Milner & Walsh (2009) showed that the feature is present but rudimentarily developed. Using μCT techniques, Milner & Walsh (2009) also investigated brain and inner ear morphology in two birds from the lower Eocene London Clay Formation, England. These two species, *Odontopteryx toliapica* (Pelagornithidae; Figure 11.6g and h) and *Prophaethon shrubsolei* (Prophaethontidae; Figure 11.6i and j) currently represent the oldest neognaths for which brain morphology is known. As with "*Numenius*" *gypsorum*, the brains (and hearing capability; Walsh *et al.*, 2009) of these taxa proved to be modern in form, except that in *Prophaethon* the wulst was far less well developed than in living relatives (e.g. red tailed tropicbird, *Phaethon rubricauda*), and the wulst of *Odontopteryx* was as poorly developed as that of "*Numenius*" *gypsorum*. The occurrence of such rudimentarily developed wulsts in these three taxa suggests that the structure was in an early stage of development in the early Tertiary, and that the first appearance of the feature and its associated functions is likely to have been towards the end of the Cretaceous.

Other Tertiary endocasts come from Neogene deposits. Mlíkovský provided brief descriptions of a relatively complete endocast and a fragment from the lower Miocene of Czechoslovakia referable to the Accipitridae (Mlíkovský, 1980), seven fragmentary endocasts from the middle Miocene of Bavaria including Pelecanidae, Phasianidae and Passeriformes (Mlíkovský, 1988), and a fragmentary specimen from the upper Pliocene of Hungary referable to the Anatidae (Mlíkovský, 1981). A complete endocast referable to an extinct species of penguin (probably *Spheniscus urbinai*) is also known from the late Miocene of Chile (Walsh, 2001). Most recently, Picasso *et al.* (2009) described a cast of the endocranium of an accipitrid from the Late Miocene of Patagonia. Brain morphology in some Pleistocene and recent species of ratites and vultures is also known (e.g. Jerison, 1973; Ashwell & Scofield, 2007; Scofield & Ashwell, 2009), but these post-Paleogene specimens show that all Neogene and younger taxa possessed entirely modern brains, little different from living members of the same genera and families.

DISCUSSION

In our introduction we highlighted the observation that, in terms of overall volume relative to body mass, the brain of a typical modern bird is larger than that of a living diapsid reptile. This size difference appears to hold true for Mesozoic archosaurs (Larsson *et al.*, 2000), although data are generally too scarce to test variation in early representatives of the group. How quickly the avian brain reached its current state of expansion is thus difficult to estimate, but we now know that bird brains were already almost modern in size and morphology by 55 Ma, and were probably little different to those of modern birds by the end of the Mesozoic (Milner & Walsh, 2009). The endocast associated with the London specimen of *Archaeopteryx* represents the only taxonomically reliable evidence of brain development in an early avialan, and demonstrates that relative brain enlargement was already well underway towards the end of the

Jurassic (Domínguez *et al.*, 2004). Did the evolution of avian flight lead to this expansion?

The flying abilities of *Archaeopteryx* and other early avialans have been hotly debated, and the taxon has been central to arguments over how flight evolved, primarily whether the ancestors of birds took to the air via a "ground up" or "trees down" route (e.g. Brugers & Chiappe, 1999; Elzanowski, 2002; Longrich, 2006). However, no matter how physically well adapted for life in the air a bird may be, its flight would be crude or impossible without sensitive somatic control. When viewed in its role as a flight computer, the development of the brain in key taxa represents important evidence beyond what can be deduced from studies of the skeleton and its mechanical flight apparatus. Domínguez *et al.* (2004) showed that, relative to body size, the brain of *Archaeopteryx* plots within the lower mass range of living birds. Moreover, the telencephalon and optic tectum of the mesencephalon were particularly well developed, exactly as would be expected in a visually oriented animal engaged in enhanced cognitive processing in an aerial environment. These authors also demonstrated a well-developed vestibular sense in this species through the form of the inner ear labyrinth. The regions that process balance and orientation information, the cerebellum and cerebellar flocculus, were also well developed. This evidence certainly lends support to the idea that *Archaeopteryx* was equipped for flight despite lacking the refined flight apparatus of living birds.

Jerison (1973) pointed out that the evolution of flight alone is unlikely to wholly account for the increase in brain size in birds. For instance, the brain of bats is not relatively larger than that of other mammals, suggesting that no significant expansion was necessary in that group to coordinate flight. Nonetheless, the brains of pterosaurs do show a similar brain/body mass ratio to *Archaeopteryx* (Witmer *et al.*, 2003), together with a similar morphological reorganization to that seen in birds (Edinger, 1941; Lewy *et al.*, 1992), and enlarged semicircular canals (Witmer, 2004). Safi *et al.* (2005) also found that extant bat species with maneuvrable flight living in complex environments have larger brains relative to body size than bats in more open environments. Their results broadly suggest that larger brains are necessary for enhanced neurological control in such environments, but that there is an energetic trade-off between the benefits of a large brain and the increased metabolic cost of its enlargement relative to the energetics of flight that in some aerial environments can lead to selection for a *decrease* in overall brain size. To our knowledge this relationship has not been tested in birds.

If an energetically expensive large brain is needed to coordinate flight, one would expect the drive to conserve energy would lead to a reduction in the brain size of secondarily flightless birds. Considering the tendency of flightless species to be larger than their volant relatives, one would also expect this effect to be compounded where evolution of a greater body mass exceeds increases in brain size. Relatively smaller brains in flightless birds were indeed found in one study (Bennett & Harvey, 1985), and Corfield *et al.* (2008) reported that the brain sizes of living palaeognaths (mostly flightless) are smaller than those of neognaths, with the notable exception of the kiwi. However, the number of flightless neognath taxa used in that study is unclear, limiting the comparability of those results for testing brain size relationships across flightless and volant living species. By comparison, Iwaniuk *et al.* (2004b) found no such relationship except in the kakapo (*Strigops habroptilus*) and extinct great auk (*Pinguinus impennis*). One possible reason for their results is that brain size reduction has lagged behind because the majority of brain growth occurs before the body has grown to full size (Iwaniuk & Nelson, 2003). Another might relate to the potential of secondarily flightless birds to retain the larger brain of their ancestors through the net energy savings that would accompany forelimb reduction. However, this is unlikely to apply to all species; the forelimbs of penguins presumably require as much energy as those of volant species in order to "fly" through the far denser medium of seawater. Nonetheless, flight muscles *and* large brains are particularly metabolically expensive (Isler & van Schaik, 2009). The observation that flightless

species have not evolved *larger* brains after being freed from the energy requirements of flight suggests that such evolutionary drivers in terrestrial environments have been less powerful than those for the evolution of flight.

If relative brain size alone is not directly related to the ability to fly, can the brain of a flier or nonflier be recognized based on morphology? Avian brain size scales primarily with body size, but the relative volume of specific regions is variable between species. As mentioned earlier, the relative expansion of the telencephalon, cerebellum, and optic tectum, together with contraction of the olfactory lobes relative to the other regions, were key steps leading to the avian brain shape we know today. Consequently, brain morphology may be more informative than overall size in terms of changes in behavior-related processing capacity (Iwaniuk & Hurd, 2005; Striedter, 2005).

It has become increasingly apparent that many nonavian theropods also possessed bird-like brains (particularly in terms of an expanded telencephalon and cerebellum) as well as more obvious bird-like attributes such as feathers (e.g. Osmólska, 2004; Kundrát, 2007; Balanoff *et al.*, 2009; Norell *et al.*, 2009). The cerebellar flocculus, important for stabilization of the visual image and normally well developed in birds, appears to be equally well developed in these theropod taxa. It is difficult to imagine a reason for such neural development in terrestrial taxa (but see Sultan, 2005), and these discoveries have consequently fueled the debate as to whether these taxa actually represent secondarily flightless birds rather than bird-like theropods (e.g. Kundrát, 2007; Kavanau, in press). One would expect a significant difference between the size and/or form of the cerebellar flocculus in volant and nonvolant species, but this has not been tested. Likewise, there appears to have been no work that has demonstrated a relationship with telencephalon size and flying ability. Thus, with the possible exception of the kiwi, the brains of flightless birds are not immediately obvious as such. Multivariate analysis of relative brain region volumes (*sensu* Iwaniuk & Hurd, 2005) may prove the best approach for distinguishing a "flying" from a "nonflying" brain. However, at present there seems to be no consistent relationship between flying ability and either brain size or morphology.

One possible reason for this apparent absence of a clear pattern might be that the neural adaptations needed for flight were already present in the flightless ancestor of Aves, and were subsequently adapted and honed for this new use (Witmer & Ridgely, 2007). If so, the regional and whole brain expansion seen in *Archaeopteryx* and Cretaceous nonavian maniraptoran theropods would have been inherited from a common ancestor. This hypothesis poses the question of what was the purpose of these adaptations in that common ancestor. One explanation for the expansion of the telencephalon might relate to drivers toward enhancement of cognition.

A wealth of data is emerging that links telencephalon size in living birds to a variety of factors including new problem-solving feeding behaviors (Lefebvre *et al.*, 1997; Nicolakakis & Lefebvre, 2000; Timmermans *et al.*, 2000; Lefebvre *et al.*, 2004; Iwaniuk & Hurd, 2005) and associated tool use (Lefebvre *et al.*, 2002; Cnotca *et al.*, 2008), and even the presence of consciousness (Eccles, 1992; Butler & Cotterill, 2006; Prior *et al.*, 2008). Apart from some striking instances of sequential and meta-tool use that exceed the abilities of nonhuman primates (Hunt & Gray, 2007; Wimpenny *et al.*, 2009), some species are thought to exhibit social learning (Tebbich *et al.*, 2001) and complex social behaviors (Clayton *et al.*, 2007; Emery *et al.*, 2007) such as cooperative hunting (Yosef & Yosef, in press). Although undoubtedly enhanced in those modern groups (especially corvids and parrots) compared with early birds and their immediate ancestors, such improvements in cognition would presumably represent a competitive advantage over less encephalized coeval species. For instance, increased encephalization has been linked to species richness (Nicolakakis *et al.*, 2003) and success during introduction to new environments (Marino, 2005; Sol *et al.*, 2002, 2005). In this context, the metabolic cost of a large brain may well be offset by the improved success in exploiting available resources, possibly providing a further explanation for the retention of large brains by flightless birds.

An arboreal lifestyle may account for the preflight expansion of the cerebellar flocculus in this hypothetical avian ancestor; visual stabilization is equally necessary in an environment of moving branches (e.g. in primates; Belton & McCrea, 2000). Expansion of the cerebellum as a whole also may relate in part to growth of specific centers that deal with snout and tongue manipulation of objects (Sultan, 2005). This would be consistent with the cognition hypothesis suggested for nonflight-related telencephalon expansion, since environmental manipulation and sensing are prerequisite information inputs for decision making. Sensory input may also have been an important factor for increases in avian brain size from the end of the Mesozoic to the Neogene; the largely Paleogene development of the wulst, and presumably its visual functions, appears to have been part a secondary pulse of telencephalic expansion. The reason for this Paleogene wulst expansion is presently unknown. To what extent the early telencephalic expansion in the ancestor of Aves was related to visual cognition is impossible to say, although *Archaeopteryx* was certainly visually oriented (Domínguez *et al.*, 2004). Garamszegi *et al.* (2002) suggested that eye (especially nocturnality and visual prey capture technique) and brain size has coevolved in birds. This relationship is logical, considering that the retina is an extension of the diencephalon, and that the optic nerve bundle should be proportionally broader in larger eyes depending on the ratio of rods to cones on the retina. However, attempts to predict visual abilities such as diurnality/nocturnality in fossil and osteological specimens so far have been inconclusive (Hall & Ross, 2007; Hall, 2008; Schmitz, 2009).

Further work is clearly needed to investigate and test the validity of these suggestions, and their bearing on the evolution and loss of flight in birds and dinosaurs.

CONCLUSIONS

Our knowledge of how the avian brain and senses function on a cellular level has advanced greatly over the past few decades, leading to a better understanding of bird behavior. However, until recently investigations of how the avian brain evolved to its present state were dependent upon a small amount of evidence from several, mostly damaged, specimens. These suggested that the avian brain had expanded from its ancestral diapsid state over a period spanning pre-Late Jurassic to early Neogene time, based on what was then known of the brain of the earliest known avialan and the fully modern brains of Miocene taxa (Jerison, 1973). The advent of noninvasive μCT approaches has greatly increased the number of taxa for which brain size and morphology are known, revealing that the brain was almost modern in size and form early in the Neogene, and constraining the timing of specific brain region development.

The senses and cognitive ability of birds are quite remarkable and offer obvious parallels for studying neurological evolution in our own group, the primates. Key questions that remain to be answered nonetheless center on the nature of the evolutionary drivers for avian brain expansion, and whether behavior (e.g. flying ability versus flightlessness) can be determined from gross features of the brain, such as form and size. These questions in turn should hold the key to finally resolving debates about the evolution of avian flight, as well as the true nature of bird-like theropods. Providing suitable new fossil specimens from important stages in avian evolution can be found, the use of new technologies such as three-dimensional visualization of X-ray and synchrotron data is likely to supply the necessary data with which to address these questions.

REFERENCES

Andersson S, Amundsen T. 1997. Ultraviolet colour vision and Ornamentation in bluethroats. *Proceedings of the Royal Society of London Series B (Biological Sciences)* **264**: 1587–1591.

Ashwell KW. S, Scofield RP. 2007. Big birds and their brains: palaeoneurology of the New Zealand moa. *Brain Behaviour and Evolution* **71**: 151–166.

Balanoff AM, Xu X, Kobayashi Y, Matsufune Y, Norrell MA. 2009. Cranial osteology of the theropod dinosaur, *Incisivosaurus gauthieri* (Theropoda: Oviraptorosauria). *American Museum Novitates* **3651**: 1–35.

Balthazart J, Taziaux M. 2009. The underestimated role of olfaction in avian reproduction? *Behavioural Brain Research* **200**(2): 248–259.

Beason RC. 2005. Mechanisms of magnetite orientation in birds. *Integrative and Comparative Biology* **45**: 565–573.

Belton T, McCrea RA. 2000. Role of the cerebellar flocculus region in cancellation of the VOR during passive whole body rotation. *Journal of Neurophysiology* **83**(3): 1599–1613.

Bennett AT, Cuthill IC. 1994. Ultraviolet vision in birds: what is its function? *Vision Research* **34**(11): 1471–1478.

Bennett PM, Harvey PH. 1985. Relative brain size and ecology in birds. *Journal of the Zoological Society of London* **207**: 151–169.

Benowitz LI, Karten HJ. 2004. Organization of the tectofugal visual pathway in the pigeon: A retrograde transport study. *Journal of Comparative Neurology* **167**(4): 503–520.

Benton MJ, Donoghue PC. J. 2007. Paleontological evidence to date the Tree of Life. *Molecular Biology and Evolution* **24**(1): 26–53.

Berkhoudt H. 1985. Structure and function of avian taste receptors. In *Form and Function in Birds, Volume 3*, King AS, McLelland J (eds). London: Academic Press; 463–496.

Bonadonna F, Spaggiari J, Weimerskirch H. 2001. Could osmotaxis explain the ability of blue petrels to return to their burrows at night? *Journal of Experimential Biology* **204**: 1485–1489.

Borror OJ, Reese CR. 1956. Vocal gymnastics of the wood thrush. *Ohio Journal of Science* **56**: 177–182.

Bowmaker JK, Martin GR. 1985. Visual pigments and oil droplets in the penguin, *Spheniscus humboldti*. *Journal of Comparative Physiology A* **156**: 71–77.

Breazile JE, Kuenzel WJ. 1993. Systema nervosum centrale. In *Nomina Anatomica Avium*, 2nd edn, Baumel J, King A, Breazile J, Evans H, Berge J (eds). Nuttall Ornithological Club: Cambridge; Publication **23**, 493–554.

Brugers P, Chiappe L. 1999. The wing of *Archaeopteryx* as a primary thrust generator. *Nature* **154**: 587–588.

Butler AB, Cotterill MJ. 2006. Mammalian and avian neuroanatomy and the question of consciousness in birds. *Biological Bulletin* **211**: 106–127.

Chatterjee S. 1991. Cranial anatomy and relationships of a new Triassic bird from Texas. *Philosophical Transactions of the Royal Society of London* **B332**: 277–342.

Clayton NS. 1998. Memory and the hippocampus in food-storing birds: a comparative approach. *Neuropharmacology* **37**: 441–452.

Clayton NS, Dally JM, Emery NJ. 2007. Social cognition by food-caching corvids. The western scrub-jay as a natural psychologist. *Philosophical Transactions of the Royal Society of London* **B362**: 507–522.

Cnotca J, Güntürkün O, Rehkämper G, Gray RD, Hunt GR. 2008. Extraordinary large brains in tool-using New Caledonian crows (*Corvus moneduloides*). *Neuroscience Letters* **433**: 241–245.

Cook RG. 2000. The comparative psychology of avian visual cognition. *Current Directions in Psychological Science* **9**(3): 83–89.

Corfield JR, Wild JM, Hauber ME, Parsons S, Kubke MF. 2008. Evolution of brain size in the palaeognath lineage, with an emphasis on New Zealand ratites. *Brain, Behaviour and Evolution* **71**: 87–99.

Cuthill IC, Partridge JC, Bennett ATD, Church SC, Hart NS, Hunt S. 2000. Ultraviolet vision in birds. In *Advances in the Study of Behaviour, Volume 29,* Slater PJB, Rosenblatt JS, Snowdon CT, Roper TJ (eds). San Diego, CA: Academic Press; 159–214.

Cunningham S, Castro I, Alley M. 2007. A new prey-detection mechanism for kiwi (*Apteryx* spp.,) suggests convergent evolution between paleognathous and neognathous birds. *Journal of Anatomy* **211**: 493–502.

De Beer G. 1954. *Archaeopteryx lithographica.* London: British Museum (Natural History).

Dechaseaux C. 1970. Moulages endocraniens d'oiseaux de l'Éocène Supérieur du Bassin de Paris. *Annales de Paléontologie* **56**: 69–72.

Delius JD, Perchard RJ, Emmerton J. 1976. Polarized light discrimination by pigeons and an electroretinographic correlate. *Journal of Comparative Physiology* **90**(6): 560–751.

Dingus L, Rowe T. 1998. *The Mistaken Extinction. Dinosaur Evolution and the Origin of Birds.* New York: W. H. Freeman and Co.

Domínguez P, Milner AC, Ketcham RA, Cookson MJ, Rowe TB. 2004. The avian nature of the brain and inner ear of *Archaeopteryx. Nature* **430**: 666–669.

Dooling RJ. 2002. *Avian Hearing and the Avoidance of Wind Turbines.* Technical Report NREL/TP-500-30844, National Renewable Energy Laboratory, Golden, Colorado. http://www.osti.gov/bridge.

Dubbeldam JL. 1989. Shape and structure of the avian brain, an old problem revisited. *Acta morphologica Neerlando-Scandinavica* **27**: 33–43.

Dubbeldam JL. 1998. Birds. In *The Central Nervous System of Vertebrates, Volume, 3.,* Nieuwenhuys R,

Ten Donkelaar HJ, Nicholson C (eds). Berlin: Springer-Verlag; 1525–1636.

Eccles JC. 1992. *Evolution of consciousness. Proceedings of the National Academy of Sciences USA* **89**: 7320–7324.

Edinger T. 1926. The brain of *Archaeopteryx. Annals and Magazine of Natural History: including Zoology, Botany and Geology* **18**: 151–156.

Edinger T. 1941. The brain of *Pterodactylus. American Journal of Science* **239**(9): 665–683.

Edinger T. 1951. The brains of the Odontognathae. *Evolution* **5**: 6–24.

Elzanowski A. 2002. Archaeopterygidae (Upper Jurassic, of Germany) In *Mesozoic Birds: Above the Heads of the Dinosaurs,* Chiappe LM, Witmer LM (eds). *Berkeley*: University of California Press; 129–159.

Elzanowski A, Galton PM. 1991. Braincase of *Enaliornis,* an early Cretaceous bird from England. *Journal of Vertebrate Paleontology* **11**: 90–107.

Emery NJ, Seed AM, von Bayern AM. P, Clayton NS. 2007. Cognitive adaptations of social bonding in birds. *Philosophical Transactions of the Royal Society of London* **B362**: 489–505.

Espaillat JE, Mason JR. 1990. Differences in taste preference between red-winged blackbirds and European starlings. *The Wilson Bulletin* **102**(2): 292–299.

Fay RR. 1988. *Hearing in Vertebrates: a Psychophysics Databook.* Winnetka, IL: Hill-Fay Associates.

Feduccia A. 1975. Morphology of the bony stapes (columella) in the Passeriformes and related groups: evolutionary implications. *University of Kansas Museum of Natural History Miscellaneous Publications* **63**: 1–34.

Feduccia A. 1999. *The Origin and Evolution of Birds,* 2nd edn. New Haven: Yale University Press.

Fleissner G, Stahl B, Thalau P, Falkenberg G, Fleissner G. 2007. A novel concept of Fe-mineral-based magnetoreception: histological and physicochemical data from the upper beak of homing pigeons. *Naturwissenschaften* **94**(8): 631–642.

Gagliardo A, Ioalé P, Bingman VP. 1999. Homing in pigeons: the role of the hippocampal formation in the representation of landmarks used for navigation. *Journal of Neuroscience* **19**(1): 311–315.

Ganchrow JR, Ganchrow D, Oppenheimer M. 1986. Chorda tympani innervation of anterior mandibular taste buds in the chicken (*Gallus gallus domesticus*). *Anatomical Record* **216**(3): 434–439.

Garamszegi LZ, Møller AP, Erritzøe J. 2002. Coevolving avian eye size and brain size in relation to prey capture and nocturnality. *Proceedings of the Royal Society of London Series B (Biological Sciences)* **269**: 961–967.

Gentle MJ, Clark JSB. 1985. The fibre spectrum of the chorda nerve in the chicken (*Gallus gallus* var. *domesticus*). *Journal of Anatomy* **140**: 105–110.

Gleich O, Dooling RJ, Manley GA. 2005 Audiogram, body mass, and basilar papilla length: correlations in birds and predictions for extinct archosaurs. *Naturwissenschaften* **92**: 595–589.

Gottschaldt KM. 1985. Structure and function of avian somatosensory receptors. In *Form and Function in Birds, Volume 3*, King AS, McLelland J (eds). London: Academic Press; 375–461.

Grubb TC. 1972. Smell and foraging in shearwaters and petrels. *Nature* **237**: 404–405.

Güntürkün O, Miceli D, Watanabe M. 1993. Anatomy of the avian thalamofugal pathway. In *Vision, Brain and Behaviour in Birds*, Zeigler HP, Bischof H-J. (eds). Massachusetts: MIT Press; 115–156.

Hagstrum JT. 2001. Infrasound and the Avian Navigational Map. *Journal of Navigation* **54**(3): 377–391.

Hainsworth FR Wolf. LL. 1976. Nectar characteristics and food selection by hummingbirds. *Oecologia* **25**: 101–113.

Hall MI. 2008. The anatomical relationships between the avian eye, orbit and sclerotic ring: implications for inferring activity patterns in extinct birds. *Journal of Anatomy* **212**: 781–794.

Hall MI, Ross CF. 2007. Eye shape and activity pattern in birds. *Journal of Zoology* **271**: 437–444.

Hart NS. 2001. The visual ecology of avian photoreceptors. *Progress in Retinal and Eye Research* **20**(5): 675–703.

Hart NS, Vorobyev M. 2005. Modelling oil droplet absorption spectra and spectral sensitivities of bird cone photoreceptors. *Journal of Comparative Physiology A* **191**: 381–392.

Håstad O, Ernstdotter E, Ödeen A. 2005. Ultraviolet vision and foraging in dip and plunge diving birds. *Biology Letters* **1**: 306–309.

Hadžiselimović H, Savković LJ. 1964. Appearance of semicircular canals in birds in relation to mode of life. *Acta Anatomica* **57**: 306–315.

Honkavaara J, Koivula M, Korpimäki E, Siitari H, Viitala J. 2002. Ultraviolet vision and foraging in terrestrial vertebrates. *Oikos* **98**: 505–511.

Hopkins MA. 1906. On the relative dimensions of the osseous semicircular canals of birds. *Biological Bulletin* **11**(5): 253–264.

Huber W. 1949. Analyse métrique du redressement de la tête chez l'embryo de poulet. *Revue Suisse de Zoologie* **56**: 286–291.

Hunt G, Gray RD. 2007. Parallel tool industries in New Caledonian crows. *Biology Letters* **3**: 173–175.

Isler K, van Schaik CP. 2009. Why are there so few smart mammals (but so many smart birds)? *Biology Letters* **5**: 125–129.

Iwaniuk AN, Hurd PL. 2005. The evolution of cerebrotypes in birds. *Brain, Behaviour and Evolution* **56**: 215–230.

Iwaniuk A, Nelson J. 2002. Can Endocranial volume be used as an estimate of brain size in birds? *Canadian Journal of Zoology* **80**: 16–23.

Iwaniuk A, Nelson J. 2003. Developmental differences are correlated with relative brain size in birds: a comparative analysis. *Canadian Journal of Zoology* **81**: 1913–1928.

Iwaniuk AN, Wylie DRW. 2006. The evolution of steropsis and the Wulst in caprimulgiform birds: a comparative analysis. *Journal of Comparative Physiology A* **192**: 1313–1326.

Iwaniuk AN, Dean KM, Nelson JE. 2004a. A mosaic pattern characterizes the evolution of the avian brain. *Proceedings of the Royal Society of London Series B (Biological Sciences)* **271**: S148–S151.

Iwaniuk AN, Nelson JE, James HF, Olson SL. 2004b. A comparative test of the correlated evolution of flightlessness and relative brain size in birds. *Journal of Zoology* **263**: 317–327.

Iwaniuk AN, Hurd PL, Wylie DR. W. 2006. Comparative morphology of the avian cerebellum: I. Degree of foliation. *Brain, Behaviour and Evolution* **68**(1): 45–62.

Iwaniuk AN, Hurd PL, Wylie DR. W. 2007. Comparative morphology of the avian cerebellum: II. Size of folia. *Brain, Behaviour and Evolution* **69**(3): 196–219.

Jerison HJ. 1968. Brain evolution and *Archaeopteryx*. *Nature* **219**: 1381–1382.

Jerison HJ. 1973. *Evolution of the brain and intelligence*. London: Academic Press.

Jones RB, Gentle MJ. 1984. Olfaction and behavioural modification in domestic chicks (*Gallus domesticus*). *Physiology and Behavior* **34**: 917–924.

Jones MP, Pierce KE, Ward D. 2007. Avian vision: a review of form and function with special consideration to birds of prey. *Journal of Exotic Pet Medicine* **16**(2): 69–87.

Kavanau L. 2010. Secondarily flightless birds or Cretaceous non-avian theropods? *Medical Hypotheses*, **74**(2): 275–276.

Kiama SG, Maina JN, Bhattacharjee J, Mwangi DK, Macharia RG, Weyrauch KD. 2006. The morphology of the pecten oculi of the ostrich, *Struthio camelus*.

Annals of Anatomy – Anatomischer Anzeiger **188**(6): 519–528.

Koch UR, Wagner H. 2002. Morphometry of auricular feathers of Barn Owls (*Tyto alba*). *European Journal of Morphology* **40**: 15–21.

Konishi M. 1970. Comparative neurophysiological studies of hearing and vocalizations in songbirds. *Journal of Comparative Physiology A* **66**(3): 257–272.

Köppl C, Gleich O. 2007. Evoked cochlear potentials in the barn owl. *Journal of Comparative Physiology A* **193**: 601–612.

Krebs JR, Clayton NS, Healy SD, Cristol DA, Patel SN, Jolliffe AR. 1996. The ecology of the avian brain: food-storing memory and the hippocampus. *The Ibis* **138**: 34–46.

Kreithen ML, Keeton WT. 1974. Detection of polarized light by the homing pigeon, *Columba livia*. *Journal of Comparative Physiology* **89**: 83–92.

Kreithen ML, Quine DB. 1979. Infrasound detection by the homing pigeon: A behavioral audiogram. *Journal of Comparative Physiology A* **129**(1): 1–4.

Kuenzi W. 1918. Versuch einer systematischen Morphologie des Gehirns der Vögel. *Revue Suisse de Zoologie* **26**: 17–112.

Kundrát M. 2007. Avian-like attributes of a virtual brain model of the oviraptorid theropod *Conchoraptor gracilis*. *Naturwissenschaften* **94**(6): 499–504.

Kurochkin EN, Saveliev SV, Postnov AA, Pervushov EM, Popov EV. 2006. On the brain of a primitive bird from the Upper Cretaceous of European Russia. *Paleontological Journal* **40**(6): 655–666.

Kurochkin EN, Saveliev SV, Dyke GJ, Pervushov EM, Popov EV. 2007. A fossil brain from the Cretaceous of European Russia and avian sensory evolution. *Biology Letters* **3**(3): 309–313.

Land MF. 1999. The roles of head movements in the search and capture strategy of a tern (Aves, Laridae). *Journal of Comparative Physiology A* **184**: 265–272.

Lambrechts MM, Hossaert-McKey M. 2006. Olfaction, volatile compounds and reproduction in birds. *Acta Zoologica Sinica* **52** (Supplement): 284–287.

Larsell O. 1967. *The Comparative Anatomy and Histology of the Cerebellum from Myxinoids Through Birds*. Minneapolis, MN: University of Minnesota Press.

Larsson HCE, Sereno PC, Wilson JA. 2000. Forebrain enlargement among nonavian theropod dinosaurs. *Journal of Vertebrate Paleontology* **20**(3): 615–618.

Lefebvre L, Whittle P, Lascaris E, Finkelstein A. 1997. Feeding innovations and forebrain size in birds. *Animal Behaviour* **53**: 549–560.

Lefebvre L, Nicolakakis N, Boire D. 2002. Tools and brains in birds. *Behaviour* **139**: 939–973.

Lefebvre L, Reader SM, Sol D. 2004. Brains, innovations and evolution in birds and primates. *Brain, Behavior and Evolution* **63**: 233–246.

Lewy Z, Milner AC, Patterson C. 1992. Remarkably preserved natural endocranial endocasts of pterosaur and fish from the late Cretaceous of Israel. *Geological Survey of Israel Current Research* **7**: 31–35.

Longrich N. 2006. Structure and function of hindlimb feathers in *Archaeopteryx lithographica*. *Paleobiology* **32**: 417–431.

Maeda K, Henbest KB, Cintolesi F, Kuprov I, Rodgers CT, Liddell PA, Gust D, Timmel CR, Hore PJ. 2008. Chemical compass model of avian magnetoreception. *Nature* **453**: 387–390.

Maier A. 1992. The avian muscle spindle. *Anatomy and Embryology* **186**(1): 1–25.

Manley GA. 1990. *Peripheral Hearing Mechanisms in Reptiles and Birds*. Heidelberg, Germany: Springer-Verlag.

Manley GA, Köppl C, Yates G.K. 1997. Activity of primary auditory neurones in the cochlear ganglion of the Emu *Dromaius novaehollandiae*: spontaneous discharge, frequency tuning, and phase locking. *Journal of the Acoustical Society of America* **101**: 1560–1573.

Marino L. 2005. Big brains do matter in new environments. *Proceedings of the National Academy of Sciences* **102**(15): 5306–5307.

Marsh OC. 1880. Odontornithes: a monograph on the extinct toothed birds of North America. *U.S. Geological Exploration of the 40th Parallel* **7**.

Martin GR. 1985. Eye. In *Form and Function in Birds*, King AS, McLelland J (eds) New York: Academic Press; 311–373.

Martin GR. 2009. What is binocular vision for? A birds' eye view. *Journal of Vision* **9**(11): 1–19.

Martin GR, Rojas LM. Ramírez Y, McNeil R. 2004. The eyes of oilbirds (*Steatornis caripensis*): pushing at the limits of sensitivity. *Naturwissenschaften* **91**: 26–29.

Martin GR, McNeil R, Marina Rojas L. 2007a. Vision and the foraging technique of skimmers (Rhincopidae). *The Ibis* **149**: 750–757.

Martin GR, Wilson K-J, Wild JM, Parsons S, Kubke MF, Corfield J. 2007b. Kiwi forego vision in the guidance of their nocturnal activities. *PLoS ONE* **2**(2): e198.

Mason JR, Clark L. 2000. The chemical senses of birds. In *Sturkie's Avian Physiology*, 5th edn, Whittow GC. (ed.). San Diego: Academic Press; 39–56.

Medina L, Reiner A. 2000. Do birds possess homologues of mammalian primary visual, somatosensory and motor cortices? *Trends in Neuroscience* **23**: 1–12.

Milner AC, Walsh SA. 2009. Avian brain evolution: new data from Palaeogene birds (Lower Eocene) from England. *Zoological Journal of the Linnean Society* **155**: 198–219.

Mlíkovský J. 1980. Zwei Vogelgehirne aus dem Miozän Böhmens. *Casopis pro mineralogii a geologii, roč* **25**: 409–413.

Mlíkovský J. 1981. Ein fossile Vogelgehirn aus dem Oberpliozän Ungarns. *Fragmenta Mineralogica et Palaeontologica* **10**: 71–74.

Mlíkovský J. 1988. Notes on the brains of the middle Miocene birds (Aves) of Hahnenberg (F.R.G.). *Casopis pro mineralogii a geologii, roč* **33**: 53–61.

Moss R, Lockie I. 1979. Infrasonic components in the song of the Capercaillie *Tetrae urogallus*. *The Ibis* **121**: 95–97.

Muller M. 1999. Size limitations in semicircular duct systems. *Journal of Theoretical Biology* **198**: 405–437.

Nebel S, Jackson DL, Elner RW. 2005. Functional association of bill morphology and foraging behaviour in calidrid sandpipers. *Animal Biology* **55**(3): 235–243.

Necker R. 2000. The somatosensory system. In *Sturkie's Avian Physiology*, 5th edn, Wittow GC (ed.). London: Academic Press; 57–69.

Necker R. 2006. Specializations in the lumbosacral vertebral canal and spinal cord of birds: evidence of a function as a sense organ which is involved in the control of walking. *Journal of Comparative Physiology A: Neuroethology, Sensory, Neural, and Behavioral Physiology* **192**(5): 439–448.

Nevitt F. 1999. Olfactory foraging in Antarctic seabirds: a species-specific attraction to krill odors. *Marine Ecology Progress Series* **177**: 235–241.

Nicolakakis N, Lefebvre L. 2000. Forebrain size and innovation rate in European birds: feeding, nesting and confounding variables. *Behaviour* **137**: 1415–1429.

Nicolakakis N, Sol D, Lefebvre L. 2003. Behavioural flexibility predicts species richness in birds, but not extinction risk. *Animal Behaviour* **65**(3): 445–452.

Norberg RA. 1978. Skull asymmetry, ear structure and function, and auditory localization in Tengmalm's Owl, *Aegolius funereus* (Linne). *Philosophical Transactions of the Royal Society of London* **B282**(991): 325–410.

Norberg RA. 2002. Independent evolution of outer ear asymmetry among five owl lineages; morphology, function and selection. In *Ecology and Conservation of Owls*, Newton I, Kavanagh R, Olsen J, Taylor I (eds). Collingwood, Victoria: CSIRO Publishing; 329–342.

Norell MA, Makovicky PJ, Bever GS, Balanoff AM, Clarke JM, Barsbold R, Rowe T. 2009. A review of the Mongolian Cretaceous dinosaur *Saurornithoides* (Troodontidae: Theropoda). *American Museum Novitates* **3654**: 1–63.

Ödeen A, Håstad O. 2003. Complex distribution of avian color vision systems revealed by sequencing the SWS1 opsin from Total DNA. *Molecular Biology and Evolution* **20**(6): 855–861.

Osmólska H. 2004. Evidence on relation of brain to endocranial cavity in oviraptorid dinosaurs. *Acta Palaeontologica Polonica* **49**: 321–324.

Papi F. 1990. Olfactory navigation in birds. *Cellular and Molecular Life Sciences* **46**(4): 352–363.

Partridge JC. 1989. The visual ecology of avian cone oil droplets. *Journal of Comparative Physiology A* **165**: 415–426.

Pearson R. 1972. *The Avian Brain*. London: Academic Press.

Picasso MBJ, Tambussi C, Dozo MT. 2009. Neurocranial and brain anatomy of a Late Miocene eagle (Aves, Accipitridae) from Patagonia. *Journal of Vertebrate Paleontology* **29**(3): 831–836.

Piersma T, van Aelst R, Kurk K, Berkhoudt H, Maas LRM. 1998. A new pressure sensory mechanism for prey detection in birds: the use of principles of seabed dynamics? *Proceedings of the Royal Society of London Series B (Biological Sciences)* **265**: 1377–1383.

Portmann A, Stingelin W. 1961. The central nervous system. In *The Biology and Comparative Physiology of Birds, Volume 2*, Marshall AJ. (ed). New York: Academic Press; 1–36.

Prior H, Schwarz A, Güntürkün O. 2008. Mirror-induced behaviour in the magpie (*Pica pica*): evidence of self-recognition. *PLoS Biology* **6**(8): e202.

Pumphrey RJ. 1948. The theory of the fovea. *Journal of Experimental Biology* **25**: 299–312.

Rehkämper G, Zilles K. 1991. Parallel evolution in mammalian and avian brains: comparative cytoarchitectonic and cytochemical analysis. *Cell and Tissue Research* **263**: 3–28.

Reiner A, Perkel DJ, Bruce LL, Butler AB, Csillag A, Kuenzel W, Medina L, Paxinos G, Shimizu T, Streidter G, Wild M, Ball GF, Durand S, Gütürkün O, Lee DW, Mello CV, Powers A, White SA, Hough G, Kubikova L, Smulders TV, Wada K, Dugas-Ford J, Husband S, Yamamoto K, Yu J, Siang C, Jarvis ED. 2004. Revised nomenclature for avian telencephalon and some related brainstem nuclei. *Journal of Comparative Neurology* **473**: 377–414.

Roper TJ. 1999. Olfaction in birds. *Advances in the Study of Behavior* **28**: 247–332.

Safi K, Seid MA, Dechmann DKN. 2005. Bigger is not always better: when brains get smaller. *Biology Letters* **1**: 283–286.

Schmitz L. 2009. Quantitative estimates of visual performance features in fossil birds. *Journal of Morphology* **270**: 759–773.

Scofield RP, Ashwell KW. S. 2009. Rapid somatic expansion causes the brain to lag behind: the case of the brain and behaviour of New Zealand's Haast's eagle (*Harpagornis moorei*). *Journal of Vertebrate Paleontology* **29**(3): 637–649.

Sillman AJ. 1973. Avian vision. In *Avian Biology, Volume 3*, Farner DS, King JR (eds), New York: Academic Press; 349–383.

Sipla JS. 2007. *The semicircular canals of birds and non-avian dinosaurs*. Unpublished PhD thesis, Stony Brook University, New York.

Skelhorn J, Rowe C. 2006. Avian predators taste-reject aposematic prey on the basis of their chemical defence. *Biology Letters* **2**: 348–350.

Smith SA, Palsek RA. 1986. Olfactory sensitivity of the Turkey Vulture (*Cathartes aura*) to three carrion-associated odorants. *The Auk* **103**: 586–592.

Sol D, Timmermans S, Lefebvre L. 2002. Behavioural flexibility and invasion success in birds. *Animal Behaviour* **63**: 495–502.

Sol D, Duncan RP, Blackburn TM, Cassey P, Lefebvre L. 2005. Big brains, enhanced cognition, and response of birds to novel environments. *Proceedings of the National Academy of Sciences USA* **102**: 5460–5465.

Stager KE. 1964. The role of olfaction in food location by the turkey vulture (*Cathartes aura*). *Los Angeles County Museum Contributions to Science* **8**: 1–63.

Stager KE. 1967. Avian olfaction. *American Zoologist* **7**: 415–419.

Stingelin W. 1957. *Vergleichend morphologische untersuchungen am vorderhirn der Vögel auf cytologischer und cytoarchitektonischer grundlage*. Basel: Verlag Helbing and Lichtenhahn.

Striedter G. 2005. *Principles of Brain Evolution*. Sunderland, MA: Sinauer Associates.

Sultan F. 2005. Why some bird brains are larger than others. *Current Biology* **15**(17): R649–R650.

Tebbich S, Taborsky M, Fessl B, Blomqvist M. 2001. Do woodpecker finches acquire tool-use by social learning? *Proceedings of the Royal Society of London Series B (Biological Sciences)* **268**: 2189–2193.

Ten Kate JH, van Barneveld HH, Kuiper JW. 1970. The dimensions and sensitivities of semicircular canals. *Journal of Experimental Biology* **53**: 501–514.

Theurich M, Langner G, Scheich H. 1984. Infrasound responses in the midbrain of the Guinea Fowl. *Neuroscience Letters* **49**: 81–86.

Timmermans S, Lefebvre L, Boire D, Basu P. 2000. Relative size of the hyperstriatum ventrale is the best predictor of feeding innovation rate in birds. *Brain, Behavior and Evolution* **56**: 196–203.

Tucker VA. 2000. The deep fovea, sideways vision and spiral flight paths in raptors. *Journal of Experimental Biology* **203**: 3745–3754.

Van Heezik YM, Gerritsen AFC, Swennen C. 1983. The influence of chemoreception on the foraging behaviour of two species of sandpiper, *Calidris alba* (Pallas) and *Calidris alpina* (L.). *Netherlands Journal of Sea Research* **17**: 47–56.

Verheyden C, Jouventin P. 1994. Olfactory behavior of foraging Procellariiformes. *The Auk* **111**: 285–291.

Vorobyev M, Osorio D, Bennett AT, Marshall NJ, Cuthill IC. 1998. Tetrachromacy, oil droplets and bird plumage colours. *Journal of Comparative Physiology* **A183** (5): 621–633.

Waldvogel JA. 1989. Olfactory orientation by birds. In *Current Ornithology, Volume 6*, Power DM (ed.). New York: Plenum; 269–321.

Walraff HG. 2004a. Navigation by homing pigeons: updated perspective. *Ethology, Ecology and Evolution* **13**: 1–48.

Walraff HG. 2004b. Avian olfactory navigation: its empirical foundation and conceptual state. *Animal Behaviour* **67**(2): 189–204.

Walsh SA. 2001. *The Bahía Inglesa Formation Bonebed: genesis and palaeontology of a Neogene Konzentrat Lagerstätte from north-central Chile*. Unpublished PhD thesis, University of Portsmouth.

Walsh SA, Barrett PM, Milner AC, Manley G, Witmer LM. 2009. Inner ear anatomy is a proxy for deducing auditory capability and behaviour in reptiles and birds. *Proceedings of the Royal Society* **B276**: 1355–1360.

Wimpenny JH, Weir AA. S, Clayton L, Rutz C, Kacelnik A. 2009. Cognitive processes associated with sequential tool use in New Caledonian crows. *PLoS ONE* **4**(8): e6471.

Wingstrand KG. 1951. *The Structure and Development of the Avian Pituitary*. Lund: Håkan Ohlssons Boktryckeri.

Witmer LM. 2004. Inside the oldest bird brain. *Nature* **430**: 619–620.

Witmer LM, Ridgely RC. 2007. Evolving an on-board flight computer: brains, ears, and exaptation in the evolution of birds and other theropod dinosaurs. *Journal of Morphology, ICVM–8 Abstracts* 1150.

Witmer LM, Ridgely RC. 2008. Structure of the brain cavity and inner ear of the centrosaurine ceratopsid *Pachyrhinosaurus* based on CT scanning and 3D visualization. In *A, New Horned Dinosaur from an Upper Cretaceous Bone Bed in Alberta*, Currie PJ (ed.). Ottawa: National Research Council Research Press; 117–144.

Witmer LM, Chatterjee S, Franzosa J, Rowe T. 2003. Neuroanatomy of flying reptiles and implications for flight, posture and behaviour. *Nature* **425**: 950–953.

Witmer LM, Ridgely RC, Dufeau DL, Semones MC. 2008. Using CT to peer into the past: 3D visualisation of the brain and ear regions of birds, crocodiles and nonavian dinosaurs. In *Anatomical Imaging: Towards a New Morphology*, Endo H, Frey R (eds). Tokyo: Springer-Verlag. 67–87.

Yosef R, Yosef N. In press. Cooperative hunting in brown-necked raven (*Corvus rufficollis*) on Egyptian mastigure (*Uromastyx aegyptius*). *Journal of Ethology.*

Young JZ. 1992. *The Life of Vertebrates*, 3rd edn. Oxford: Oxford University Press.

12 Evolving Perceptions on the Antiquity of the Modern Avian Tree

JOSEPH W. BROWN[1] AND M. VAN TUINEN[2]

[1]University of Michigan Museum of Zoology, Ann Arbor, USA
[2]University of North Carolina at Wilmington, Wilmington, USA

The age of the modern bird (Neornithes) origin and radiation has been greatly debated for over a century and remains contentious. Times of origin and diversification among neornithine birds have been interpreted variably from paleontological data, and more recently from analysis of molecular data. An apparent conflict (albeit not always clearly defined) between these two types of data is often highlighted (Brown *et al.*, 2008; van Tuinen *et al.*, 2006). We are now at a crucial time where molecular information can be constrained explicitly in a probabilistic framework by both the fossil record and its uncertainty in order to further investigate this apparent discrepancy between molecular and fossil chronograms. In addition, an overwhelming benefit of modern molecular approaches is to provide divergence time estimates where paleontological information is lacking or uncertain. In this chapter, we highlight (i) the history of opinions on the antiquity of the neornithine tree, culminating in the oft-heated "rock–clock" debate, (ii) current state-of-the-art modeling of historically variable molecular substitution rates (i.e. "relaxed molecular clocks"), and (iii) limitations (and future prospects) of present molecular phylogenetic dating techniques.

UNDERSTANDING OF AVIAN EVOLUTIONARY HISTORY PRIOR TO CLADISTIC THOUGHT (PRE-1960S): FOSSILS INTERPRET AN ANCIENT ORIGIN OF MODERN BIRDS

In the 19th century, comparison of known fossil extinct birds (*Archaeopteryx*, *Ichthyornis*, *Hesperornis*) to extant birds implied an ancient origin of modern birds. Early workers were impressed with the similarities between ratites (e.g. Ostrich, Emu, Kiwi) and dinosaurs. For example, Huxley (1868) famously pointed out that the dinosaur *Compsognathus* represents an intermediate stage towards modern birds. He suggested a scenario in which early ratites descended from *Compsognathus* and gave rise to modern birds. Huxley viewed the divergence of birds and dinosaurs as taking place in the late Paleozoic and he considered *Archaeopteryx* to be more derived than some living ratites. One year earlier, Cope (1867) argued for a diphyletic origin of birds. The flightless penguins and ratites were more closely related to dinosaurs, while modern birds evolved independently from *Archaeopteryx* through pterosaurian ancestry. Similarly, Vogt (1880) suggested that dinosaurs led to ratites, but he considered the ancestor of *Archaeopteryx* and nonratite birds to be a lizard. This evolutionary scenario was followed by Mivart (1881). Subsequent to the first findings of *Hesperornis*, several workers emphasized the anatomical similarities between this flightless lineage and the living ratites, particularly in lacking a keel on the sternum. Wiedersheim (1884, 1885) set out a detailed description of his view on the separate origin of the flightless ratites and other modern birds. According to his view, one lineage consisted of modern volant birds and *Archaeopteryx* sharing a common scaled-lizard-like ancestor with pterosaurs. The other lineage included ratites and *Hesperornis* as direct descendants of dinosaurs. These two lineages converged somewhere in the late Paleozoic or early Triassic in a common "Ur-ancestor." Lowe (1928) considered ratites as primarily flightless and he proposed that ratites were ancestral to *Archaeopteryx*, *Ichthyornis*, and modern volant birds. Thus, the perception of an ancient (and sometimes dual) origin of modern birds (Paleozoic or Triassic) was not an isolated one, but shared by many authors until the middle of the 20th century (see also; Lindsay, 1885; Lucas, 1916; von Steinmann, 1922; Heilman, 1926; Holmgren, 1955; Verheyen, 1960; von Blotzheim, 1960; Friant, 1968). These scenarios are certainly beyond the current discussion, but they illustrate the propensity of early workers to emphasize character similarities in avian classifications and to view modern birds as having an ancient origin among reptiles.

DUAL PERSPECTIVES ON THE HISTORICAL DEPTH OF THE AVIAN TREE: CLADISTIC REINTERPRETATION OF FOSSILS AND THE "ROCK–CLOCK" DEBATE

Two paradigm shifts occurred in the 1960s that advanced our understanding of avian evolutionary history. These shifts were the development of cladistics in classification (Meise, 1963; Hennig, 1966), and the molecular clock in genetics (Zuckerkandl & Pauling, 1962, 1965). Although initial use was infrequent, both methods have become increasingly popular since the 1990s

through the development of faster computers, algorithms, and polymerase chain reaction (PCR). In the 1990s, major gaps between modern birds and *Archaeopteryx* were filled in with the findings of many Enantiornithines, *Confuciusornis*, and *Patagopteryx* (Chiappe, 1995; Chapter 3, this volume). Soon thereafter, major divisions among modern birds were becoming defined by molecular data (Sibley & Ahlquist, 1990; Groth & Barrowclough, 1999; van Tuinen *et al.*, 2000) and the phylogenetic history of modern and extinct birds elucidated (Chiappe & Dyke, 2002, 2006). Cladistic reinterpretation has removed the majority of Cretaceous "modern birds" from Neornithes, while other putatively "modern" birds are too fragmentary to be of cladistic use (Dyke & van Tuinen, 2004). Thus, upon reevaluation of the fossil evidence, Neornithes was reinterpreted as a much younger taxon, with diversification largely being limited to the Cenozoic (the extreme viewpoint represented by Feduccia, 1995, 2003). The published exception is that of *Vegavis iaai*, a Cretaceous member of modern waterfowl (crown Anatoidea; Clarke *et al.*, 2005). This fossil, dated at ca. 66–68 Ma (million years ago), constrains a minimally Late Cretaceous divergence of crown Anseriformes, Galloanseres, Neognathae and Neornithes.

The first DNA-sequence-based avian chronograms were also reconstructed during the mid-1990s (Cooper & Penny, 1997; Hedges *et al.*, 1996). Following the development of sophisticated, user-friendly software (see below), several divergence-time studies followed in quick succession (Kumar & Hedges, 1998; Waddell *et al.*, 1999; Haddrath & Baker, 2001; van Tuinen & Hedges, 2001; Paton *et al.*, 2002; Baker *et al.*, 2006, 2007; Ericson *et al.*, 2006; Pereira & Baker, 2006a, 2006b, 2008; Slack *et al.*, 2006; van Tuinen *et al.*, 2006; Brown *et al.*, 2007, 2008; Pereira *et al.*, 2007; Wright *et al.*, 2008). Utilizing a variety of methods and fossil calibrations, all mitochondrial studies estimated Mid- to Early Cretaceous divergences (100–140 Ma) between Paleognathae and Neognathae (van Tuinen, 2009). In addition, of constructed chronograms for 12 bird orders, nine support a Cretaceous crown ordinal divergence (Ratitae, Galliformes, Anseriformes,

Charadriiformes, Passeriformes, Falconiformes, Psittaciformes, Strigiformes, Apodiformes), with younger divergence times for the remaining investigated orders (Piciformes, Sphenisciformes, Columbiformes) (for a summary, see Hedges & Kumar, 2009). Some of these times may not reflect ordinal diversification due to uncertainty regarding monophyletic status (e.g. Falconiformes, Gruiformes, Ratitae) but the majority of these ordinal ages signify extensive gaps in the fossil record. It is presently unclear to what extent taxon sampling (Linder *et al.*, 2005; van Tuinen *et al.*, 2006; Hug & Roger, 2007), fossil sampling (Hug & Roger, 2007), rate heterogeneity (Pereira & Baker, 2006a), genomic sampling (van Tuinen & Hadly, 2004), incorrect modeling of rates and substitution patterns (Brown *et al.*, 2007; Brown *et al.*, 2008; Svennblad, 2008), incorrect fossil constraints (Brown *et al.*, 2007; Ksepka, 2009), or true lack of fossils are responsible for these tremendous gaps.

Exegesis of the relevant literature on fossil and molecular divergence time estimates culminated in the (unfortunately) so-called "rocks vs. clocks" debate (Benton, 1999; Easteal, 1999), where (generally speaking) one source of historical information was regarded as vastly superior to the alternative. No attempt will be made here to chronicle this supposed debate, as it is presently clear that paleontological and molecular data are largely complementary (van Tuinen *et al.*, 2006): fossils necessarily post-date divergence (speciation) events (being products of such events), while molecular estimates (which, generally speaking, ignore coalescent times) generally pre-date divergence events (Magallón, 2004; Brown *et al.*, 2008). Thus, estimates from paleontological vs. genetic data likely act to bracket actual speciation (divergence) times of neornithine taxa. In this same vein, the perceived "gaps" inferred from fossil vs. molecular data are almost entirely dependent upon hypotheses derived from information extrapolated from genetic sequence data of extant taxa (since fossil age is unlikely to change appreciably upon re-evaluation). So, given that current (mostly mitochondrial) molecular estimates are unapologetically ancient, how might this perceived temporal pattern be altered?

Unless lineage effects on genomic rates are significantly different between birds and mammals, both accuracy and precision of divergence time estimates will likely increase in birds with analysis of more nuclear data (van Tuinen & Hadly, 2004a, b). Nuclear-based chronograms are now becoming available and provide useful comparison to existing mitochondrial estimates. These data are extremely desirable for two important reasons: (i) generally evolving much more slowly than mitochondrial DNA, nuclear DNA sequences are free of substitutional saturation for much longer time periods, which is particularly germane to estimating the age of deep nodes in the avian evolutionary tree; and (ii) the nuclear genome offers a nearly unlimited number of unlinked, independent loci with which more general divergence time estimates can be made (while, because of physical linkage, the 37 genes of the mitochondrial genome constitute a single, potentially idiosyncratic, "superlocus"). Whether these data reduce the gap between rocks and clocks is the focus of current research. The number of nuclear loci in use is currently small, algorithms are not always appropriate (Brown *et al.*, 2007), and nuclear data are often perceived as complementary instead of alternative source of historical information. Sequencing of a plethora of nuclear genes for all major nonpasserine families (Ericson *et al.*, 2006; Hackett *et al.*, 2008) and improved modeling of rates among birds with many fossil constraints will reinvigorate investigation into the antiquity of modern birds. It has already been pointed out that fossil gaps are not universally distributed in birds but appear to be limited to the stem group of modern bird orders and families (van Tuinen *et al.*, 2006).

EXTRACTION OF PHYLOCHRONOLOGICAL INFORMATION FROM MOLECULAR SEQUENCE DATA: EVOLUTION OF OUR UNDERSTANDING OF THE "EVOLUTION OF THE RATE OF EVOLUTION"

Evolutionary biologists are ultimately interested in biological diversity – how it is generated, how it is maintained, and how it is lost. Clearly our understanding of biodiversity would be incomplete without a temporal perspective. How long, for example, does it take for a clade to reach size x? When did novel adaptive innovation y evolve? What geophysical/environmental phenomena likely triggered speciation event z? It is here that the "molecular clock" proves itself an extremely useful concept, as it allows elucidation of not only the timing of macroevolutionary events, but also the extent of their temporal clustering – what we might dub the "phylochronological" signal. Extraction of this signal enables us to construct more informed hypotheses regarding the processes and mechanisms of diversification. Indeed, for taxa with poor or absent fossil records, a molecular clock may be the only means by which to infer phylochronological patterns. However, despite its utility, the "molecular clock" concept has required considerable retooling to accommodate the heterogeneity ubiquitous to large molecular data sets (Bromham & Penny, 2003; Magallón, 2004; Rutschmann, 2006; Welch & Bromham, 2005). On the whole we can perceive a trend to making molecular clock models more general by relaxing simplifying assumptions of previous implementations.

Molecular clock theory was borne of the pioneering work of Emile Zuckerkandl and Linus Pauling (Zuckerkandl & Pauling, 1962, 1965), and Emanuel Margoliash (1963). Working with mammalian protein sequences, these authors remarked that sister lineages contained very similar numbers of amino acid substitutions. From these empirical observations they posited that although molecular substitution is best regarded a stochastic process, over long periods of time it can be considered approximately constant at rate λ. This simple yet powerful hypothesis yields testable predictions, which, if passed, enables us to interpret a molecular phylogeny in terms of absolute time rather than a simple nesting of clades. Assuming a strict molecular clock, dating of nodes in a tree is a trivial task. The time to the most recent common ancestor, t_{MRCA}, of two taxa separated by genetic distance d is calculated as:

$$t_{MRCA} = \frac{d}{2\lambda}$$

(the coefficient 2 is required because both lineages undergo substitutional accumulation in time t). This equation assumes, of course, that the (constant) rate of molecular evolution λ is known. The value of λ usually comes from the calibration of genetic distances with the fossil record (see below), although occasionally dates of biogeographic events are used in place of fossil calibrations. Extending this to multiple taxa is straightforward, since all lineages within the tree are assumed to share the same value of λ, although there will generally be a need to correct for stochastic deviations from ultrametericity (i.e. that all terminal branches in the phylogram line up at the present). Unfortunately, most present day molecular data sets reject the economy of the strict molecular clock. Avian dating studies wishing to employ a global molecular clock therefore require data pruning, either via "gene-shopping" (Hedges *et al.*, 1996; Kumar & Hedges, 1998) or "taxon-shopping" (van Tuinen & Dyke, 2004; van Tuinen & Hedges, 2001), to obtain a matrix that will not reject a molecular clock. This is a reasonable practice if one believes that the majority of genes and taxa conform to expectation. However, this is unreasonable if one wishes to retain all hard-earned data within an analysis, or if one believes that the processes of substitution are more heterogeneous. Indeed, on the timescale of neornithine evolution one might predict that the signal of a strict molecular clock would decay due to stochastic variations alone. We will briefly summarize here the most popular approaches to estimating divergence times with nonclock-like data.

Overdispersed clocks

Rejection of a molecular clock is typically considered as evidence for rate variation across lineages. However, as Gillespie & Langley (1979) argue, the molecular clock hypothesis (as commonly employed) actually consists of two constituent assumptions: (i) that substitution rates are constant, and (ii) that substitutions occur according to a Poisson process. An alternative interpretation of lineage-specific variability in the number of substitutions could thus be to question the second assumption; specifically, whether the variance afforded through a Poisson distribution (where the variance is equal to the mean) of substitutions through time adequately describes the variability we observe in empirical data sets. This interpretation is validated through use of alternative Gaussian (Cutler, 2000) and negative binomial (Bedford *et al.*, 2008) distributions which adequately describe lineage-specific substitution counts where a Poisson distribution fails. An interesting corollary of this alternative constant-rate high-variance molecular clock hypothesis is the absence of any assumed correlation of rate with phylogeny (see below); in an overdispersed clock long branches need not be clustered on a tree, because "rate of evolution" is not assumed to be a heritable trait (a strict molecular clock, in contrast, carries with it the implicit assumption that substitution rate is entirely heritable). Indeed, application of Cutler's (2000) method (as implemented in the program dating5) to avian mtDNA sequence data revealed distinct temporal diversification patterns not revealed through other methods (Brown *et al.*, 2008). Whereas autocorrelated methods (see below) generally infer gradual patterns of diversification, the overdispersed reconstruction infers both short periods of extensive diversification and long periods of stasis. Despite the perceived promise and interpretive simplicity of an overdispersed clock, existing analytical programs are limited and thus are rarely used.

Local clocks

The more common approach to the rejection of a Poisson-distributed molecular clock is to assume among-lineage rate heterogeneity. The most straightforward way to extend the original molecular clock concept to accommodate rate variation across a tree is to employ "local" molecular clocks (Yoder & Yang, 2000). Here, regions of a tree are assumed to evolve according to a strict Poisson-distributed molecular clock, but different clades can have different rates λ_i. This is certainly a better description of empirical data, as different clades of birds display markedly different rates of

molecular evolution as revealed through trends in branch length heterogeneity in reconstructed phylograms (e.g. Hackett *et al.*, 2008). However, the number of individual local clocks and their discrete placement in the tree is inherently subjective. Although various local clock models can be compared statistically (in the program PAML; Yang, 2007), the local clock approach has largely been abandoned for approaches that let the data themselves indicate where changes in evolutionary rate likely occur within a tree.

Rate smoothing

The local clock approach above assumes a few (potentially great) discrete changes in the Poisson rate of substitution λ_i across a tree. A more general approach is to allow an arbitrary number of such changes in λ_i, but to "smooth" transitions in rate to minimize large changes. Two general approaches exist, these differing in the direction of smoothing; it is debatable which direction is optimal. On the one hand, sister lineages are by definition the exact same age, and because they share a recent common ancestor they are likely of similar size, life history traits, DNA repair efficiency, etc. – characteristics that are thought to influence substitution rates. It therefore seems sensible to focus on sister lineages when minimizing deviations from a molecular clock. This is the approach that PATHd8 (Britton *et al.*, 2007) takes, and can be regarded as a smoothed local-clock approach. Here, path lengths are averaged successively from the tips of a tree back through internal nodes. The averaged sister pathlengths are assumed to obey a strict Poisson molecular clock, although different sister-pairs can have different rates because of branch length differences or reference to simple fossil-imposed age constraints. The simplicity of the calculations involved allow for the dating of very large trees (hundreds of taxa), very quickly (typically $\ll 1$ s of computation). However, when applied to avian data (Ericson *et al.*, 2006), the approach infers divergence time estimates that are strikingly younger than those from alternative more rigorous approaches (Brown *et al.*, 2007, 2008), suggesting that the method may be overly simplistic. Indeed, the current implementation of PATHd8 has been demonstrated to be statistically biased, generating overly young and precise divergence time estimates for even relatively simple simulated sequences (Svennblad, 2008). More work is required to see if this method can be rescued, but at the moment it is best considered with caution.

The alternate direction of smoothing, from ancestor to descendant branches, may therefore be more reasonable; indeed, the evolution of rate variation from ancestor to descendant branches is the very process we are trying to understand. If the trait "rate of molecular substitution" is heritable to any degree (for instance, because of inheritance of DNA-repairing enzymes), then smoothing in this direction can be expected to extract more meaningful information. The program r8s (Sanderson, 2003) takes such an autocorrelated-rates approach, but penalizes rates that change too quickly across the tree in a fashion akin to smoothing in regression analysis. In nonparametric rate smoothing (NPRS; Sanderson, 1997) optimal rates and dates are inferred by simply minimizing the penalty function. However, NPRS is generally not recommended for most data sets as it tends towards overfitting, inferring large fluctuations in rate where short branch lengths are located. The alternative semiparametric penalized likelihood (PL; Sanderson, 2002) approach is an extension to NPRS which involves a smoothing parameter that controls the relative contributions of rate smoothing and data-fitting; large values of the smoothing parameter favor minimizing rate changes over data-fitting (tending towards a molecular clock), while small values of the smoothing parameter tend towards NPRS. The optimal smoothing value is determined through a data-driven sequence-based cross-validation procedure. Application of this method to avian data matrices has yielded reasonably consistent divergence time estimates that generally agree with more realistic, computationally intensive approaches (Haddrath & Baker, 2001; Paton *et al.*, 2002; Harrison *et al.*, 2004; Baker *et al.*, 2005; Brown *et al.*, 2008). Nevertheless, the method (as currently implemented) has the drawback that a single global smoothing parameter controls the extent of rate change across an entire tree. This assumption may be unreasonable for trees that potentially span tens or hundreds of millions of years if constraints on substitution rate variability have changed appreciably over evolutionary time. Likewise, the autocorrelation assumption that forms the basis of r8s smoothing has recently come under question (Drummond *et al.*, 2006), and for birds specifically (Brown *et al.*, 2008), although the extent of the decay of autocorrelation of rates is likely dependent upon taxon sampling and tree age.

Modeling "relaxed" clocks

A more rigorous approach to dating nonclock-like molecular genetic sequence data is to explicitly model rate heterogeneity itself. Although much more computationally time-consuming than the methods above, these model-based approaches have two basic advantages: (i) they extract more biologically interpretable information, and (ii) the relative fit of alternative candidate relaxed-clock models to empirical sequence data can be computed statistically using existing tools. Modeling rate heterogeneity generally takes one of two forms: (i) modeling the *process* of rate change across a tree, or (ii) modeling the *product* of such rate changes. The former methods require assumptions about how rate change proceeds over evolutionary time. The most highly utilized model of this type was developed through the work by Jeff Thorne and colleagues on modeling the "the rate of evolution of the rate of evolution" (Thorne *et al.*, 1998; Kishino *et al.*, 2001; Thorne & Kishino, 2002). Implemented in the popular MCMC program Multidivtime (Thorne, 2003), this model implicitly assumes an autocorrelated process of rate evolution from ancestor to descendent branches. Specifically, the model assumes that the substitution rate at the descendent branch conforms to a lognormal distribution, the mean of which is equal to the logarithm of the rate at its ancestral branch. The variance of this lognormal distribution is determined by both a sampled autocorrelation parameter v and the inferred length of time separating the two nodes. An intuitively satisfying property of this approach is that sister branches, while being autocorrelated (to some degree) in rate to their common ancestral branch, can nevertheless potentially differ considerably from one another. This model has been applied extensively to avian data matrices (Pereira & Baker, 2006a,b , 2008; Slack et al., 2006; Baker et al., 2007; Brown et al., 2007, 2008; Pereira et al., 2007, and is largely responsible for the statistically robust Cretaceous molecular time-scale that has been emerging over the past decade (Figure 12.1). Nevertheless, Multidivtime is starting to show its age, being limited to a simple substitution model (F84 + G), which is inappropriate for large taxon and/or character samples, and simple "hard" age calibrations (upper, minimum, or fixed; see below).

Alternate approaches to modeling autocorrelated-rates across a tree are based upon the Ornstein–Uhlenbeck (OU) process. Also called a "mean-reverting process," here rates can change across a tree in a near-Brownian fashion, although an equilibrium rate is enforced through use of a "spring" restraint that "pulls" a rate (either down or up) towards the equilibrium rate with a force proportional to how far it is removed from the mean; in effect, the model penalizes extreme rates. Unlike the autocorrelated lognormal model of Multidivtime, the OU model (and its variants; see below) possesses a stationary distribution (i.e. the mean and variance do not change over time or across the tree), and hence is a very different conceptual take on the process of rate evolution; whereas in Multidivtime a branch-specific rate is explicitly tied to its ancestral branch rate, in methods employing the OU process all rates are instead tied to the same underlying equilibrium rate. The idea of the reality of an "underlying equilibrium rate" is a deep and provocative assumption about the process of rate evolution, and in a way the OU approach can be thought of as modeling the distribution of rates around an "absolute" molecular clock. An early implementation of the OU process for phylochronological analysis (in the program PhyBayes; Aris-Brosou & Yang, 2002, 2003) was shown to be flawed through inappropriate priors

overly influencing the results (Welch et al., 2005); in particular, the priors were biased to infer higher rates of substitution near the root of the tree. However, the recent implementation of the "CIR" model (essentially a "squared-OU" model, which preserves rate positivity and avoids the prior bias above) in the program PhyloBayes overcomes many of these problems, and is well supported by empirical data (Lepage et al., 2007). Despite the promise and success of this approach, its very recent development has meant that avian molecular genetic data have yet to be analyzed in this way.

While the models above represent autocorrelated rate evolution as a continuous process, it is also possible to build piecewise relaxed clock models. For example, the compound Poisson stochastic process approach of Huelsenbeck et al. (2000) assumes that substitutions generally occur according to the standard Poisson-distributed molecular clock, but that changes in rate λ_i occur along the tree according to an independent Poisson process. Using standard MCMC machinery, divergence times and their associated credibility intervals can be estimated while accommodating uncertainty in all other model parameters, including the frequency and degree of discrete rate changes. This model can be thought of as a generalized local clock approach, but has an advantage over the subjective procedure above in that the number, degree, and location of the inferred shifts in substitution rate are data-driven instead of investigator-proclaimed. Unfortunately, a lack of development beyond its introductory paper (Huelsenbeck et al., 2000), together with a lack of available software (but see Himmelmann & Metzler, 2009), has meant that this straightforward and biologically interpretable approach has yet to realize its potential.

In contrast to the models above, the second class of models, those that model the *product* of rate heterogeneity, do not make any explicit assumptions about how rate changes. Rather, these models make assumptions about the shape of the resulting distribution of rates, and assume that branch-specific rates are each drawn independently from this distribution.

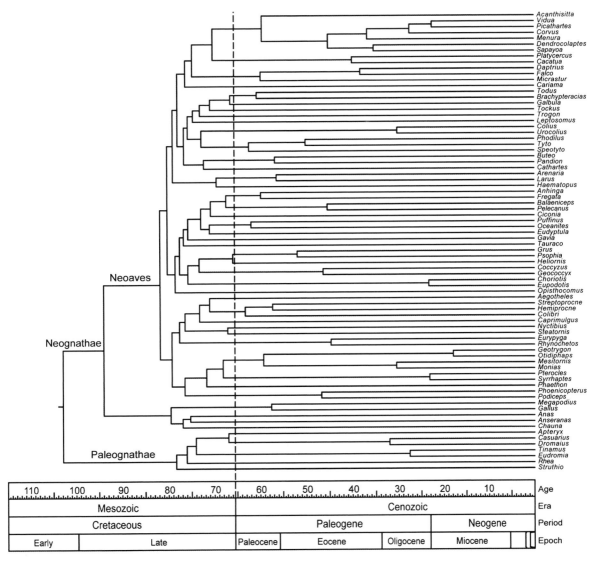

Fig. 12.1   Molecular genetic evolutionary timescale of neornithine diversification inferred from the analysis of data from Hackett *et al.* (2008) utilizing a Bayesian uncorrelated lognormally-distributed relaxed molecular clock in BEAST (Brown *et al.*, unpublished data). The vertical dashed line identifies the Cretaceous–Paleogene (K–Pg) boundary.

Using MCMC methodologies, proposed branch rates are accepted at a frequency that is proportional to their posterior probabilities. Unlike the other models above (but similar to the overdispersed clock), these models make no assumptions regarding an autocorrelation of substitution rates across a tree, and so are frequently referred to as "uncorrelated" models. There is good reason to question the autocorrelation assumption; even if "rate of evolution" *is* heritable, the accumulation of stochastic variation over millions of years may mean that autocorrelation decays to zero along the branches separating the nodes in a tree (Drummond *et al.*, 2006). Regardless, relaxing the autocorrelation assumption means that autocorrelation itself can be tested; if rates are indeed autocorrelated (and sufficient signal is present in the data) then the sampled rates should reflect that pattern. Autocorrelation has only been evaluated once in a broad-scale sample of Neornithes (Brown *et al.*, 2008), and was rejected. However, because of the time-dependency of autocorrelation decay, recovery of a genuine signal of autocorrelation will likely require a more dense taxon sampling than has been performed previously so that time intervals between nodes can be minimized.

Among these uncorrelated models, the uncorrelated lognormal model has enjoyed the most use to date, and has been shown to be superior to an uncorrelated exponential model of rate variation for a number of data sets (Drummond *et al.*, 2006), including birds (Brown *et al.*, 2008). The flexibility of the lognormal distribution means that it is able to be fit to a broad range of rate distribution shapes, and explains why it is implemented in several Bayesian relaxed clock applications, including BEAST (Drummond *et al.*, 2006; Drummond & Rambaut, 2007), MCMCtree (Rannala & Yang, 2007; Yang, 2007), and PhyloBayes (Lartillot & Philippe, 2004; Lepage *et al.*, 2007). Each of these packages has its own advantages. MCMCtree, for example, explicitly allows for potential error in fossil calibration ages (e.g. from stratum misidentification) by adding non-zero probability tails to otherwise "hard" fossil constraints (Yang & Rannala, 2006). The benefit of using PhyloBayes is that it implements seven different clock models, enabling a researcher to statistically compare alternative models using the same statistical machinery rather than relying on indirect comparisons across software packages/implementations (Lepage *et al.*, 2007). Finally, in addition to flexible xml-coding support, which allows for the construction of arbitrarily complex models, BEAST is unique in that of all the relaxed clock methods available, it is the only one that does not require a fixed tree topology. This inclusion of topological uncertainty is especially appealing for avian studies, where higher level relationships are still unsettled.

In summary, there are currently a number of approaches readily available to researchers for phylochronological reconstruction using nonclock-like molecular genetic sequences (Table 12.1), although none of them can be considered a panacea (see below). These approaches run the gamut from quick-and-dirty "corrections" to an imperfect clock (e.g. PATHd8) to sophisticated descriptions of either the process of rate evolution itself (e.g. CIR model) or the product of such evolution (e.g. uncorrelated lognormal model). Given the breadth of choices available, the ideal course of action would be to test several distinct approaches to see if the phylochronological signal is consistent across model/method assumptions (Linder *et al.*, 2005; Britton *et al.*, 2007; Hug & Roger, 2007; Lepage *et al.*, 2007; Brown *et al.*, 2008; Hipsley *et al.*, 2009): concordant results across methods would lend additional credence to resulting inferences, whereas dissonance could help identify potential model assumption violations. For example, a recent comparison of five dating methods on the Neornithes tree using mtDNA revealed broadly consistent origin estimates for the major clades (Brown *et al.*, 2008), while also calling into question the appropriateness of one method (PATHd8) for the particular data set.

LIMITATIONS OF CURRENT MOLECULAR PHYLOGENETIC DATING TECHNIQUES

A model represents a conceptual understanding of how "nature" influences physical entities to generate the distribution of empirical observations. Models can be constructed from empirical (inductive) or theoretical (deductive) expectations, with the ideal situation being a motivated iteration between the two sources of understanding (Box, 1976). However, a model should not endeavor to "fit an elephant" (that is, try to describe reality in its entirety; Steel, 2005), but instead attempt to extract information from the *salient* components of the underlying process,

Table 12.1 A comparison of available programs for the dating of non-clocklike molecular genetic sequences.

Program	Clock approach	Statistical inference	Input data	Multiple partitions	Age constraints	Topology	Notes
PATHd8	Rate smoothing (sister branches)	N/A	Phylogram	No	Minimum, maximum, and fixed[1]	Fixed	Extremely fast for even large trees; however, current implementation delivers overly young and precise estimates
r8s	Rate smoothing (ancestor-descendent)	Penalized likelihood	Phylogram	No	Minimum, maximum, and fixed	Fixed	Optimal smoothing determined via sequence-based cross-validation
dating5	Overdispersed clock	Maximum likelihood	Phylogram	No	Minimum, maximum, and fixed	Fixed	Software has not seen recent development
PAML	Local clocks	Maximum likelihood	Nucleotide or amino acid sequences	Yes	Minimum, maximum, and fixed	Fixed	Location of discrete clocks is user-defined
HLS2000[†]	Compound Poisson process	Bayesian MCMC	DNA sequences	No	Fixed*	Fixed	Software has not seen recent development
Multidivtime[‡]	Autocorrelated model	Bayesian MCMC	Nucleotide or amino acid sequences	Yes	Minimum, maximum, and fixed	Fixed	Nucleotide substitution model limited to f84, which is likely too simplistic for most large/old trees
PhyBayes	OU process	Bayesian MCMC	Nucleotide sequences	Yes	N/A[¶]	Fixed	Software has not seen recent development
MCMCtree[§]	Uncorrelated lognormal	Bayesian MCMC	Nucleotide or amino acid sequences	Yes	Probability distributions	Fixed	Explicitly accommodates potential error in calibration ages
PhyloBayes	Various autocorrelated and uncorrelated models	Bayesian MCMC	Nucleotide or amino acid sequences	No	N/a[¶]	Fixed	Seven clock models can be compared within the same software
BEAST	Uncorrelated lognormal/exponential	Bayesian MCMC	Nucleotide or amino acid sequences	Yes	Probability distributions	Estimated	Xml-coding support allows for the construction of arbitrarily complex models of sequence evolution

* A fixed-age constraint is required.
[†]This program was not given a proper name, so the initials of the authors are used. Not much is known about this program as it is not distributed and has not enjoyed use beyond the original study. Although a number of possible extensions are discussed by the authors, these apparently have not been implemented.
[‡]The initial steps of analysis require the PAML package.
[§]Part of the PAML package.
[¶]Estimates relative ages only. Absolute ages are generated through scaling relative ages to a fixed age constraint; if more than one constraint is available an average date across all constraints is typically estimated.

formalized with estimable parameters. The idea of *saliency* should be recognized as a relative concept; with greater thought, and a broader collection of empirical observations, our idea of what constitutes a "salient" component of a process continues to evolve, leading to a richer understanding of the sources of variation. Such is the condition of our understanding of the processes of molecular evolution. Larger molecular genetic data matrices (in terms of both taxon and especially character sampling) have afforded an increased power to identify more subtle (but increasingly important) sources of variation. Consequently, several simplifying assumptions in our standard modeling of the molecular genetic evolutionary process are currently being challenged, and may eventually translate to improvements in the extraction of phylochronological signal or potentially identify biases of past methods.

Molecular substitution models

Molecular evolution is typically modeled as a continuous-time Markovian substitution process. This conceptual framework, originally constrained for practical reasons (Felsenstein, 1981), carries with it several explicit and implicit assumptions: (i) *stationarity* (the probabilities of stochastic substitution do not change through time, or are in equilibrium); (ii) *homogeneity* (the equilibrium character frequencies and substitution-rate matrix are identical across lineages); (iii) *time-reversibility* (the process of substitution looks the same both forwards and backwards in time); and (iv) *independence* (all sites within an alignment are considered identical and independently distributed (i.i.d.) realizations of the same evolutionary process). Although held by all of the relaxed clock methods above, none of these assumptions is likely to strictly hold true (and indeed empirical data exist to contest each of them), however, since all models are wrong we should concern ourselves with what is *importantly* wrong (Box, 1976).

Of these assumptions, *independence* is unique in that it focuses along a genetic sequence (rather than across a tree like the remaining assumptions). This assumption is actually a composite assumption: (i) sites evolve independently from one another, and (ii) all sites evolve according to the same underlying process (in practice, the same substitution model). Strict violation of the first component is ensured through the physical linkages between nucleotides, although inclusion of molecular markers from disparate regions of the genome (say, different chromosomes) can represent "more independent" information. Violation of the second component is readily apparent through inspection of the characteristics of various character classes (e.g. genes, coding/noncoding regions, codon positions, etc.), which often differ considerably in terms of nucleotide composition and levels of polymorphism. Failure to accommodate for this will necessarily lead to a compromised inference (where, for example, relative rate parameters and equilibrium character state frequencies are averages over potentially distinct genomic regions). Nevertheless, the *independence* assumption is also unique in that its violation is all but solved. For example, the introduction of among-site gamma-distributed rate heterogeneity enormously increases the fit of models to empirical data (Yang, 1996). More generally, recently developed mixed (Lartillot & Philippe, 2004; Pagel & Meade, 2004) and partitioned (e.g. Nylander *et al.*, 2004) models allow heterogeneity in the substitutional process across sites and loci. Partitioned models are available in a number of relaxed clock methods, although they are most flexible in BEAST.

The remaining model assumptions above reflect expectations of the uniformity of the molecular substitution process(es) over both time and lineages. The adoption of the *time-reversibility* assumption is due almost entirely to numerical convenience: it both reduces the required number of substitution parameters to be estimated (as compared to the more general model; Rodríguez *et al.*, 1990), and allows for efficient computation of the likelihood of an unrooted tree via Felsenstein's "pulley principle" (Felsenstein, 1981). However, reasons to settle for simpler reversible models are rapidly dematerializing: (i) Bayesian MCMC-sampling methods, together with increasingly powerful and available computational resources, can easily accommodate the relatively small increase in the number of estimable parameters, and (ii) signal from large present day empirical data matrices are revealing support for irreversibility (Squartini & Arndt, 2008), overturning earlier conclusion from studies (Yang, 1994) that may simply have suffered from a lack of power.

The final two assumptions, *homogeneity* and *stationarity*, are tightly related in that violation of one typically involves violation of the other. Of all the assumptions, these two are most likely to influence phylochronological inference through biasing both topology and branch length estimation. Moreover, the strict validity of these assumptions over evolutionary time spans, such as the diversification history of Neornithes, is dubious. Indeed, clear evidence for the violation of one or both of these assumptions is the observed empirical base compositional biases across lineages that cannot be described by stochastic variation alone. While such compositional biases can potentially be masked through data filtering (e.g. translating to amino acids for coding sequences, or employing R–Y coding), a more satisfying approach that makes use of more evolutionary information is to model compositional changes themselves. Fortunately, several nonhomogeneous/nonstationary models exist which do just that (e.g. Galtier & Gouy, 1998), for example by allowing base composition to change across a tree according to a compound piecewise-constant Poisson stochastic process (Blanquart & Lartillot, 2006), similar to the compound Poisson relaxed molecular clock model (Huelsenbeck *et al.*, 2000) above. A distinct violation of the stationarity assumption involves the concept of "heterotachy," where site-specific substitution rates change in different parts of the tree (Lopez *et al.*, 2002). Thankfully, substitution models now exist that accommodate heterotachous evolution (Tuffley & Steel, 1998; Kolaczkowski & Thornton, 2008; Pagel & Meade, 2008; Wu *et al.*, 2008). In summary, it is not yet known whether biologically realistic violations of these substitution model assumptions will surface as

significant biases to phylochronological signal. However, we have at our disposal solutions to each of these problems, all that remains is to graft constituent models together (e.g. Blanquart & Lartillot, 2008), identify the salient features, and apply the resulting models to the problem of phylochronological inference.

Modeling of rate evolution

Besides concerns regarding the suitability of the level of sophistication of existing molecular substitution models, we must also consider whether the correct components of evolutionary rate heterogeneity are indeed being modeled. It is unclear, for example, if substitution rate evolution is best considered autocorrelated in time (such that "rate" is a heritable character), or if an "episodic" clock (Gillespie, 1984) is a better description of empirical data. Correlated rate models, if valid, enable greater inferential precision because rate/date estimation at a given node can make use of not only local but also distant evolutionary information (Lepage *et al.*, 2007). The uncorrelated models above are episodic clocks that offer no explanation of *why* rates vary. A distinct type of episodic clock involves punctuated (or speciational) molecular evolution, where substitution rates are elevated during speciation, with the result that the lengths of the branches (in terms of the expected number of substitutions per site) in a clade of a tree are positively correlated to the number of speciation events (Pagel *et al.*, 2006). Such a scenario could explain stark branch length differences between the speciose Passeriformes and depauperate Pelecaniformes in reconstructed phylograms (Hackett *et al.*, 2008). Indeed, punctuated *morphological* evolution has been inferred in birds (Paleognaths; Cubo, 2003). However, a signal of punctuated *molecular* substitution rates (which could potentially mislead phylogenetic dating) was not found in a recent study of Neornithes (Brown *et al.*, 2008), although identification of this kind of signal would surely benefit greatly by increased taxon sampling. Another potential explanatory variable to consider is effective population size, which has been shown to be correlated (negatively) with substitution rate in a range of eukaryotic taxa (Bedford *et al.*, 2008). Additionally, effective population size strongly determines the rate of lineage sorting, and since lineage sorting post-dates speciation it makes sense to estimate divergence times and effective population sizes simultaneously (Liu & Pearl, 2007; Rannala & Yang, 2003). Given that many avian species differ greatly in effective population size (think of rails versus gulls), this may be a worthwhile avenue of research to pursue. A final consideration involves the perceived time-dependency of molecular substitution rates (Ho & Larson, 2006; Ho *et al.*, 2005); here, extant population-level polymorphisms that would not persist over evolutionary timescales (i. e. mutations that do not become substitutions) bias reconstruction methods into inferring that substitution rates are higher in the present than they were in the past. The influence of this pervasive phenomenon on divergence time estimation has not yet been fully investigated, and not at all in birds.

Age constraints

One of the most compelling developments in recent relaxed clock model implementations is the ability to construct age probability distributions for fossil-calibrated nodes. Previous molecular dating techniques (e.g. r8s, Multidivtime, and PATHd8) allowed only the enforcement of "hard" age constraints: (i) absolute minimum (i.e. that the speciation event represented by the calibrated node *must* pre-date the fossil; the fossil, of course, being a product of the speciation event), (ii) absolute maximum (information which, strictly speaking, cannot come from the fossil record), or (iii) fixed ages (i.e. the fossil perfectly represents the age of the node without error). The new probability distributions available in BEAST (Drummond & Rambaut, 2007) and MCMCtree (Yang & Rannala, 2006) offer two main advantages over these simple constraints: (i) additional information (e.g. from models of fossil preservation) can be incorporated into the calibration, effectively lending more credence to the fossil record,

and (ii) uncertainty in the age of the fossil itself can be accommodated. However, the same flexibility that makes these distributions so attractive unfortunately also makes them inherently subjective. Although rightly considered with enthusiasm (Ho, 2007; Ho & Phillips, 2009), there is presently no rigorous protocol for determining the optimal shape (e.g. Gaussian, lognormal, uniform, exponential, etc.; Figure 12.2) and breadth of these distributions, which makes direct comparisons across studies difficult. A joint collaboration of paleontologists and molecular phylogeneticists working on this problem would allow greater extraction of phylochronological signal from the fossil record, and subsequently generate better divergence time estimates.

Study design

Finally, improvements in divergence time estimation will require systematic attention to sampling with respect to which loci, taxa, and fossil calibrations should be included in a given study. For example, over long evolutionary timescales mtDNA can be expected to exhibit substitutional saturation, which may bias relaxed clock studies through an underestimation of branch lengths deep in the tree (consequently underestimating the ages of deeper nodes). Under this scenario, it may be desirable to utilize slower evolving nuclear introns, or to mask saturation through translating nucleotide sequences to amino acids (if coding) or otherwise through R–Y recoding (Woese *et al.*, 1991). The extent of taxon sampling has been found to be influential in molecular dating (e.g. Linder *et al.*, 2005), presumably due to node-density effects (Venditti *et al.*, 2006), where more substitutions are discovered (making branch lengths longer) in regions of the tree with higher taxon sampling. Lastly, while it is generally a good strategy to incorporate calibration information from as many fossils as possible (Bremer *et al.*, 2004; Hug & Roger, 2007), it is imperative that these fossils are scrutinized closely, as one incorrectly dated or taxonomically misdiagnosed fossil can potentially invalidate an entire analysis. It thus seems prudent to test

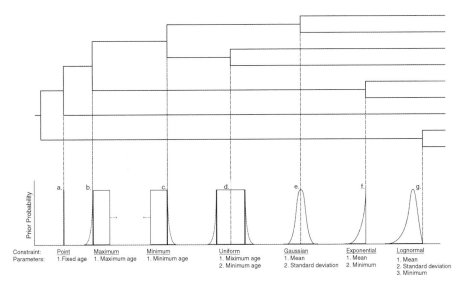

Fig. 12.2 Alternative temporal calibration constraints for use in molecular divergence time estimation. (a) At the extreme, the age of a fossil may be used as a point estimate for the divergence of two taxa. Here, the age of the node is treated as known without error; hence, potential divergence times above or below the point estimate receive zero prior probability, and so are not considered in the estimation process. This is generally poor practice, as (i) potential errors in phylogenetic placement and geological dating of the fossil are ignored, and (ii) diagnosable fossils are unlikely to temporally correspond precisely to cladogenesis. Alternatively, fossils can be considered as (b) maximum or (c) minimum ages. (Strictly speaking, fossil evidence can only provide minimum ages, not maximum ages, as the absence of (earlier) evidence cannot necessarily be interpreted as evidence of (earlier) absence.) Typically, such calibrations are implemented as "hard" constraints (bold lines), where prior divergence times on one side of the constraint are equiprobable (arrows) while all values on the other side of the constraint receive zero prior probability. However, such calibrations can also be implemented as "soft" constraints (Yang & Rannala, 2006), where exponential distributions constituting some proportion of the overall prior density (say, 2.5%) extend beyond the "hard" bounds (dashed curves). A soft prior gives small but nonzero probability to ages beyond that of the fossil and hence accommodate potential errors in geologic dating or phylogenetic placement, potentially leading to the identification of inappropriate/invalid constraints. Minimum and maximum constraints can be combined into a single calibration. (d) When considerable uncertainty is present this is typically modeled with a uniform (equiprobable) distribution (with either "hard" or "soft" bounds). (e) In situations where existing evidence suggests some time periods are more probable *a priori* than others (say, biogeographic events), a more appropriate modeling may take the form of a Gaussian (normal) distribution. (f) The form of the exponential distribution is such that the mode (highest probability) of the distribution is the age of the fossil itself; ages younger than the fossil receive zero prior probability (i.e. a "hard" bound). Application of an exponential constraint is useful when (i) a fossil is thought to temporally correspond closely to the relevant cladogenetic event, or (ii) where additional information (say, fossil preservation curves) can be used to inform prior construction. (g) Perhaps the most appropriate distribution to model uncertainty regarding a cladogenetic event is the lognormal distribution. This distribution has a "hard" minimum at the age of the fossil, a lag until the mode of the distribution (i.e. sampled diagnosable fossils are *expected* to post-date the cladogenetic event), and has a long ("soft") tail that allows for the (small but nonzero) possibility that the divergence event occurred much earlier than what the fossil would suggest. As with the Gaussian (standard deviation) and exponential (where mean = standard deviation) distributions, the breadth of the lognormal distribution is considerably malleable to the information at hand. Nevertheless, at the moment construction of lognormal (and other) constraints is more of an art than a science; scientists would do best to investigate the sensitivity of divergence time inferences to changes in temporal prior constraints.

suites of calibrations for dating consistency (Near & Sanderson, 2004; Near *et al.*, 2005), although it should be kept in mind that exceptionally "good" fossils (i.e. those that are especially old, or more closely approximate the age of the node they are meant to date) are likely to appear "inconsistent."

In conclusion, we are at a very exciting stage in molecular phylogenetic systematics; not only are we well aware of the potential unsuitability of assumptions made by early relaxed clock approaches, but (more importantly) we have a firm grasp on what *further* tribulations may be lurking in the future. The widespread adoption of Bayesian philosophies over the past decade in particular has ushered in a new paradigm for methodological implementation, *complecto errorem* (embrace uncertainty), where uncertainty in "nuisance" parameters (model components that are essential for a salient description of the evolutionary process, but are otherwise not of direct interest) can be integrated out, rendering conclusions that are compelling to a degree that have not been heretofore possible. Moreover, the rapidly decreasing costs associated with molecular genetic sequencing means that it will soon be possible to interrogate enormous amounts of data for subtle signals of past molecular substitution rate evolution. With such information in hand we can expect more accurate, precise, and consistent phylochronological inferences, which in turn will better enable us to understand and appreciate the dynamics of neornithine diversification.

ACKNOWLEDGMENTS

JWB thanks A. Drummond for helpful discussions, J. Johnson and J. McCormack for comments on earlier drafts of the manuscript, and G. Ligeti and S. Youth for encouragement throughout.

REFERENCES

Aris-Brosou S, Yang Z. 2002. Effects of models of rate evolution on estimation of divergence dates with special reference to the metazoan 18S ribosomal RNA phylogeny. *Systematic Biology* **51**: 703–714.

Aris-Brosou S, Yang Z. 2003. Bayesian models of episodic evolution support a late Precambrian explosive diversification of the Metazoa. *Molecular Biology and Evolution* **20**: 1947–1954.

Baker AJ, Huynen LJ, Haddrath O, Millar CD, Lambert DM. 2005. Reconstructing the tempo and mode of evolution in an extinct clade of birds with ancient DNA: The giant moas of New Zealand. *Proceedings of the National Academy of Sciences USA* **102**: 8257–8262.

Baker AJ, Pereira SL, Haddrath OP, Edge K-A. 2006. Multiple gene evidence for expansion of extant penguins out of Antarctica due to global cooling. *Proceedings of the Royal Society of London B (Biological Sciences)* **273**: 11–17.

Baker AJ, Pereira SL, Paton TA. 2007. Phylogenetic relationships and divergence times of Charadriiformes genera: multigene evidence for the Cretaceous origin of at least 14 clades of shorebirds. *Biology Letters* **3**: 205–209.

Bedford T, Wapinski I, Hartl DL. 2008. Overdispersion of the molecular clock varies between yeast, drosophila and mammals. *Genetics* **179**: 977–984.

Benton MJ. 1999. Early origins of modern birds and mammals: molecules vs. morphology. *BioEssays* **21**: 1043–1051.

Blanquart S, Lartillot N. 2006. A Bayesian compound stochastic process for modeling nonstationary and nonhomogeneous sequence evolution. *Molecular Biology and Evolution* **23**: 2058–2071.

Blanquart S, Lartillot N. 2008. A site- and time-heterogeneous model of amino acid replacement. *Molecular Biology and Evolution* **25**: 842–858.

Box GEP. 1976. Science and statistics. *Journal of the American Statistical Association* **71**: 791–799.

Bremer K, Friis EM, Bremer B. 2004. Molecular phylogenetic dating of asterid flowering plants shows early Cretaceous diversification. *Systematic Biology* **53**: 496–505.

Britton T, Anderson CL, Jacquet D, Lundqvist S, Bremer K. 2007. Estimating divergence times in large phylogenetic trees. *Systematic Biology* **56**: 741–752.

Bromham L, Penny D. 2003. The modern molecular clock. *Nature Reviews Genetics* **4**: 216–224.

Brown JW, Payne RB, Mindell DP. 2007. Nuclear DNA does not reconcile "rocks" and "clocks" in Neoaves: a comment on Ericson *et al. Biology Letters* **3**: 257–259.

Brown JW, Rest JS, García-Moreno J, Sorenson MD, Mindell DP. 2008. Strong mitochondrial DNA support for a Cretaceous origin of modern avian lineages. *BMC Biology* **6**: 6.

Chiappe LM. 1995. The first 85 million years of avian evolution. *Nature* **378**: 349–355.

Chiappe LM, Dyke GJ. 2002. The Mesozoic radiation of birds. *Annual Review of Ecology and Systematics* **33**: 91–124.

Chiappe LM, Dyke GJ. 2006. The early evolutionary history of birds. *Journal of the Palaeontological Society of Korea* **22**: 133–151.

Clarke JA, Tambussi CP, Noriega JI, Erikson GM, Ketcham RA. 2005. Definitive fossil evidence for the extant avian radiation in the Cretaceous. *Nature* **433**: 305–308.

Cooper A, Penny D. 1997. Mass survival of birds across the Cretaceous-Tertiary boundary: molecular evidence. *Science* **275**: 1109–1113.

Cope ED. 1867. An account of the extinct reptiles which approached the birds. *Proceedings of the Academy of Natural Sciences of Philadelphia* **1867**: 234–235.

Cubo J. 2003. Evidence for speciational change in the evolution of ratites (Aves: Palaeognathae). *Biological Journal of the Linnean Society* **80**: 99–106.

Cutler DJ. 2000. Estimating divergence times in the presence of an overdispersed molecular clock. *Molecular Biology and Evolution* **17**: 1647–1660.

Drummond AJ, Rambaut A. 2007. BEAST: Bayesian evolutionary analysis by sampling trees. *BMC Evolutionary Biology* **7**: 214.

Drummond AJ, Ho SYW, Phillips MJ, Rambaut A. 2006. Relaxed phylogenetics and dating with confidence. *PLoS Biology* **4**: e88.

Dyke GJ, van Tuinen M. 2004. The evolutionary radiation of modern birds (Neornithes): reconciling molecules, morphology and the fossil record. *Zoological Journal of the Linnean Society* **141**: 153–177.

Easteal S. 1999. Molecular evidence for the early divergence of placental mammals. *BioEssays* **21**: 1052–1058.

Ericson PGP, Anderson CL, Britton T, Elzanowski A, Johansson US, Källersjö M, Ohlson JI, Parsons TJ, Zuccon D, Mayr G. 2006. Diversification of Neoaves: integration of molecular sequence data and fossils. *Biology Letters* **4**: 543–547.

Feduccia A. 1995. Explosive evolution in tertiary birds and mammals. *Science* **267**: 637–638.

Feduccia A. 2003. "Big bang" for Tertiary birds? *Trends in Ecology and Evolution* **18**: 172–176.

Felsenstein J. 1981. Evolutionary trees from DNA sequences: a maximum likelihood approach. *Journal of Molecular Evolution* **17**: 368–376.

Friant M. 1968. Sur les ceintures des membres des Oiseaux. *Acta Anatomica* **69**: 262–273.

Galtier N, Gouy M. 1998. Inferring pattern and process: maximum-likelihood implementation of a nonhomogeneous model of DNA sequence evolution for phylogenetic analysis. *Molecular Biology and Evolution* **15**: 871–879.

Gillespie JH. 1984. The molecular clock may be an episodic clock. *Proceedings of the National Academy of Sciences USA* **81**: 8009–8013.

Gillespie JH, Langley CH. 1979. Are evolutionary rates really variable? *Journal of Molecular Evolution* **13**: 27–34.

Groth JG, Barrowclough GF. 1999. Basal divergences in birds and the phylogenetic utility of the nuclear RAG–1 gene. *Molecular Phylogenetics and Evolution* **12**: 115–123.

Hackett SJ, Kimball RT, Reddy S, Bowie RCK, Braun EL, Braun MJ, Chojnowski JL, Cox WA, Han K-L, Harshman J, Huddleston CJ, Marks BD, Miglia KJ, Moore WS, Sheldon FH, Steadman DW, Witt CC, Yuri T. 2008. A phylogenomic study of birds reveals their evolutionary history. *Science* **320**: 1763–1768.

Haddrath O, Baker AJ. 2001. Complete mitochondrial DNA genome sequences of extinct birds: ratite phylogenetics and the vicariance biogeography hypothesis. *Proceedings of the Royal Society of London B (Biological Sciences)* **268**: 939–945.

Harrison GL, McLenachan PA, Phillips MJ, Slack KE, Cooper A, Penny D. 2004. Four new avian mitochondrial genomes help get to basic evolutionary questions in the late Cretaceous. *Molecular Biology and Evolution* **21**: 974–983.

Hedges SB, Kumar S (eds). 2009. *The Timetree of Life.* Oxford: Oxford University Press.

Hedges SB, Parker PH, Sibley CG, Kumar S. 1996. Continental breakup and the ordinal diversification of birds and mammals. *Nature* **381**: 226–229.

Heilman G. 1926. *The Origin of Birds.* London: Witherby.

Hennig W. 1966. *Phylogenetic Systematics.* Urbana: University of Illinois Press.

Himmelmann L, Metzler D. 2009. TreeTime: an extensible C + + software package for Bayesian phylogeny reconstruction with time-calibration. *Bioinformatics* **25**: 2440–2441.

Hipsley C, Himmelmann L, Metzler D, Muller J. 2009. Integration of Bayesian molecular clock methods and fossil-based soft bounds reveals early Cenozoic origin of African lacertid lizards. *BMC Evolutionary Biology* **9**: 151.

Ho SYW. 2007. Calibrating molecular estimates of substitution rates and divergence times in birds. *Journal of Avian Biology* **38**: 409–414.

Ho SYW, Larson G. 2006. Molecular clocks: when times are a-changin'. Trends in Genetics **22**: 79–83.

Ho SYW, Phillips MJ. 2009. Accounting for calibration uncertainty in phylogenetic estimation of evolutionary divergence times. *Systematic Biology* **58**: 367–380.

Ho SYW, Phillips MJ, Cooper A, Drummond AJ. 2005. Time dependency of molecular rate estimates and systematic overestimation of recent divergence times. *Molecular Biology and Evolution* **22**: 1561–1568.

Holmgren N. 1955. Studies on the phylogeny of birds. *Acta Zoologica* **36**: 243–328.

Huelsenbeck JP, Larget B, Swofford DL. 2000. A compound Poisson process for relaxing the molecular clock. *Genetics* **154**: 1979–1892.

Hug LA, Roger AJ. 2007. The impact of fossils and taxon sampling on ancient molecular dating analyses. *Molecular Biology and Evolution* **24**: 1889–1897.

Huxley TH. 1868. On the animals which are most nearly intermediate between birds and reptiles. *Annals and Magazine of Natural History* **4**: 66–75.

Kishino H, Thorne JL, Bruno WJ. 2001. Performance of a divergence time estimation method under a probabilistic model of rate evolution. *Molecular Biology and Evolution* **18**: 352–361.

Ksepka DT. 2009. Broken gears in the avian molecular clock: new phylogenetic analyses support stem galliform status for Gallinuloides wyomingensis and rallid affinities for Amitabha urbsinterdictensis. *Cladistics* **25**: 173–197.

Kolaczkowski B, Thornton JW. 2008. A mixed branch length model of heterotachy improves phylogenetic accuracy. *Molecular Biology and Evolution* **25**: 1054–1066.

Kumar S, Hedges SB. 1998. A molecular timescale for vertebrate evolution. *Nature* **392**: 917–920.

Lartillot N, Philippe H. 2004. A Bayesian mixture model for across-site heterogeneities in the amino-acid replacement process. *Molecular Biology and Evolution* **21**: 1095–1109.

Lepage T, Bryant D, Philippe H, Lartillot N. 2007. A general comparison of relaxed molecular clock models. *Molecular Biology and Evolution* **24**: 2669–2680.

Linder HP, Hardy CR, Rutschmann F. 2005. Taxon sampling effects in molecular clock dating: an example from the African Restionaceae. *Molecular Phylogenetics and Evolution* **35**: 569–582.

Lindsay B. 1885. On the avian sternum. *Proceedings of the Zoological Society of London* **53**: 684–716.

Lopez P, Casane D, Philippe H. 2002. Heterotachy, an important process of protein evolution. *Molecular Biology and Evolution* **19**: 1–7.

Lowe PR. 1928. Studies and observations bearing on the phylogeny of the ostrich and its allies. *Proceedings of the Zoological Society of London* **16**: 185–247.

Lucas FA. 1916. On the beginnings of flight. *American Museum Journal* **16**: 5–11.

Margoliash E. 1963. Primary structure and evolution of cytochrome c. *Proceedings of the National Academy of Sciences USA* **50**: 672–679.

Meise W. 1963. Verhalten der staussartigen Vögel und monophylie der Ratitae. *Proceedings of the XIII International Ornithological Congress* **1**: 115–125.

Mivart G. 1881. A popular account of chameleons. *Nature* **24**: 309–312.

Near TJ, Sanderson MJ. 2004. Assessing the quality of molecular divergence time estimates by fossil calibrations and fossil-based model selection. *Philosophical Transactions of the Royal Society of London B (Biological Sciences)* **359**: 1477–1483.

Near TJ, Meylan PA, Shaffer HB. 2005. Assessing concordance of fossil calibration points in molecular clock studies: an example using turtles. *American Naturalist* **165**: 137–146.

Nylander JA, Ronquist AF, Huelsenbeck JP, Nieves-Aldrey JL. 2004. Bayesian phylogenetic analysis of combined data. *Systematic Biology* **53**: 47–67.

Pagel M, Meade A. 2008. Modelling heterotachy in phylogenetic inference by reversible-jump Markov chain Monte Carlo. *Philosophical Transactions of the Royal Society of London B (Biological Sciences)* **363**: 3955–3964.

Pagel M, Venditti C, Meade A. 2006. Large punctuational contribution of speciation to evolutionary divergence at the molecular level. *Science* **314**: 119–121.

Paton T, Haddrath O, Baker AJ. 2002. Complete mitochondrial DNA genome sequences show that modern birds are not descended from transitional shorebirds. *Proceedings of the Royal Society of London B (Biological Sciences)* **269**: 839–846.

Pereira SL, Baker AJ. 2006a. A mitogenomic timescale for birds detects variable phylogenetic rates of molecular evolution and refutes the standard molecular clock. *Molecular Biology and Evolution* **23**: 1731–1740.

Pereira SL, Baker AJ. 2006b. A molecular timescale for galliform birds accounting for uncertainty in time estimates and heterogeneity of rates of DNA substitutions across lineages and sites. *Molecular Phylogenetics and Evolution* 38: 499–509.

Pereira SL, Baker AJ. 2008. DNA evidence for a Paleocene origin of the Alcidae (Aves: Charadriiformes) in the Pacific and multiple dispersals across northern oceans. *Molecular Phylogenetics and Evolution* 46: 430–445.

Pereira SL, Johnson KP, Clayton DH, Baker AJ. 2007. Mitochondrial and nuclear DNA sequences support a Cretaceous origin of Columbiformes and a dispersal-driven radiation in the Paleogene. *Systematic Biology* 56: 656–672.

Rannala B, Yang Z. 2007. Inferring speciation times under an episodic molecular clock. *Systematic Biology* 56: 453–466.

Rodríguez F, Oliver JL, Marín A, Medina JR. 1990. The general stochastic model of nucleotide substitution. *Journal of Theoretical Biology* 142: 485–501.

Rutschmann F. 2006. Molecular dating of phylogenetic trees: A brief review of current methods that estimate divergence times. *Diversity and Distributions* 12: 35–48.

Sanderson MJ. 1997. A nonparametric approach to estimating divergence times in the absence of rate constancy. *Molecular Biology and Evolution* 14: 1218–1231.

Sanderson MJ. 2002. Estimating absolute rates of molecular evolution and divergence times: a penalized likelihood approach. *Molecular Biology and Evolution* 19: 101–109.

Sanderson MJ. 2003. r8s: inferring absolute rates of molecular evolution and divergence times in the absence of a molecular clock. *Bioinformatics* 19: 301–302.

Sibley CG, Ahlquist JE. 1990. *Phylogeny and Classification of Birds.* Haven: Yale University Press.

Slack KE, Jones CM, Ando T, Harrison GL, Fordyce RE, Penny D. 2006. Early penguin fossils, plus mitochondrial genomes, calibrate avian evolution. *Molecular Biology and Evolution* 23: 1144–1155.

Squartini F, Arndt PF. 2008. Quantifying the stationarity and time reversibility of the nucleotide substitution process. *Molecular Biology and Evolution* 25: 2525–2535.

Steel M. 2005. Should phylogenetic models be trying to "fit an elephant"? *Trends in Genetics* 21: 307–309.

Svennblad B. 2008. Consistent estimation of divergence times in phylogenetic trees with local molecular clocks. *Systematic Biology* 57: 947–954.

Thorne JL. 2003. MULTIDISTRIBUTE. Available from the author (http://statgen.ncsu.edu/thorne/multidivtime.html).

Thorne JL, Kishino H. 2002. Divergence time and evolutionary rate estimation. *Systematic Biology* 51: 689–702.

Thorne JL, Kishino H, Painter IS. 1998. Estimating the rate of evolution of the rate of evolution. *Molecular Biology and Evolution* 15: 1647–1657.

Tuffley C, Steel M. 1998. Modeling the covarion hypothesis of nucleotide substitution. *Mathematical Biosciences* 147: 63–91.

Van Tuinen M. 2009. Birds (Aves). In *The Timetree of Life*, Hedges SB, Kumar S (eds). Oxford: Oxford University Press; 409–411.

Van Tuinen M, Dyke GJ. 2004. Calibration of galliform molecular clocks using multiple fossils and genetic partitions. *Molecular Phylogenetics and Evolution* 30: 74–86.

Van Tuinen M, Hadly EA. 2004a. Calibration and error in placental molecular clocks: a conservative approach using the Cetartiodactyl fossil record. *Journal of Heredity* 95: 200–208.

Van Tuinen M, Hadly EA. 2004b. Error in estimation of rate and time inferred from the early amniote fossil record and avian molecular clocks. *Journal of Molecular Evolution* 59: 267–276.

Van Tuinen M, Hedges SB. 2001. Calibration of avian molecular clocks. *Molecular Biology and Evolution* 18: 206–213.

Van Tuinen M, Sibley CG, Hedges SB. 2000. The early history of modern birds inferred from DNA sequences of nuclear and mitochondrial ribosomal genes. *Molecular Biology and Evolution* 17: 451–457.

Van Tuinen M, Stidham TA, Hadly EA. 2006. Tempo and mode of modern bird evolution observed with large-scale taxonomic sampling. *Historical Biology* 18: 205–221.

Venditti C, Meade A, Pagel M. 2006. Detecting the node-density artifact in phylogeny reconstruction. *Systematic Biology* 55: 637–643.

Verheyen RA. 1960. Les nandous (Rheiformes) sont apparentes aux tinamous (Tinamidae/Galliformes). *Le Gerfaut* 50: 289–293.

Vogt C. 1880. Archaeopteryx macroura, an intermediate form between birds and reptiles. *The Ibis* 22: 434–456.

Von Blotzheim U. 1960. Zur morphologie und ontogenese von schultergürtel, sternum und becken von Struthio, Rhea und Dromiceius: ein beitrag zur phylogenese der Ratiten. *Proceedings of the 7th International Ornithological Congress* 1: 240–251.

Von Steinmann G. 1922. Laufvögel und flügvögel. *Anatomische Anzeiger* 55: 239–244.

Waddell PJ, Cao Y, Hasegawa M, Mindell DP. 1999. Assessing the Cretaceous superordinal divergence times within birds and placental mammals by using whole mitochondrial protein sequences and an extended statistical framework. *Systematic Biology* **48**: 119–137.

Welch JJ, Bromham L. 2005. Molecular dating when rates vary. *Trends in Ecology and Evolution* **20**: 320–327.

Welch J, Fontanillas E, Bromham L. 2005. Molecular dates for the "Cambrian explosion": the influence of prior assumptions. *Systematic Biology* **54**: 672–678.

Wiedersheim R. 1884. Die stammesentwicklung der Vögel. *Biologische Zentrallblatt* **3**: 654–668; 688–695.

Wiedersheim R. 1885. Ueber die vorfahren der heutigen Vögel. *Humboldt* **4**: 213–225.

Woese C, Achenbach L, Rouviere P, Mandelco L. 1991. Archaeal phylogeny: reexamination of the phylogenetic position of *Archaeoglobus fulgidus* in light of certain composition-induced artifacts. *Systematic and Applied Microbiology* **14**: 364–371.

Wright TF, Schirtzinger EE, Matsumoto T, Eberhard JR, Graves GR, Sanchez JJ, Capelli S, Muller H, Scharpegge J, Chambers GK, Fleischer RC. 2008. A multilocus molecular phylogeny of the parrots (Psittaciformes): support for a Gondwanan origin during the Cretaceous. *Molecular Biology and Evolution* **25**: 2141–2156.

Wu J, Susko E, Roger AJ. 2008. An independent heterotachy model and its implications for phylogeny and divergence time estimation. *Molecular Phylogenetics and Evolution* **46**: 801–806.

Yang Z. 1994. Estimating the pattern of nucleotide substitution. *Journal of Molecular Evolution* **39**: 105–111.

Yang Z. 1996. Among-site rate variation and its impact on phylogenetic analyses. *Trends in Ecology and Evolution* **11**: 367–372.

Yang Z. 2007. PAML 4: Phylogenetic analysis by maximum likelihood. *Molecular Biology and Evolution* **24**: 1586–1591.

Yang Z, Rannala B. 2006. Bayesian estimation of species divergence times under a molecular clock using multiple fossil calibrations with soft bounds. *Molecular Biology and Evolution* **23**: 212–226.

Yoder AD, Yang Z. 2000. Estimation of primate speciation dates using local molecular clocks. *Molecular Biology and Evolution* **17**: 1081–1090.

Zuckerkandl E, Pauling L. 1962. Molecular disease, evolution, and genic heterogeneity. In *Horizons in Biochemistry*, Kasha M, Pullman B. (eds). New York: Academic Press; 189–225.

Zuckerkandl E, Pauling L. 1965. Evolutionary divergence and convergence in proteins. In *Evolving Genes and Proteins*, Bryson V, Vogel HJ (eds). New York: Academic Press; 97–166.

13 Major Events in Avian Genome Evolution

CHRIS L. ORGAN AND SCOTT V. EDWARDS

Harvard University, Cambridge, USA

Birds have long captured the attention of naturalists for their beauty, the range of their behaviors, and of course, their mastery of the air. With the growth of experimental science and modern biology, birds remained the focus of important biological research including work on how the vertebrate brain evolved (Jarvis *et al.*, 2005) and the effect of environmental fluctuations on speciation (Grant & Grant, 2002). The steep rise in genetics research made possible by the polymerase chain reaction (PCR), capillary sequencing, and bioinformatics in the 1980s and 1990s made the comparative study of birds even more tractable (Edwards *et al.*, 2005). Within three years of the first draft publication of the human genome (IHGSC, 2001), the first draft of the chicken (*Gallus gallus*) genome was published (Hillier *et al.*, 2004). Since then, over 20 mammal genomes have been sequenced (though many at low 2× coverage), resulting in a robust comparative framework for studying the genomics of mammals (O'Brien *et al.*, 1999; Green, 2007). By contrast, the diapsid (bird–reptile) side of the amniote tree remains undersampled. Despite the greater diversity of Reptilia compared to Mammalia (ca. 17,000 reptile species: ca. 10,000 in birds and ca. 7000 nonavian reptiles; compared with ca. 5000 species of mammals), as of this writing only three species – *Gallus* (Hillier *et al.*, 2004), *Taeniopygia* (Warren et al., 2010), and *Meleagris* (Dalloul et al., 2010) have been sampled from which to understand the genome

structure and biology of the avian, let alone the more basal reptilian, genome. But the genome of the green anole (*Anolis carolinensis*) is currently being sequenced and the available avian genomes have already provided numerous insights into the structure and evolution of the diapsid genome in general and the bird genome in particular (Ellegren, 2005, 2007, 2008).

The variation of genomic traits among nonavian reptiles is still largely unstudied (but see Shedlock *et al.*, 2007). However, there are many reasons to predict that, despite the morphological and behavioral diversity of birds, genomic characters will show less variability than those seen in mammals. In fact, the uniformity of several aspects of the avian genome is already well documented, such as genome size and karyotype (Burt *et al.*, 1999; Burt, 2002; Gregory, 2002; Organ *et al.*, 2008). The variation in traits such as total genome size and intron size have been examined across birds with varying degrees of taxon sampling and phylogenetic rigor, and some quantitative studies have shown that genome size is more constrained across birds than are other traits, such as basal metabolic rate, and shows signatures of stabilizing selection (Waltari & Edwards, 2002). Alternatively, work by Oliver *et al.* (2007), explains broad patterns of genome size variation across taxa by invoking a proportional model of genome evolution in which small genomes undergo the least amount of evolutionary change (a pattern consistent across

Living Dinosaurs: The Evolutionary History of Modern Birds, First Edition. Edited by Gareth Dyke and Gary Kaiser.
© 2011 John Wiley & Sons, Ltd. Published 2011 by John Wiley & Sons, Ltd.

eukaryotes). The proportional model is essentially a neutral model of genome evolution (Lynch & Conery, 2003; Lynch, 2007a) and suggests that simple explanations for genome size variation cannot be regarded as either neutral or selective *a priori*.

Genomics will increasingly play a larger role in population, organismal, ecological, and evolutionary biology in birds and other animals. Headway has already been made in this direction with the development of genome-wide reference markers for phylogeographic analysis (Backström *et al.*, 2008). And aside from the ability of genomics to inform other disciplines, research along the continuum between genetics and genomics is of inherent interest for understanding avian biology. For example, natural selection produces adaptive phenotypes which are based upon genetic variants and developmental networks, and the record of these adaptations can sometimes be recovered in genome sequences. This makes genomics an unprecedented goldmine for evolutionary biologists to discover broad and universal patterns of evolution – no small feat in the messy world of biology. Phylogenetic inference has been, and will continue to be, advanced by genomics, but genomicists are not just limited to questions of relatedness: they also probe other evolutionary questions. For example, the relationship between gene duplication and the immergence of novel protein function remains of paramount interest.

Even now comparative genomicists can piece together the major events that have shaped the avian genome, though we run the risk of oversimplifying our inferences due to undersampling of taxa. The following characteristics of the avian genome represent major evolutionary shifts from the ancestral amniote condition: small and relatively invariant genome size, a paucity of repetitive elements, a dispersion of the karyotype through an increase in microchromosomes, the stabilization of a Z–W sex chromosome system, high isochore structure, and fewer and shorter genes (Figure 13.1). We present some of these features mapped onto the most current phylogeny of extant birds (Hackett *et al.*, 2008) using parsimony (Maddison & Maddison, 2008). Many of these characters are interrelated, sometimes

causally so, as in the relationship between low repetitive element density and small genome size. Although many questions remain, the emerging view suggests that the avian genome is streamlined with minimal amounts of noncoding sequence and a high concentration of genes. Of several trends in the transition from the ancestral amniote to the avian genome, the extinction of repetitive element lineages within the avian genome and its effects on genome size are among the most prominent.

DEMISE OF REPETITIVE ELEMENTS AND SHRINKING GENOME SIZE

Compared with other animal groups, genome size is less variable within amniotes, whereas the genomes of amphibians and lungfish can vary by an order of magnitude (Gregory, 2007). The largest amniote genome described so far belongs to the red viscacha rat (*Tympanoctomys barrerae*; 8.4 pg). The smallest amniote genome belongs to the pheasant (*Phasianus colchicus*; 0.97 pg). Indeed, no other group of amniotes has smaller or less variable genome size than birds (Tiersch & Wachtel, 1991; Gregory, 2002). When mapped onto the avian tree using a simple method like parsimony there is a strong signal of genome size homogeneity (Figure 13.2), as suggested earlier by Waltari & Edwards (2002). Paleognaths possess the largest genomes among birds (e.g. Ostrich, *Struthio camelus*, has a genome size of 2.16 pg). Some have suggested that the large genomes of basal birds are less the result of their phylogenetic position and more a consequence of their flightlessness (Hughes, 2000), a hypothesis based on pairwise rather than phylogenetic comparisons. Indeed, the fact that the Galliformes, a clade within the secondmost basal group of birds, the neognaths, contains the bird with the smallest genome on record (the common pheasant noted above) could suggest adaptive evolution in genome size along the line to paleognaths, although most Galliformes species are poor flyers. A recent comparative analysis, although limited in its taxon sampling within birds, found no

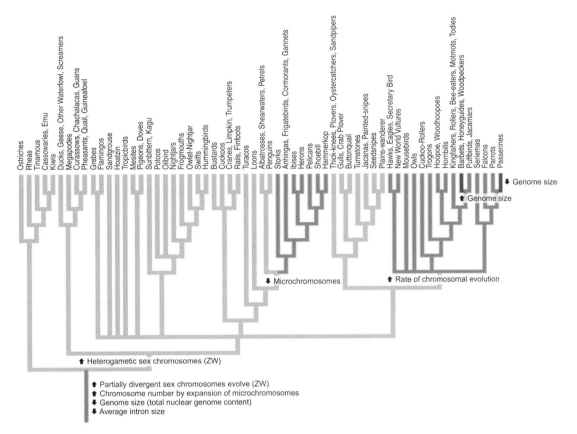

Fig. 13.1 Diagram of various events in the evolution of the avian genome. The tree is based on a phylogenetic analysis of 19 nuclear genes (Hackett *et al.*, 2008). Branch shading simply highlights areas of genomic change.

evidence for differences in genome size evolving between paleognaths and neognaths (Organ & Shedlock, 2009).

The primary mechanism accounting for the streamlined genomes of birds is the lower density of active and recently extinct repetitive elements (Shedlock, 2006). Repetitive elements are stretches of DNA that repeat in their sequence, either tandemly or dispersed in the genome, and include both transposable elements and simple repeats, such as microsatellites. Transposable elements (transposons) are stretches of DNA capable of moving to different locations within the genome of a single cell by an RNA intermediate using reverse transcriptase (endogenous retroviruses; retroelements) or by a cut-and-paste DNA mechanism using transposase (Kid-

well, 2005). Repetitive elements play important roles within genomes aside from selfishly making copies of themselves, and indeed appear to have been critical for genome evolution (Shedlock & Okada, 2000; Kazazian, 2004; Shedlock *et al.*, 2004). One important function recently discovered is the co-option of transposable elements into regulatory elements (Fableta *et al.*, 2007). The diversity of transposable elements apparently also determines, in part, the expansion in gene families because gene duplication by retroposition appears rare in the chicken genome (Hillier *et al.*, 2004). Long interspersed nuclear elements (LINEs) found abundantly in mammal genomes are thought to be responsible for the reverse transcription of retrotransposed genes, whereas the prevailing active element in birds (chicken repeat 1; CR1) lacks a mechanism to copy polyadenylated mRNA (Haasa *et al.*, 2001; Hillier *et al.*, 2004). Most of the repetitive element signal within the *Gallus* genome, including CR1, comes from extinct and decaying elements (Hillier *et al.*, 2004), although differences in repetitive content between *Gallus* and *Taeniopygia*, such as the presence of short interspersed elements in the later but not the former (Warren *et al.*, 2010), suggests active repetitive element evolution in crown-clade Aves.

There are multiple explanations for the observed diversity of genome size within animals, ranging from emergent proportional models in which larger genomes evolve at faster rates (Oliver *et al.*, 2007), to nearly neutral models in which population-level dynamics determine whether drift or selection predominates in shaping genome size (Lynch, 2007). Purifying (stabilizing) selection has also been proposed, in which selection acts against too much "junk DNA" (Waltari & Edwards, 2002; Knight *et al.*, 2005). Stabilizing selection is usually invoked to explain the variation in genome size among avian clades. A different type of mechanism, one more in line with adaptationist scenarios, is suggested to be responsible for the reduction of avian genome size from a larger ancestral amniote genome. This scenario invokes positive selection acting on genome and nucleus or cell size to maximize efficiency of cellular and organism-level traits, such as metabolic rate (Cavalier-Smith, 1978, 1985; Kozlowski *et al.*, 2003; Vinogradov & Anatskaya, 2006). But adaptationist explanations remain a common problem in evolutionary biology (Gould & Lewontin, 1979). The small genomes of birds are perhaps too easy to interpret in adaptive terms – in this case, that small genome/cell size is an adaptation associated with elevated metabolism required for flight (Hughes & Hughes, 1995; Gregory, 2002; Andrews *et al.*, 2008). The essential idea is that genome size affects the size of the nucleus and cell, and therefore the capacity (due to surface area to volume ratios) to diffuse everything from nutrients to dissolved gases across cell membranes.

Recent work on genome size evolution in dinosaurs showed that the typical "avian" streamlined genome likely arose within saurischian dinosaurs long before birds and volant flight evolved (Organ *et al.*, 2007). Nevertheless, Organ *et al.* (2007) hypothesize that there may be an adaptive explanation for genome size reduction in saurischian and theropod dinosaurs. Indeed, accumulating evidence suggests that an adaptive explanation might have traction. Inference and analysis of genome size in pterosaurs – which together with bats and birds represent all vertebrate lineages to evolve powered flight – suggests that these flying reptiles evolved small genomes (Organ & Shedlock, 2009). More comparative support for this trend comes from bats, which have smaller genomes than related mammals and from the correlation of wing morphology (a biomechanical proxy for flying ability) with genome size in birds (Andrews *et al.*, 2008). The hypothesis that genome size is modulated, at least in part, by physiological factors associated with cell size is not without its critics (Lynch, 2007), and rightly so given the heavy reliance on correlations to support it. Experimental research on the connection between physiology and genome size is needed. It is likely given our current understanding of biology in general, that multiple factors (proportional rates, drift, and selection) have been important in shaping genome size in birds. But size is only one aspect of genome architecture. How a genome is packaged into chromosomes, its karyotype, is another important aspect of genome architecture and the subject of the next section.

EXPANDING KARYOTYPE AND SEX CHROMOSOMES

The structure and organization of an organism's chromosomes (its karyotype) is important because it provides not only the physical location of genes, but their spatial relationships as well. A striking feature of the avian genome is the large number of microchromosomes (Figures 13.3–13.4). While the large number of microchromosomes in birds is striking compared with mammals, most reptile genomes are composed of numerous microchromosomes, which, like those found in birds, are structured differently than macrochromosomes

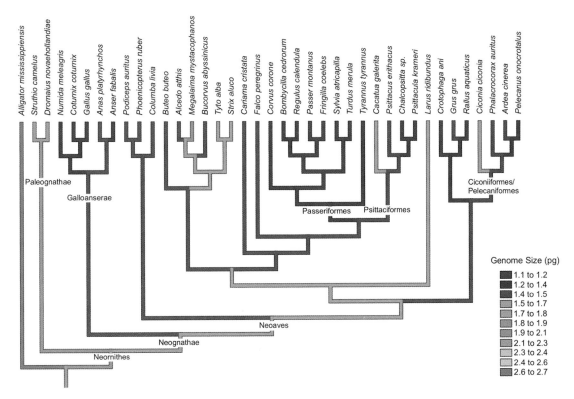

Fig. 13.2   Diagram of genome size variation within modern birds (Neornithes). Data were obtained from the animal genome size database (Gregory, 2007) and mapped onto a pruned version of the most recent phylogenetic framework for Neornithes (Hackett *et al.*, 2008) used squared-change parsimony (Maddision & Maddison, 2008). [This figure appears in color as plate 13.2]

(Burt, 2002). For example, despite similar gross features, such as centromeres and telomeres, avian microchromosomes undergo higher rates of point substitutions (Axelsson *et al.*, 2005) and more readily blend parental genetic information through higher recombination rates, nearly five times the rate of recombination of mammalian chromosomes (Rodionov *et al.*, 1992a,b; Backström *et al.*, 2006). Microchromosomes are also GC-rich, suggesting they contain a higher density of genes than do macrochromosomes (Hillier *et al.*, 2004). Associated with the increased density of genes is the reduction of intergenic DNA and repetitive elements.

Although the compliment of chromosomes in birds is less variable than in other amniote groups, chromosomal rearrangements are common (Christidis, 1990). For example, bird species with the same ploidy, say 2n = 80, display abundant chromosomal inversions and translocations as well as shifting locations of the centromere from the middle (metacentric) to the ends of the chromosomal arms (teleocentric). Hexamer repeats in the form (TTAGGG)n are common in the telomeres of vertebrate chromosomes, but occur at interstitial and centromeric regions on avian macrochromosomes, suggesting that ancestral chromosomes fused together. Avian microchromosomes are especially enriched with (TTAGGG)n repeats, which may contribute to the higher rates of recombination observed in microchromosomes (Nanda *et al.*, 2002). Some species,

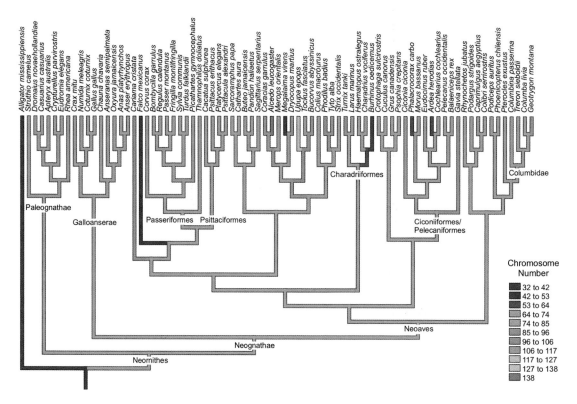

Fig. 13.3 Diagram of karyotype variation within modern birds (Neornithes). Conspecific karyotype data used for Asian barbets (*Megalaima*) (Kaul & Ansari, 1981), owls (*Strix*) (Takagi & Sasaki, 1974), and cormorants (*Phalacrocorax*) (Ebied *et al.*, 2005). Other karyotype data were obtained from the literature (Benirschke, 1977; Waldrigues & Ferrari, 1982; Christidis, 1990; Qingsong *et al.*, 1995; Nishida *et al.*, 2008). Karyotype was mapped onto a pruned version of the most recent phylogenetic framework for Neornithes (Hackett *et al.*, 2008) used squared-change parsimony (Maddision & Maddison, 2008). [This figure appears in color as plate 13.3]

such as barred antshrike (*Thamnophilus doliatus*) and yellow-legged buttonquail (*Turnix tanki*) show highly derived karyotpic organization in that all of their chromosomes are telocentric (Christidis, 1990). The stone curlew (*Burhinus oedicnemus*) has the smallest number of chromosomes in any bird known, while accipiters (bird-eating hawks in the order Falconiformes) have between 66 and 68 chromosomes and only six to 12 microchromosomes (Christidis, 1990; Rodionov, 1997). The accipitrid condition is in stark contrast with other bird groups. Although striking differences between micro- and macrochromo-

somes exist in species like the estrildid finch (*Pytilia phoenicoptera*) and the Crimson Rosella parrot (*Platycercus elegans*), there is often a gradation between the two categories, as is seen in species like the House Sparrow (*Passer domesticus*) (Christidis, 1990).

The ancestral karyotype of birds, inferred by comparative mapping studies, contained around 20 microchromosomes (Burt, 2002), while a subsequent analysis using Bayesian phylogenetic comparative methods reconstructed 26 microchromosomes with a 95% credibility interval between 13 and 39 (Organ *et al.*, 2008). This

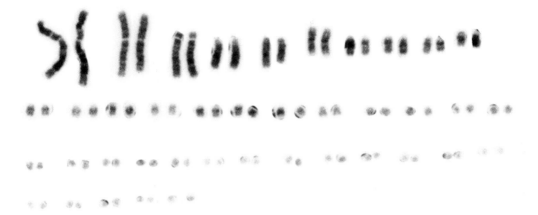

Fig. 13.4 Karyotype of female emu (*Dromaius novaehollandiae*; 2n = 80; 20 macrochromosomes and 60 microchromosomes) showing a gradation among chromosome sizes. Note the many microchromosomes characteristic of diapsid genomes.

agreement suggests that the ancestral bird karyotype was not much different from those of other reptiles and somewhere after the bird–crocodilian split, the accumulation of microchromosomes resulted in an expanded karyotype. Fission of macrochromosomes and microchromosomes best explain the mechanistic origin of the avian karyotype (Burt, 2002). However, fusion appears to have played a role in chromosomal evolution as well (Shibusawa *et al.*, 2004), and some groups such as Asian barbets (*Megalaima*) and kingfishers (*Alcedo*) show high rates of chromosomal evolution. Long stretches of conserved synteny occur between chickens and humans, suggesting a low rate of chromosomal translocations compared with higher rates of intrachromosomal rearrangements, such as inversions (Hillier *et al.*, 2004).

In addition to the reorganization of the genome through microchromosomes, the avian karyotype evolved to play an important role in the life history of birds with the rise of sex chromosomes (Ellegren & Parsch, 2007). While reptiles display an array of sex determining mechanisms, all birds possess the Z–W system of genotypic sex determination (Organ & Janes, 2008). It is thought that sex chromosomes evolve from pairs of ancestral autosomes that gained one or more sex-specific genes (Charlesworth *et al.*, 2005). Once established, a lack of recombination results in the accumulation of deleterious mutations and degradation of the heterogametic sex chromosome (W in female birds or Y in male mammals, for example).

Sex chromosomes in derived birds (Neoaves) have regions specific to the Z or W chromosome (Lawson-Handley *et al.*, 2004), but ratites possess only partially diverged sex chromosomes that contain pseudoautosomal regions, in which recombination occurs between the Z and W chromosomes (Janes *et al.*, 2009). However, the diverged sex chromosomes found in most birds may have evolved by widespread convergence rather than being representative of the ancestral state in Neoaves, which has been dated to 102–170 Ma, long after the avian divergence from crocodilians (225–245 Ma) (Lawson-Handley *et al.*, 2004). This evidence suggests that the temperature-dependent system seen in crocodilians, which lack sex chromosomes, may be representative of the condition in the common ancestor of crocodilians and birds. Sex chromosomes and genotypic sex determination allow random allocation of 50:50 sex determination to offspring, regardless of environmental fluctuations and may therefore have played an important role in the rapid radiation of birds into numerous and varied niches by the Late Cretaceous/early Paleocene. Specific genes governing sex determination in birds, such as the mammalian *SRY* gene, are unknown.

SHORT GENES, MULTIGENE FAMILIES, AND ISOCHORES

The average avian gene appears to be smaller than the average mammalian gene (Hughes & Hughes, 1995). Although the size of exons are similar between birds and mammals, the average intron (blocks of DNA cleaved out during processing into messenger RNA) is smaller, which results in smaller genes (Figure 13.5). Smaller introns likely arose in nonavian dinosaurian ancestors

(Waltari & Edwards, 2002). Although the total gene count in *Gallus* is estimated at roughly 20,000 (assembly v2.1), the gene count of *Taeniopygia* is estimated at 17,500 (Warren *et al.*, 2010) and *Meleagris* at 16,000 (Dalloul *et al.*, 2010). These studies have also shown that gene families have differentially expanded/contracted along the different lineages leading to these three species.

The nucleotides that form the base composition of the genome are not evenly distributed within chromosomes, but occur in blocks of GC-rich regions (Bernardi, 2000). GC-rich regions are also unevenly spread across chromosomes, with microchromosomes more GC-rich than macrochromosomes, at least based on the genome of *Gallus* (Figure 13.5; Hillier *et al.*, 2004). With only three bird genomes fully sequenced it is perhaps too early to draw

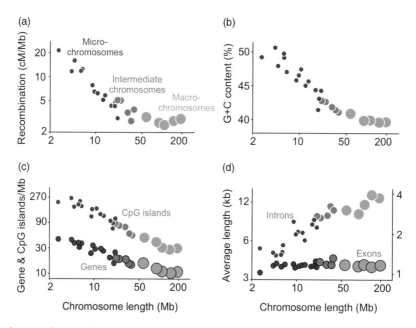

Fig. 13.5 Distribution of genome features across chromosome size classes (macro-, intermediate, and mirco-chromosomes) redrawn from Hillier *et al.* (2004). Small dark circles represent microchromosomes, medium gray circles are intermediately sized chromosomes, and large light gray circles are macrochromosomes. (A) Recombination rate is higher in microchromosomes; (B) G + C content is also higher on microchromosomes; (C) genes and CpG islands are more abundant on microchromosomes; and (D) genes on microchromosomes are shorter than those on larger chromosomes owing to shorter intron sizes. In (C) and (D) the black outlines denote genes and exons respectively.

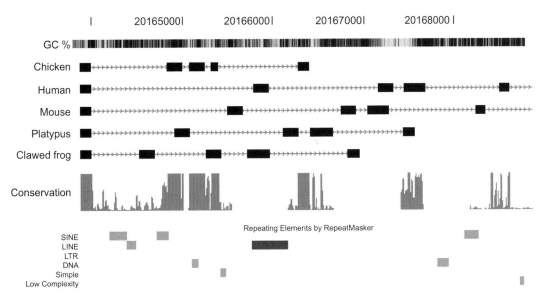

Fig. 13.6 An example of a chicken gene (*Rhodopsin*) compared with the orthologous (homologous across species) genes obtained with the UCSC Genome Browser (Karolchik *et al.*, 2003). RefSeq genes are displayed. Boxes denote exons. For repetitive elements, light gray rectangles are from human and black (CR-1) are from the chicken sequence data as reported on the genome browser by RepeatMasker (Smit *et al.*, 1996–2004). Note the paucity of repetitive elements in the chicken compared with the human. GC content is determined in a five base pair window and conservation plots are drawn from the chicken sequence.

conclusions about how the average gene has evolved in birds compared with the average gene in mammals. The chicken genome project found that, compared with humans, the average gene in *Gallus* (homologous genes among lineages, or orthologous) is under strong purifying selection ($d_N/d_S = 0.06$). Similar findings have been reported for *Meleagris* and *Taeniopygia* (Dalloul *et al.*, 2010) – presumably because there is a relationship between gene concentration on microchromosomes and the intensity of purifying selection. As in mammals (Clark *et al.*, 2003) genes involved with reproduction and the immune system appear to evolve particularly quickly, and likely for adaptive reasons (Heger & Ponting, 2007). Immune genes are rapidly evolving likely because of an arms race among organisms and pathogens, while genes associated with reproduction may be under the influence of sexual selection.

Most genes belong to gene families, and the size of the average gene family in the chicken is smaller than in mammals (Hillier *et al.*, 2004; Organ *et al.*, 2010). In addition, the vast majority of gene duplicates in *Gallus* appear to have arisen recently, producing an exponentially distributed pattern of gene family ages (Organ *et al.*, 2010), a pattern consistent within other Eukaryotes (Lynch & Conery, 2000). This pattern implies that most genes are pseudogenized through the accumulation of deleterious mutations soon after duplicating (assuming a relatively constant rate of gene duplication) and are ultimately eliminated from the genome. Moreover, nonprotein coding RNA paralogs are dispersed throughout the chicken genome and are associated with a low density (compared with mammals) of nonprotein coding RNA pseudogenes (Hillier *et al.*, 2004). Taken together, these findings suggest that coding and noncoding genes do not duplicate by similar mechanisms in birds and mammals. For example, duplications in birds and mammals may differ in the prevalence of mechanisms such as unequal crossover and retrotransposition.

Genome-wide, the expansion and contraction of gene families and the degree to which they are subject to selection provides important insights into mechanisms other than nucleotide substitution contributing to genomic and phenotypic evolution. In addition, there is an ongoing debate about whether sequence variation in coding regions or changes in the expression and regulation of genes contributes more to phenotypic evolution and adaptation (Hoekstra & Coyne, 2007). Recent research suggests that the gain and loss of large stretches of DNA and chromosomal segments (Biémont, 2008), including the gain and loss of whole genes (Hahn *et al.*, 2007) may also contribute substantially to phenotypic variation. Animal genomes appear to be much more fluid than previously thought and in the near future comparative genomics at the population level will likely both complicate and clarify our picture of how genomes contribute to phenotypic adaptation in birds.

PHYLOGENOMICS

The rapid accumulation of sequence data is already illuminating hitherto obscured regions of avian genomic biology. These new data will also facilitate the phylogenomics of birds, which has already revealed insights into the pattern of avian evolution, including surprising relationships such as the sisterhood of passerines and parrots (Hackett *et al.*, 2008). The phylogenomic analysis of Hackett *et al.* (2008) also suggests that patterns of character evolution were more complex than previously thought, such as diurnal behavior independently evolving several times within birds. Although such studies do not challenge deep splits within the avian tree, such as the Paleognathae–Neognathae split or the split between Galloanserae and Neoaves higher in the tree, other well-established groups such as the Falconiformes (Falconidae and Accipitridae) are not supported by multigene phylogenetic inference. These insights have been hard won because convergence at the morphological level obfuscates phylogenetic relationships. Moreover, the rapid divergence of birds in the Cretaceous created patterns of genetic coalescence from which species-level phylogeny is difficult to discern, though conceptual and algorithmic advances are making headway (Edwards, 2009).

CONCLUSIONS

Although we have outlined some major events during the evolution of the avian genome, at the time of this writing it is far too early to develop a comprehensive theory on the natural history of genome evolution in birds. Such a synthesis must at minimum wait for the publication of the first non-avian reptile genome from *Anolis carolinensis*. The rewriting of the avian phylogenetic tree is just one way in which the comparative genomics of birds is advancing ornithology and systematics. In addition, insights into the structure, function, and diversity of avian genomes will be essential to advance our understanding of avian biology in general.

ACKNOWLEDGMENTS

We would like to thank Rebecca Kimball for supplying the avian tree file used in this chapter and Dan Janes for supplying the karyotype micrograph.

REFERENCES

Andrews CB, Mackenzie SA, Gregory TR. 2008. Genome size and wing parameters in passerine birds. *Proceedings of the Royal Society, Series B (Biological Sciences)* **276**: 55–61.

Axelsson E, Webster MT, Smith NGC, Burt DW, Ellegren H. 2005. Comparison of the chicken and turkey genomes reveals a higher rate of nucleotide divergence on microchromosomes than macrochromosomes. *Genome Research* **15**: 120–125.

Backström N, Brandström M, Gustafsson L, Qvarnström A, Cheng H, Ellegren H. 2006. Genetic mapping in a natural population of collared flycatchers (*Ficedula albicollis*): conserved synteny but gene order rearrangements on the avian Z chromosome. *Genetics* **174**: 377–386.

Backström N, Fagerberg S, Ellegren H. 2008. Genomics of natural bird populations: a gene-based set of reference markers evenly spread across the avian genome. *Molecular Ecology* **17**: 964–980.

Benirschke RJ. 1977. Karyological difference between *Sagittarius* and *Cariana* (Aves). *Cellular and Molecular Life Sciences* **33**: 1021–1022.

Bernardi G. 2000. Isochores and the evolutionary genomics of vertebrates. *Gene* **241**: 3–17.

Biémont C. 2008. Within-species variation in genome size. *Heredity* **101**: 297–298.

Burt DW. 2002. Origin and evolution of avian microchromosomes. *Cytogenetics and Genome Research* **96**: 97–112.

Burt DW, Bruley C, Dunn IC, Jones CT, Ramage A, Law AS, Morrice DR, Paton IR, Smith J, Windsor D, Sazanov A, Fries R, and Waddington D. 1999. The dynamics of chromosome evolution in birds and mammals. *Nature* **402**: 411–3.

Cavalier-Smith T. 1978. Nuclear volume controlled by nucleoskeletal DNA, selection for cell volume and cell growth rate, and the solution of the DNA C-value paradox. *Journal of Cell Science* **34**: 247–278.

Cavalier-Smith T. 1985. Cell volume and the evolution of eukaryotic genome size. In *The Evolution of Genome Size*, Cavalier-Smith T (ed). Chichester: John Wiley & Sons; 104–184.

Charlesworth D, Charlesworth B, Marais G. 2005. Steps in the evolution of heteromorphic sex chromosomes. *Heredity* **95**: 118–128.

Christidis L. 1990. Chordata 3B. Aves. In *Animal Cytogenetics*, Bernard J, Kayano H, Levan A (eds). Berlin: Gebrüder Borntraeger; 4.

Clark AG, Glanowski S, Nielsen R, Thomas PD, Kejariwal A, Todd MA, Tanenbaum DM, Civello D, Lu F, Murphy B, Ferriera S, Wang G, Zheng X, White TJ, Sninsky JJ, Adams MD, Cargill M. 2003. Inferring nonneutral evolution from human-chimp-mouse orthologous gene trios. *Science* **302**: 1960–1963.

Dalloul, R.A., Long, J.A., Zimin, A.V., Aslam, L., Beal, K., Ann Blomberg, L., Bouffard, P., Burt, D.W., Crasta, O., Crooijmans, R.P.M.A., Cooper, K., Coulombe, R.A., De, S., Delany, M.E., Dodgson, J.B., Dong, J.J., Evans,

C., Frederickson, K.M., Flicek, P., Florea, L., Folkerts, O., Groenen, M.A.M., Harkins, T.T., Herrero, J., Hoffmann, S., Megens, H.-J., Jiang, A., de Jong, P., Kaiser, P., Kim, H., Kim, K.-W., Kim, S., Langenberger, D., Lee, M.-K., Lee, T., Mane, S., Marcais, G., Marz, M., McElroy, A.P., Modise, T., Nefedov, M., Notredame, C., Paton, I.R., Payne, W.S., Pertea, G., Prickett, D., Puiu, D., Qioa, D., Raineri, E., Ruffier, M., Salzberg, S. L., Schatz, M.C., Scheuring, C., Schmidt, C.J., Schroeder, S., Searle, S.M.J., Smith, E.J., Smith, J., Sonstegard, T.S., Stadler, P.F., Tafer, H., Tu, Z., Van Tassell, C.P., Vilella, A.J., Williams, K.P., Yorke, J.A., Zhang, L., Zhang, H.-B., Zhang, X., Zhang, Y.& Reed, K.M. 2010. Multi-Platform Next-Generation Sequencing of the Domestic Turkey (*Meleagris gallopavo*): Genome Assembly and Analysis. *PLoS Biology* **8**: e1000475.

Ebied AM, Hassan HA, Almaaty AHA, Yaseen AE. 2005. Karyotypic characterization of ten species of birds. *Cytologia* **70**: 181–194.

Edwards SV. 2009. Is a new and general theory of molecular systematics emerging? *Evolution* **63**: 1–19.

Edwards SV, Jennings WB, Shedlock AM. 2005. Phylogenetics of modern birds in the era of genomics. *Proceedings of the Royal Society, Series B (Biological Sciences)* **272**: 979–992.

Ellegren H. 2005. The avian genome uncovered. *Trends in Ecology and Evolution* **20**: 180–186.

Ellegren H. 2007. Molecular evolutionary genomics of birds. *Cytogenetic and Genome Research* **117**: 120–130.

Ellegren H. 2008. Sequencing goes 454 and takes large-scale genomics into the wild. *Molecular Ecology* **17**: 1629–1631.

Ellegren H, Parsch J. 2007. The evolution of sex-biased genes and sex-biased gene expression. *Nature Reviews Genetics* **8**: 689–698.

Fableta M, Rebolloa R, Biémonta C, Vieira C. 2007. The evolution of retrotransposon regulatory regions and its consequences on the *Drosophila melanogaster* and *Homo sapiens* host genomes. *Gene* **390**: 84–91.

Gould SJ, Lewontin RC. 1979. The spandrels of San Marco and the Panglossion paradigm: a critique of the adaptationist programme. *Proceedings of the Royal Society of London, Series B (Biological Sciences)* **205**: 581–598.

Grant PR, Grant BR. 2002. Unpredictable evolution in a 30-year study of Darwin's finches. *Science* **296**: 707–711.

Green P. 2007. 2x genomes – Does depth matter? *Genome Research* **17**: 1547–1549.

Gregory TR. 2002. A bird's-eye view of the C-value enigma: genome size, cell size, and metabolic rate in the class Aves. *Evolution* **56**: 121–130.

Gregory TR. 2007. *Animal Genome Size Database*. http://www.genomesize.com/. [Accessed April 15, 2007.]

Haasa NB, Grabowskia JM, Northa J, Moranb JV, Kazazian HHJ, John BE. 2001. Subfamilies of CR1 non-LTR retrotransposons have different 50 UTR sequences but are otherwise conserved. *Gene* **265**: 175–183.

Hackett SJ, Kimball RT, Reddy S, Bowie RCK, Braun EL, Braun MJ, Chojnowski JL, Cox WA, Han K-L, Harshman J, Huddleston CJ, Marks BD, Miglia KJ, Moore WS, Sheldon FH, Steadman DW, Witt CC, Yuri T. 2008. A phylogenomic study of birds reveals their evolutionary history. *Science* **320**: 1763–1768.

Hahn MW, Han MV, Han SG. 2007. Gene family evolution across 12 *Drosophila* genomes. *PLoS Genetics* **3**: 2135–2146.

Heger A, Ponting CP. 2007. Evolutionary rate analyses of orthologs and paralogs from 12 *Drosophila* genomes. *Genome Research* **17**: 1837–1849.

Hillier LW, Miller W, Birney E, *et al.* 2004. Sequence and comparative analysis of the chicken genome provide unique perspectives on vertebrate evolution. *Nature* **432**: 695–716.

Hoekstra HE, Coyne JA. 2007. The locus of evolution: evo devo and the genetics of adaptation. *Evolution* **61**: 995–1016.

Hughes AL. 2000. *Adaptive Evolution of Genes and Genomes*. Oxford: Oxford University Press.

Hughes AL, Hughes MK. 1995. Small genomes for better flyers. *Nature* **377**: 391.

IHGSC (International Chicken Genome Sequencing Consortium). 2001. Initial sequencing and analysis of the human genome. *Nature* **409**: 860–921.

Janes DE, Ezaz T, Edwards SV. 2009. Recombination and nucleotide diversity in the sex chromosomal pseudoautosomal region of the emu, *Dromaius novaehollandiae*. *Journal of Heredity* **100**: 125–136.

Jarvis ED, Güntürkün O, Bruce L, Csillag A, Karten H, Kuenzel W, Medina L, Paxinos G, Perkel DJ, Shimizu T, Striedter G, Wild JM, Ball GF, Dugas-Ford J, Durand SE, Hough GE, Husband S, Kubikova L, Lee DW, Mello CV, Powers A, Siang C, Smulders TV, Wada K, White SA, Yamamoto K, Yu J, Reiner A, Butler AB. 2005. Avian brains and a new understanding of vertebrate evolution. *Nature Reviews Neuroscience* **6**: 151–159.

Karolchik D, Baertsch R, Diekhans M, Furey TS, Hinrichs A, Lu YT, Roskin KM, Schwartz M, Sugnet CW, Thomas DJ, Weber RJ, Haussler D, and Kent WJ. 2003. The UCSC Genome Browser Database. *Nucleic Acids Research* **31** (1): 51–54.

Kaul D, Ansari HA. 1981. Chromosomal polymorphism in a natural population of the northern green barbet, *Megalaima zeylanica caniceps* (Franklin) (Piciformes: Aves). *Genetica* **54**: 241–245.

Kazazian HHJ. 2004. Mobile elements: drivers of genome evolution. *Science* **303**: 1626–1632.

Kidwell MG. 2005. The evolution of genomic parasites. In: *The Evolution of the Genome*, Gregory TR (ed.). Boston: Elsevier Academic Press: 165–213.

Knight CA, Molinari NA, et al. 2005. The large genome constraint hypothesis: evolution, ecology and phenotype. *Annals of Botany* **95**: 177–190.

Kozlowski J, Konarzewski M, Petrov DA. 2003. Cell size as a link between noncoding DNA and metabolic rate scaling. *Proceedings of the National Academy of Sciences* **100**: 14080–14085.

Lawson-Handley L, Ceplitis H, Ellegren H. 2004. Evolutionary strata on the chicken Z chromosome: Implications to sex chromosome evolution. *Genetics* **167**: 367–376.

Lynch M. 2007. Colloquium papers: the frailty of adaptive hypotheses for the origins of organismal complexity. *Proceedings of the National Academy of Sciences* **104**: 8597–8604.

Lynch M. 2007. *The Origins of Genome Architecture*. Sunderland; Sinauer Associates.

Lynch M, Conery JS. 2000. The evolutionary fate and consequences of duplicate genes. *Science* **290**: 1151–1155.

Lynch M, Conery JS. 2003. The origins of genome complexity. *Science* **302**: 1401–1404.

Maddison WP, Maddison DR. 2008. *Mesquite: A Modular System for Evolutionary Analysis, Version 2.5*. http://mesquiteproject.org.

Nanda I, Schrama D, Feichtinger W, Haaf T, Schartl M, Schmid M. 2002. Distribution of telomeric (TTAGGG)n sequences in avian chromosomes. *Chromosoma* **111**: 215–227.

Nishida C, Ishijima J, Kosaka A, Tanabe H, Habermann FA, Griffin DK, Matsuda Y. 2008. Characterization of chromosome structures of Falconinae (Falconidae, Falconiformes, Aves) by chromosome painting and delineation of chromosome rearrangements during their differentiation. *Chromosome Research* **16**: 171–181.

O'Brien SJ, Menotti-Raymond M, Murphy WJ, Nash WG, Wienberg J, Stanyon R, Copeland NG, Jenkins NA,

Womack JE, Graves JA. 1999. The promise of comparative genomics in mammals. *Science* **286**: 458–481.

Oliver MJ, Petrov D, Ackerly D, Falkowski P, Schofield OM. 2007. The mode and tempo of genome size evolution in eukaryotes. *Genome Research* **17**: 594–601.

Organ CL, Janes DE. 2008. Evolution of sex chromosomes in Sauropsida. *Integrative and Comparative Biology* **48**: 512–519.

Organ CL, Shedlock AM. 2009. Palaeogenomics of pterosaurs and the evolution of small genome size in flying vertebrates. *Proceedings of the Royal Society, Biology Letters* **5**: 47–50.

Organ CL, Shedlock AM, Meade A, Pagel M, Scott V. Edwards SV. 2007. Origin of avian genome size and structure in nonavian dinosaurs. *Nature* **446**: 180–184.

Organ CL, Moreno RG, Edwards SV. 2008. Three tiers of genome evolution in reptiles. *Integrative and Comparative Biology* **48** (4): 494–504.

Organ CL, Rasmussen M, Baldwin MW, Kellis M, Edwards SV. 2010. A phylogenomic approach to the evolutionary dynamics of gene duplication in birds. In *Evolution After Gene Duplication*, Dittmar K, Liberles D (eds). Chichester: John Wiley & Sons; 253–268.

Qingsong K, Zhiping L, Jin'en Z. 1995. Preliminary study on the karyotype of *Ardea purpucea* and *Ardea cinerea*, Ciconiiformes. *Journal of Forestry Research* **6**: 65–67.

Rodionov V. 1997. Evolution of avian chromosomes and linkage groups. *Russian Journal of Genetics* **33**: 25–738.

Rodionov V, Chelysheva LA, Solovei IV, Myakoshina YA. 1992a. Chiasmata distribution in lampbrush chromosomes of the chicken *Gallus gallus* domesticus: recombination hot spots and their possible significance for correct disjunction of homologous chromosomes in the first meiotic division. *Russian Journal of Genetics* **28**: 151–160.

Rodionov V, Myakoshina YA, Chelysheva LA, Solovei IV, Gaginskaya ER. 1992b. Chiasmata on lampbrush chromosomes of *Gallus gallus* domesticus: a cytogenetic study of recombination frequency and linkage group lengths. *Russian Journal of Genetics* **28**: 53–63.

Shedlock AM. 2006. Phylogenomic investigation of CR1 LINE diversity in reptiles. *Systematic Biology* **55**: 902–911.

Shedlock AM, Okada N. 2000. SINE insertions: Powerful tools for molecular systematics. *Bioessays* **22**: 148–160.

Shedlock AM, Takahashi K, Okada N. 2004. SINEs of speciation: Tracking lineages with retroposons. *Trends in Ecology and Evolution* **19**: 000–000.

Shedlock AM, Botka CW, Zhao S, Shetty J, Zhang T, Liu JS, Deschavanne PJ, Edwards SV. 2007. Phylogenomics of non-avian reptiles and the structure of the ancestral amniote genome. *Proceedings of the National Academy of Sciences* **104**: 2767–2772.

Shibusawa M, Nishibori M, Nishida-Umehara C, Tsudzuki M, Masabanda J, Griffin DK, Matsuda Y. 2004. Karyotypic evolution in the Galliformes: An examination of the process of karyotypic evolution by comparison of the molecular cytogenetic findings with the molecular phylogeny. *Cytogenetic and Genome Research* **106**: 111–119.

Smit AF, Hubley R, Green P. 1996–2004. *RepeatMasker Open–3.0.* http://www.repeatmasker.org.

Takagi N, Sasaki M. 1974. A phylogenetic study of bird karyotypes. *Chromosoma* **46**: 91–120.

Tiersch TR, Wachtel SS. 1991. On the evolution of genome size of birds. *Journal of Heredity* **82**: 363–368.

Vinogradov AE, Anatskaya OV. 2006. Genome size and metabolic intensity in tetrapods: a tale of two lines. *Proceedings of the Royal Society, Series B (Biological Sciences)* **273**: 27–32.

Waldrigues A, Ferrari I. 1982. Karyotypic study of cuculiform birds. I. Karyotype of the smooth-billed ani (Crotophaga ani). *Brazilian Journal of Genetics* **1**: 121–129.

Waltari E, Edwards SV. 2002. Evolutionary dynamics of intron size, genome size, and physiological correlates in archosaurs. *American Naturalist* **160**: 539–552.

Warren, W.C., Clayton, D.F., Ellegren, H., Arnold, A.P., Hillier, L.W., Kunstner, A., Searle, S., White, S., Vilella, A.J., Fairley, S., Heger, A., Kong, L., Ponting, C.P., Jarvis, E.D., Mello, C.V., Minx, P., Lovell, P., Velho, T.A.F., Ferris, M., Balakrishnan, C.N., Sinha, S., Blatti, C., London, S.E., Li, Y., Lin, Y.-C., George, J., Sweedler, J., Southey, B., Gunaratne, P., Watson, M., Nam, K., Backstrom, N., Smeds L., Nabholz, B., Itoh, Y., Whitney, O., Pfenning, A.R., Howard, J., Volker, M., Skinner, B.M., Griffin, D.K., Ye, L., McLaren, W. M., Flicek, P., Quesada, V., Velasco, G., Lopez-Otin, C., Puente, X.S., Olender, T., Lancet, D., Smit, A.F.A., Hubley, R., Konkel, M.K., Walker, J.A., Batzer, M.A., Gu, W., Pollock, D.D., Chen, L., Cheng, Z., Eichler, E. E., Stapley, J., Slate, J., Ekblom, R., Birkhead, T., Burke, T., Burt, D., Scharff, C., Adam, I., Richard, H., Sultan, M., Soldatov, A., Lehrach, H., Edwards, S.V., Yang, S.-P., Li, X., Graves, T., Fulton, L., Nelson, J., Chinwalla, A., Hou, S., Mardis, E.R. & Wilson, R.K. 2010. The genome of a songbird. *Nature* **464**: 757–762.

14 Bird Evolution Across the K–Pg Boundary and the Basal Neornithine Diversification

BENT E. K. LINDOW

Natural History Museum of Denmark, Copenhagen, Denmark

As discussed in earlier chapters, the clade of "modern birds" (Neornithes) encompasses the approximately 9600 species of birds currently alive and their immediate fossil ancestors. Monophyly of the group with regards to earlier, archaic clades of Mesozoic birds appears well-established by phylogenetic analyses (Cracraft, 1986; Cracraft *et al.*, 2004).

Several recent studies indicate that the many events concerning the origin, early evolution and radiation of the modern birds (Neornithes) took place around the time of the mass extinction at the Cretaceous–Paleogene boundary (Dyke & van Tuinen, 2004; Ericson *et al.*, 2006; Slack *et al.*, 2006; van Tuinen *et al.*, 2006). Despite an extremely poor fossil record with hardly any phylogenetically informative material for the latest Cretaceous and Paleocene (Hope, 2002; Mayr, 2005a; but see Stidham, 2002, 2008), a number of hypotheses on the impact of the K–Pg-boundary events on neornithine evolution have been proposed (Feduccia, 1995, 2003; Retallack, 2004; Robertson *et al.*, 2004).

This chapter provides a brief review of the Late Cretaceous and early Paleogene (Paleocene and early Eocene Epochs) fossil record of modern birds, then proceeds to a discussion of our current knowledge about the shape of the basal radiations within the neornithine tree based on both molecular and morphological phylogenetic analyses. Subsequently, the proposed hypotheses of the clade's survival across the K–Pg boundary of the group are discussed separately, as the current fossil record of neornithine birds is too inadequate for linking the mass extinction to patterns of origin and diversification. Finally, the current knowledge of the approximate temporal pattern of major radiation events within Neornithes is synthesized in the light of combined evidence from phylogenetic analyses and the fossil record.

THE CRETACEOUS AND EARLY PALEOGENE FOSSIL RECORD OF MODERN BIRDS

Although a number of fragmentary and isolated avian fossils from the Cretaceous have previously been assigned to Neornithes, the taxonomic status of most of this material is very equivocal at best (Hope, 2002). Database analysis of Cretaceous neornithine bird-fossil material, combined with studies of its preservation and its patchy distribution, implies that neornithines were relatively uncommon in the Late Cretaceous world (Fountaine *et al.*, 2005). This argument contrasts with preliminary reports by Stidham [2002, 2008], which indicate that neornithine birds from the latest Cretaceous of the Western Interior

Living Dinosaurs: The Evolutionary History of Modern Birds, First Edition. Edited by Gareth Dyke and Gary Kaiser.
© 2011 John Wiley & Sons, Ltd. Published 2011 by John Wiley & Sons, Ltd.

Seaway of North America were very diverse, both taxonomically and ecologically. Nonetheless, the discoveries of the fossil anseriforms (waterfowl) *Teviornis gobiensis* and *Vegavis iaai* from the latest Cretaceous of the Mongolia and Antarctica, respectively, have provided definite and properly phylogenetically constrained evidence of the presence of neornithine birds prior to the K–Pg boundary (Kurochkin *et al.*, 2002; Clarke *et al.*, 2005). In addition, a number of reasonably complete, putative fossil modern birds have been reported, deriving from the latest Cretaceous of Antarctica. These include a charadriiform (shorebird) referred to the family Burhinidae and the alleged gaviid (loon) *Polarornis gregorii* (Case, 2001; Chatterjee, 2002; Cordes, 2002). However, the former specimen still awaits formal, proper publication and the phylogenetic affinities of the latter are equivocal at best (Dyke & van Tuinen, 2004; Mayr, 2004). Another putative gaviiform is the late Cretaceous *Neogaeornis wetzeli*, represented by an isolated tarsometatarsus from the late Cretaceous of Chile (Olson, 1992).

In the first epoch of the Paleogene system, the Paleocene (65.5–55.8 Ma; dates for this and all subsequent epoch and stages from Gradstein *et al.*, 2004), the fossil record improves vastly in terms of number of taxa, although little of the material is preserved as reasonably complete specimens. Globally, members of the following groups have been described with certainty: Lithornithidae ("stonebirds"; extinct, volant Palaeognathae; Houde, 1988; Mayr, 2007a), Gastornithidae (giant, flightless, and possibly meat-eating birds related to the Anseriformes; Buffetaut, 1997; Mayr, 2007a), Presbyornithidae (extinct waterfowl; Olson, 1994; Benson, 1999), Cariamae (the clade including the extant seriemas; Mayr, 2007a), Phorusrhacidae (extinct meat-eating "terror-cranes"; Alvarenga & Höffling, 2003), Messelornithidae (extinct crane-like birds; Mourer-Chauviré, 1995a), Odontopterygiformes (extinct bony-toothed birds; Harrison, 1985; Bourdon, 2005; Bourdon *et al.*, 2005), Sphenisciformes (penguins; Tambussi *et al.*, 2005; Slack *et al.*, 2006; Tambussi & Acosta Hospitaleche, 2007), Strigiformes (owls; Rich & Bohaska, 1976;

Mourer-Chauviré, 1994; Mayr, 2007a). In addition, there are a number of taxa whose exact phylogenetic positions remain uncertain: for example, members of the "form family" Graculavidae (Olson & Parris, 1987); the Remiornithidae (extinct birds possibly of palaeognathous affinities; Mayr, 2005a); *Walbeckornis* (Mayr, 2007a) and a new taxon of a large ground-dwelling bird (most likely a palaeognath; Buffetaut & de Ploeg, 2008).

By the time of the early and middle Eocene Epoch (55.8–48.6 and 48.6–37.2 Ma, respectively), the fossil record of modern birds suddenly improves drastically in terms of both quantity and quality, chiefly due to the presence of a number of deposits in Europe and North America with exceptional preservation qualities (Fossil-Lagerstätten *sensu* Seilacher *et al.*, 1985). These include the lower Eocene London Clay (UK) and Fur (Denmark) Formations; the middle Eocene Green River Formation (North America) and Messel (Germany) deposits and the middle–upper Eocene Geiseltal deposits (Germany) (Mayr, 2005a; Lindow & Dyke, 2006). The number of known clades is too large to list here; it has been estimated that at least 55 family-level clades of birds are known from the Paleogene of Europe and another 25 in the early Eocene of North America (James, 2005). Interested readers are referred to review papers such as Mayr (2005a) and Lindow & Dyke (2006).

The range of fossil and living clades of birds known today will probably continue to be extended for years to come. Paul (2003) conducted a historical, bibliographic study comparing "palaeontological effort" (collecting effort/number of specimens described) to the publication date of the earliest stratigraphic record of fossil families. The results indicate that within the clade, families of birds still have relatively incompletely known stratigraphic ranges, which are still being extended back in time (Paul, 2003). Similar "collector curve" studies corroborate this observation, but have noted that new fossils appear to be "filling known gaps" and are "not creating new ones", i.e. ranges are not being extended, at least for Mesozoic groups (Fountaine *et al.*, 2005) (see Dyke & Gardiner, Chapter 5, this volume).

However, the geographic range of fossil birds is being extended, especially into the Southern Hemisphere (Fountaine et al., 2005). Interestingly, a statistically significant shift in preservation environments of avian fossils between the Cretaceous and the Paleogene has been noted (Dyke et al., 2007); a database investigation of fossil specimens has shown that neornithine birds are much more prevalent in aquatic environments than their nonneornithine predecessors, which could indicate a biological shift in habitat.

These analyses are especially relevant in relation to future efforts to discover new localities for well-preserved fossil bird material; especially in the light of the apparent need for the occurrence of rare depositional environments (e.g. Solnhofen, Liaoning, Fur, Green River, and Messel), which allow researchers to gain adequate anatomical information in order to identify the taxonomy of the specimens and faithfully reconstruct their true phylogenetic relationships.

As has been noted by several authors, the neornithine fossil record is seriously skewed towards the Northern Hemisphere, especially Europe and North America, for historical reasons (Cracraft, 2001). However, it has repeatedly been pointed out that the closest contemporary recent relatives of the Paleogene fossil birds of the Northern Hemisphere are to be found in the Southern Hemisphere, a fact which has often been interpreted as support for a Gondwanan origin of Neornithes as a whole (Mourer-Chauviré, 1995b; Mayr, 2005a). Other studies, which have combined information derived from both phylogeny and recent biogeographical distribution patterns of Neornithes, suggest that the clade originated and began to diversify during or before the breakup of Gondwanaland in the Early Cretaceous (Hedges et al., 1996; Cracraft, 2001; Ericson, 2008). An equally valid interpretation is that this distribution can be attributed to the displacement to the Southern Hemisphere at the end of the Paleogene of groups originating in the Northern Hemisphere, due to the disappearance of tropical or subtropical environments (Fain & Houde, 2004). Indeed, because little sampling effort in deposits of this age (and older) has taken place in the Southern

Hemisphere, hypotheses suggesting that some, or all, of the modern avian radiation may have taken place outside of the well-sampled Northern Hemisphere regions cannot readily be tested at present (Lindow & Dyke, 2006).

THE PATTERN OF NEORNITHINE DIVERSIFICATION: PHYLOGENETIC ANALYSES

Before any well-founded hypotheses of evolution, diversification, and lineage dynamics of a given avian clade can be made, the exact pattern of relationships within it must be established through phylogenetic analyses (Dyke, 2003a). The interrelationships of the higher-order clades within Neornithes have been largely unresolved for a long time, despite the efforts of both morphological and molecular studies, which has been pointed out by several authors (Cracraft, 2001; Dyke & van Tuinen, 2004; Fain & Houde, 2004; Edwards et al., 2005). However, two recent analyses with a wide array of genomic sampling have yielded very promising results (Ericson et al., 2006; Hackett et al., 2008).

Morphological Analyses

As noted by Mayr & Clarke (2003) there are comparatively few well-sampled morphological phylogenetic analyses of neornithine birds, whereas the relationships among extinct groups diverging earlier than the clade Neornithes are relatively well-understood (e.g. Chiappe & Dyke, 2002, 2006).

Phylogenetic analyses of Neornithes based on morphology have usually been restricted to studies of individual groups, such as Charadriiformes (Mickevich & Parenti, 1980; Chu, 1995; all based upon modified versions of the character matrix in Strauch (1978) non-cladistic study), Larinae (Chu, 1998) and Galliformes (Dyke et al., 2003).

Despite calls for the use and inclusion of well-preserved bird fossils in phylogenetic analyses based on morphological data to help resolve issues of relationships within neornithines (Dyke & van Tuinen, 2004; Lindow & Dyke, 2006), fossils have rarely been used on a large scale, despite the fact that a number of almost complete and relatively well-preserved ones exist (e.g. many of the taxa listed in Mayr, 2005a). Use of fossil taxa is usually restricted to outgroups (Mayr & Clarke, 2003; Livezey & Zusi, 2006, 2007) or included in analyses restricted to less inclusive clades, often at "ordinal" level (e.g. Ericson, 1997; Livezey, 1997; Dyke & Gulas, 2002; Mayr, 2002, 2003, 2005b; Dyke, 2003b; Gulas-Wroblewski & Wroblewski, 2003; Mayr & Mourer-Chauviré, 2004). However, recent phylogenetic analyses have employed several fossil and extant taxa, but unfortunately only sampled for single extinct clades (Odontopterygiformes or Sphenisciformes) and had somewhat skewed taxonomic distributions (e.g. Bourdon, 2005; Bourdon *et al.*, 2005; Slack *et al.*, 2006).

Indeed, fossil Paleogene neornithine taxa appear to have been almost completely ignored in the few wide-sampled morphological analyses of Neornithes. Instead, the importance of Mesozoic non-neornithine birds as outgroups for character polarization has been oft promoted (Livezey & Zusi, 2001, 2007). This is unfortunate as previous studies of other groups have shown that solely utilizing fossils as outgroups will not resolve problems of ingroup relationships and help recover correct topologies; fossils with basal positions within the clades to be analysed must also be included (Donoghue *et al.*, 1989).

The absence of characters drawn from good, well-represented fossil material, has meant that many years of anatomical work has proved unsuccessful in unravelling the relationships of extant birds on the basis of osteological characters (Cracraft *et al.*, 2004; Mayr, 2005a). In particular the clade Neoaves (the "neoavian comb" of Cracraft *et al.*, 2004) remains largely unresolved (see Livezey, Chapter 4, this volume).

Relatively recently, however, Livezey & Zusi (2006, 2007) published a very exhaustive phylogenetic analysis of 150 taxa of Neornithes and 2954 characters. Amongst other results, their analysis recovered a monophyletic Palaeognathae (with

the exception of the fossil *Lithornis*; see below); a monophyletic Neognathae and a monophyletic "Galloanserae" as sister-group to a well-supported (bootstrap, 99%; Bremer support, 18) monophyletic Neoaves. Interestingly, while the clade Neoaves appears relatively well supported within the analysis, many of its subclades and successive clades are in fact quite weakly supported (i.e., bootstrap, <50%; Bremer support, 1–3; Livezey & Zusi, 2007, figure 10). Despite this large sample of taxa, few fossils were included. A number of theropod dinosaur taxa and Mesozoic avians were employed as outgroups. The extinct palaeognath *Lithornis* appears to be the only fossil Paleogene avian taxon included in the analysis *a priori*. Interestingly, it was recovered as the sister group to all other Neornithes, contrasting with previous studies which have considered it to be part of the Palaeognathae (Houde, 1988; Dyke, 2003c; Leonard *et al.*, 2005). Furthermore, the extinct orders Dinornithiformes and Aepyornithiformes were included in the analysis, but only placed *a posteriori* within a backbone-constrained most parsimonious tree, due to initial analytical problems apparently caused by missing data for these two taxa (Livezey & Zusi, 2007).

This large morphologically based analysis has been critically reviewed by Mayr (2007b). Several examples of incorrect, doubtful, or unacceptably ill-assumed generalized character scoring in the analysis of Livezey & Zusi (2006, 2007) not withstanding, Mayr (2007b) correctly showed that very large amounts of homoplastic characters in a morphological data set swamps and overrules relatively few phylogenetically informative characters, resulting in the weakly supported and practically unresolved nodes mentioned above. He advocated that future morphological analyses should concentrate on identifying and utilizing a smaller number of key apomorphic characters, instead of just increasing the amount of characters in data sets. Nonetheless, Mayr (2007b) also acknowledged that the large morphological analysis did in fact manage to recover a number of novel clades also discovered in the molecular analyses of Fain & Houde (2004) and Ericson *et al.* (2006).

Despite the above taxonomic shortcomings and minor discrepancies between the various morphological analyses, an overall, repeated pattern of divisions within Neornithes can be summarized as follows (Mayr & Clarke, 2003; Cracraft *et al.*, 2004; Slack *et al.*, 2006; Livezey & Zusi, 2007): (i) A division between the clades Palaeognathae (ratites, tinamous, and extinct lithornithids and Neognathae (all other groups); (ii) one or more divisions involving waterfowl (Anseriformes), landfowl (Galliformes), and extinct bony-toothed birds (Odontopterygiformes) (exact topology of this clade unresolved; see Bourdon, 2005; Mayr, 2008) and Neoaves (all other groups); and (iii) a mostly unresolved clade Neoaves.

MOLECULAR ANALYSES

Untill recently molecular analyses have also failed to adequately resolve the "neoavian comb" within Neornithines (Johansson *et al.*, 2001; Poe & Chubb, 2004; Sorenson *et al.*, 2003; Dyke & van Tuinen, 2004; Edwards *et al.*, 2005). However, one recent analysis using β-fibrinogen recovered two previously unrecognized parallel clades within Neoaves, "Metaves" and "Coronaves" (Fain & Houde, 2004). An even more recent analysis using a wider array of gene regions (c-*myc*, RAG-1, myoglobin, β-fibrinogen, and ornithine decarboxylase) managed to resolve Neoaves further and corroborate the monophyly of several groups previously recovered in independent morphological and molecular studies (Ericson, *et al.*, 2006). However, the above-mentioned monophyletic clades "Metaves" and "Coronaves" collapsed when the β-fibrinogen data were excluded.

The inability to adequately resolve Neoaves reflects short internodes resulting from rapid radiation at the base of modern birds; in turn these short internodes are responsible for confounding attempts at resolving polytomies adequately (Johansson *et al.*, 2001; Feduccia, 2003; Sorenson *et al.*, 2003; Poe & Chubb, 2004; Fain & Houde, 2004; Edwards *et al.*, 2005; Hackett *et al.*, 2008).

Results of several molecular analyses support the same overall topology as described above for morphological analyses: (i) a deeply rooted division between Palaeognathae and Neognathae; (ii) a well-established and relatively deeply rooted division between Galloanserae (Galliformes + Anseriformes) and Neoaves; and (iii) a rapid, almost simultaneous radiation between the remaining groups within Neoaves (Ericson, 2008).

THE K–Pg EXTINCTION EVENT AND HYPOTHESES OF ITS IMPACT ON NEORNITHINE EVOLUTION

The Cretaceous–Paleogene (K–Pg) mass extinction took place at a geologically and environmentally complicated time in Earth history (Archibald, 1997). Although the exact order of events and the weight of various proposed mechanisms behind global mass extinction has been and continues to be debated (e.g. Sharpton & Ward, 1990; Ryder *et al.*, 1996), a general overview of events for the Cretaceous–Paleogene extinction event can be summarized as follows: first, there was a major marine regression, which resulted amongst others things, in terrestrial habitat fragmentation (Archibald, 1997). Second, a phase of continental flood basalts, the Deccan traps in India injected large amounts of SO_2 and CO_2 into the atmosphere (Courtillot *et al.*, 1996; Chenet *et al.*, 2008). Third, there were one or more bolide impacts, one of which occurred in the Caribbean region (Alvarez *et al.*, 1980; Alvarez, 1983; Hildebrand *et al.*, 1991). Observational evidence suggests that the bolide impact resulted in a short, sharp "impact winter" caused by clouds of impact ejecta darkening the atmosphere (Wolfe, 1991; Wolfe & Russell, 2001) along with increased rates of acid rain (Prinn & Fegley, 1987; Sigurdsson *et al.*, 1992; Retallack, 2004). In the minutes and hours immediately after the bolide impact, a "short-term infrared thermal event" has been hypothesized, where particles of ejecta sent into suborbital trajectories heated up to incandescent levels and bathed the Earth in lethal infrared radiation and causied global wildfires (Melosh *et al.*, 1990; Robertson *et al.*, 2004). Finally, there was a brief, intense greenhouse warming with increased levels of precipitation (Wolfe 1990; Wolfe & Russell, 2001; Arens & Jahren, 2000). This sequence of events in turn led to major simultaneous disruptions of global terrestrial and marine ecosystems and mass extinction (Norris, 2001; Wolfe & Russell, 2001).

As noted by several authors, the best sampled terrestrial K–Pg boundary record is found in the Western Interior of North America, and almost all hypotheses on the nature and effect of the extinction event on terrestrial organisms (including birds) have been based on this record (Archibald & Bryant, 1990; Sheehan & Fastovsky, 1992; Archibald, 1997; Wolfe & Russell, 2001; Retallack, 2004; Robertson *et al.*, 2004). Naturally the extremely limited, regional nature of this record calls for a cautionary approach to correlating cause and effect of the mass extinction on a global scale (Archibald & Bryant, 1990; Sheehan & Fastovsky, 1992; Archibald, 1997; Robertson *et al.*, 2004).

Several studies have highlighted the necessity of making analyses of the ecological selectivity of the extinction event, i.e. examining *both* which taxa became extinct *and* which survived, in order to understand the nature of the event and test associated hypotheses. This has been studied for both marine and terrestrial sections (e.g. Kitchell *et al.*, 1986; Sheehan & Hansen, 1986; Gallagher, 1991; Rhodes & Thayer, 1991; Sheehan & Fastovsky, 1992; Sheehan *et al.*, 1996; Retallack, 2004; Robertson *et al.*, 2004).

Birds have figured in relatively few scenarios considering selective extinction and survival at the Cretaceous–Paleogene boundary, probably due to their extremely limited fossil record. The latest Cretaceous and Paleocene avian fossil record is so scant that it is not even certain whether the earlier diverging Mesozoic groups of non-neornithine birds disappeared before or at the boundary, or perhaps survived across it (Chiappe, 1995; Padian & Chiappe, 1998; Chiappe & Dyke, 2002); although preliminary reports of chronostratigraphic data from North America suggest that their diversity was declining or they had disap-

peared before the end of the Maastrichtian (Stidham, 2002, 2008).

Based strictly on the fossil record of birds, and ignoring molecular studies, Feduccia [1995, 2003] proposed that birds underwent a massive extinction event at the Cretaceous–Paleogene boundary. His scenario proposes that all avian groups basal to the neornithine birds became extinct, and that only two clades within the Neornithes, the paleognaths and the "transitional shorebirds" survived an evolutionary bottleneck, before embarking on a rapid diversification in the Paleocene. However, the assumption on which this line of reasoning is based is flawed, as has been noted by several authors. First, the evidence for a morphological grade of "transitional shorebirds" (Feduccia, 1976, 1977, 1978; Olson & Feduccia, 1980) has been repeatedly criticized for its deeply flawed methodology and assumptions (Cracraft, 1981, 1988; Livezey, 1997; van Tuinen *et al.*, 2003). Second, proper cladistic analyses based on morphology (McKitrick, 1991; Ericson, 1997; Livezey, 1997) and molecular studies of mitochondrial genomes have also soundly refuted the possibility of shorebirds being basal to other neornithines (Paton *et al.*, 2002). Third, cladistic analyses have found no synapomorphies between a key taxon to the "transitional shorebird hypothesis", the extinct anseriform *Presbyornis*, and charadriiforms and/or phoenicopterids (Ericson, 1997; Livezey, 1997); the similarities between them are plesiomorphic character states for neognathous birds (Ericson, 1997). Fourth, discoveries of phylogenetically well-constrained latest Cretaceous crown-group anseriforms have destroyed another central tenet of the hypothesis, namely that these evolved from transitional shorebirds after the Cretaceous (Clarke *et al.*, 2005). Finally, there is not enough fossil evidence in the latest Cretaceous, which could indicate that lineages of non-neornithine birds became extinct at the K–Pg boundary, but it is just as likely that they disappeared some time before the event (Chiappe, 1995; Padian & Chiappe, 1998).

Retallack (2004) attempted to explain the selective extinction of large dinosaurian herbivores and carnivores at the K–Pg boundary through the destructive effects of increased acid rain due to bolide impact and atmospheric shock (Prinn & Fegley, 1987) alongside aerosols and carbonic acid from volcanic eruptions coupled with the detritus-buffered survival hypothesis (Sheehan & Hansen, 1986; Sheehan & Fastovsky, 1992). There is solid, independent evidence for a dramatic increase in pH values from pedo-, chemo-, and bioassays from the latest Cretaceous and earliest Paleogene of North America. A comparative bioassay of organismic pH-tolerance levels indicates that in North America, terrestrial pH levels were suppressed to between 5.5 and 4, supported by the observations of severe extinctions of non-marine molluscs, while amphibians and fish survived relatively unscathed (Retallack, 2004). Severe acid rain could damage plant ecosystems through leaf browning and fall, in turn killing off the large herbivores needing large quantities of green plant material (Prinn & Fegley, 1987). This in turn would lead to the extinction of the large and middle-sized carnivorous forms preying upon these. On the other hand, taxa dependant on the detritus foodchain, which has forest-soil leaf litter, decaying plant material, and fungi, and includes annelids, molluscs, and arthropods at is base, would be buffered against extinction (Sheehan & Hansen, 1986). The fossil record adequately supports that terrestrial insectivorous and omnivorous mammals along with turtles, crocodiles, and champsosaurs (based on the detritus foodchain in streams) favourably survived the extinction event (Sheehan & Hansen, 1986).

While the acid-rain–detritus-buffering mechanism adequately explains the extinction of large herbivores and carnivores and survival of the above-mentioned vertebrates, it cannot adequately address the selective survival of neornithine birds in contrast with their enantiornithine cousins and small theropod dinosaurs, although neornithines are specifically discussed by Retallack (2004). The neornithines considered by him are the "transitional charadriiforms", which as mentioned above is a nonexistent taxonomic group based on faulty reasoning.

Nonetheless, the hypothesis can still be tested by adopting a very conservative estimate of which neornithine lineages were definitely present in the latest Cretaceous based on the above-mentioned phylogenetic analyses and the known fossil record: phylogenetically well-constrained anseriforms (Clarke et al., 2005) positively infers the existence of Galliformes (the sister-taxon of Anseriformes); Neoaves (the sister-taxon of "Galloanserae") and Palaeognathae (the sister taxon of Neognathae) and quite likely also Odontopterygiformes (Bourdon, 2005; Mayr, 2008).

Palaeognathae, Galliformes, Anseriformes, and stem Neoaves would be selectively able to survive the extinction event as they are coupled to the terrestrial or aquatic detritus-buffered foodchains. However, the problem is that two other groups equally likely to be coupled to the detritus foodchain became extinct: enantiornithine birds and small theropod dinosaurs (Sheehan & Fastovsky, 1992). While preliminary reports indicate that enantiornithines might have become extinct some time before the K–Pg boundary (Stidham, 2002, 2008), countless studies have now shown that the theropod dinosaurs had the same physiological and anatomical adaptations as birds and for ecological purposes can be considered small, feathered, nonflying birds. Therefore birds and their small dinosaurian theropod cousins become an obstacle to the acid-rain–detritus-feeding selective survival/extinction hypothesis, although it may adequately explain the survival of other groups.

Based on the hypothesis that heated, re-entering ejecta grains from the Chicxulub impact bathed the Earth in intense infrared (IR) radiation (Schmitz, 1988; Melosh et al., 1990), killing unsheltered organisms on a global scale within minutes and a few hours immediately after the bolide impact, and causing global wildfires (Robertson et al., 2004).

The physical evidence for the global dust cloud in the marine record is a worldwide "basal layer" of iridium-enriched fine-grained clay with elevated levels of concentrated metal trace elements hypothesized to derive from atmospheric fallout of impact-related ejecta grains (Melosh, et al., 1990). However, this was questioned earlier by Schmitz (1988) who used major element analysis to show that the metal elements of the "basal layer" are most likely to derive from local seawater and have been concentrated in the "basal layer" as a precipitation horizon associated with algal decomposition on the seafloor at the end of the Maastrichtian. Furthermore, the absence of physical evidence in the terrestrial record of North America for the infrared radiation-inducted global wildfires has been noted; the amounts of charcoal in the boundary layers are in fact well below (or are absent) the background levels seen in latest Cretaceous of North America (Belcher *et al.*, 2003).

Robertson *et al.* (2004) argued that the terrestrial animal groups which survived were the ones that were able to find shelter in water, burrows, wood, or beneath rocks: fishes, amphibians, turtles, crocodilians, champsosaurs, neornithine birds, and mammals. Animals not able to submerge themselves in water or burrow would not survive the intense, short-term thermal shock.

However, the presence and survival of several neornithine bird groups across the K–Pg boundary actually provides serious problems for the hypothesis of "thermal death by IR" as clearly evidenced in the convoluted discussion by Robertson *et al.* (2004). Once more, the list of putative neornithine groups present in the Late Cretaceous by Robertson *et al.* (2004) is ill-founded. It is partially based on a number of dubious taxa of isolated elements from Hope (2002) and associated ghost-lineages based on the preliminary phylogenetic hypotheses of Cracraft (2001) along with "shorebirds similar to charadriiformes" (Robertson *et al.*, 2004), who appear to be a legacy of the above-mentioned flawed hypothesis of Feduccia (1995). Again, this particular hypothesis can be tested by considering the four or five neornithine lineages that were definitely present in the latest Cretaceous: palaeognaths, anseriforms, galliforms, the neoavian lineage, and odontopterygiforms.

While the anseriforms (and possibly also the odontopterygiforms, assuming the latter had already evolved an aquatic or semi-aquatic ecology at the time) could survive by diving according to Robertson *et al.* (2004), there remains a serious respiratory problem to be dealt with. Although it is

true that ectothermic vertebrates (fishes, amphibians, crocodiles, champsosaurs, and turtles) could remain submerged for a relatively long time (hours) due to their low metabolism, out of harms' reach of the IR, as contended by Robertson *et al.* (2004), birds (and the semi-aquatic mammals for that matter) are in trouble. As clearly stated by Robertson *et al.* (2004), the birds are unable to hold breath more than one minute, after which they need to resurface to replenish their air supply. However, when resurfacing, they would be subject to the IR. If this went on for hours, as posited by the hypothesis, the birds would be in dire straits due to the metabolic need of having to stay surfaced after repeated, prolonged dives. During this time they still receive the full blast of the IR. In an ill-argued attempt to circumvent this problem Robertson *et al.* (2004) suggests that as their plumage would get singed, water would penetrate it, adding to their protection and allowing the birds to seek shelter "under rocks or in agitated shallow water". Given the globally destructive conditions of the initial hypothesis, it is more likely that the anseriforms and birds with similar ecological preferences also would have been killed off by the IR.

As correctly pointed out by Robertson *et al.* (2004) galliforms and palaeognaths pose a much more serious obstacle to the hypothesis, as their recent representatives are all nonswimming and nonburrowing. No reasonable explanation is given for galliform survival, while it is suggested that the small size of earliest known palaeognaths (the much later Paleogene Lithornithidae of Houde, 1998) could indicate "they might have burrowed" (Robertson *et al.*, 2004). However, this is an unfounded suggestion without any supporting morphological evidence whatsoever.

Generally the proposed extinction/survival hypothesis is deeply flawed; the IR thermal radiation model is simply *too* lethal. Using the assumptions by Melosh *et al.* (1990), Robertson *et al.* (2004) describe the global flux of thermal radiation as around $10 \, \text{kW m}^{-2}$, "power levels comparable to those obtained in a domestic oven set at 'broil'" and lasting several hours. While they explicit state that it is the effects of lethal IR and *not* the associated global wildfires they are analysing, these two factors are so intimately associated, that their joint effect must be considered (Belcher *et al.*, 2003). Although Robertson *et al.* (2004) go on to suggest that semi-aquatic neornithine birds were able to survive and find shelter from the IR "in lakes, marshes or swamplands having dense sheltering vegetation unlikely to burn fully" and later restate that "Dense marsh vegetation common in the Cretaceous probably did not burn completely", the initial conditions set by the hypothesis makes this "survival strategy" rather inadequate. High temperatures and burning vegetation would quickly force the sheltering animals out into the open, where they would have been roasted by IR. As noted above, staying continuously submerged for a longer period is not an option for birds.

Summing up, Robertson *et al.* (2004) need a string of dubious or completely unfounded *ad hoc* explanations to explain the selective survival of the neornithine lineages. Together with the above-mentioned geological and sedimentological evidence (Schmitz, 1988; Belcher *et al.*, 2003, and references therein), the survival of particularly galliforms and palaeognaths indicates that a phase of thermal IR probably did not happen or (if it did) was not important enough to be a major factor in extinction or survival at the K–Pg boundary.

Without proposing any specific hypothesis, but noting that larger organisms are more prone to extinction during times of ecological crisis, Hone *et al.* (2008) suggested that a trend towards smaller size in ornithuromorph birds (including Neornithes) might have aided their survival during the K–Pg extinction event.

TIMING BASAL NEORNITHINE DIVERSIFICATION EVENTS

The early pattern of radiation within Neornithes, especially with regard to the temporal resolution of events, has been subject to much debate and investigation. Most of the debate resolves around the discrepancies between the fossil record, which preserves little evidence of neornithine diversity pre-dating the K–Pg boundary (at least in the Northern Hemisphere) and molecular studies, which imply both origins and diversification deep into the Cretaceous (Hedges *et al.*, 1996; Cracraft, 2001; Dyke & van Tuinen, 2004). An important point in the debate, which has been largely overlooked by advocates of both morphology and molecular studies, is the need to adequately distinguish between *origin* and *diversification* (Dyke & van Tuinen, 2004; van Tuinen *et al.*, 2006).

Hard evidence in the shape of the presence of undoubted anseriforms in the latest Cretaceous implicates the existence of at least three other lineages of Neornithes in the Cretaceous: Galliformes (the sister-taxon of Anseriformes); Neoaves (the sister-taxon of "Galloanserae"), and Palaeognathae (the sister taxon of Neognathae) (Clarke *et al.*, 2005). In addition, phylogenetic analyses (Bourdon, 2005; Bourdon *et al.*, 2005) and morphological studies (Mayr, 2008) indicate that Odontopterygiformes are placed outside Neoaves, possibly as the sister group of Anseriformes to the exclusion of Galliformes. This means that the bony-toothed birds were also present in the latest Cretaceous.

Although partially based on the incorrect identification of fossil taxa, Cracraft (1986) was among the first to propose a Cretaceous origin of the Neornithes based on a proper phylogenetic analysis of morphological characters. Combining the then known fossil record of three avian clades (Strigiformes, Caprimulgiformes, and Apodiformes) in a statistical fossil gap analysis (Marshall, 1997), Bleiweiss (1998) showed that these three orders most likely originated within a short time during the early Paleocene.

Studies using a "molecular clock" approach have also repeatedly placed the divergence and origin of modern birds within the Cretaceous (Hedges *et al.*, 1996; Cooper and Penny, 1997; Paton *et al.*, 2002). However, as has been pointed out several times, much of the calibration has been made uncritically using fossils, which are scanty and/or of extremely equivocal taxonomic affinities (Chiappe & Dyke, 2002; Dyke & van Tuinen, 2004). Also, some genes, such as RAG-1, clearly behave in a nonclock-like way (Paton *et al.*, 2002).

One example of this practice is the study of Cooper & Penny (1997), which employed a molecular clock approach using 12S rRNA and the c-*mos* nuclear gene to propose the origin of several neornithine clades of birds in the Cretaceous, 120–140 Ma. However, among the fossil specimens they used to calibrate their data (Cooper & Penny, 1997) was the above-mentioned "gaviiform" *Polarornis* (Chatterjee, 2002), whose affinities are equivocal at best (Dyke & van Tuinen, 2004; Mayr, 2004), and Paleocene "transitional charadriiforms" (Olson, 1985), whose taxonomic identification is based on flawed methodological assumptions (Cracraft, 1981, 1988). Furthermore, their corrected parsimony tree (Cooper & Penny, 1997) is at odds with accepted phylogenies of modern birds, in that galliform birds are the sister group of all other neornithine taxa, including ratites.

Other studies have tried to circumvent the problems with the immediate fossil record, by using extremely old, fossil-supported external calibration points such as the 310 Ma split between synapsid (mammals) and diapsid (birds) reptiles (Hedges *et al.*, 1996). However, as noted by Padian & Chiappe (1998), the age of bird diversification was calculated against an incorrect date (Late Triassic) for crown-group mammals (Middle to Late Jurassic), and is therefore incorrect by a difference of about 60 Myr.

Also, the date of 310 Ma has been shown to be based on: (i) a very uncertain identification of the Carboniferous amniote fossil *Hylonomus* as a diapsid; (ii) failure to include substantial precision errors of geological ages, which of course would be magnified in subsequent nodes calculated on the basis of this date; (iii) failure to correctly employ standard errors; i.e. using error bars instead of 95% and 99% confidence intervals. Including correctly calculated estimates of error in these studies would have yielded a much larger interval of 132–67 Ma (95% confidence) for the neornithine diversification (Graur & Martin, 2004). Also, molecular clock methods have been shown to overestimate divergence dates due to statistical test failure to properly exclude genes with nonclock-like behaviour and incorrect arithmetic treatment of skewing in clock estimates (Benton & Ayala, 2003).

A solution to the above mentioned questions on origin and diversification and the discrepancies between "molecular clock" dating and the fossil record has been provided by the recent analysis of Ericson *et al.* (2006). As noted above, their analysis of several genes managed to both resolve the clade Neoaves better than previous analyses and recover several clades within Neoaves, which had previously been proposed by independent analyses, both morphological and molecular. More importantly, for the first time the topology of the resulting tree was calibrated by the use of phylogenetically well-documented and well-constrained fossils. The results indicated that while some neoavian lineages were present in the latest Cretaceous, the main higher-level diversification phase within the clade took place during the Paleocene, after the K–Pg boundary. Although specifics of this analysis have been criticized and the same data have been used to support an extensive diversification at the end of the Cretaceous (Brown *et al.*, 2007), the initial assumptions and results have been adequately defended (Ericson *et al.*, 2007).

These results fit surprisingly well with the previously mentioned conclusions of the fossil gap analysis of Bleiweiss (1998), which was centred on the neoavian clades Strigiformes, Caprimulgiformes, and Apodiformes. Furthermore, fossil evidence in the shape of phylogenetically well-constrained fossil penguins (Sphenisciformes; members of Neoaves; Slack *et al.*, 2006) indicate that the main phase of rapid neoavian diversification must have taken place by the earliest Paleocene.

CONCLUSIONS

In recent years the early evolution of modern birds (Neornithes) has been the focus of intense scientific research and debate, with many and varied research efforts in a number of fields, including paleontological, morphological, and molecular studies. Investigations have especially centred on such issues as the relationships within and between various clades within Neornithes; the origin of the clade as a whole; and the timing and pattern of different radiations within the clade, chiefly that of the subclade Neoaves. Molecular clock studies have long advocated a date of origin for modern birds in the Cretaceous Period, and recent fossil evidence has unequivocally confirmed the presence of members of the Neornithes in the latest Cretaceous (Clarke *et al.*, 2005), just before the great mass extinction at the K–Pg boundary, some 65 Ma. Also, a combination of several molecular analyses of relationship calibrated against phylogenetically well-documented and well-constrained fossils have managed to temporally constrain the radiation of one of the main clades within Neornithes, the Neoaves (Ericson *et al.*, 2006). While the clade already started to

differentiate within the latest Cretaceous, the main, explosive phase of diversification is now regarded as having taken place during the Paleocene Epoch, possibly within as little as 5 Myr of the K–Pg boundary mass extinction (Ericson *et al.*, 2006). The exact dates of origination and diversification differ between studies and are still subject to much debate (Ericson *et al.*, 2006, 2007; Slack *et al.*, 2006; Brown *et al.*, 2007), chiefly due to lack of well-preserved Late Cretaceous fossils to calibrate the earliest divergences against. Nonetheless, the main origination and/or diversification phases within Neornithes as recognized by current morphological and molecular consensus can be summarized as follows (Figure 14.1):

1 division between Palaeognathae and Neognathae (ca. 100 Ma; Ericson *et al.*, 2006; Slack *et al.*, 2006; van Tuinen *et al.*, 2006);

2 one or more divergences resulting in the origin of Galliformes, Anseriformes and probably

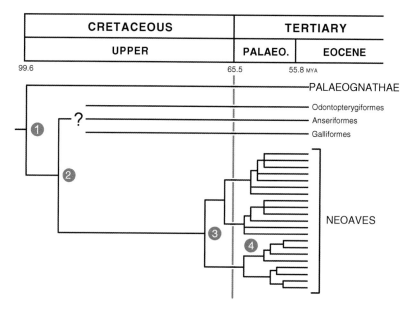

Fig. 14.1 Conceptual diagram synthesizing current molecular and morphological models of origin and diversification of neornithine lineages during the late Cretaceous and earliest Paleogene. Grey circles denote main phases of origin and diversification: 1, Palaeognath/Neognath division; 2, divergence of anseriform–galliform–odontopterygiform complex; 3, initial neoavian diversification; 4: main, extensive neoavian diversification. Vertical grayshade line: K–Pg boundary. (Based on Bourdon, 2005; Ericson *et al.*, 2006; Slack *et al.*, 2006; van Tuinen *et al.*, 2006; Mayr, 2008; ages from Gradstein *et al.*, 2004.)

Odontopterygiformes (95–90 Ma; Ericson *et al.*, 2006; Slack *et al.*, 2006; van Tuinen *et al.*, 2006);

3 initial basal diversification of Neoaves in the latest Cretaceous (75–65 Ma; Ericson *et al.*, 2006);

4 main, extensive diversification within Neoaves after the K–Pg boundary (65–55 Ma; Ericson *et al.*, 2006).

The survival of several lineages across the mass extinction event (aftermath of Phase 3), and especially the rapid and extensive diversification (Phase 4) event taking place immediately after the mass extinction event, are especially intriguing from a research viewpoint considering extinction recovery and subsequent ecological niche adaptation and diversification. This pattern compares closely to the "alternative explosive model" and overall conclusions of van Tuinen *et al.* (2006) based on molecular clock analyses, as well as divergence models 3 and 4 of Penny & Philips (2004), suggesting that: (i) the effects of the K–Pg mass extinction was relatively limited on terrestrial vertebrates compared to marine; and (ii) that neornithine birds moving into new niches may be competing with and displacing (juvenile) pterosaurs and/or small dinosaurs.

Slack *et al.* (2006) suggested that smaller pterosaurs might have been losing niche-space to neornithine birds, but noted the need for additional fossil data. However, database analysis coupled with the rock record indicate that the alleged decline in pterosaur diversity in the Late Cretaceous is a result of taphonomic bias, and there is no evidence for competitive replacement of pterosaurs by birds (Butler *et al.*, 2008). Furthermore, morphological analyses do not support competitive exclusion between pterosaurs and Mesozoic birds, indicating that the massive radiation amongst Neoaves immediately after K–Pg extinction event cannot be explained by birds taking up niches previously occupied by pterosaurs (McGowan & Dyke, 2007).

The dynamics, if any, of the replacement of enantiornithines with neornithines is also an interesting area for further study. Although Hope (2002) and Stidham (2002, 2008) presented data which may suggest that enantiornithines were becoming less diverse and perhaps displaced by more taxonomically and ecologically diverse neornithines at the end of the Cretaceous, confirmation of this will require denser fossil sampling across the boundary and especially identification of geological deposits with high potential for the preservation of articulated, well-preserved avian fossil material.

The documented survival of several lineages through the K–Pg boundary events suggests that modern birds may actually be relatively good at surviving in times of ecological stress and are able to rapidly expand and diversify into new niches within a short time. Unfortunately, the issue of the exact response of Neornithine birds to the K–Pg boundary mass extinction remains largely unanswered. The hypotheses so far presented have been based on the assumed ecological adaptations of a phylogenetically flawed clade of "transitional shorebirds" (Feduccia, 1995, 2003; Retallack, 2004; Robertson *et al.*, 2004). Furthermore, as discussed above, the selective survival of terrestrial birds contrasted with their small theropod dinosaur and/or enantiornithine cousins presents some problems to the selective extinction/survival hypothesis involving acid rain (Retallack, 2004) and a major obstacle to the infrared radiation hypothesis of Robertson *et al.* (2004).

However, given a more detailed fossil record comparable to that of mammals, birds may actually become an important asset for understanding the exact extinction mechanisms for terrestrial vertebrates at the K–Pg boundary. Based on the currently published fossil record it is impossible to determine the exact dynamics and effects of the events at the K–Pg boundary on neornithine and non-neornithine birds (Chiappe, 1995; Cracraft, 2001) although promising, but unpublished data from the latest Cretaceous of North America (Stidham, 2002, 2008) may be able to remedy this situation in the near future.

Similarly, the reason for the rapid post-K–Pg boundary diversification and expansion is unknown and a dearth of well-preserved fossil material from the Paleocene Epoch currently makes it impossible to shed much light on this important phase of neornithine radiation. The discovery of one or more earliest Paleocene Fossil-Lagerstätten comparable to the ones known from the early and middle Eocene is needed in order to document exactly which lineages were evolving and how.

ACKNOWLEDGMENTS

Many heartfelt thanks to Gareth Dyke and Gary Kaiser for inviting me to write this chapter; as well as reviewing it. The author was funded by a post-doctoral grant from the Carlsberg Foundation.

REFERENCES

Alvarenga HMF, Höfling E. 2003. Systematic revision of the Phorusrhacidae (Aves: Ralliformes). *Papéis Avulsos de Zoologia* **43**: 55–91.

Alvarez LW, Alvarez W, Asaro F, Michel HV. 1980. Extraterrestrial cause for the Cretaceous–Paleogene boundary extinction. *Science* **208**: 1095–1108.

Alvarez LW. 1983. Experimental evidence that an asteroid impact led to the extinction of many species 65 million years ago. *Proceedings of the National Academy of Sciences* **80**: 627–642.

Archibald JD. 1997. Extinction, Cretaceous. In *Encyclopedia of Dinosaurs*, Currie PJ, Padian K (eds). San Diego: Academic Press; 221–230.

Archibald JD, Bryant LJ. 1990. Differential *Cretaceous/Tertiary* extinctions of nonmarine vertebrates: evidence from northeastern Montana. *Geological Society of America Special Paper* **247**: 549–562.

Arens NC, Jahren AH. 2000. Carbon isotope excursion in atmospheric CO_2 at the Cretaceous–Paleogene boundary: evidence from terrestrial sediments. *Palaios* **15**: 314–322.

Belcher CM, Collinson ME, Sweet AR, Hildebrand AR, Scott AC. 2003. Fireball passes and nothing burns – the role of thermal radiation in the Cretaceous–Paleogene event: evidence from the charcoal record of North America. *Geology* **31**: 1061–1064.

Benson RD. 1999. *Presbyornis isoni* and other late Paleocene Birds from North Dakota. *Smithsonian Contributions to Paleobiology* **89**: 253–259.

Benton MJ, Ayala FJ. 2003. Dating the tree of life. *Science* **300**: 1698–1700.

Bleiweiss R. 1998. Fossil gap analysis supports early Paleogene origin of trophically diverse avian orders. *Geology* **26**: 323–326.

Bourdon E. 2005. Osteological evidence for sister group relationship between pseudo-toothed birds (Aves: Odontopterygiformes) and waterfowls (Anseriformes). *Naturwissenschaften* **92**: 586–591.

Bourdon E, Boya B, Iarochène M. 2005. Earliest African neornithine bird: a new species of Prophaethontidae (Aves) from the Paleocene of Morocco. *Journal of Vertebrate Paleontology* **25**: 157–170.

Brown JW, Payne RB, Mindell DP. 2007. Nuclear DNA does not reconcile "rocks" and "clocks" in Neoaves: a comment on Ericson *et al. Biology Letters* **3**: 257–259.

Buffetaut E. 1997. New remains of the giant bird *Gastornis* from the Upper Paleocene of the eastern Paris Basin and the relationships between *Gastornis* and *Diatryma. Neues Jahrbuch für Geologie und Paläontologie Monatshefte* **1997** (3): 179–190.

Buffetaut E, G. de Ploeg. 2008. A large ostrich-like bird from the Late Palaeocene of the Paris Region. In *Programme & Abstracts, SVPCA 2008*, Dyke G, Naish D, Parkes M (eds). Dublin: University College Dublin and the National Museum of Ireland; 16–17.

Butler R, Barrett P, Nowbath S, Upchurch P. 2008. Estimating the effects of the rock record on pterosaur diversity patterns: implications for hypotheses of bird/pterosaur competitive replacement. *Journal of Vertebrate Paleontology* **28** (Supplement 59A).

Case JA. 2001. Latest Cretaceous records of modern birds from Antarctica: center of origin or fortuitous occurrence? *PaleoBios* **21**. (Supplement 40).

Chatterjee S. 2002. The morphology and systematics of *Polarornis*, a Cretaceous loon (Aves: Gaviidae) from Antarctica. In *Proceedings of the 5th Symposium of the Society of Avian Paleontology and Evolution*, Zhou Z, Zhang F (eds), Beijing, 1–4 June 2000. Beijing: Science Press; 125–155.

Chenet A-L, Fluteau F, Courtillot V, Gérard M, Subbarao KV. 2008. Determination of rapid Deccan eruptions across the Cretaceous-Paleogene boundary using paleomagnetic secular variation: results from a 1200-meter thick section in the Mahabaleshwar escarpment. *Journal of Geophysical Research* **113**: B04101, doi: 10.1029/2006JB004635.

Chiappe LM. 1995. The first 85 million years of avian evolution. *Nature* **385**: 349–355.

Chiappe LM, Dyke GJ. 2002. The Mesozoic radiation of birds. *Annual Review of Ecology and Systematics* **33**: 91–124.

Chiappe LM, Dyke GJ. 2006. The early evolutionary history of birds. *Journal of the Paleontological Society of Korea* **22**: 133–151.

Chu PC. 1995. Phylogenetic reanalysis of Strauch's osteological data set for the Charadriiformes. *Condor* **97**: 174–196.

Chu PC. 1998. A phylogeny of the gulls (Aves: Larinae) inferred from osteological and integumentary characters. *Cladistics* **14**: 1–43.

Clarke JA, Tambussi CP, Noriega JI, Erickson GM, Ketcham RA. 2005. Definitive fossil evidence for the extant avian radiation in the Cretaceous. *Nature* **433**: 305–308.

Cooper A, Penny D. 1997. Mass survival of birds across the KT boundary: molecular evidence. *Science* **275**: 1109–1113.

Cordes AH. 2002. A new Charadriiform avian specimen from the early Maastrichtian of Cape Lamb, Vega Island, Antarctic Peninsula. *Journal of Vertebrate Paleontology* **22**: 46A.

Courtillot V, Jaeger J-J, Yang Z, Féraud G, Hofmann C. 1996. The influence of continental flood basalts on mass extinctions: where do we stand? *Geological Society of America Special Paper* **307**: 513–525.

Cracraft J. 1981. Toward a phylogenetic classification of the recent birds of the world (Class Aves). *The Auk* **98**: 681–714.

Cracraft J. 1988. The major clades of birds. In *The Phylogeny and Classification of the Tetrapods, Volume 1: Amphibians, Reptiles, Birds*, Benton MJ (ed.). Oxford: Clarendon Press; 10–12.

Cracraft J. 1986. The origin and early diversification of birds. *Paleobiology* **12**: 383–399.

Cracraft J. 2001. Avian evolution, Gondwana biogeography and the Cretaceous-Paleogene mass extinction event. *Proceedings of the Royal Society Series B (Biological Sciences)* **268**: 459–469.

Cracraft J, Barker FK, Braun MJ, Harshman J, Dyke GJ, Feinstein J, Stanley S, Cibois A, Schikler P, Beresford P, García-Morena J, Sorenson MD, Yuri T, Mindell DP. 2004. Phylogenetic relationships among modern birds (Neornithes): toward an avian tree of life. In *Assembling the Tree of Life*, Cracraft J, Donoghue MJ (eds). New York: Oxford University Press; 468–489.

Donoghue MJ, Doyle JA, Gauthier J, Kluge AG, Rowe T. 1989. The importance of fossils in phylogeny reconstruction. *Annual Review of Ecology and Systematics* **20**: 431–460.

Dyke GJ. 2003a. "Big bang" for Paleogene birds? *Trends in Ecology and Evolution* **18**: 441–442.

Dyke GJ. 2003b. The phylogenetic position of *Gallinuloides* Eastman (Aves: Galliformes) from the Paleogene of North America. *Zootaxa* **199**: 1–10.

Dyke GJ. 2003c. The fossil record and molecular clocks: basal radiations within the Neornithes. In *Telling the Evolutionary Time. Molecular Clocks and the Fossil Record*, Donoghue PJ, Smith MP (eds). London: Taylor and Francis; 263–277.

Dyke GJ, Gulas BE. 2002. The fossil galliform bird *Paraortygoides* from the Lower Eocene of the United Kingdom. *American Museum Novitates* **3360**: 1–14.

Dyke GJ, van Tuinen M. 2004. The evolutionary radiation of modern birds (Neornithes. reconciling molecules, morphology and the fossil record. *Zoological Journal of the Linnean Society* **141**: 153–177.

Dyke GJ, Gulas BE, Crowe TM. 2003. Suprageneric relationships of galliform birds (Aves, Galliformes. a cladistic analysis of morphological characters. *Zoological Journal of the Linnean Society* **137**: 227–244.

Dyke GJ, Nudds RL, Benton MJ. 2007. Modern avian radiation across the Cretaceous-Paleogene boundary. *The Auk* **124**: 339–341.

Edwards SV, Jennings WB, Shedlock AM. 2005. Phylogenetics of modern birds in the era of genomics. *Proceedings of the Royal Society Series B (Biological Sciences)* **272**: 979–992.

Ericson PGP. 1997. Systematic relationships of the Palaeogene family Presbyornithidae (Aves: Anseriformes). *Zoological Journal of the Linnean Society* **121**: 429–483.

Ericson PGP. 2008. Current perspectives on the evolution of birds. Contributions to Zoology **77**: 109–116.

Ericson PGP, Anderson CL, Britton T, Elzanowski A, Johansson US, Källersö M, Olson J.L., Parsons TJ. Zuccon, D. Mayr, G. 2006. Diversification of Neoaves: integration of molecular sequence data and fossils. *Biology Letters* **2**: 543–547.

Ericson PGP, Anderson CL, Mayr G. 2007. Hangin' on to our rocks 'n clocks: a reply to Brown *et al*. *Biology Letters* **3**: 260–261.

Fain MG, Houde P. 2004. Parallel radiations in the primary clades of birds. *Evolution* **58**: 2558–2573.

Feduccia A. 1976. Osteological evidence for the shorebird affinities of the flamingos. *The Auk* **93**: 587–601.

Feduccia A. 1977. Hypothetical stages in the evolution of modern ducks and flamingos. *Journal of Theoretical Biology* **67**: 715–721.

Feduccia A. 1978. *Presbyornis* and the evolution of ducks and flamingos. *American Scientist* **66**: 298–304.

Feduccia A. 1995. Explosive evolution in Paleogene birds and mammals. *Science* **267**: 637–638.

Feduccia A. 2003. "Big bang" for Paleogene birds? *Trends in Ecology and Evolution* **18**: 172–176.

Fountaine TMR, Benton MJ, Dyke GJ, Nudds RL. 2005. The quality of the fossil record of Mesozoic birds. *Proceedings of the Royal Society of London, Series B (Biological Sciences)* **272**: 289–294.

Gallagher WB. 1991. Selective extinction and survival across the *Cretaceous/Tertiary* boundary in the northern Atlantic coastal plain. *Geology* **19**: 967–970.

Gradstein FM, Ogg JG, Smith AG, Agterberg FP, Bleeker W, Cooper RA, Davydov V, Gibbard P, Hinnov LA, House MR, Lourens L, Luterbacher HP, McArthur J, Melchin MJ, Robb LJ, Shergold J, Villeneuve M, Wardlaw BR, Ali J, Brinkhuis H, Hilgen FJ, Hooker J, Howarth RJ, Knoll AH, Laskar J, Monechi S, Plumb KA, Powell J, Raffi I, Röhl U, Sadler P, Sanfilipo A Schmitz B, Shackleton NJ, Shields GA, Strauss H, Van Dam J, Van Kolfschoten T, Veizer, J, Wilson D. 2004. *A Geologic Time Scale 2004*. Cambridge: Cambridge University Press.

Graur D, Martin W. 2004. Reading the entrails of chickens: molecular timescales of evolution and the illusion of precision. *Trends in Genetics* **20**: 80–86.

Gulas-Wroblewski BE, Wroblewski AF-J. 2003. A crown-group galliform bird from the Middle Eocene Bridger Formation of Wyoming. *Palaeontology* **46**: 1269–1280.

Hackett SJ, Kimball RT, Reddy S, Bowie RCK, Braun EL, Braun MJ, Chojnowski JL, Cox WA, Han K-L, Harshman J, Huddleston CJ, Marks BD, Miglia KJ, Moore WS, Sheldon FH, Steadman DW, Witt CC, Yuri T. 2008. A phylogenomic study of birds reveals their evolutionary history. *Science* **320**: 1763–1768.

Harrison CJO. 1985. A bony-toothed bird (Odontopterygiformes) from the Palaeocene of England. *Tertiary Research* **7**: 23–25.

Hedges SB, Parker PH, Sibley CG, Kumar S. 1996. Continental breakup and the diversification of birds and mammals. *Nature* **381**: 226–229.

Hildebrand AR, Penfield GT, Kring DA, Pilkington M, Zanoguera AC, Jacobsen SB, Boynton WV. 1991. Chicxulub Crater; a possible *Cretaceous/Tertiary* boundary impact crater on the Yucatan Peninsula, Mexico. *Geology* **19**: 867–871.

Hone DWE, Dyke GJ, Haden M, Benton MJ. 2008. Body size evolution in Mesozoic birds. *Journal of Evolutionary Biology* **21**: 618–624.

Hope S. 2002. The Mesozoic Radiation of the Neornithes. In *Mesozoic Birds: Above the Heads of Dinosaurs*, Chiappe LM, Witmer LM (eds). Berkeley: University of California Press; 339–388.

Houde P. 1988. *Paleognathous Birds from the Early Paleogene of the Northern Hemisphere*. Cambridge: Nuttall Ornithological Club, Publication 22.

James HF. 2005. Paleogene fossils and the radiation of modern birds. *The Auk* **122**: 1049–1054.

Johansson US, Parsons TJ, Irestedt M, Ericson PGP. 2001. Clades within the "higher land birds", evaluated by nuclear DNA sequences. *Journal of Zoology and Systematic Evolutionary Research* **39**: 37–51.

Kitchell JA, Clark DL, Gombos AM. 1986. Biological selectivity of extinction: a link between background and mass extinction. *Palaios* **1**: 504–511.

Kurochkin EN, Dyke GJ, Karhu AA. 2002. A new presbyornithid bird (Aves, Anseriformes) from the Late Cretaceous of Southern Mongolia. *American Museum Novitates* **3386**: 1–11.

Leonard L, van Tuinen M, Dyke GJ. 2005. A new specimen of the fossil Palaeognath *Lithornis* from the earliest Palaeogene of Denmark. *American Museum Novitates* **3491**: 1–11.

Lindow BEK, Dyke GJ. 2006. Bird evolution in the Eocene: climate change in Europe and a Danish fossil fauna. *Biological Reviews of the Cambridge Philosophical Society* **81**: 483–499.

Livezey BC. 1997. A phylogenetic analysis of basal Anseriformes, the fossil *Presbyornis*, and the interordinal relationships of waterfowl. *Zoological Journal of the Linnean Society* **121**: 361–428.

Livezey BC, Zusi RL. 2001. Higher-order phylogenetics of modern Aves based on comparative anatomy. *Netherlands Journal of Zoology* **51**: 179–205.

Livezey BC, Zusi RL. 2006. Higher-order phylogeny of modern birds (Theropoda, Aves: Neornithes) based on comparative anatomy. I. Methods and characters. *Bulletin of the Carnegie Museum of Natural History* **37**: 1–556.

Livezey BC, Zusi RL. 2007. Higher-order phylogeny of modern birds (Theropoda, Aves: Neornithes) based on comparative anatomy. II. Analysis and discussion. *Zoological Journal of the Linnean Society* **149**: 1–95.

Marshall CR. 1997. Confidence intervals on stratigraphic ranges with nonrandom distribution of fossil horizons. *Paleobiology* **23**: 165–173.

Mayr G. 2002. A new specimen of *Salmila robusta* (Aves: Gruiformes: Salmilidae n. fam.) from the Middle Eocene of Messel. *Paläontologische Zeitschrift* **76**: 305–316.

Mayr G. 2003. Phylogeny of Early Paleogene swifts and hummingbirds (Aves: Apodiformes). *The Auk* **120**: 145–151.

Mayr G. 2004. A partial skeleton of a new fossil loon (Aves, Gaviiformes) from the early Oligocene of Germany with preserved stomach content. *Journal of Ornithology* **145**: 281–286.

Mayr G. 2005a. The Paleogene fossil record of birds in Europe. *Biological Reviews of the Cambridge Philosophical Society* **80**: 1–28.

Mayr G. 2005b. The postcranial osteology and pylogenetic position of the Middle Eocene *Messelastur gratulator* Peters, 1994 – a morphological link between owls (Strigiformes) and falconiform birds? *Journal of Vertebrate Paleontology* **25**: 635–645.

Mayr G. 2007a. The birds from the Paleocene fissure filling of Walbeck (Germany). *Journal of Vertebrate Paleontology* **27**: 394–408.

Mayr G. 2007b. Avian higher-level phylogeny: well-supported clades and what we can learn from a phylogenetic analysis of 2954 morphological characters. *Journal of Zoological Systematics and Evolutionary Research* **46**: 63–72.

Mayr G. 2008. A skull of the giant bony-toothed bird *Dasornis* (Aves: Pelagornithidae) from the Lower Eocene of the Isle of Sheppey. *Palaeontology* **51**: 1107–1116.

Mayr G, Clarke JA. 2003. The deep divergences of neornithine birds: a phylogenetic analysis of morphological characters. *Cladistics* **19**: 527–553.

Mayr G, Mourer-Chauviré C. 2004. Unusual tarsometatarsus of a mousebird from the Paleogene of France and the relationships of *Selmes* Peters, 1999. *Journal of Vertebrate Paleontology* **24**: 366–372.

McGowan AJ, Dyke GJ. 2007. A morphospace-based test for competitive exclusion among flying vertebrates: did birds, bats and pterosaurs get in each other's space? *Journal of Evolutionary Biology* **20**: 1230–1236.

McKitrick MC. 1991. Phylogenetic analysis of avian hindlimb structure. *Miscellaneous Publications Museum of Zoology, University of Michigan* 179: 1–85.

Melosh HJ, Schneider NM, Zahnle KJ, Latham D. 1990. Ignition of global wildfires at the Cretaceous/Tertairy boundary. *Nature* **343**: 251–254.

Mickevich MF, Parenti LR. 1980. Review of the "The phylogeny of the Charadriiformes (Aves. a new estimate using the method of character compatibility analysis". *Systematic Zoology* **29**: 108–113.

Mourer-Chauviré C. 1994. A large owl from the Palaeocene of France. *Palaeontology* **37**: 339–348.

Mourer-Chauviré C. 1995a. The Messelornithidae (Aves: Gruiformes) from the Paleogene of France. *Courier Forschungsinstitut Senckenberg* **181**: 95–105.

Mourer-Chauviré C. 1995b. Dynamics of the avifauna during the Paleogene and the early Neogene of France. Settling of the recent fauna. *Acta Zoologica Cracoviensis* **38**: 325–342.

Norris RD. 2001. Impact of K–T boundary events on marine Life. In *Palaeobiology II*, Briggs DEG, Crowther PR (eds). Oxford: Blackwell Science; 229–231.

Olson SL. 1985. The fossil record of birds. In *Current Ornithology Volume VIII*, Farner DS, King JR, Parks KC (eds). New York: Plenum Press; 80–238.

Olson SL. 1992. *Neogaeornis wetzeli* Lambrecht, a Cretaceous loon from Chile (Aves: Gaviidae). *Journal of Vertebrate Paleontology* **12**: 122–124.

Olson SL. 1994. A giant *Presbyornis* (Aves: Anseriformes) and other birds from the Paleocene Aquia Formation of Maryland and Virginia. *Proceedings of the Biological Society of Washington* **107**: 429–435.

Olson SL, Feduccia A. 1980. *Presbyornis* and the origin of the Anseriformes (Aves: Charadriomorphae). *Smithsonian Contributions to Zoology* **323**: 1–24.

Olson SL, Parris DC. 1987. The Cretaceous birds of New Jersey. *Smithsonian Contributions to Paleobiology* **63**: 1–22.

Padian K, Chiappe LM. 1998. The origin and early evolution of birds. *Biological Reviews of the Cambridge Philosophical Society* **73**: 1–42.

Paton T, Haddrath O, Baker AJ. 2002. Complete mitochondrial DNA genome sequences show that modern birds are not descended from transitional shorebirds. *Proceedings of the Royal Society, Series B (Biological Sciences)* **269**: 839–846.

Paul CRC. 2003. Ghost ranges. In *Telling the Evolutionary Time. Molecular Clocks and the Fossil Record*, Donoghue PJ, Smith MP (eds). London: Taylor and Francis; 91–106

Penny D, Phillips MJ. 2004. The rise of birds and mammals: are microevolutionary processes sufficient for macroevolution. *Trends in Ecology and Evolution* **19**: 516–522.

Poe S, Chubb AL. 2004. Birds in a bush: five genes indicate explosive evolution of avian orders. *Evolution* **58**: 404–415.

Prinn RG, Fegley Jr., 1987. Bolide impacts, acid rain, and biospheric traumas at the Cretaceous–Tertiary boundary. *Earth and Planetary Science Letters* **83**: 1–15.

Retallack GJ. 2004. End-Cretaceous acid rain as a selective extinction mechanism between birds and dinosaurs. In *Feathered Dragons: Studies on the Transition from Dinosaurs to Birds*, Currie PJ, Koppelhus EB, Shugar MA (eds). Bloomington: Indiana University Press; 35–64.

Rhodes MC, Thayer CW. 1991. Mass extinctions: ecological selectivity and primary production. *Geology* **19**: 877–880.

Rich PV, Bohaska DJ. 1976. The world's oldest owl: a new strigiform from the Paleocene of southwestern Colorado. In *Collected Papers in Avian Paleontology Honoring the 90th Birthday of Alexander Wetmore*, Olson SL (ed.). Washington, DC: Smithsonian Contributions to Paleobiology; 87–93.

Robertson DS, McKenna MC, Toon OB, Hope S, Lillegraven AJ. 2004. Survival in the first hours of the Cenozoic. *GSA Bulletin* **116**: 760–768.

Ryder G, Fastovsky D, Gartner S. 1996. *The Cretaceous–Tertiary Event and other Catastrophes in Earth history*. Geological Society of America, Special Paper 307.

Schmitz B. 1988. Origin of microlayering in worldwide distributed Ir-rich marine *Cretaceous/Tertiary* boundary clays. *Geology* **16**: 1068–1072.

Seilacher A, Reif WE, Westphal F. 1985. Sedimentological, ecological and temporal patterns of fossil Lagerstätten. *Philosophical Transactions of the Royal Society of London* **B311**: 5–23.

Sharpton VL, Ward PD. 1990. *Global Catastrophes in Earth History: An Interdisciplinary Conference on Impacts, Volcanism, and Mass Mortality*. Geological Society of America, Special Paper 247.

Sheehan PM, Hansen TH. 1986. Detritus feeding as a buffer to extinction at the end of the Cretaceous. *Geology* **14**: 868–870.

Sheehan PM, Fastovsky DE. 1992. Major extinctions of land dwelling vertebrates at the Cretaceous–Tertiary boundary, eastern Montana. *Geology* **20**: 556–560.

Sheehan PM, Coorough PJ, Fastovsky DE. 1996. Biotic selectivity during the K/T and Late Ordovician extinction events. *Geological Society of America Special Paper* **307**: 477–489.

Sigurdsson H, d'Hondt S, Carey S. 1992. The impact of the *Cretaceous/Tertiary* bolide on evaporite terrane and generation of major sulfuric acid aerosol. *Earth and Planetary Science Letters* **109**: 543–559.

Slack KE, Jones CM, Ando T, Harrison GLA, Fordyce RE, Arnason U, Penny D. 2006. Early penguin fossils, plus mitochondrial genomes, calibrate avian evolution. *Molecular Biology and Evolution* **23**: 1144–1155.

Sorenson MD, Oneal E, García-Moreno J, Mindell DP. 2003. More taxa, more characters: the hoatzin problem is still unresolved. *Molecular Biology and Evolution* **20**: 1484–1499.

Stidham TA. 2002. Extinction and survival of birds (Aves) at the Cretaceous-Paleogene boundary: evidence from Western North America. *Geological Society of America Abstracts* **34**: 355.

Stidham TA. 2008. Extinction of birds at the end of the Cretaceous: chronostratigraphic data. *Journal of Vertebrate Paleontology* **28**: 147A–148A.

Strauch JG. 1978. The phylogeny of the Charadriiformes (Aves): a new estimate using the method of character compatibility analysis. *Transactions of the Zoological Society of London* **34**: 263–345.

Tambussi C, Acosta Hospitaleche C. 2007. Antarctic birds (Neornithes) during the Cretaceous-Eocene times. *Revista de la Asociación Geológica Argentina* **62**: 604–617.

Tambussi CP, Reguero MA, Marenssi SA, Santillana SN. 2005. *Crossvallia unienwillia*, a new Spheniscidae (Sphenisciformes, Aves) from the late Paleocene of Antarctica. *Geóbios* **38**: 557–675.

Van Tuinen M, Paton T, Haddrath O, Baker A. 2003. "Big bang" for Paleogene birds? a reply. *Trends in Ecology and Evolution* **18**: 442–443.

Van Tuinen M, Stidham TA, Hadly EA. 2006. Tempo and mode of modern bird evolution observed with large-scale taxonomic sampling. *Historical Biology* **18**: 205–221.

Wolfe JA. 1990. Paleobotanical evidence for a marked temperature increase following the Cretaceous/Tertiary boundary. *Nature* **343**: 152–154.

Wolfe JA. 1991. Paleobotanical evidence for a June "impact winter" at the Cretaceous/Tertiary boundary. *Nature* **352**: 420–423.

Wolfe JA, Russell DA. 2001. Impact of K–T boundary events on terrestrial life. In *Palaeobiology II*, Briggs DEG, Crowther PR (eds). Oxford: Blackwell Science; 232–234.

15 Functional and Phylogenetic Diversity in Marine and Aquatic Birds

GARY KAISER

Royal British Colombia Museum, Victoria, Canada

The various groups of aquatic and marine birds have long been familiar to naturalists and most have been formally recognized as discrete taxa since the 19th century (e.g. Garrod, 1873; Fürbringer, 1888; Beddard, 1898). Although these birds constitute a major portion of the diversity among the higher taxa of Class Aves and are well-represented among fossils since the Paleogene (Olson, 1985; Warheit, 2002), they are unknown from the Cretaceous except for the fossils *Vegavis* and *Teviornis* (Kurochkin *et al.*, 2002; Dyke & van Tuinen, 2004; Clarke *et al.*, 2005; see Lindow, Chapter 14, this volume). Living representatives include ducks and their relatives (Anseriformes); loons (Gaviiformes); grebes (Podicipediformes); penguins (Sphenisciformes); tube-nosed seabirds or petrels (Procellariiformes); cormorants, pelicans, gannets, frigatebirds, and tropicbirds (Pelecaniformes); auks, gulls, terns, phalaropes, and a variety of other shorebirds (Charadriiformes); and the rails and sungrebes (Ralliformes). The appearance of these taxa in 21st-century morphology-based phylogenies (e.g. Livezey & Zusi, 2007) and, with some anomalies, biomolecular phylogenies (e.g. Hackett *et al.*, 2008) support their biological significance and implied evolutionary relationships.

As discussed in earlier chapters, until the end of the 20th century, ornithologists were unable to reach a general agreement on the classification of birds and, therefore, were unable to develop a plausible story for the evolutionary relationships among seabirds (Raikow, 1974). Occasionally some particular similarity appeared as a phenetic argument in support of an evolutionary hypothesis (e.g. Verheyen, 1958) but none of these gained widespread acceptance. Part of the difficulty may have stemmed from a superficial similarity imposed on all members of the group by life in aquatic habitats, just as a comparable level of general similarity characterizes most small forest birds or the long-legged waders that stalk through marshes and grasslands. However, unlike the rather conservative terrestrial groups, marine and aquatic birds exhibit a variety of unique structural adaptations that help them cope, not only with food handling, but also with both the ability to move through two very different media and to search for resources across very large distances. The purpose of this chapter is to review some of those functional adaptations in the light of modern phylogenetic hypotheses and recently discovered fossil forms.

A FRAMEWORK FOR MARINE BIRD CLASSIFICATION

Biomolecular phylogenies

The application of biomolecular techniques to the construction of avian phylogenies has offered solutions to some long-standing puzzles over the relationships among marine or aquatic groups (e.g. Friesen *et al.*, 1996; Kennedy & Page, 2002; Pereira & Baker, 2008). However, even these technically sophisticated approaches have been faced with unexpected difficulties. Marine and aquatic birds are exceptionally mobile and well able to undertake migrations from one hemisphere to another (Shaffer *et al.*, 2006). Such behavior introduces the prospect of an important role for processes of speciation affected by such dispersal and, thereby, greatly increases the complexities of interpreting bio-molecular analyses. Colonial seabirds (e.g. *Oceanodroma castro, O. leucorhoa*) are exceptional among vertebrates in providing evidence for peripatric, parapatric, and sympatric models of speciation processes (Friesen, 2007).

The biomolecular tapestry of Sibley & Ahlquist (1990) was the first attempt to use genetic material (nuclear DNA) as the basis for classification of the whole of Class Aves (see Livezey, Chapter 4, this volume). Although it was based solely on physical characteristics of the DNA molecule (i.e. the difference in melting points between pure and mixed samples), it was sufficiently sensitive to distinguish many of the traditional orders and families. The resulting tapestry of relationships scattered the marine and aquatic birds among other clades in unexpected ways, linking waterfowl (Anseriformes) to land fowl (Galliformes) in a basal branch of the Subclass Eoaves and placing many of the remaining waterbirds among more terrestrial taxa in an all-inclusive Order Ciconiiformes (Figure 15.1). Subsequently both the methodology and the taxon sampling approach used by Sibley and Ahlquist were called into question (Houde, 1987a; Sarich *et al.*, 1989; Lanyon, 1992; Harshman, 1994) and the influence of their ideas on understanding of the evolution of birds declined.

Sibley & Ahlquist's (1990) treatment of Pelecaniformes was particularly controversial because it fragmented that order placing frigatebirds near the crown, close to loons and petrels, but assigning the pelicans to a position among the storks, a location that agreed with a suggested relationship to the shoebill stork (*Balaeniceps rex*) (Cottam, 1957; Mayr, 2003; see Livezey, Chapter 4, this volume). The remaining pelecaniform birds were placed near the grebes and herons (Ardeiformes) (Figure 15.1) but split into two groups, with the tropicbirds described as a sister clade to a Parvorder Sulida (cormorants, anhingas, and boobies). A more recent biomolecular phylogeny (Hackett *et al.*, 2008) also recognizes a close relationship between the storks and pelicans, but places both

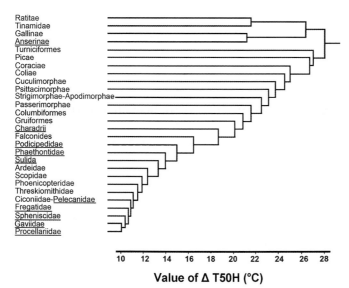

Value of Δ T50H (°C)

Fig. 15.1 Higher taxa in the biomolecular tapestry of Sibley & Ahlquist (1990) based on melting temperature changes between pure and mixed samples of nuclear DNA. The names of orders that include marine and aquatic birds are underlined.

Major Branch   Avian Orders

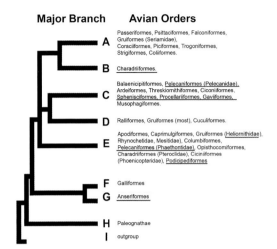

A   Passeriformes, Psittaciformes, Falconiformes, Gruiformes (Seriamidae), Coraciiformes, Piciformes, Trogoniformes, Strigiformes, Coliiformes.

B   Charadriiformes.

C   Balaenicipitiformes, Pelecaniformes (Pelecanidae), Ardeiformes, Threskiornithiformes, Ciconiiformes, Sphenisciformes, Procellariiformes, Gaviiformes, Musophagiformes.

D   Ralliformes, Gruiformes (most), Cuculiformes.

E   Apodiformes, Caprimulgiformes, Gruiformes (Heliornithidae), Rhynochetidae, Mesitidae), Columbiformes, Pelecaniformes (Phaethontidae), Opisthocomiformes, Charadriiformes (Pteroclidae), Ciconiiformes (Phoenicopteridae), Podicipediformes.

F   Galliformes

G   Anseriformes.

H   Paleognathae

I   outgroup

Fig. 15.2 Avian phylogeny derived from characteristics of nuclear DNA by Hackett *et al.* (2008). The names of orders that include marine and aquatic birds are underlined.

with other pelecaniform birds (Figure 15.2). In that phylogeny, only the tropicbirds lack a close relationship to other pelecaniform birds.

The nearly linear array of higher taxa in the Sibley & Ahlquist (1990) tapestry (Figure 15.1) immediately encourages correlations with anticipated trends in avian evolution. For instance, within the Neoaves, weight increases as one moves from the small forest birds in the basal clades towards the crown clades that hold such large birds as penguins and loons (Maurer, 1998). That trend has important implications for several seemingly unrelated aspects of avian biology. Small forest birds with altricial young and elaborate nests appear close to the basal groups while many of the crown clades have more precocious young and build minimal nests. The same forest birds also use low-speed, highly maneuverable flight, implying that the long-range, high-speed flight of large aquatic and marine birds is advanced. That arrangement has parallels in the history of human flight in which slow, low-powered machines preceded the development of sophisticated long-range aircraft – but history has not been kind to the tapestry.

Sibley & Ahlquist's (1990) pioneering effort demonstrated that biomolecular approaches to avian phylogeny could have much to offer but ultimately methodological weaknesses, especially the lack of an appropriate outgroup, discouraged widespread application of the tapestry (Houde, 1987a; Sarich *et al.*, 1989; Lanyon, 1992; Harshman, 1994). Recent studies of mitochondrial and nuclear DNA (e.g. Fain & Houde, 2004; Hackett *et al.*, 2008) have further undermined its importance even though they have been less comprehensive in their scope.

The very large genomic analysis of Hackett *et al.* (2008) comes closest to achieving the goals of Sibley & Ahlquist's (1990) pioneering effort. It also identifies the early separation of the Galloanserinae from other neognathous birds, a dichotomy that was perhaps the single most important and novel contribution of Sibley & Ahlquist (1990; Figure 15.2). At higher taxonomic levels, the phylogeny of Hackett *et al.* (2008) shares some structural features with morphology-based phylogenies, including the recent work of Livezey & Zusi (2007). As in all biomolecular studies, the dependence on DNA from living species isolates the proposed phylogeny from the fossil record.

Aside from basal branches for the Paleognathae and Galloanserinae, Hackett *et al.* (2008) arrange the remaining lineages into five large clusters: (A) forest birds, including hawks and owls; (B) charadriiform birds; (C) pelecaniform birds, herons, ibises, storks, petrels, penguins, and loons; (D) cranes, rails, and cuckoos; and (E) caprimulgiforms and apodiforms with the sunbittern (*Eurypyga* sp.) and kagu (*Rhynochetos jubatus*). Unfortunately a large, unresolved polytomy leaves other important lineages, such as pigeons, flamingos, grebes, tropicbirds, and the hoatzin (*Opisthocomus hoazin*) in an undifferentiated cluster. In addition some of the relationships in the main body of the phylogeny are unexpected, even anomalous. Cuckoos appear in a lineage with rails, and the kagu with the potoo (*Nyctibius* sp.) while the bulk of the pelecaniform birds are isolated from the tropicbirds. By creating numerous exceptions and increasing the number of times a particular characteristic might have arisen, these anomalies make it very difficult to postulate general evolutionary themes or patterns across this phylogeny. Fortunately, for the purposes of this chapter, many marine and aquatic birds are contained in a single large lineage that has some similarities with the arrangement proposed by Livezey & Zusi (2007).

Morphology-based phylogenies

Marine and aquatic birds are distinctive because significant changes in morphology have accompanied adaptation to their preferred habitats. Consequently a morphology-based analysis such as the recent phylogeny from Livezey & Zusi (2007) holds great promise as a tool for understanding the sequence of some evolutionary events in the history of birds. Unlike molecular-based hypotheses, Livezey & Zusi (2007) are able to link their phylogeny to the fossil record by rooting it in the bird-like characteristics of the Theropoda. They use 2954 morphological characteristics to arrange 33 orders of living birds in five large monophyletic groups: Cohort Paleognathae, Subcohort Galloanserae, Division Natatores, Subdivision Terrestrornithes, and Subdivision Dendrornithes (Figure 15.3). The location of aquatic and marine

birds among the basal groups echoes their placement in the early attempt by Max Fürbringer (1888) to base a phylogeny for birds on measured distances between clusters of characters.

In the Livezey & Zusi (2007) phylogeny, a series of stepwise changes lead from the ratites in the basal clades to small forest birds (e.g. Coraciiformes, Piciformes, Passeriformes) near the crown. The progress of those changes, from groups dominated by large birds that are often marine or aquatic, to groups dominated by very small forest birds implies the existence of broad evolutionary trends and responses to adaptive pressures that we should be able to track through avian history. Each order and family represents a unique interpretation of those trends in the face of separate environmental factors faced by the modern group's ancestors. Comparably polarized events are also implied by the biomolecular phylogeny of Hackett *et al.* (2008) that stands in stark contrast to the opposite arrangement in the Sibley & Ahlquist (1990) biomolecular tapestry (Maurer, 1998). In both the Livezey & Zusi (2007) and the Hackett *et al.* (2008) phylogenies, many taxa among the early clades include marine or aquatic birds (Figure 15.4), suggesting that there is a general

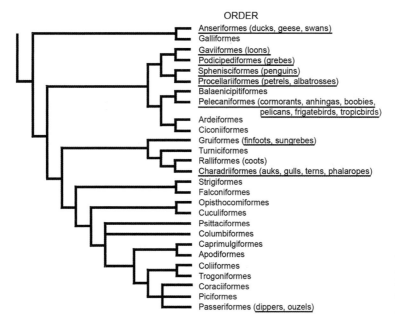

Fig. 15.3 Avian phylogeny derived from the analysis of 2954 morphological characteristics (Livezey & Zusi, 2006, 2007). The names of orders that include marine and aquatic birds are underlined.

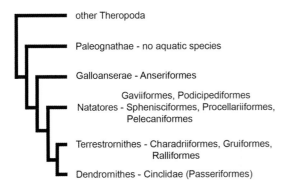

other Theropoda

Paleognathae - no aquatic species

Galloanserae - Anseriformes

Gaviiformes, Podicipediformes
Natatores - Sphenisciformes, Procellariiformes,
Pelecaniformes

Terrestrornithes - Charadriiformes, Gruiformes,
Ralliformes

Dendrornithes - Cinclidae (Passeriformes)

Fig. 15.4 Groups of diving birds within the major divisions of the phylogeny from Livezey & Zusi (2007).

trend from large birds at the base to small birds at the crown with significant consequences for the evolution of flight capabilities, reproductive strategies, and other aspects of avian biology (Figure 15.3).

Although there are no marine or specifically aquatic species among either the known fossils of theropod dinosaurs or among the living Paleognathae, eight orders of specialized marine and aquatic birds appear near the base of the Neognathae (Livezey & Zusi, 2007). One aquatic order, Anseriformes (ducks, geese, and related birds), is placed basally within the Galloanserinae, while most other marine or aquatic groups appear in the appositely named Division Natatores (Figure 15.4). The Charadriiformes (auks, gulls, and shorebirds) and the Ralliformes (rails) are placed in the more crownward Terrestrornithes, a transitional taxon that includes a mix of terrestrial and aquatic forms. Rather than being anomalies, the Charadriiformes and Ralliformes suggest that the aquatic lifestyle represents a successful strategy, worthy of imitation through parallel evolution. The crown group, Dendrornithes contains 14 orders, numerous families, and thousands of species of small forest birds. It contains no marine members but the relatively unspecialized dippers or ouzels (Cinclidae) are aquatic, wing-propelled divers (Kingery, 1996).

The extreme dispersal that gives rise to exceptional speciation processes (Friesen, 2007) may complicate both the biomolecular and morphological analyses of seabird phylogeny. Dispersal aggravates the problems created by the general scarcity of avian fossils and obscures the identity of preferred habitats or even likely breeding ranges for long extinct lineages.

The problem is exacerbated in Mesozoic birds, such as the flightless Hesperornithiformes (see O'Connor *et al.*, Chapter 3, this volume), both by the great age of the fossils and by their tenuous relationship to living groups (Olson, 1985; Houde, 1987b; Hope, 2002). In addition, time has been sufficient for great changes in the shapes of continents and presumably in the ocean currents that are so important in determining the distribution of prey species. Fossil forms from post-Cretaceous deposits existed in a more familiar geography and are more likely to be related to extant groups. Some, such as the flightless mancalline auks and plotopterids (Warheit, 2002; Mayr, 2005) appear to have bred along the coasts where their fossilized remains are found. The giant soaring birds among the Pelagornithidae, or pseudo-toothed birds, were much more mobile and may have shared their behaviour with modern oceanic soaring birds; their breeding grounds could have been located almost anywhere on the globe.

The biology of the Pelagornithidae poses several challenges (Stillwell *et al.*, 1997; see Bourdon, Chapter 8, this volume). In many ways their evolution appears to parallel that of much smaller, but extant, albatrosses and the two groups may have shared a similar life style. The distinctive bony processes that make up the "toothed" jaws in the Pelagornithidae appear adapted for holding slippery prey such as squid or small fish prey seized from the surface; albatrosses use a hook at the tip of the beak for the same purpose. However, most pelagornithids were very much larger than the extant albatrosses. They were so large that their wings appear to contradict aerodynamic theory about functional limits to the size of soaring birds (Pennycuick, 1987, 2002, 2008). Their taxonomy is also more problematic than most other recent fossil groups. Traditionally, they have been placed within the Pelecaniformes as the family Pelagornithidae (Fürbringer, 1888; McKee, 1985;

Goedert, 1989; Gonzalez-Barbra, 2002), but many are represented by only a few fragments (e.g. *Cyphornis*, Wetmore, 1928) and may never be confidently assigned to a living taxon. Recently, very complete material from North Africa and Europe suggests that the pseudo-toothed birds represent a sister lineage, Odontopterigiformes, to Anseriformes (Bourdon, 2005, Chapter 8, this volume; Mayr, 2008).

The proposal to place the pelagornithids so deep among basal phylogenies raises questions about their breeding strategies. The breeding strategy of the albatross requires a relatively large egg (i.e. about 35% of adult body weight; Carey *et al.*, 1980) and extended periods of intense parental care (Warham, 1996). None of the other basal lineages (Anseriformes, Gaviiformes, Podicipediformes) lay large eggs and their parental care is much less intense than more crownward groups. Breeding season in anseriforms is followed by a flightless moult period; albatrosses use a more gradual process to replace flight feathers with only minor effects on aerial efficiency.

STRUCTURAL CHARACTERISTICS

Most marine and aquatic birds form a remarkably cohesive group in terms of their basic body structure in that they are all relatively large birds with a superficial resemblance to ducks. Their necks are as long as those of most other birds but often appear shorter because they are folded when the bird is at rest. Their bodies are densely muscled and somewhat elongated with the hind limbs so far back that walking on land is difficult or impossible. Most species can stand erect but many cannot walk more than a few steps and terrestrial locomotion is limited to sliding on the breast. The skeleton of aquatic birds is usually robust with large surfaces for the attachment of powerful flight muscles. Among diving species and the truly oceanic birds, the density of the body is increased by limited pneumatization of the bones. In addition, the skeletons do not exhibit an exceptional degree of fusion. Most groups appear to find the flexibility of independent bones, connected by extensive developments of

relatively heavy connective tissue, preferable to the rigidity of ankylosis, even though fusion reduces overall body weight.

There are some skeletal adaptations unique to diving species. For instance, the tarsometatarsus may be laterally compressed and rather bladelike, supposedly to ease its passage through the water while the bird is swimming (Storer, 1960). Grebes carry that compression to an extreme and the flattened bone may help the lower limb act as a kind of wing, facilitating a unique lift-based swimming technique (Johansson & Norberg, 2001). Diving species also have exceptionally elongated ribs that extend caudally, almost to the cloaca. The function of such ribs is not clear but they may be important in strengthening the abdominal wall, helping to keep its shape during deep dives. Closer to the spinal column, the ribs are often supported by long and well-developed uncinate processes.

The variety of skeletal adaptations in the Pelecaniformes makes them exceptional among aquatic and marine groups. In the pelicans and frigatebirds, the bones are greatly inflated and highly pneumatized. As a result, they are unusually large but very thin-walled. This same characteristic appears among the fossil pelagornithids and is one of the reasons that their fossils are so fragmentary. Unlike most other seabirds, modern pelicans show a considerable degree of skeletal fusion. In particular, the pelvic bones and the synsacrum extend forward to capture additional vertebrae in the "lumbar" region, stiffening the lower spinal column (Figure 15.5). In the pelican's pectoral girdle, the enlarged and inflated furcula is thoroughly fused to a relatively small sternum, creating a very small keel. The pectoral girdle of the frigatebird is similar (Figure 15.6). The skeletons of the tropicbirds, gannets, and cormorants are not so specialized and are superficially similar to those of other aquatic birds such as loons or grebes.

The Pelecaniformes are also the only group of seabirds whose skulls lack one of the characteristic features of marine birds. In all of the other orders, the skull has a pair of large troughs between the dorsal edges of the orbits and the mid-line. In life, those troughs house the salt glands, modified tear glands, that help the bird maintain its osmotic

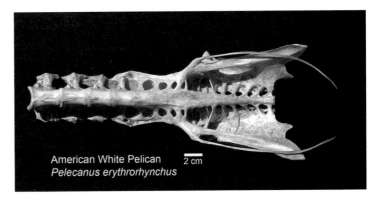

American White Pelican
Pelecanus erythrorhynchus
2 cm

Fig. 15.5 Pelvic region of the spinal column in the American white pelican (*Pelecanus erythrorhynchus*) showing the fusion of additional vertebrae to the anterior portion of the synsacrum.

balance by selectively excreting excess salt and are, therefore, essential for life at sea (Nishimura & Fan, 2002; Hughes, 2003; Laverty & Skadhauge, 2008). In the Pelecaniformes, the salt glands are housed within the orbits of the eyes and the roof of the skull is smooth (Figure 15.7).

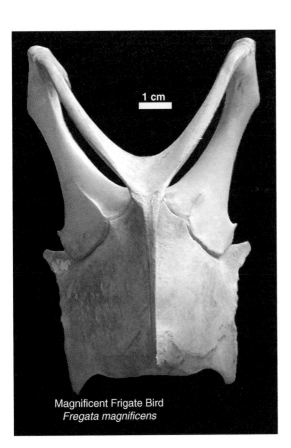

1 cm

Magnificent Frigate Bird
Fregata magnificens

Fig. 15.6 Ventral surface of the pectoral girdle in the magnificent frigatebird showing the very short sternum, fused furcula, and massive coracoids.

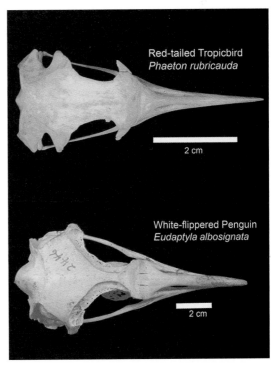

Red-tailed Tropicbird
Phaeton rubricauda
2 cm

White-flippered Penguin
Eudaptyla albosignata
2 cm

Fig. 15.7 Salt glands are carried within the orbits of pelecaniform birds (top) but in dorsal troughs in more typical seabirds such as the sphenisciform (bottom).

Externally, the Pelecaniformes are different from other seabirds in having less waterproof and less dense plumage. Most live in the tropics or the warm temperate zone but loose plumage does not prevent cormorants from occupying icy waters in the North Pacific and North Atlantic oceans. A downy layer may keep the skin dry and water trapped by larger feathers is warmed by body heat, as it is in human divers using wetsuits. Cormorant feathers shed water readily and they are able to take off quickly from a partially submerged position. However, prolonged submersion may wet the feathers and cormorants are noted for spending long periods at roosts with their wings spread for drying.

Wetting of the feathers is one explanation for the scarcity of reports of swimming by frigatebirds. However, it is more likely that they are trying to avoid the gradually accumulated weight of dried salt. On the Galapagos Islands, tourists are often taken to see frigatebirds bathing and swimming in a large freshwater lake on San Cristobal Island.

Members of the Pelecaniform birds also fail to develop a brood patch, and incubate their eggs and brood their young with heat from their feet (Siegel-Causey, 1997). Their eggs may be large (except in the cormorants) but have rather small yolks and produce altricial, rather helpless young. Procellariiform and alcid eggs are large and range from 35% to 41% yolk by volume but pelecaniform eggs range from 17% to 27% yolk (Jones, 1979; Carey et al., 1980). Consequently, newly hatched procellariiform and alcid young are generally much more developed than pelecaniform hatchlings and some alcid young are genuinely precocious (Gaston, 2004).

One widespread skeletal feature, that historically has had little value in taxonomy (Gadow, 1933), distinguishes members of the Neoaves from both the Paleognathae and the ancestors of birds among theropod dinosaurs (Kaiser, 2007). Thoracic hypapophyses are most prominent in diving birds, first appear among the Anseriformes and later in the Natatores and to varying degrees in diving birds among the Terrestrornithes. They are absent in the Galliformes and

many soaring birds but their presence in small raptors, swifts, and other aerobatic birds in the Dendrornithes suggests that they play a role in vigorous flapping flight. Typically, hypapophyses are small in terrestrial birds but unusually long examples that occur in the mousebird (Coliiformes) are an exception. In those birds, the hypapophses may support the internal organs when the birds hang upside down or roost in dense communal masses.

Hypapophyses develop as laterally compressed ventral processes of thoracic vertebrae that extend into the body cavity, separating the lungs. Larger examples may flare laterally (loons) (Figure 15.8). In falcons (Falconidae) and some other birds, they may extend forward and backward to fuse with adjacent hypapophyses, creating a supporting strut along a fused section of vertebrae (Figure 15.9). Usually they are present only on anterior thoracic vertebrae but in cormorants they appear on vertebrae fused to the synsacrum. Typically, the most anterior hypapophysis serves as a point of attachment for the ancinus longus muscle of the neck where its elongated structure may be useful to birds that dart their heads forward to catch prey (Gadow, 1933). In the Cassin's auklet (Ptychoramphus aleuticus), and other auks, a ligament anchors the tips of most posterior hypapophysis to vertebrae in the synsacrum, limiting dorso-ventral bending of the lower spinal column (Figure 15.10).

Among marine and aquatic birds, the development of thoracic hypapophyses varies both between groups and within groups. Among the petrels, the largest and most elaborate hypapophyses occur in the diving-petrels (Pelecanoididae), smaller ones in diving shearwaters (Puffinus sp.), but there are none at all in albatrosses. Among pelecaniform birds, hypapophyses are typically small or absent. Among Charadriiformes, they are particularly large and elaborate in auks but small or absent in gulls, terns, and other nondiving charadriiforms.

One characteristic that is supposed to distinguish birds from their ancestors among the dinosaurs is the presence of a plow-shaped pygostyle. It carries the large fleshy bulb to which the tail

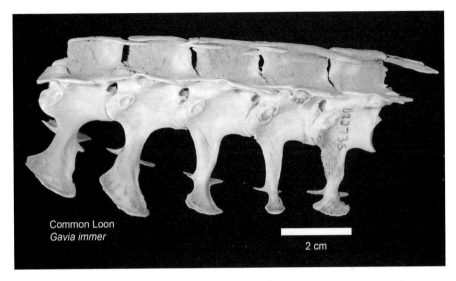

Fig. 15.8 Independent thoracic vertebrae with very large hypapophyses in the common loon (*Gavia immer*).

feathers are attached and plays a key role in the controlled movement of those feathers. Flightlessness appears to have been accompanied by the loss of a well-formed pygostyle among the living paleognaths (including Tinamiformes) but it is usually a distinctive feature in neornithine birds. The Podicipediformes are an exception. Their tails have undergone extreme reduction;

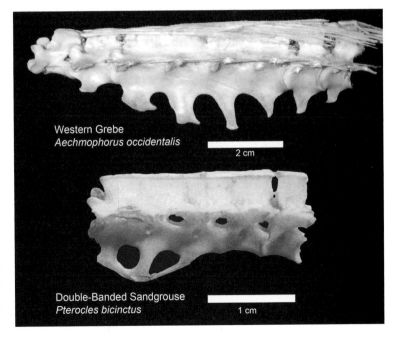

Fig. 15.9 Fused series of thoracic vertebrae with simple hypapophyses in the western grebe (*Aechmophorus occidentalis*) and hypapophyses with fused tips in the double-banded sandgrouse (*Pterocles bicinctus*).

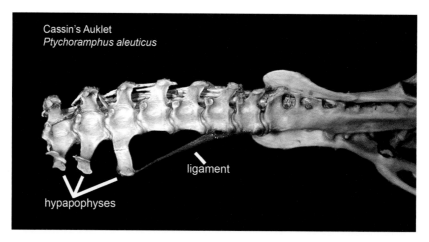

Fig. 15.10   Central thoracic region in the Cassin's auklet (*Ptychorhamphus aleuticus*) showing the ligament connecting the anterior hypapophyses to the ventral spinal process of the synsacrum (Kaiser, 2007).

the tail rectrices are essentially indistinguishable from adjacent contour feathers and the pygostyle is a featureless nub at the end of the spinal column.

The shape of the tailbones, particularly the pygostyle, varies among other groups of marine and aquatic birds. In species that generate aerial speed (anseriforms, gaviiforms, podicipediforms, and many charadriiforms) or great aerial efficiency (procellariiforms) the tail fan and its supporting skeleton are greatly reduced. Among these birds, the tail's most important flight function may be service as a splitter-plate (Maybury & Rayner, 2001). A splitter-plate helps to maintain the laminar flow of air around the body and reduces energy expenditure by reducing turbulence and delaying the detachment of vortices. In more aerobatic species such as frigatebirds or storm-petrels (Hydrobatidae), the tails play an important role in maneuver. The feathers are long and the pygostyle and its neighboring bones are large.

THE AQUATIC HABIT

Life on the water offers many advantages to birds – providing that they can keep warm and stay dry. Aquatic ecosystems are exceptionally complex and contain a wide variety of foraging opportu-

nities for those birds that can reach them. Floating is less costly in terms of energy expenditure than standing and water creates a defensive moat that keeps purely terrestrial predators at a distance, isolating nests and young, and offering a fairly safe place to sleep. For species that can dive beneath the surface, water may even provide a temporary hiding place from aerial predators.

Open stretches of water are typically subject to stronger winds than occur over land, making it much easier for birds to achieve sufficient ground speed for a take-off and when the flight is over, water offers a soft landing, especially for young birds just learning to fly. When birds need to replace worn feathers, some water birds drop all of their flight feathers at once yet are still able to find sufficient food. With such advantages, it is not surprising that so many different lineages include water birds. Not all water bodies are equally hospitable, but although the high seas can be particularly hostile, a great many birds earn a living on them.

The term "seabird" is used to describe the specialized subgroup of aquatic birds that live on the oceans and derive their livelihood from them. In spite of their amazing diversity and long reign as the dominant form of life on the planet, dinosaurs remained entirely terrestrial and never made a home in the oceans. Perhaps they were unable to displace their contemporaries among the giant marine reptiles of the Mesozoic. Nonetheless avian descendants of the Theropoda learned to exploit oceanic habitats, and continue to do so with truly spectacular success.

Most mammals find it difficult to cope with the hostile conditions of marine habitats, even along the shore. The ability of seabirds to breed and raise young in the face of fierce oceanic storms at isolated and dangerous locations left early ornithologists facing miserable working conditions. A few of those naturalists were intrepid adventurers who seem to have enjoyed the adventure of their hardships, but others transferred personal feelings about working conditions to their subjects and portrayed seabirds as ornithological outcasts, clinging grimly to life on the edges of the world (e.g. Bent, 1946). It has only been since improvements in transportation and communication made ornithologists relatively comfortable on even the most remote nesting areas that we have come to view seabirds as profoundly well adapted and successful organisms that thrive on a rich array of resources (Ricklefs, 1990).

The success of seabirds has depended on their ability to overcome the series of challenges posed by excess salt, low temperatures, and prey hidden in vast areas of trackless habitat (Warham, 1996; Gaston, 2004). Specialized tear glands excrete salt allowing seabirds to maintain their osmotic balance (Hughes, 2003; Laverty & Skadhauge, 2008). Dense waterproof plumage insulates them from the cold and makes it possible for many species to nest at high latitudes and exploit the unusually rich polar seas. Flight is usually the key to finding scattered patches of prey. Seabirds that specialize in aerial performance often feed on the surface but many others are able to dive and pursue prey deep into the ocean (Warham, 1996; Gaston, 2004). Some parts of the sea are so rich that even flightless species and those that go through a flightless molt period can make a successful living.

BEHAVIORAL DISTINCTIONS

The term "aquatic bird" can be applied to any species that lives in or on water but "seabird" refers specifically to those birds that spend a significant part of their lives in saltwater habitats. The term is not meant to describe a natural cluster of related groups but includes representatives from seven discrete orders. Among the aquatic birds, only the Heliornithidae and Fulicidae (order Ralliformes) are strictly aquatic but the other orders are not necessarily uniformly marine in their habits. The intensity of the commitment to marine habitats varies even among families, but for convenience, seabirds can be divided into three functional groups depending on how completely they are committed to life at sea.

1 Marginal or temporary seabirds live on or near the seashore. Marginal seabirds include many Charadriiformes such as the Laridae (gulls and terns; *Creagrus, Anous,* and *Proscelsterna* are genuinely pelagic), nine families of wading birds such as Scolopacidae (sandpipers) and Charadriidae (plovers). Temporary seabirds may spend much of their lives on saltwater but typically return to freshwater to breed and raise their young in an environment with fewer physiological challenges. All of the Gaviiformes (loons), some Podicipediformes (grebes), some Charadriiformes, such as Phalaropidae (phalaropes), and some Anseriformes, such as the Mergini (sea ducks), fall into this category (Johnsgard, 1987);

2 Coastal seabirds are more or less restricted to nearshore waters and shallow seas over the continental shelf. Most of the Pelecaniformes (pelicans, cormorants, frigatebirds, tropicbirds, and gannets) can be included in this group, but many pelicans and cormorants, and all of the anhingas live only in freshwater habitats (Carboneras, 1992; Elliott, 1992; Orta, 1992a,b) while the frigatebirds are more oceanic and wander far out to sea. The marine forms in this category may spend their whole lives on saltwater but typically return to nocturnal roosts on land and never wander too far from sources of freshwater;

3 Oceanic seabirds, such as Sphenisciformes (penguins), Alcidae (auks: Charadriiformes), and Procellariiformes (petrels) are fully adapted to life on the high seas and may spend much of

their lives far from shore, especially outside of the breeding season (Warham, 1996; Gaston, 2004). The Order Procellariiformes is the most structurally and behaviorally diverse taxon among the truly oceanic birds. It is also the most globally distributed group of birds (Figure 15.11) and includes diving-petrels, storm-petrels, gadfly petrels (*Pterodroma* sp.), prions (*Pachyptila* sp.), shearwaters, fulmars (*Fulmarus* sp.), and albatrosses (*Diomedea* sp.). Penguins are less diverse in appearance and all can be described as massive, flightless birds, more at home in the water than on land (see Ksepka & Ando, Chapter 6, this volume). Penguins are typical of the south polar seas but a few species range as far north as the Equator. The centre of diversity for the petrels is in the Southern Hemisphere but many storm-petrels and fulmars breed in the Northern Hemisphere where their ranges overlap areas occupied by auks (Figure 15.11). The Alcidae are restricted to north temperate and Arctic seas but look somewhat like small penguins. Like penguins, they use their wings for underwater locomotion. The extinct great auk (*Pinguinus impennis*) and

some seemingly ancestral forms (e.g. Mancallinae) were flightless (Olson & Hasegawa, 1979) but all extant species of auk can fly.

LOCOMOTION AND FORAGING

Diving

The strongest influence on the overall design of the skeleton in marine and aquatic birds is the bird's method of collecting food, especially if it requires locomotion underwater. Moving through a medium as dense as water requires large muscles and the support of a robust skeleton.

Foot-propelled divers have an elongated body reflecting the in-line position of two large muscle masses, one for the wings and aerial flight, and one for the hind limbs and underwater locomotion. The leg action of swimming birds is similar to that of walking in terrestrial birds. It is hardly surprising then the pelvic bones of birds that can achieve great speed underwater, such as loons, grebes, and cormorants, have hips that are somewhat reminiscent of those found in running ratites (and

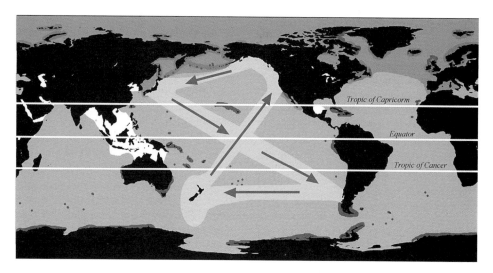

Fig. 15.11 Global occurrence of Procellariiformes and Alcidae. Breeding areas are shown in darker tones. Occasionally, members of the Procellariiformes may wander into an uncolored area but it is not a regular part of their habitat. The large figure eight, in the Pacific Ocean, indicates the approximate track of the annual migration by the sooty shearwater (*Puffinus griseus*). [This figure appears in color as Plate 15.11].

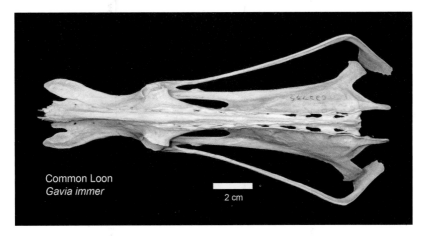

Fig. 15.12 The narrow pelvic skeleton of a common loon (*Gavia immer*) is typical of other foot-propelled diving birds such as grebes and cormorants.

theropod dinosaurs). The hind limbs are mounted near the anterior end of greatly elongated pelvic bones that extend back towards the tail (Figure 15.12).

In Ratites and the great majority of swimming birds, the movement of the legs is through the saggital plane. In such animals, running or rapid swimming would seem to pose a risk to eggs being carried in the abdomen and may be one of the factors limiting egg size in such birds (Dyke & Kaiser, 2010). A different posture may allow loons and grebes to carry larger eggs safely while achieving high speeds. In the water, loons and grebes splay their legs to the side, swimming with strokes closer to the saggital plane (Johansson & Norberg, 2001; Hertel & Campbell, 2007). Interestingly, the strokes are synchronized and not alternating as in other swimming birds.

Wing-propelled divers can make do with much smaller pelvic muscles than foot-propelled divers and a concentration of flight muscles on the thorax gives them a rotund profile. Auks and diving-petrels are dependent on their wings for most of their locomotion and rarely walk. On land, they prefer to slide on their bellies, pushed along by their rather small legs. The hind limbs are not as fragile as they look and in many species they are used to excavate extensive nesting burrows. Because both aerial and underwater locomo-

tion is achieved by using the wings, there is no risk of interference with unlaid eggs, and the eggs of auks can be exceptionally large (Gaston & Jones, 1998).

Penguins have retained an elongated shape because they are both flightless, wing-propelled divers and pedestrians (see Ksepka & Ando, Chapter 6, this volume). They need exceptionally muscular hind limbs for walking, jumping, and scrambling over uneven surfaces. Penguin legs only look short because they are kept beneath the abdominal skin with the knees alongside the belly. For their body size, penguins do not lay exceptionally large eggs. The penguin-like great auk also laid a modest egg compared to flying auks (Gaston & Jones, 1998).

Foot-propelled diving is likely the most primitive method of underwater locomotion and is presumably a direct extension of terrestrial locomotion. The Mesozoic giant, *Hesperornis* was a foot-propelled diver but there are no known users of underwater wing-propulsion from the Mesozoic. Foot-propulsion appears in all of the basal groups such as the diving-ducks, loons, and grebes, but many of the modern foot-propelled divers have been reliably reported as using their wings for additional thrust. All modern phylogenies imply that wing-propelled diving evolved independently as the main form of underwater propulsion in at least four and perhaps five lineages, depending on the nature of the relationship between Sphenisciformes and Procellariiformes. Both of those orders contain only wing-propelled divers (although shearwaters use all four limbs; Warham, 1996) and these groups may have shared a common ancestor with similar capabilities. Wing-propulsion is also used by gannets (Carboneras, 1992), tropicbirds (Brewer & Hertel, 2007), and perhaps to a limited extent by cormorants. Wing propulsion is the only form of underwater locomotion used by auks and, in emergencies, by other charadriiform birds such as gulls, sandpipers, and plovers.

The dipper in the family Cinclidae (Passeriformes) is the only aquatic bird to lack webbed or fringed toes and must use its wings to reach the bottom of shallow streams. Once submerged, it clings to the substrate with sharp claws and forages for bottom-dwelling organisms not unlike its relatives on dry land (Kingery, 1996). It is perhaps the least specialized of all aquatic birds.

Wing-propelled divers typically have an elongated and strengthened humerus that is reflected in high values for the brachial index (ratio of humerus length to ulna length or BI). The dipper has a BI of 0.9, typical of its near relatives, but it may not need a particularly robust humerus because it makes only shallow dives and does not use its wings for extended pursuit of its prey (Kingery, 1996). Among more typical wing-propelled divers the BI is much higher, ranging in living auks from 1.2 to 1.6, in diving-petrels from 1.2 to 1.3, and in penguins from 1.2 to 1.5. Among flightless auks, the BI ranges to 2.5 in the fossil Mancallidae and to 1.9 in the recently extinct great auk. Shearwaters appear to be more specialized for soaring than diving but they can reach depths of 70 m (Weimerskirch & Sagar, 1996) and have a BI between 1.0 and 1.1 while their nondiving relatives have a BI between 0.9 and 1.0. In rails, unexpectedly high values of BI, about 1.25, in both foot-propelled divers such as coots (*Fulica* sp.) and nondiving species (*Porzana* sp.), may be more related to wing shape and rather galliform-like flight than to diving technique (Figure 15.13).

Pelecaniform birds either collect their prey from the surface or make spectacular plunges into the sea (Brewer & Hertel, 2007). Their values of BI are not exceptional, suggesting that aerial flight has been the most important influence on their wing structure. Frigatebirds are strictly surface feeders and never alight on the water, let alone plunge into it, suggesting that the unusually fusion of their furcula and sternum is entirely related to aerial flight. Their BI of 0.8 is lower than other pelecaniforms but comparable to that of other highly maneuverable birds (Rayner & Dyke, 2002). Pelicans are broad-winged soaring birds that also have the furcula fused to the sternum as in the frigatebird, but are not nearly so aerobatic. Their BI is close to 0.9. In boobies, cormorants, and tropicbirds, that occasionally use their wings for underwater propulsion, the BI ranges from 0.9 to 1.0. Only the gannets, which regularly use their wings underwater, have a BI between 1.1 and 1.2.

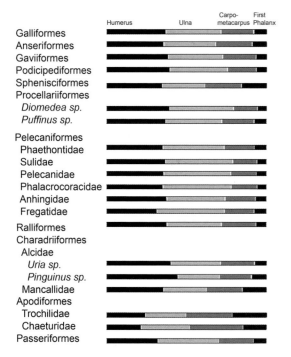

Fig. 15.13 Approximate lengths of the skeletal elements in the wings of various birds.

Important changes in the shape of the humerus are very rare in birds and represent a commitment to very specific locomotory techniques, as in the thick, stubby humerus of hummingbirds (Trochilidae) whose flight depends on very rapid wing beats. The humerus is also modified in the truly specialized wing-propelled divers, such as penguins, diving-petrels, and auks but it is not simply elongated. It is widened and slightly arched, presumable to increase its ability to cope with the stresses of wing-driven propulsion underwater (Figure 15.14). Comparable adaptations occur in the extinct plotopterids and mancallid auks (Olson & Hasegawa, 1979). Because the same structural adaptations are found in several unrelated lineages, it seems likely that there are limited opportunities, within the design constraints of the avian wing, to overcome the biomechanical challenges imposed by this form of locomotion.

Aerial flight

Aerial locomotion has also had profound effects on the functional biology of marine and aquatic birds, especially those that live on the high seas. The adaptive advantages of flight efficiency and great range are obvious for species that live in an environment whose resources are as patchily and as sparsely distributed as those of the open ocean.

Exceptionally efficient flight has appeared in a variety of forms among several families, including the Procellariiformes, the Pelecaniformes, the Charadriiformes, and the Pelagornithes of the early Eocene (Stillwell *et al.*, 1996; see Bourdon, Chapter 8, this volume). The advantages of speed to birds that live at sea are more difficult to appreciate but speed is important for such temporary seabirds as ducks, loons, and grebes, and extreme speeds have become the definitive characteristic of the Alcidae (Table 15.1; Kaiser, 2007).

Through much of the 20th century, air speed and its related energy consumption were difficult to measure and the ability of seabirds to maintain such a sustained effort baffled ornithologists. At the same time, the misleading simplicity of the auks' aerial performances frequently led to their capabilities being derided or dismissed as an evolutionary waypoint to flightlessness (Bent, 1946; Storer, 1960). The full appreciation of the effect of speed on avian ecology needed to wait for the application of more sophisticated technology (Elliott *et al.*, 2004) and, more importantly, the development of testable theories of avian flight (Rayner, 1988, 1993; Pennycuick, 2008).

Recent radar observations of flight speeds have revealed that few birds travel at the values predicted by mathematical models; large birds tend to fly close to speeds requiring maximum

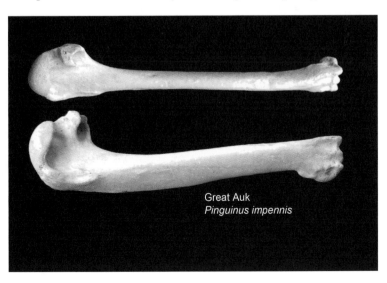

Fig. 15.14 Ventral (bottom) and posterior (top) views of the humerus of the great auk (*Pinguinus impennis*), a large and flightless auk that, like all its smaller relatives, used wing propulsion underwater.

Great Auk
Pinguinus impennis

Table 15.1 Flight speeds of marine birds from the coast of British Columbia. Most observations were recorded in May or June while birds were commuting between foraging areas and nests. Most were within 10 m of the sea surface and winds were calm. Where samples are adequate, values are shown ±1 SD.

Species	Flight speed (m s^{-1})	Sample size (n)	Wing loading (N m^{-2})	Egg to body ratio (%)
Barrows goldeneye (*Bucephala clangula*)*	18.9	1		9.3
Common goldeneye (*B. clangula*)*	16.4	1	150.7	7.1
Harlequin duck (*Histrionicus histrionicus*)*	18.1 ± 4.2	12		8.5
Common merganser (*Mergus merganser*)*	16.4	6	217.7	6.7
	19.4 ± 3.9	11		
Common loon (*Gavia immer*)	18.3†	3	195.2	5.8
	21.1 ± 1.9*	22		
Red-throated loon (*G. stellata*)*	17.8 ± 3.6	11	169.1	6.5
Pacific loon (*G. pacifica*)*	19.2 ± 4.2	13		5.3
Red-necked grebe (*Podiceps grisegena*)‡	10.8 ± 0.8	64	128.1	3.2
Common murre (*Uria lomvia*)*	19.7 ± 2.8	20	175.9	11.1
Pigeon guillemot (*Cepphus columba*)*	18.9 ± 3.1	16	100.2	11.8
Rhinoceros auklet (*Cerorhinca monocerata*)*	17.8 ± 1.9	9	127.6	14.6
Marbled murrelet (*Brachyramphus marmoratus*)	23.6 ± 3.3	100	134.3	16.3
Pre-laying*	22.5 ± 3.6	1435		
Incubation*	21.1 ± 4.2	1435		
Nestling care*	21.4 ± 3.9	30		
Early winter§				
Arriving†	18.3	424		
Departing†	29.2	487		

Sources for flight speed: *Elliott & Kaiser (2009), † Burger (2001), ‡ Blake & Chan (2006), § Elliott *et al.* (2004).

power while small birds show greater variety, often traveling at speeds that offer maximum range (Bruderer & Boldt, 2001; Alerstam *et al.*, 2007). Bruderer & Boldt (2001) attributed this phenomenon to evolutionary forces counteracting speeds at both ends of the scale. Small birds, with low wing loading, avoid slow speeds that would be adversely affected by wind, while large birds, with high wing loading, avoid speeds high enough to interfere with maneuverability or control. Alerstam *et al.* (2007) took a different approach, concluding that wing loading accounts for about 50% of the variation in flight speed while phylogeny accounted for another 35%, possibly through differences in flight mode. In effect behavioral adaptations for flight appear to override simplistic interpretations of flight energetics.

In both of the above studies, the focus was on air speeds achieved by land birds during migration. Only a few water birds were included and neither project examined the flight of oceanic birds. In a sample from British Columbia of marine and aquatic birds that use continuous flapping flight (Table 15.1), the average speed of the auks is more than 10 k h^{-1} faster than that of the ducks, loons, and grebes even though their wing loading is lower. Such a large discrepancy suggests that the extra effort is of significant advantage to an auk. It may reflect the auk's need to have as long a foraging period as possible at several feeding areas in spite of a long commute to its nest, however, that does not explain the continued use of high speeds outside the breeding season (Table 15.1). Perhaps those speeds reflect the urgency of trips between localized foraging opportunities at rapids and other short-term current systems, created by tide cycles.

Among the auks, none flies faster than the marbled murrelet (*Brachyramphus marmoratus*), a 200-g resident of the North Pacific that often nests 70 km inland and at elevations over 1000 m. Among birds restricted to flapping flight, the marbled murrelet lays the largest egg for its size (38.5 g; Zimmerman & Hipfner, 2007) and has an egg to body ratio of 16.3% (Table 15.1). Air speeds and egg to body values for other marine and aquatic taxa in the same habitats are much lower. After the murrelet egg hatches, each parent makes daily trips of up to 100 km carrying fish between marine foraging areas and the nest (Hull *et al.*, 2001; Kaiser, 2007). They are capable of making those trips at speeds of up to 181 km h^{-1} (Elliott *et al.*, 2004) in spite of small wings (wing loading 135 N m^{-2}, aspect ratio 10.6).

The key to the murrelet's success in applying speed to the challenges of its lifestyle may lie in the abundance and richness of its preferred prey. In British Columbia, it often feeds on the Pacific sand lance (*Ammodytes hexapterus*) whose caloric density (5000 cal g^{-1}) (Vermeer & Devito, 1986) is close to that of butter. It also feeds on herring, anchovies, and other oily, schooling fish. Such a diet allows the murrelet to devote a huge proportion of its daily energy budget to flight, expending most of that allocation on thrust. As a consequence, the additional energetic costs of generating sufficient lift to carry heavy eggs or fish, or to fly uphill remain relatively small (Kaiser, 2007). Applications of aerial speed to life-history phenomena may be a logical consequence of dependence on such energetically expensive flight.

Somewhat surprisingly, none of the flighted members of the basal clades are tentative fliers with weak abilities. Even the Tinamiformes (Palaeognathae) and Galliformes (Galloanserinae) that are essentially pedestrian animals depend on vigorous flapping to escape predators. The take-off of the tinamous is not nearly as explosive as that of the Galliformes (Stegmann, 1978) but both groups use a burst of energy to get airborne and glide with a few additional flaps to a landing site. None of the other living paleognaths are flighted but members of the sister clade to the Galliformes, the Anseriformes, are as adept at the use of muscle power for sustained flight as members of any crown clade and include many birds capable of long-distance migrations.

The flight of the Anseriformes is similar to that of their neighbors in the Natatores, the Gaviiformes, and the Podicipediformes. The members of these groups are heavily muscled water birds that forage underwater and have robust skeletons. As in the auks discussed above, they employ rapid, energetically expensive beats of relatively small wings to achieve high speeds. Wing strokes in cruising flight are surprisingly shallow. Body weight is an important aerodynamic factor in achieving their flight speed (Rayner, 1988; Tennekes, 1998; Pennycuick, 2008) and it also allows them to carry sufficient fuel in the form of body fat to carry that body on transcontinental migrations. Although such body shapes and flight styles seems to leave these birds dependent on abundant, energy-rich foods, it is part of a very successful life-strategy found in several other lineages within the Natatores and the Terrestrornithes: boobies, gannets, diving-petrels, as well as auks. Even the entirely flightless penguins might be considered an extreme example of this same strategy.

Within the Natatores and the Terrestrornithes there are also groups that have developed specialized flight styles based on energy-saving techniques that allow them to search for food across great distances. Large, terrestrial species, with long, wide wings (e.g. many storks, cranes, pelicans, and anhingas) use thermal soaring that exploits updrafts generated when solar energy heats the air. Many of these birds also exploit updrafts created by interactions between the wind and topography. Over water, useful updrafts are usually not as large but frigatebirds have such low wing loading that they are able to exploit them. Another group of marine birds, the Procellariiformes is not dependent on updrafts for soaring and appears to have developed the most efficient flight of all. The albatrosses in particular use finely tuned movements of their long narrow wings to exploit air movement over oceanic waves and are able to fly across vast distances with minimal effort (Pennycuick, 2002, 2008).

Shearwaters and most of the pelecaniform birds take their prey underwater while the frigatebirds and albatrosses specialize in surface items. Among the relatives of these large marine birds are a handful of much smaller species (storm-petrels, terns, and phalaropes) that have developed specialized flight styles and foraging techniques associated with feeding from the water's surface.

Other new flight styles appear as one moves crownward in the Livezey & Zusi (2007) phylogeny, into the entirely terrestrial and more arboreal Dendrornithes. More birds tend to have smaller sizes, lower wing loading, and shorter, wider wings (lower aspect ratios). In terms of flight performance, those changes imply decreases in speed, range, and efficiency with a resulting tendency away from aquatic or marine habitats in favor of three-dimensional, terrestrial, and forested habitats in which patches of food and other resources are more densely clustered but are both vertically and horizontally distributed. Most members of the crown clades can be described as small forest birds. Their flight is very different from that of marine and aquatic birds, exhibiting an exceptional degree of maneuverability and aerobatics that are useful in the complex environment of a forest canopy.

Pronation and wing flexion appear to play a much larger role in the flight of the crown clades than in the basal groups. Even giants in this group, such as the Great Hornbill (*Buceros bicornis*), have wings that are highly flexible in two directions (Kemp, 2001). They are capable of pronating to the point where the tips touch and they can curl their wings inwards, to form a tube around the body during the recovery stroke. By increasing the area of the lifting surface, pronation facilitates rapid vertical take-offs and gentle vertical landings. The value of rolling the wing around the body is not so clear but perhaps the wing experiences less drag when the feathers are drawn through the air than when a stiff, oar-like structure is pushed through it. Aquatic and marine birds use some pronation but few are capable of a standing take-off. Species with high wing loading, such as loons, depend on rapid wing beats and a long taxiing run for take-off, while soaring birds, such as albatrosses depend on the ability of their long, stiff wings to exploit passing breezes.

Generally, marine and aquatic birds are known for flying in long straight lines or sweeping curves; few are capable of aerobatics. Their aerial maneu-

verability may be limited by their tail, which is usually too short to serve either as an effective rudder or for auxiliary lifting. When these birds need to prevent a stall and maintain aerial stability at low speeds, they deploy their webbed feet that usually offer a larger surface area than the tail feathers. During cruising flight, the short tail may be most useful as a splitter plate, helping to extend the laminar flow of air passing over and under the body and reducing the amount of drag from turbulence (Maybury & Rayner, 2001). Highly coordinated and controlled aerobatics are used by such long-tailed species as the frigatebird and storm-petrel. Their abilities imply that any neural connections required for such flight evolved early in avian history and were eventually coordinated with wing movement (Maynard-Smith, 1952). Perhaps the movements of aerial control, seen in aerobatic species, originated in the simple comfort movements and stretches that birds need to preen all parts of the body.

For most of the marine and aquatic birds that fly on stiff, narrow wings, there are advantages to having a relatively long humerus (and high values of BI). A longer humerus, associated with a suite of functional adaptations in the wing and shoulder, helps to reduce wing inertia (Rayner & Dyke, 2002). Consequently we find many values of BI > 1.0 in Livezey & Zusi's (2007) Natatores, with values of BI commonly exceeding 1.1 in diving birds. Not surprisingly the values reach extremes in wing-propelled divers that "fly" in a much denser medium: diving-petrels 1.23, auks 1.32, and penguins 1.36 (Nudds *et al.*, 2004) (Figure 15.15).

Wing flexibility is closely related to the value of the brachial index (Rayner & Dyke, 2002; Nudds *et al.*, 2004). A short humerus lies subparallel to the axis of the body and increases the length of the arc traced by the wing tip during elevation and depression of the wing. In most forest birds of the Dendrornithes, the BI is < 1.0, reaching an extreme in the swifts (e.g. *Apus apus* 0.48). The BI is also exceptionally low in the highly aerobatic frigatebird (*Fregata minor* 0.76). The frigatebird is so lightly built that, in spite of extremely narrow wings, it attains a wing loading of only $40\,\mathrm{N\,m^{-2}}$,

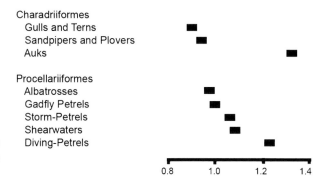

Fig. 15.15 Brachial Indices: relative bone lengths in the forelimb of various orders of birds (Nudds *et al.* 2004; Nudds, 2007).

comparable to that of the broader winged, terrestrial soaring birds such as the turkey vulture (*Cathartes aura*).

Recent additions to the fossil record suggest that the sustained energetic expenditure associated with the vigorous flapping flight may have arisen early among the ornithurine birds. Five specimens of a small aquatic bird, *Gansus*, from Lower Cretaceous deposits in China offer evidence that aquatic birds have an exceptionally long history (You *et al.*, 2006). The head of *Gansus* is missing so that little can be determined of its food habits but large webbed feet and the presence of hypapophyses on the thoracic vertebrae support the idea that this bird foraged underwater. Its rather modern-looking wings suggest that it was capable of the sustained flight needed to move between water bodies. *Gansus* had no external digits interrupting the outline of the wing, its shoulder is much like that of a modern bird, and the sternal keel is large and robust (You *et al.*, 2006). Two of the specimens include both humerus and ulna to give BI values of 0.92 and 0.98 (supplementary data to You *et al.*, 2006). Such values would be low for a wing-propelled diver but are within the range of modern foot-propelled diving birds (Nudds *et al.*, 2004).

CONCLUSIONS

Except for the Anseriformes, represented by *Vegavis* (Clarke *et al.*, 2005) and the presbyornithid *Teviornis* (Kurochkin *et al.*, 2002; Dyke & van Tuinen, 2004), there are no uncontroversial fossils for the living groups of marine or aquatic birds dating back to the Mesozoic and early stages in the evolution of these groups. However, living marine and aquatic birds share a suite of general characteristics that should help us recognize such fossils if or when they are found. Some of these are the same characteristics that distinguish their lineages from those of the crown clades in the morphology-based phylogeny of Livezey & Zusi (2007).

Our ability to conceive of a variety of trends that range in a rational way across the phylogeny of Livezey & Zusi (2007) implies some biological basis for the evolutionary story implicit in that phylogeny but such trends are merely argument and not proof. The morphological characters are linked to specialized functional adaptations that they support and many of the trends cannot be tracked across independently constructed biomolecular phylogenies. Unfortunately, where a trend has been clear, as in the case of the weight increase across the Sibley & Ahlquist (1990) tapestry (Maurer, 1998), the phylogeny has been weak (Houde, 1987a; Sarich *et al.*, 1989; Lanyon, 1992; Harshman 1994). Where the biomolecular analyses have been strong, as in the recent work of Hackett *et al.* (2008), relationships to morphological or behavioral characteristics remain obscure.

In spite of the seemingly indefatigable logic behind modern biomolecular analyses, there may not be great cause for concern that biomolecular phylogenies often show only general agreement with the morphology-based variety. Generally, the biomolecular analyses identify the same families and orders that have been a familiar part of avian taxonomy for more than a century but the approach seems to struggle with the resolution of events in deep time. The millions of years that have passed since modern birds appeared in the Early Cretaceous offers a great opportunity for significant segments of the genetic record to have disappeared without a trace and for sections to have become a palimpsest, overwritten repeatedly by similar or recurring molecular events. Biomolecular chemistry is a relatively new science dependent on observations mediated by specialized technology and sophisticated mathematics. We should not be too surprised that the evolutionary history we perceive through studies of DNA proves as difficult to interpret and as contradictory as that from our glimpse of the fragmentary and sparse fossil record.

ACKNOWLEDGMENTS

M. C. E. McNall of the Royal British Columbia Museum provided access to the avian skeletal collection for the measurements used to determine the brachial index.

REFERENCES

Alerstam T, Rosén M, Bäckman J, Ericson PGP, Hellgren O. 2007. Flight speeds among bird species: Allometric and phylogenetic effects. *PloS Biology* **5**: 1657–1662.

Beddard FE. 1898. *The Structure and Classification of Birds*. London: Longmans Green and Company.

Bent AC. 1946. *Life Histories of North American Diving Birds: Order Pygopodes*. New York: Dodd, Mead and Company.

Blake RW, Chan KH. 2006. Flight speeds of seven bird species during chick-rearing. *Canadian Journal of Zoology* **84**: 1047–1052.

Brewer ML, Hertel F. 2007. Wing morphology and flight behavior of pelecaniform seabirds. *Journal of Morphology* **268**: 866–877.

Bourdon E. 2005. Osteological evidence for sister group relationship between pseudotoothed birds (Aves: Odontopterygiformes) and waterfowls (Anseriformes). *Naturwissenschaften* **92**: 586–5591.

Bruderer B, Boldt A. 2001. Flight characteristics of birds: 1 radar measurements of speeds. *The Ibis* **143**: 176–204.

Burger AE. 2001. Using radar to estimate populations and assess habitat associations of marbled murrelets. *Journal of Wildlife Management* **65**: 696–715.

Carey C, Rahn H, Parisi P. 1980. Calories, water, lipid and yolk in avian eggs. *Condor* **82**: 335–343.

Carboneras C. 1992. Family Sulidae (boobies and gannets). In *Handbook of Birds of the World, Volume 1*, del Hoyo J, Elliott A, Sargatal J (eds). Barcelona: Lynx Edicions; 312–325.

Clarke JA, Tambussi CP, Noriega JI, Erickson GM, Ketcham RA. 2005. Definitive fossil evidence for the extant avian radiation in the Cretaceous. *Nature* **433**: 305–308.

Cottam PA. 1957. The pelecaniform characteristics of the skeleton of the shoebill stork, *Balaeniceps rex. Bulletin of the British Museum (Natural History) Zoology* **5**: 49–72.

Dyke GJ, Kaiser GW. 2010. Cracking a developmental constraint: egg size and bird evolution. *Records of the Australian Museum* **62** (1): 207–216.

Dyke GJ, van Tuinen M. 2004. The evolutionary radiation of modern birds (Neornithes): Reconciling molecules, morphology and the fossil record. *Zoological Journal of the Linnean Society* **141**: 153–177.

Elliott A. 1992. Pelecanidae (Pelicans). In *Handbook of Birds of the World, Volume 1*, del Hoyo J, Elliott A, Sargatal J (eds). Barcelona: Lynx Edicions; 290–311.

Elliott KH, Kaiser GW. 2009. Flight speeds of some British Columbia birds. *British Columbia Birds* **19**: 1–5.

Elliott KH, Hewett M, Kaiser GW, Blake, RW. 2004. Flight energetics of the Marbled Murrelet, *Brachyramphus marmoratus. Canadian Journal of Zoology* **82**: 644–652.

Fain MG, Houde P. 2004. Parallel radiations in the primary clades of birds. *Evolution* **58**: 2558–2573.

Friesen VL. 2007. New roles for molecular genetics in understanding seabird evolution, ecology, and conservation. *Marine Ornithology* **35**: 89–96.

Friesen VL, Baker AJ, Piatt, JF. 1996. Phylogenetic relationships within the Alcidae (Charadriiformes: Aves) inferred from total molecular evidence. *Molecular Biology and Evolution* **13**: 359–367.

Fürbringer M. 1888. *Untersuchungen zur Morphologie und Systematik der Vögel, zugleich ein Beitrag zur Anatomie der Stüz - und Bewegungsorgane*. Amsterdam: Van Holkema.

Gadow HF. 1933. *The Evolution of the Vertebral Column.* Cambridge: Cambridge University Press.

Garrod AH. 1873. On some points in the anatomy of *Steatornis. Proceedings of the Zoological Society of London* **1873**: 526–535.

Gaston AJ. 2004. *Seabirds: A Natural History.* New Haven: Yale University Press.

Gaston AJ, Jones IL. 1998. The Auks: Alcidae. Oxford: Oxford University Press.

Goedert JL. 1989. Giant late Eocene marine birds (Pelecaniformes: Pelagornithidae) from northwestern Oregon. *Journal of Paleontology* **63**: 939–944.

Gonzalez-Barbra G, Schwennicke T, Goedert JL, Barnes LG. 2002. Earliest Pacific Basin record of the Pelagornithidae (Aves: Pelecaniformes). *Journal of Vertebrate Paleontology* **22**: 722–725.

Hackett SJ, Kimball RT, Reddy S, Bowie RCK, Braun EL, Braun MJ, Chojnowski JL, Cox WA, Han K-L, Harshman J, Huddleston CJ, Marks BD, Miglia KJ, Moore WS, Sheldon FH, Steadman DW, Witt CC, Yuri T. 2008. A phylogenomic study of birds reveals their evolutionary history. *Science* **320**: 1763–1768.

Harshman J. 1994. Reweaving the tapestry: what can we learn from Sibley and Ahlquist (1990)? *The Auk* **111**: 377–388.

Hertel F, Campbell KE. 2007. The antitrochanter in birds: form and function in balance. *The Auk* **124**: 789–805.

Hope S. 2002. The Mesozoic radiation of the Neornithes. In *Mesozoic Birds: Above the Heads of Dinosaurs,* Chiappe LM, Witmer LM (eds). Berkeley: University of California Press; 339–388.

Houde P. 1987a. Critical-evaluation of DNA hybridization studies in avian systematics. *The Auk* **104**: 17–32.

Houde P. 1987b. Histological evidence for the systematic position of *Hesperornis* (Odontornithes: Hesperornithiformes). *The Auk* **104**: 125–129.

Hughes MR. 2003. Regulation of salt gland, gut and kidney interactions. *Comparative Biochemistry and Physiology* **A136**: 507–524.

Hull CL, Kaiser GW, Lougheed C, Lougheed L, Boyd S, Cooke F. 2001. Variation in commuting distance of Marbled Murrelets (*Brachyramphus marmoratus*): ecological and energetic consequences of nesting further inland. *The Auk* **118**: 1036–1046.

Johansson LC, Norberg UML. 2001. Lift based paddling in diving grebe. *Journal of Experimental Biology* **204**: 1687–1696.

Johnsgard PA. 1987. *Diving birds of North America.* Lincoln: University of Nebraska Press.

Jones PJ. 1979. Variability of egg size and composition in the great white pelican (*Pelecanus onocrotalus*). *The Auk* **96**: 407–8.

Kaiser GW. 2007. *The Inner Bird: Anatomy and Evolution.* Vancouver: University of British Columbia Press.

Kemp J. 2001. Family Bucerotidae (Hornbills). In *Handbook of Birds of the World, Volume 1,* del Hoyo J, Elliott A, Sargatal J (eds), Barcelona: Lynx Edicions; 436–523.

Kennedy M, Page RDM. 2002. Seabird supertrees combining partial estimates of procellariiform phylogeny. *The Auk* **119**: 88–108.

Kingery HE. 1996. The American dipper *Cinclus mexicanus;* Number 229. In *Birds of North America Online,* Poole A. (ed.). Ithaca: Cornell Laboratory of Ornithology. [Retrieved from http://bna.birds.cornell.edu/bna/species/229 doi: 10.2173/bna.229]

Kurochkin EN, Dyke GJ, Karhu AA. 2002. A new Presbyornithid bird (Aves, Anseriformes) from the Late Cretaceous of Southern Mongolia. *American Museum Novitates* **3386**: 1–11.

Lanyon SM. 1992. Review of Sibley and Ahlquist, 1990. *Condor* **94**: 304–307.

Laverty G, Skadhauge E. 2008. Adaptive strategies for post-renal handling of urine in birds. *Comparative Biology and Physiology* **A149**: 246–254.

Livezey BC, Zusi RL. 2006. Higher-order phylogeny of modern birds (Theropoda, Aves: Neornithes) based on comparative anatomy: – 1. Methods and characters. *Bulletin of Carnegie Museum of Natural History* **37**: 1–544.

Livezey BC, Zusi RL. 2007. Higher-order phylogeny of modern birds (Theropoda, Aves: Neornithes) based on comparative anatomy: II. Analysis and discussion. *Zoological Journal of the Linnean Society* **149**: 1–95.

Maurer BA. 1998. The evolution of body size in birds. I. Evidence for non-random diversification. *Evolutionary Ecology* **12**: 925–34.

Maybury WJ, Rayner JMV. 2001. The avian tail reduces body parasitic drag by controlling flow separation and vortex shedding. *Proceedings of the Royal Society of London Series B (Biological Sciences)* **268**: 1405–10.

Maynard-Smith J. 1952. The importance of the nervous system in the evolution of flight. *Evolution* **6**: 127–29.

Mayr G. 2003. The phylogenetic affinities of the shoebill (*Balaeniceps rex*). *Journal of Ornithology* **144**: 157–75.

Mayr G. 2005. Tertiary plotopterids (Aves, Plotopteridae) and a novel hypothesis on the phylogenetic relationships of penguins (Spheniscidae). *Journal of Zoological Systematics and Evolutionary Research* **43**: 61–71.

Mayr G. 2008. A skull of the giant bony-toothed bird *Dasornis* (Aves: Pelagornithidae) from the Lower Eocene of the Isle of Shepley. *Paleontology* **51**: 1107–16.

McKee JWA. 1985. A pseudodontorn (Pelecaniformes: Pelagornithidae) from the Middle Pliocene of Hawera, Taranaki, New Zealand. New Zealand. *Journal of Zoology* **12**: 181–184.

Nishimura H, Fan Z. 2002. Sodium and water and urine concentration in avian kidney. In *Osmoregulation and Drinking in Vertebrates*, Hazon N, Flik G (eds). Cambridge: Bios Science Publishers, SEB Symposium Series Volume 54; 129–151.

Nudds RL, Dyke GJ, Rayner JMV. 2004. Forelimb proportions and the evolutionary radiation of the Neornithes. *Proceedings of the Royal Society of London Series B (Biological Sciences)* **271**: S324–S327.

Olson SL. 1985. The fossil record of birds. In *Current Ornithology VIII*, Farner DS, King JR, Parks KC (eds). New York: Plenum Press; 80–238.

Olson SL, Hasegawa Y. 1979. Fossil counterparts of giant penguins from the North Pacific. *Science* **206**: 688–689.

Orta J. 1992a. Family Phaethontidae (Tropicbirds). In *Handbook of Birds of the World, Volume 1*, del Hoyo J, Elliott A, Sargatal J (eds). Barcelona: Lynx Edicions; 280–289.

Orta J. 1992b. Family Phalacrocoracidae (Cormorants), Family Anhingidae (Darters), Family Fregatidae (Frigatebirds). In *Handbook of Birds of the World, Volume 1*, del Hoyo J, Elliott A, Sargatal J (eds). Barcelona: Lynx Edicions; 326–374.

Pennycuick CJ. 1987. Flight of seabirds. In *Seabirds: Feeding Biology and Role in Marine Ecosystems*, Croxall JP (ed). Cambridge: Cambridge University Press; 43–62.

Pennycuick CJ. 2002. Gust soaring as a basis for flight of petrels and albatross (Procellariiformes). *Avian Science* **2**: 1–12.

Pennycuick CJ. 2008. *Modelling the Flying Bird*. Amsterdam: Elsevier, Theoretical Ecology Series Volume 5.

Pereira SL, Baker AJ. 2008. DNA evidence for a Paleocene origin of the Alcidae (Aves: Charadriiformes) in the Pacific and multiple dispersals across northern oceans. *Molecular Phylogenetics and Evolution* **46**: 430–445.

Raikow RJ. 1974. Problems in avian classification. In *Current Ornithology II*, Johnston RF (ed). New York: Plenum Press; 187–212.

Rayner JMV. 1988. Form and function in avian flight. In *Current Ornithology V*, Johnston RF (ed). New York: Plenum Press; 1–66.

Rayner JMV. 1993. On aerodynamics and the energetics of vertebrate flapping flight. *Contemporary Mathematics* **141**: 351–97.

Rayner JMV, Dyke GJ. 2002. Origins and evolution of diversity in the avian wing. In *Vertebrate Biomechanics and Evolution*, Bels VL, Gasc J-P, Casinos A (eds). Oxford: Bios Scientific Publishers; 297–317.

Ricklefs RE. 1990. Seabird life histories and the marine environment: some speculations. *Colonial Waterbirds* **13**: 1–6.

Sarich VM, Schmid CW, Marks J. 1989. DNA hybridization as a guide to phylogenies: a critical analysis. *Cladistics* **5**: 3–32.

Sibley CG, Ahlquist JE. 1990. *Phylogeny and Classification of Birds: A Study in Molecular Evolution*. New Haven: Yale University Press.

Siegel-Causey D. 1997. Phylogeny of the Pelecaniformes: molecular systematics of a primitive group. In *Avian Molecular Evolution and Systematics*, Mindell DP (ed). San Diego: Academic Press; 159–172.

Shaffer SA, Tremblay Y, Weimerskirch H, Scott D, Thompson DR, Sagar PM, Moller H, Taylor GA, Foley DG, Block BA, Costa DP. 2006. Migratory shearwaters integrate oceanic resources across the Pacific Ocean in an endless summer. *Proceedings of the National Academy of Sciences* **103**: 12799–12802.

Stegmann BC. 1978. *Relationships of the Superorders Alectoromorphae and Charadriomorphae (Aves): a Comparative Study of the Avian Hand*. Cambridge: Nuttall Ornithological Club, Publication 17.

Stillwell JD, Jones CM, Leavy RH, Harwood DM. 1997. First fossil bird from East Antarctica. *Antarctic Journal of the United States* **33**: 1–7.

Storer RW. 1960. Evolution in the diving birds. *International Ornithological Congress* **12**: 694–707.

Tennekes H. 1998. *The Simple Science of Flight: From Insects to Jumbo Jets*. Cambridge: MIT Press.

Verheyen R. 1958. Contribution a la Systèmatique des Alciformes. *Bulletin Royal Belgian Institute of Natural Sciences* **34**: 1–22.

Vermeer K, Devito. K. 1986. Size, caloric content, and association of prey fishes in meals of nestling rhinoceros auklets. *Murrelet* **67**: 1–9.

Warham J. 1996. *The Behaviour, Population Biology, and Physiology of the Petrels*. London: Academic Press.

Warheit KI. 2002. The seabird fossil record and the role of paleontology in understanding seabird community structure. In *Biology of Marine Birds*, Schreiber EA, Burger J (eds). Boca Raton: CRC Press; 17–55.

Weimerskirch H, Sagar PM. 1996. Diving depths of Sooty Shearwaters *Puffinus griseus*. *The Ibis* **138**: 786–94.

Wetmore A. 1928. *The Systematic Position of the Fossil Bird Cyphornis magnus*. Ottawa: Canada Department of Mines, Geological Series No. 48:

You HL, Lamanna MC, Harris JD, Chiappe LM, O'Connor JK, Ji SA, Lü JC, Yuan C-X, Li DQ, Zhang X, Lacovara KJ, Dodson P, Ji Q. 2006. A nearly modern amphibious bird from the Early Cretaceous of northwestern China. *Science* **312**: 1640–1643.

Zimmerman K, Hipfner JM. 2007. Egg size, eggshell porosity, and incubation period in the marine bird family Alcidae. *The Auk* **124**: 307–315.

Part 4 The Future: Conservation and Climate Change

16 The State of the World's Birds and the Future of Avian Diversity

GAVIN H. THOMAS

Imperial College London, Ascot, and University of Bristol, UK

"...current and predicted environmental per-turbations form a double-edged sword that will slice into both the legacy and future of evolution." Myers and Knoll, 2001

The future of avian diversity hangs in the balance as continued and escalating human activity impacts on populations, species, and ecosystems on a global scale. Extinctions have been documented for 134 bird species since the year 1500 (Bird-Life International, 2008), while a further four species persist only in captivity. At least 15 species regarded as Critically Endangered are, in all likelihood, extinct (Butchart *et al.*, 2006a), bringing the total known probable extinctions in the past 500 years to 153 (BirdLife International, 2008). Many of these, as well as many prehistoric extinctions, were driven by human activities (Chatterjee, 1997). At ca. 31 species per million per years, the recent rate of extinction far outstrips the estimated one species per million per year background rate in the fossil record (Pimm *et al.*, 1995; Pimm *et al.*, 2006). Yet, the rate of known extinctions is a substantial underestimate of the true rate due largely to the increases in the numbers of described species, particularly during the 19th century (Pimm *et al.*, 2006). Despite conservation efforts that have rescued 15 species from the brink of extinction since the late 20th century (Butchart *et al.*, 2006b), the rate of extinction is expected to increase in the coming decades and by the end of the 21st century could rise to 1500 species per million per year as human activities drive rapid climate change, continued habitat loss, and increases in the number and abundance of invasive species (Pimm *et al.*, 2006; Ehrlich & Pringle, 2008).

The loss of species, along with an even larger proportional loss of populations (Hughes *et al.*, 1997), increased invasions by alien species, biotic mixing or homogenization, and reduction or loss of major biomes, have been described as first-order effects of future changes to biodiversity (Myers, 1985, 1996; Myers & Knoll, 2001). Myers (1985) was among the first to draw attention to the much wider impacts of species loss on the ecological and evolutionary processes that operate within communities and ecosystems. Potential effects on process include the disruption of gene-flow due to the fragmentation of species ranges, reductions in genetic diversity due to population declines, and the disruption of ecosystem interactions due to biotic interchanges (Myers & Knoll, 2001). Alarmingly, many of these expected impacts of biodiversity loss will be mirrored and often exacerbated by the effects of climate change (Walther *et al.*, 2002; Parmesan, 2006; Brook *et al.*, 2008). For example, just as local extinction can result in the loss of functional groups (Şekercioğlu *et al.*, 2004; Şekercioğlu, 2006b) and breakdown of interactions within communities (Petchey *et al.*, 2008), climate change can cause a mismatch between the phenologies of predator and prey (e.g. Both *et al.*, 2009) or drive species turnover and community reassembly due to range shifts (Schaefer *et al.*, 2008). At the same time, reduced genetic diversity or constraints on phenotypic plasticity may inhibit species' responses to climate change (Visser, 2008).

In this chapter I review our current understanding of the threats to and drivers of recent and projected avian extinctions, and discuss how differential responses to global change may shape the future of avian diversity. My initial focus is on comparative studies at broad taxonomic scales that: (i) show that extinction risk is not randomly distributed across the avian tree of life; (ii) demonstrate that extinction results in the disproportionate loss of evolutionary history; and (iii) have identified commonalities and idiosyncrasies in the biological traits of threatened species. Variation in extinction risk is a product of both species' intrinsic biology that has an evolutionary history (phylogenetic effects) and threat processes that vary in their spatial distribution (geographic effects). The effects of global change also vary spatially and I will highlight global studies that show that future broad-scale patterns of avian diversity are likely to depend on geographic gradients in land conversion driven by both direct anthropogenic impacts and climate change (Davies *et al.*, 2006; Jetz *et al.*, 2007). The way in which bird species' are able respond to climate change will determine their future survival and contribute to the stability or breakdown of community dynamics. I will discuss studies that show that birds are not tracking climate change fast enough, and how limits to phenotypic plasticity and evolutionary change at a genetic level contribute to a bleak picture of future avian diversity.

EXTINCTION RISK IS NOT PHYLOGENETICALLY RANDOM

Today, 1226 (12%) of the worlds bird species are threatened with extinction (BirdLife International, 2008), primarily from habitat loss but also from exploitation and the impacts of invasive

species. This is not a random set of species. Bennett & Owens (1997) calculated the probability of a family of a given species richness containing the number of threatened species identified in the 1994 world list of threatened birds (Collar *et al.*, 1994). Several families have a higher proportion of threatened species than expected by chance alone, including parrots (Psittacidae), pheasants (Phasianidae), albatrosses (Procellariidae), and pigeons (Columbidae). Subsequent analyses on revised data have not altered the general pattern (e.g. Russell *et al.*, 1998; Bennett & Owens, 2002; Bennett *et al.*, 2005). The clustering of threat is also observed on phylogenies of regional avifauna: British birds that are of conservation concern due to recent population declines are phylogenetically clustered (Thomas, 2008). The non-random taxonomic or phylogenetic distribution of threat has two important evolutionary implications: first, the extinction of threatened species will result in an uneven loss of evolutionary history; and second, it indicates that some species may be evolutionarily predisposed to elevated extinction risk.

LOSS OF EVOLUTIONARY HISTORY

Not all species are equal with respect to the amount of evolutionary history that they represent (Faith, 1992; Purvis *et al.*, 2000). Some species branch-off early in the tree of life, have few or no close relatives and hence are evolutionarily distinct. For example, the rifleman (*Acanthisitta chloris*) and the rock wren (*Xenicus gilviventris*) share a branch of the passerine phylogeny that diverged from the rest of the order perhaps 82 Ma (Barker *et al.*, 2004) and represent a large amount of unique evolutionary history. In contrast, the majority of the speciose *Zosterops* clade diverged around 2 Ma (Moyle *et al.*, 2009) and most of the evolutionary history of the clade is shared across many species. Indeed, most species do have close relatives and contribute relatively little unique evolutionary history to the tree of life (Nee & May, 1997; Purvis *et al.*, 2000). Because of this, if extinction of species was random with respect to phylogeny (the "field of bullets" scenario, Raup, 1992) then the loss of evolutionary history would be proportionally lower than the loss of species (Nee & May, 1997). But as we have seen above, threatened birds tend to be taxonomically or phylogenetically clumped and the loss of evolutionary history is far from random. Several studies based on taxonomic data have shown that the amount of unique avian evolutionary history that is currently imperiled is disproportionately large. If all currently threatened species were to go extinct then we would lose a disproportionate number of genera (Russell *et al.*, 1998; Purvis *et al.*, 2000) with 38 more threatened avian genera than expected by chance alone (Purvis *et al.*, 2000). The disjunction between numbers of threatened species and numbers of threatened genera extends to higher taxonomic levels. Older tribes of both the New World and global avifauna have a higher proportion of threatened species than younger tribes leading to a proportionally higher loss of evolutionary history (Gaston & Blackburn, 1997). Brooks *et al.* (2005) reported that in addition to the unexpectedly high numbers of threatened genera, there are also more threatened families (five) than predicted at random (fewer than two). They also report that there have been more confirmed extinctions of whole genera in the past 500 years than expected by chance. Estimates of current and possible future extinction rates within avian orders are predicted to result in the loss of entire orders within centuries (McKinney, 1998). Phylogenetic approaches to the loss of evolutionary history are more powerful than taxonomic methods because they take into account the full hierarchical branching structure and variation in branch length of phylogenetic trees, however, we currently lack a complete species level phylogeny for the world's birds. Nonetheless, von Euler (2001) used estimates for the ages of taxa at each taxonomic level (Sibley & Ahlquist, 1990; Sibley & Monroe, 1990, 1993) to generate an approximation of the avian phylogenetic tree. Based on these estimates he quantified the expected loss of evolutionary history measured as the amount of branch length or "phylogenetic diversity" (PD; Faith, 1992). The estimated loss of

PD, based on the numbers of extinct, critically endangered, endangered, and vulnerable species was consistently higher than expected but only marginally so. For example, across all threat categories, 9.93% of PD would be lost compared to 9.01% if threat was distributed randomly. However, it is likely that the loss of (and difference from random) phylogenetic diversity would be considerably larger based on the true phylogeny of birds. This is because the taxonomic approach used by von Euler inevitably underestimates the extent of tree imbalance since all examples at a given taxonomic level are assumed to be equal in age.

Taken together, the examples above strongly indicate that we stand to lose unexpectedly large amounts of avian evolutionary history. Yet, while species are a transparent and tangible measure of biodiversity, measures of evolutionary history are perhaps less so. Indeed, exactly what phylogenetic branching structure measures (PD in particular) has been debated (Faith, 2002). Phylogenetic diversity is most frequently considered to be a measure of feature diversity (Faith, 1992; Crozier, 1997; Mooers *et al.*, 2005; Forest *et al.*, 2007). This means simply that it captures some aspect of the phenotypic or genetic distinctness of the species that it represents and in doing so encompasses a more holistic view of biodiversity than species counts. This assumes that species traits evolve at a constant rate across the phylogenetic tree so that feature diversity accumulates proportionally with time (or whatever units branch lengths are measured in for the phylogeny in question). This assumption will hold for some traits in some taxa, but not all traits in all taxa since many traits are not neutral (Diniz-Filho, 2004). Alternate measures that explicitly model or measure phenotypic divergence have been argued to be more appropriate if the aim is to capture the processes that generate feature diversity. Owens & Bennett (2000b) devised a simple metric of relative phenotypic divergence based on the method of independent contrasts (Felsenstein, 1985). They examined the degree to which diversity in avian clutch size is under threat and identified 21 avian families that contained significantly more threatened diversity in clutch size than expected by chance. The mesites (Mesitornithidae) had the highest clutch size diversification under threat score and although more than half the families listed contained fewer than 10 species, speciose families such as pheasants (Phasianidae) and rails (Rallidae) were also among the 21. Owens & Bennett (2000b) stressed that PD would generally do a better job of describing overall evolutionary distinctness than a single trait, however, Faith (2002) argued that not only does PD better capture evolutionary distinctness but it is also more effective as a metric of phenotypic diversity.

An alternate view of the information content of phylogenetic trees is that long branches are relicts or evolutionary dead-ends not worthy of conservation, whereas speciose groups characterized by short branches indicate future evolutionary potential (Erwin, 1991). On this basis, the rapid and recent diversification of *Zosterops* (Moyle *et al.*, 2009) would be taken as evidence that they will continue to diversify in the future. However, as Krajewski (1991) points out there is no phylogenetic theory that suggests that recent speciation equates to future radiation and such inference is little more than speculation. Although phylogeny may not be an informative measure of evolutionary potential, the concept of conserving evolutionary processes has a role in the future of avian diversity and I return to this point at the end of the chapter.

Leaving aside the arguments on the precise meaning of PD, it is clear that it reflects evolutionary history and that it is likely that it retains at least some information on the evolutionary processes of lineage and phenotypic diversification (Crozier, 1997). Moreover, conserving the distinctness of species offers novel ways of generating interest in conserving evolutionary processes. A particularly compelling approach treats phylogenetic diversity as a measure of evolutionary heritage that can be used to measure the biodiversity harbored within countries or other geopolitical units (Frankel, 1974; Mooers & Atkins, 2003; Mooers *et al.*, 2005). Applied to the birds of Indonesia, Mooers and Atkins (2003) showed that 500 Myr of unique, threatened avian evolutionary heritage are held within the countries borders.

Several metrics for quantifying the evolutionary distinctness of individual species and combining these with measures of threat have been described (Redding & Mooers, 2006; Faith, 2008). These approaches were taken to the public arena when, in 2007, the Zoological Society of London launched the Evolutionary Distinct Globally Endangered (EDGE) program. This program is aimed at raising awareness and funds for the conservation of species that might otherwise be overlooked by more conventional conservation management approaches. It was applied initially to the most unusual and threatened mammals (Isaac *et al.*, 2007) and amphibians, with EDGE birds in preparation subsequently included in 2009.

EVOLUTIONARY PREDISPOSITION TO THREAT

Analyses of the threat to avian evolutionary history provide alternate measures of the potential losses of biodiversity but yield little information on the mechanisms that underlie extinction risk. However, the taxonomic bias in extinction risk suggests that there may indeed be common factors that make certain species predisposed to elevated extinction risk (Fisher & Owens, 2004). This is because closely related species share more evolutionary history than more distantly related species do and consequently they tend to be more similar in their life-histories, morphology, ecology, and behavior (Felsenstein, 1985; Harvey & Pagel, 1991). Understanding how species' traits relate to level of threat can reveal the general processes that determine variation in extinction risk, identify species in need of conservation action, and potentially indicate where future conservation concern may lie (Fisher & Owens, 2004).

Numerous mechanisms and potential intrinsic biological traits have been proposed to explain interspecific variation in extinction risk (Bennett & Owens, 2002; Fisher & Owens, 2004). These include variation in body size (Pimm *et al.*, 1988), life history (e.g. reduced fecundity, Pimm *et al.*, 1988), diet (Terborgh, 1974), habitat preferences or specificity (Bibby, 1995), population size or geographic range size and setting (Pimm *et al.*, 1988; Gaston, 1994a, b), migratory behavior (Pimm *et al.*, 1988), and the intensity of sexual selection (Møller, 2000). Explorations of the links between species traits and extinction risk, while accounting for phylogeny, have helped to elucidate which of these are the most important in birds. But, as we will see, the traits that are important for one family or clade are often not so for other clades and patterns across the whole class are not necessarily mirrored at finer taxonomic or geographic scales.

At a global, class-wide scale, several traits have been identified as being important in predisposing some species to increased extinction risk. Life history, and in particular variation in clutch size, seems to be particularly important. Species with a low intrinsic rate of population increase due to low fecundity will take longer to recover from a decline in density and are therefore expected to be more prone to extinction from stochastic demographic events (Pimm *et al.*, 1988). Bird species with low fecundity (small clutch sizes) are indeed especially susceptible to extinction (Bennett & Owens, 1997). The expected role of body size variation in determining variation in the risk of extinction is less transparent because body size may be either positively or negatively related to other traits that would be expected to increase the risk of extinction (Gaston & Blackburn, 1995). For example, large body size may be associated with low abundance, indicating that large bodied species may be more at risk. Alternately, small body size may be associated with susceptibility to environmental perturbation (Gaston & Blackburn, 1995). Two class-wide studies (Gaston & Blackburn, 1995; Bennett & Owens, 1997) have shown that large body size is associated with elevated risk of extinction. Gaston & Blackburn (1995) showed that threatened species are, on average, larger than nonthreatened species, and Bennett & Owens (1997) found a positive correlation between body size and extinction risk. However, although reanalysis has confirmed the link between both body size and clutch-size and extinction risk (Bennett & Owens, 2002; Bennett

et al., 2005), more recent class-wide studies have found that clutch-size, but not body size, is important (Morrow & Pitcher, 2003). In the same study, Morrow and Pitcher also examined whether increased intensity of sexual selection was associated with increased extinction risk. They used four measures of sexual selection (social mating system, sexual size dimorphism, sexual dichromatism, and testis size) to capture different aspects of both pre- and post-mating selection. Only testis size, a measure of post-mating sexual selection, was associated with extinction risk. Species with large testis were more threatened than species with small testis and this remained the case when body size and clutch size were included as correlates. It should be noted that these results where based on analyses that accounted for the effects phylogeny but did not hold when using the raw data and may have overcompensated for phylogenetic effects (see Purvis, 2008). However, Morrow and Pitcher's results are supported by evidence that sexually dimorphic species have higher extinction rates following introductions to islands than sexually monomorphic species (Sorci *et al.*, 1988; McLain *et al.*, 1995, 1999).

Each of the class-wide analyses discussed above focused exclusively on the intrinsic biology of species. However, while some species with, for example, small clutch sizes are prone to extinction risk, some are not. Why is this so? One explanation is that intrinsic biological traits make a species more susceptible to certain types of threat, but if the species does not encounter that particular type of threat then it will probably not be at greater risk of extinction. This idea was examined in detail by Owens & Bennett (2000a). They suggested that species with slow life-histories (e.g. small clutch size or large body size) may be particularly vulnerable to threats from human persecution or introduced predators, whereas species that are ecologically highly specialized (narrow range in diet or habitat preferences) will be most susceptible to threats imposed by habitat loss or degradation. These predictions held up remarkably well (Figure 16.1; Owens & Bennett, 2000a; Bennett & Owens, 2002; Bennett *et al.*, 2005). Where the

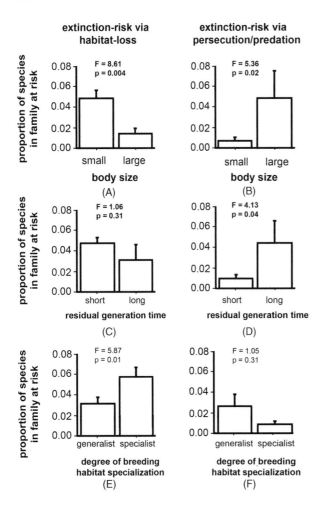

Fig. 16.1 The effects of interactions between types of threat and species' ecology on extinction risk across families of birds. Proportion of species at risk of extinction among families against: (A) body size threatened by habitat loss and (B) persecution or predation; (C) residual generation time threatened by habitat loss and (D) persecution or predation; and (E) degree of breeding habitat specialization threatened by habitat loss and (F) persecution or predation. (From Owens & Bennett, 2000a.)

main threat was habitat loss, bird families with on average small body sizes and highly specialized breeding habitat requirements had a higher proportion of species at risk than families with large average body sizes or generalist habitat preferences. There was no significant difference in risk between families with short and long generation times. Where the main threat was from persecution or introduced predators, families with on average large body sizes and long generation times had much higher proportions of threatened species than families with small average body sizes or short generation times. There was no significant difference in risk between generalist and specialist families.

Revealing general ecological processes is important in our understanding of the mechanisms of threat but has a limited value for practical conservation (Fisher & Owens, 2004). To do that, more focus is needed on particular clades, regions, or both (Fisher & Owens, 2004). Studies that have taken this approach demonstrate the idiosyncratic roles of biology and threat processes in determining population trends and extinction risk. Long et al. (2007) focused on the order Anseriformes and found that two intrinsic factors, small population size and small global range size, and two extrinsic factors, the amount of wetland lost within a species' range (measured as the increase in area of agricultural land) and the total number of different threat processes operating, were the key determinants of recent population declines. Krüger and Radford (2008) found that extinction risk was higher in Accipitridae species with large body sizes, low reproductive rate, specialized habitats, low plumage polymorphism, and less acrobatic displays. Some studies have focused on restricted geographic rather than taxonomic scope. For example, Prinzing *et al.* (2002) found that long-distance migration and small body size were the strongest correlates of population declines in European birds (although they focus discussion on the lack of a convincing relationship with indices of sexual selection). Gage *et al.* (2004) found that extinction risk among neotropical birds was highest among species that occupy fewer zoogeographical regions, have more limited elevation ranges, are smaller bodied, live at higher altitude, use only a single microhabitat, and occupy the edges of undisturbed habitat. When restricted further so that analyses were carried out on each of four zoogeographical regions, they found that limited elevation range was associated with extinction risk in the central Andes and northern Andes; extinction risk decreased with the number of microhabitats in central South America and southern Amazonia but increased in the northern Andes; higher altitude was associated with elevated extinction risk in central South America; and large bodied species were more at risk in southern Amazonia.

Analyses that are restricted both taxonomically and geographically may have the most practical value. Among North American shorebirds (Charadriiformes) the migration route and level of threat on the nonbreeding grounds were the strongest correlates of population decline, suggesting that conservation of stopover sites is likely to play an important role in halting recent declines (Thomas *et al.*, 2006). In gamebirds (Galliformes) intrinsic traits including latitudinal range, body mass, elevation range, and habitat use, and variables reflecting human impacts, including human population density, total food consumption, and diet composition influence extinction risk (Keane *et al.*, 2005). These factors are largely consistent within different Galliform clades (Cracidae and

Phasiandae) and geographic regions (Africa, Asia, Latin America, and the Caribbean).

What is striking from this brief overview of comparative analyses of extinction risk and population trends in birds is just how much variation there is in the intrinsic correlates of imperilment between clades and at different taxonomic scales. Of equal importance is the remarkably low explanatory power of many of the models (Bennett & Owens, 2002). Even with carefully justified explanatory variables, comparative analyses can often explain only a small proportion of the variation in extinction risk or population trend. For example, the best overall model in Gage *et al.*'s (2004) analysis of neotropical birds (across all zoogeographic regions) explained < 10% of the variation in extinction risk. Even with a very large number of potential explanatory variables (26, including five different measures of size), the final model of Krüger & Radford's (2008) analyses of extinction risk in hawks (Accipitridae) explained just 14.8% of the variation in extinction risk. Why do analyses of species' traits generally do so badly at explaining extinction risk or population trends? The most likely explanation may not be that the wrong traits have been identified, rather that species decline partly because of what they are and partly because of where they are. To fully understand variation in threat we need to consider the geographic distribution of current avian diversity and how this will be impacted by current and projected changes in human activities.

THE GEOGRAPHIC DISTRIBUTION OF THREATENED BIRDS

The breeding ranges of all extant birds have recently been compiled to produce maps of their global distributions (Orme *et al.*, 2005; Hawkins *et al.*, 2007; Jetz *et al.*, 2007). These have been used to identify the hotspots of total, threatened, and range-restricted (or endemic) avian species richness (Figure 16.2; Orme *et al.*, 2005). The geographic location of the different types of hotspot are not congruent (Orme *et al.*, 2005). Although

(A)

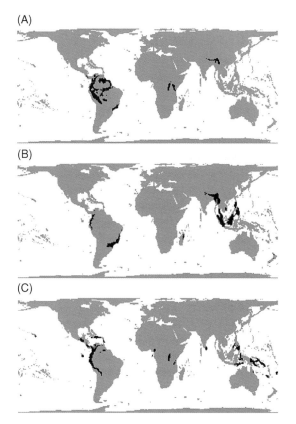

(B)

(C)

Fig. 16.2 Avian biodiversity hotspots. Hotspots of: (A) species richness; (B) threatened species richness; and (C) endemic species richness. Hotspots (shown in black) are the richest 2.5% of grid cells for each category. (From Orme *et al.*, 2005.)

some species richness hotspots are also threatened species richness hotspots (specifically hotspots in the Andes, the Himalayas, and the Atlantic coastal forests), the overall congruence between the two is remarkably low: the hotspots of species richness contained 49% (4371 species) of the world's extant birds but only 27% (293) of the threatened species. This type of disjunction had been discussed previously by Jetz & Rahbek (2001) with respect to African birds: patterns of endemic or threatened species richness are the product of entirely (endemic species) or primarily (threatened species) small-ranged species, whereas patterns in total species richness are driven by large-ranged

species. This occurs despite the fact that most species have small ranges: for subSaharan African birds, the widest ranging quartile of species accounts for 70.5% of occurrence records across 1° grid cells, whereas the smallest ranged species accounted for only 1.3% of occurrences (Jetz & Rahbek, 2001). Consequently, the overlap with human impacts may differ between total, threatened, and range-restricted species richness.

Davies *et al.* (2006, 2007) explored the correlations between both total and species richness and a range of environmental and human impact variables. They found that, on a global scale, measures of human impact were better predictors of the number of threatened species per grid cell (controlling for overall species richness) than were measures of ecology (Davies *et al.*, 2006). Population density, agricultural area and GDP all have strong impacts on the number of threatened species globally, whereas the only major ecological predictor was elevation range (a proxy for topographic variability). The Normalized Difference Vegetation Index (NDVI), a measure of productivity, had a lesser role. In contrast, when analyzed separately for each biogeographic realm (Olson *et al.*, 2001) the results are rather idiosyncratic. but in general indicate that ecology and environment are more important than human impact. For example, NDVI is a major predictor of threatened species richness in the Afrotropics, whereas in IndoMalaya temperature is more important. Human population density is also a significant correlate of total richness, albeit of secondary importance to ambient energy and temperature (Davies *et al.*, 2007). The correlation between threatened species richness and human impacts is particularly concerning. Although we cannot be certain of cause and effect, it is clearly a plausible inference that recent human impacts are responsible for the high proportion of threatened species in many parts of the world.

Perhaps a greater cause for alarm is the continued and escalating impact that humans are having globally. The analyses outlined above were based on recent measures of human populations, wealth, technology, and agricultural practice. Yet human impacts are projected to be even more widespread in the future and will encompass changes in land use and climate (Mitchell *et al.*, 2004; Carpenter *et al.*, 2005; Van Vuuren *et al.*, 2006; Lee & Jetz, 2008). These changes are expected to hit not just threatened species but also species that are currently considered to be of least concern. Thomas *et al.* (2004) used projections of species' distributions under future climate scenarios to assess extinction risks for a range of taxa in selected regions around the world. They modeled the association between a species' current climate and its distribution to generate a 'climate envelope' that is assumed to remain unchanged in the future. Of the species and regions that they analyzed, ca. 18% would be lost under minimal climate warming scenarios, compared to ca. 24% under moderate and ca. 35% under maximum-change scenarios. For birds, in which endemics from Mexico, Europe, Queensland, and South Africa were analyzed, the estimates ranged from no extinctions forecast (South Africa) to a loss of ca. 85% of species (Queensland). Jetz *et al.* (2007) explored future distributions of all the world's birds under projected climate and land-use change. Future changes in species ranges under alternate global change scenarios were estimated, and even under the most benevolent projections at least 400 species by 2050, and 900 species by 2100, were predicted to suffer > 50% range contractions (Figure 16.3). The most dramatic declines are forecast among small-ranged tropical endemics that will be impacted by large-scale anthropogenic land conversion (Figure 16.3). Climate change is forecast to impact mainly on high-latitude species, though the effects are likely to be less severe in the short to mid-term than those predicted for land-use change in the tropics (Figure 16.3).

ECOSYSTEM CONSEQUENCES OF AVIAN DECLINES

Current global human distributions and future projected anthropogenic impacts on biodiversity will, if unchecked, have a dramatic impact on the avian distributions. Yet it is not just numbers of species that will be affected. Birds provide a range of ecosystem services and functions that are perhaps the most diverse of all vertebrate groups yet their role has frequently been overlooked (Table 16.1, Şekercioğlu *et al.*, 2004; Şekercioğlu, 2006a,b). Birds provide both regulatory roles in ecosystems, including seed dispersal (Armesto & Rozzi, 1989; Cordeiro & Howe, 2003), pollination (Feinsinger *et al.*, 1982; Hadley & Betts, 2009), pest control (Mols & Viser, 2002; Van Bael *et al.*, 2003; Perfecto *et al.*, 2004), and carcass disposal (DeVault, 2003), and have supporting roles in nutrient decomposition (Sanchez-Pinero & Polis, 2000) and ecosystem engineering (Daily *et al.*, 1993; Casas-Criville & Valera, 2005; Valdivia-Hoeflich *et al.*, 2005). We know that just as species richness is not randomly distributed around the world, nor is phenotypic and functional diversity. Nectarivores fulfill important roles as pollinators in many systems, particularly in Australia, New Zealand, and Oceania (Ford, 1985), and are particularly species rich in the neotropics (Kissling *et al.*, 2009). The distribution of avian body sizes varies substantially, with more small-bodied species towards the tropics and in areas with high temperature (Olson *et al.*, 2009). Clutch size is typically higher in the Northern than the Southern Hemisphere (Jetz *et al.*, 2008). As extinctions occur at different rates in different regions, both species and functional diversity will be lost, with consequences for communities and ecosystems (Table 16.1). Because frugivores are important seed dispersers their loss may inhibit the ability of dependent plant species to withstand the predicted large-scale changes in land use in this region (see Jetz *et al.*, 2007). Şekercioğlu *et al.* (2004) explored the global consequences of loss of functional diversity by addressing the current and future distributions of threatened birds across a range of ecological and geographical groupings. They generated three possible scenarios for future extinctions. In the first, best case, scenario they assumed that current and future conservation practice will see no change in the status of currently threatened species but will be enough to prevent any new species being elevated to threatened or higher categories of conservation concern.

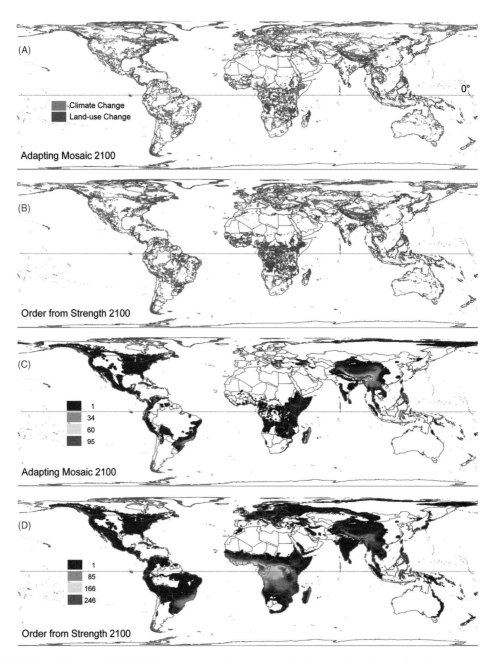

Fig. 16.3   Projected impacts of global change on geographic patterns of avian species richness. Land cover conversion in 2100 due to climate and land-use based on projections of (A) the environmentally proactive Adapting Mosiac and (B) the environmentally reactive Order from Strength models. Species richness of birds with projected range declines of ≥ 50% on a 0.5° grid for (C) the Adapting Mosiac and (D) Order from Strength models. (From Jetz *et al.*, 2007.) [This figure appears in color as Plate 16.3.]

Table 16.1 Ecological and economical contributions of avian functional groups. From Şekercioğlu *et al.* (2004) and Şekercioğlu (2006a).

Functional group	Ecological process	Ecosystem service and economic benefits	Consequences of loss of functional group
Frugivores	Seed dispersal	Removal of seeds from parent tree; escape from seed predators; improved germination; increased economic yield; increased gene flow; recolonization and restoration of disturbed ecosystems	Disruption of dispersal mutualisms; reduced seed removal; clumping of seeds under parent tree; increased seed predation; reduced recruitment; reduced gene flow and germination; reduction or extinction of dependent species
Nectarivores	Pollination	Outbreeding of dependent and/or economically important species	Pollinator limitation; inbreeding and reduced fruit yield; evolutionary consequences; extinction
Scavengers	Carrion consumption	Removal of carcasses; leading other scavengers to carcasses; nutrient recycling; sanitation	Slower decomposition; increases in carcasses; increases in undesirable species; disease outbreaks; changes in cultural practices
Insectivores	Predation on insects	Control of insect populations; reduced plant damage; alternate to pesticides	Loss of natural pest control; pest outbreaks; crop losses; trophic cascades
Piscivores	Predation on fishes and invertebrates; production of guano	Controlling unwanted species; nutrient deposition around rookeries; soil formation in polar environments; indicators of fish stocks; environmental monitors	Loss of guano and associated nutrients; impoverishment of associated communities; loss of socioeconomic resources and environmental monitors; trophic cascades
Raptors	Predation on vertebrates	Regulation of rodent populations; secondary dispersal	Rodent pest outbreaks; trophic cascades; indirect effects
All species	Various	Environmental monitoring; indirect effects; birdwatching tourism; reduction of agricultural residue; cultural and economic uses; ecosystem engineering	Losses of socioeconomic resources and environmental monitors; unpredictable consequences

In the second, intermediate, scenario they compared the threatened species lists from 1994 (Collar *et al.*, 1994) to 2003 (International Union for Conservation of Nature and Natural Resources, 2003) and assume that the rate at which additional species were elevated to threatened during this period will be maintained in the future. The final, worst case, scenario takes the rate inferred in scenario two and adds a further 1% increase in rate per decade so that the rate at which species go extinct gradually increases throughout the 21st century. Şekercioğlu *et al.* (2004) argue that species that are classified as endangered, critically endangered, or extinct in the wild sum to

only 0.025% of the global bird population and therefore contribute little to ecosystem process and function compared to nonthreatened species. They regard such species as functionally extinct and species that have undergone recent population or range contraction to be functionally deficient. On this basis, and under the three future scenarios, by 2100 6–14% of bird species will be extinct, 7–25% will be functionally extinct, and a staggering 13–52% will be functionally deficient (Figure 16.4). Species in some habitats, including those that are marine or forest dwelling, and some regions, particularly New Zealand, Oceania, Malagasy, and South Polar will lose functional

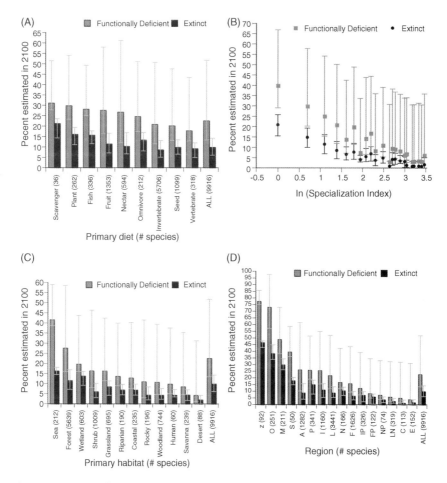

Fig. 16.4   Predicted percentages of extinct and functionally deficient bird species for scenario 2 (intermediate; see text for details). (A) Distribution based on primary diet. (B) Effects of the degree of specialization derived from the product of habitats used and food types consumed. (C) Distribution based on primary habitat. (D) Distribution based on biogeographic region: A, Austral; C, Cosmopolitan; E, Eastern Hemisphere; F, Afrotropical; I, Indomalayan; M, Malagasy; N, Nearctic; L, neotropical; O, Oceania; P, Palearctic; S, South Polar; and Z, New Zealand. (From Şekercioğlu *et al.*, 2004.)

diversity above the global average (Figure 16.4C and D). Similarly, and perhaps more seriously, some functional groups will fare particularly poorly including frugivores, herbivores, nectarivores, piscivores, and scavengers (Figure 16.4a). In addition, some 40% of species that are specialized in a single guild are predicted to be functionally deficient by 2100 (Figure 16.4b). The impacts on ecosystem and community function could be severe (Table 16.1).

Functional extinction of bird species is of serious importance for a particular ecosystem or community only if that function cannot be filled by other members of the community (Walker, 2003). However, there is evidence that avian communities have very low levels of functional redundancy. That is, if a species is lost, its function is lost with it unless species from neighboring communities can successfully invade the depauperate community. The seeds of *Leptonychia usambarensis*, a tree endemic to the eastern Usambara Mountains of Tanzania, are dispersed by several bird species (particularly stripe-cheeked greenbul *Andropadus milanjensis* and Shelley's greenbul *Andropadus masukuensis*) in continuous expanses of rainforest. However, where the forest has become fragmented, the birds were rare or even absent and juvenile recruitment of the *Leptonychia* was reduced (Cordeiro & Howe, 2003). Among British birds, functional diversity (measured according to size, diet, foraging strata, habitat, and time) in a given $10\,km^2$ assemblage is typically lower than expected by chance (Petchey *et al.*, 2007), suggesting that function within communities is replaceable. However, over a 20-year period, changes in species richness were matched by proportionally similar changes in functional diversity (Petchey *et al.*, 2007). This implies that there is little redundancy in the functional composition of British avian communities and that loss of function is buffered by community reassembly, presumably from neighboring assemblages. Whether this would hold globally or at a finer (community rather than assemblage) scale is not clear. It seems reasonable to expect that the overall lack of functional redundancy will be a general phenomenon, but less likely that this will be so for irreplaceability. Across the neotropics many species of hummingbird have coevolved with the plants on which they feed to the extent that many species of plant are dependent on hummingbirds for pollination (Stiles, 1981). The three-wattled bellbird *Procnias tricarunculata* disperses seeds of *Ocotea endresiana*, a shade-tolerant neotropical tree, to canopy gaps where seedling survival is higher than for seeds dispersed over much shorter distances by other bird species (Wenny & Levey, 1998). Similarly, the southern cassowary *Casuarius casuarius* is the only frugivore large enough to act as a long-distance disperser for many plants in the Queensland rainforest (Stocker & Irvine, 1983). These examples are only the tip of the iceberg for potential loss of irreplaceable avian functional diversity on a global scale (Şekercioğlu *et al.*, 2004; Şekercioğlu, 2006a,b). It is then all the more concerning to consider that the responses of many species to climate change may have similar

effects on community structure and function. In the next section I review the evidence for avian responses to recent climate change, particularly with respect to range movements and phenology, and consider their impacts on evolutionary and ecological processes.

RESPONSES TO CLIMATE CHANGE IN BIRDS

The rapidity of recent and projected climate change poses new challenges to the worlds biota (Root *et al.*, 2003). Limits to bird ranges may often be set by physiological tolerance to and energy expenditures that compensate for, temperature (Root, 1988a,b). As global temperature warms, there are three ways in which species' can respond. They can track changes in climate by shifting their ranges to suitable habitats elsewhere; they can show phenotypically plastic responses (i.e. changes in phenotypic traits without changes in genotype); or, they can show microevolutionary (genetic) change (Holt, 1990; Davis *et al.*, 2005; Visser, 2008). The former will result in species turnover and community reassembly and therefore alter the balance of ecosystem and community dynamics. The latter two can mitigate the effects of climate change such that ecological communities may remain intact. But, if the phenologies of different species respond at different rates then interspecific interactions will break down (Harrington *et al.*, 1999; Root *et al.*, 2003; Both *et al.*, 2009). Consequently, community and ecosystem structure and function will depend critically on how different species respond to climate change. There is widespread evidence that birds are responding to climate change in both distribution and phenology.

The first observations of range change in birds have been summarized by Parmesan (2006) who highlighted studies showing northward range shifts of bird range in Iceland, Finland, and Britain in the 1930s and 1940s (Salomonsen, 1948; Kalela, 1949, 1952; Gudmundsson, 1951; Harris, 1964). Northward shifts averaging $18.9\,km$ have been observed recently across 20 species of British birds between 1968–1972 and 1988–1991

(Thomas & Lennon, 1999), a period over which temperature increased in Great Britain (Conway, 1998). Range shifts are also known to have occurred in polar and tropical regions. In Antarctica, Adélie (*Pygoscelis adeliae*) and emperor (*Aptenodytes forsteri*) penguins have responded to changes in sea-ice extent with both species now essentially absent from the northernmost parts of their recent ranges (Croxall *et al.*, 2002; Ainley *et al.*, 2003). In the tropics, the rufous hummingbird *Selasphorus rufus* wintered in Mexico as recently as the 1970s with fewer than 30 sightings per year along the Gulf Coast of the USA, yet by 1996 there were a total of 1643 documented winter occurrences in this region (Hill *et al.*, 1998).

However, the observed distributional changes in birds may not be proceeding rapidly enough to track climate change. Devictor *et al.* (2008) developed a measure, the community temperature index (CTI), to quantify the balance within an assemblage between species whose ranges are typically associated with high temperature and species whose ranges are typically associated with low temperature. A high CTI indicates that there is a relatively high proportion of species associated with high compared to low temperatures. Among assemblages of French breeding birds, CTI was found to have increased since 1989 in association with increases in temperature. However, the increases in CTI corresponded to a community-wide northward shift in species' ranges of 91 km, compared to an equivalent northward shift in temperature of 273 km. The 182 km lag in species ranges suggests that French breeding birds are not tracking climate change fast enough (Devictor *et al.*, 2008).

Avian phenology may also be responding too slowly to a warming climate. Global climate change has been shown to affect the phenology, distributions, and life histories of many organisms (Rohde, 1992; Post *et al.*, 2001; Parmesan & Yohe, 2003; Parmesan, 2006), particularly birds (reviewed by Crick, 2004; Sparks & Mason, 2004). The best-known example of advancement in phenology is based on laying date data spanning 57 years (92,828 records from 1939 to 1995) from the British Trust for Ornithol-

ogy nest record scheme (Crick *et al.*, 2003). These data show that the laying date of 31 out of 36 species was related to temperature or rainfall, with 19 species (53%) showing long-term trends in laying date that corresponded with climate change (Crick *et al.*, 1997; Crick & Sparks, 1999). Using UK-specific climate scenarios, Crick & Sparks (1999) predicted that for 27 species included in their study, the laying date would advance by an average of 8 days, and a maximum of 18 days, by 2080. A similar study of nest records from North America for the tree swallow *Tachycineta bicolor* showed that temperature predicted a 9-day advancement in laying date between 1959 and 1991 (Dunn & Winkler, 1999). Changes in migration have been more extensively documented. The average arrival and departure dates of 20 migrant species in Oxfordshire, UK have advanced by 8 days in the past 30 years (Cotton, 2003). Arrival dates in the UK had advanced in 17 out of 20 species compared to advances in departure dates in 15 out of 20. The overall time spent at the breeding grounds remained essentially unchanged during the 30-year period. Cotton (2003) suggests that arrival time advanced with increasing winter temperatures in subSaharan Africa, while departure time advanced with elevated summer temperatures in the Oxfordshire breeding grounds. Hüppop & Hüppop (2003) found that all but one species out of 25 bird species (24 passerines and the woodcock) that pass through the island of Helgoland (southeastern North Sea) showed a trend towards earlier mean spring passage times since 1909. The trend was statistically significant in seven species that are short–medium distance migrants and in ten species that are long-distance migrants. The earlier mean spring passage times coincide with higher North Atlantic Oscillation (NAO) indices.

There are numerous other examples of phenological responses attributed to climate change in birds (Crick, 2004). However, there is also evidence that these changes are much slower than those at lower trophic levels, causing a mismatch between the phenologies of some bird species and their food (Harrington *et al.*, 1999; Both *et al.*, 2009). Several studies of insectivorous birds have shown that the phenology of the off-spring food supply has advanced more rapidly than breeding phenology (Visser *et al.*, 1998; Both & Visser, 2005; Pearce-Higgins *et al.*, 2005). A long-term study of great tits (*Parus major*) on the Hoge Veluwe (Netherlands) showed that there was strong directional selection for advancement in the date of egg-laying due to temperature increases that resulted in advances in vegetation phenology and annual peak dates of caterpillar biomass (Visser *et al.*, 1998). Yet during this period the egg-laying date remained unchanged. At the same location, the spring arrival of migrant pied fly-catchers (*Ficedula hypoleuca*) remained unchanged but arrival was earlier than the peak laying date and laying date advanced by an average of 10 days. But, as with great tits, this still lagged behind the advancement of spring on the breeding grounds (Both & Visser, 2001). There is, however, variation in the responses of the same species in different parts of their range, since the laying date of great tits of Wytham Wood, Oxford, UK has tracked advancement in peak caterpillar biomass (Cresswell & McCleery, 2003).

At a population level, the most severe consequences of the disjunction between the timing of peak food supplies and the timing of breeding is rapid population decline. This is evident in some Dutch populations of the pied flycatcher where declines in abundance of 90% have been recorded (Both *et al.*, 2006). At an ecosystem level, because climate change can be expected to influence all species, variation in response between species may impact on all levels and interactions (Harrington *et al.*, 1999). Both *et al.* (2009) explored variation in phenology at four trophic levels: (i) budburst in pedunculate oak trees *Quercus robur*; (ii) hatching date of caterpillars that feed on oak buds; (iii) laying and hatching date in four passerine species (coal tit *Parus ater*, blue tit *Cyanistes caeruleus*, great tit *Parus major*, and pied flycatcher *Ficedula hypoleuca*) whose offspring depend on caterpillars for food; and (iv) peak food requirements date (estimated as 12 days from hatching) in sparrowhawks *Accipiter nisus* that prey on passerine birds. Both *et al.* (2009) found that budburst advanced only slightly (0.17

days per year) between 1988 and 2005, while both caterpillar (0.75 days per year) and passerine (0.36–0.5 days per year) hatching dates advanced significantly between 1985 and 2005, but raptor hatching dates did not change. Caterpillar hatching dates were closely correlated with budburst, and passerine hatching was correlated with caterpillar hatching. However, in both cases the slope of < 1 implies that congruence between phenologies declined over the period of studies. Both *et al.* (2009) suggest that phenological responses to climate change are slower at higher trophic levels, resulting in a disjunction between peaks of food demands and peaks of food availability.

Why are avian phenological responses lagging behind those of their prey? A trait that is phenotypically plastic with respect to temperature should show a near-instantaneous response to climate change (Visser, 2008). For example, if great tits respond directly to changes in the advancement of peak caterpillar biomass then phenotypically plastic changes in the timing of reproduction would be exactly concurrent. However, phenotypic responses can be hampered if the environment in which selection occurs differs from the environment in which, for example, reproductive decisions are made (Visser *et al.*, 1998). In Dutch great tits, the relationship between peak of food availability and the cue for reproduction appears to have changed, implying that selection and decision making do not take the same cues (Visser *et al.*, 1998). Among migratory pied flycatchers, phenotypic plasticity in laying date is constrained by spring arrival: the timing of the former relies on environmental cues on the breeding grounds, whereas the timing of the latter is dependent on environmental cues on the wintering grounds (Both & Visser, 2001).

If differences in phenological responses of birds and their prey are usually explained by constraints on plasticity (Visser *et al.*, 1998; Both & Visser, 2001; Both *et al.*, 2009), then adequate responses will depend on microevolutionary change (Visser, 2008). Yet, the very few studies of genotypic responses to climate change in birds have found no evidence that birds are responding to climate change at the genetic level (Pulido, 2007;

Garant *et al.*, 2008; Teplitsky *et al.*, 2008). Body size declines have been associated with climate change in several bird species (Yom-Tov, 2001; Yom-Tov *et al.*, 2006; Teplitsky *et al.*, 2008) but there is no evidence for genetic change. Between 1958 and 2004, mean body mass of individually marked red-billed gulls (*Larus novaehollandiae scopulinus*) from New Zealand declined at a rate of 0.1 SD per generation and correlated significantly with increased temperature (Teplitsky *et al.*, 2008). Using an "animal model" (Kruuk, 2004), phenotypic mass was broken down into additive genetic, random, and other fixed effects to estimate the genetic component of body mass change. In contrast to the mean phenotypic body mass, there was no relationship between the genetic component of body mass change and declines in temperature (Teplitsky *et al.*, 2008). Similarly, the population of great tits (*Parus major*) in Wytham Woods, Oxford showed significant changes in mean phenotype of reproductive traits, including laying date, clutch size, and egg mass collected from 1965 to 1988 compared to the warmer period from 1989 to 2004 (Garant *et al.*, 2008). There was also significant additive genetic variance and heritability in these traits within each period. However, comparisons of matrices of additive genetic variances and covariances (the **G** matrix, Lande, 1979) between the two periods revealed no change in the genotypic component of reproductive traits (Garant *et al.*, 2008).

The lack of evidence for microevolutionary responses to climate change, and the constraints on phenotypic plasticity imposed by differential cues in reproductive traits show that ecosystem dynamics may be disrupted and they also have implications for community structure and function, particularly in migratory birds. Migratory species are likely to be particularly affected by climate change (Both & Visser, 2001) because it may influence their breeding and migration phenology, migration distance (Visser *et al.*, 2009), as well as both breeding and wintering ranges. Communities of migratory birds may therefore be particularly impacted by climate change. Resident species may benefit from warmer winters due in particular to reduced energy demands, and this

may be detrimental to long-distance migrants due to increased competition.

Geographic variation in the number of long-distance migrants in Europe was found to decrease with increasing winter temperature and decreasing spring temperature (Lemoine & Böhning-Gaese, 2003). Lemoine & Böhning-Gaese (2003) used this relationship to predict and test how community composition changed between censuses in 1980–81 and 1990–92 in the Lake Constance region of central Europe. They found, consistent with the model, that the proportion of long-distance migrants decreased while the proportion of short-distance migrants and resident species increased with warmer winter and colder spring temperatures. Extending the predictive model to 21 sites across Europe revealed similar trends (Lemoine *et al.*, 2006). Taken together, these studies imply that long-distance migrants may be particularly susceptible to the effects of climate change. However, it is not clear whether species turnover due to range changes (community reassembly, Schaefer *et al.*, 2008) or phenotypic adaptation will generally govern responses of bird communities to climate change. Distinguishing between these two possibilities is important because species turnover affects community dynamics often in unpredictable ways and may impact on function in a similar way to species extinction (see Table 16.1). In contrast, community composition will remain essentially intact if species respond by adaptation. Using spatial variation in the number of species that potentially migrate (i.e. they are migratory in part of their range) and those that actually migrate in a community, Schaefer *et al.* (2008) showed that spatially the number and proportion of migratory species decreased with decreasing temperature in the coldest month and increasing spring temperature. The spatial models imply that a 1° increase in the temperature of the coldest month results in a decline of 1.3 migratory species (corresponding to 1.78% of the number of migrants). This decline corresponds to a 2.18% decline in the proportion of migratory activity but only a 0.20% decline in the proportion of migratory propensity. Declines in migratory propensity imply species turnover and were less important than declines in migratory activity that imply phenotypic adaptation. When compared to an expected 3.4° increase in temperature under a moderate climate change scenario (A2, Mitchell *et al.*, 2004) these models imply that climate change will have only a minor impact on Europe-wide community composition of migrant bird species (Schaefer *et al.*, 2008). It is encouraging that species turnover and community reassembly will apparently have only a very minor impact on migrant bird communities. However, the lagged-response of some species, including long-distance migrants (Both *et al.*, 2006), and communities (Devictor *et al.*, 2008), and the absence of evidence for microevolutionary responses to climate change (Garant *et al.*, 2008; Teplitsky *et al.*, 2008) suggests that there may still be a significant impact on the functioning of these communities.

A FUTURE FOR AVIAN DIVERSITY?

Taken together, there is little evidence that birds are able to respond, either through phenotypic plasticity or via genetic adaptation, sufficiently rapidly to cope with global change (Visser, 2008). Avian diversity in the future is threatened by a double-edged sword of habitat loss and climate change (Myers & Knoll, 2001). Habitat loss will drive contraction and fragmentation of species ranges and population. As populations decline so too does genetic diversity while gene flow between populations will be disrupted as habitats become more fragmented (Templeton *et al.*, 2001). Reduced genetic diversity reduces evolutionary potential simply because there is less variation in genetic material on which selection can act (Frankham, 2005). We have seen that there is little evidence to suggest that there is adequate genotypic evolution in response to climate change and that there are limits on the extent to which phenotypic plasticity can mitigate these effects. This suggests first, that our current estimates of the number of threatened bird species are very likely to be significant underestimates since they do not

account for projected climate or land-use changes; and second, that the time to extinction for many currently imperiled species may be substantially faster than current estimates suggest. We are already faced with the huge challenge of conserving threatened species but even the idea that we can save communities and ecosystems in their current state is itself under threat (Hannah *et al.*, 2002). The impacts of projected extinctions acting differentially on functional groups, the effects of mistiming of phenology on trophic interaction networks, and species turnover in the face of climate-driven range shifts will all impact not just on the numbers of birds but on the structure and function of the communities and ecosystems in which they live.

At a global scale, the projected impacts of land use and climate change on land cover are only poorly predicted by past human impact (Lee & Jetz, 2008). Current conservation priorities therefore provide poor targets for future threats and there is a pressing need for revised reserve-based conservation planning (Lee & Jetz, 2008). New conservation targets are challenging because often we do not adequately understand the species' responses to environmental perturbation. Much value will therefore be placed on understanding and conserving evolutionary and ecological processes rather than species or habitats *per se.* Unfortunately, while we are making progress in our ability to predict future extinctions, we remain poor at predicting future evolutionary processes (Woodruff, 2001; Barraclough & Davies, 2005). Where and what to conserve therefore remains unclear. However, both future projections (Jetz *et al.*, 2007) and the current distribution of anthropogenic threats (Davies *et al.*, 2006) suggest that areas that have been identified as rich in endemic species and possible historical centers of clade origin and speciation (Croziat *et al.*, 1974; Ricklefs & Schluter, 1993; Jetz *et al.*, 2004) will be among the hardest hit by continued negative human impacts on biodiversity. While a proliferation of vacant niches due to extinction and the formation of barriers to gene flow (and increased potential for reproductive isolation) due habitat fragmentation may promote speciation in the future (Templeton et al., 2001), the lack of genetic diversity will very likely inhibit it (Jablonski, 2001). If areas or clades that have an elevated propensity to speciate are lost then future diversity will be impoverished not only by impending extinctions but also by a decline in speciation rate (Rosenzweig, 2001). Avian diversity is, in common with the rest of world's biota, facing a bleak and uncertain future.

REFERENCES

Ainley DG, Ballard G, Emslie SD, Fraser WR, Wilson, PR, Woehler EJ. 2003. Adelie penguins and environmental change. *Science* **300**: 429–430.

Armesto JJ, Rozzi R. 1989. Seed dispersal syndromes in the rain forest of Chiloé: evidence for the importance of biotic dispersal in a temperate rain forest. *Journal of Biogeography* **16**: 219–226.

Barker FK, Cibois A, Schilker P, Feinstein J, Cracraft J. 2004. Phylogeny and diversification of the largest avian radiation. *Proceedings of the National Academy of Sciences USA* **101**: 11040–11045.

Barraclough TG, Davies TJ. 2005. Predicting future speciation. In Phylogeny and Conservation, Purvis A, Gittleman JL, Brooks TM (eds). Cambridge: Cambridge University Press; 400–418.

Bennett PM, Owens IPF. 1997. Variation in extinction risk among birds: chance or evolutionary predisposition? *Proceedings of the Royal Society of London Series B (Biological Sciences)* **264**: 401–408.

Bennett PM, Owens IPF. 2002. The Evolutionary Ecology of Birds. Oxford: Oxford University Press.

Bennett PM, Owens IPF, Nussey D, Garnett ST, Crowley GM. 2005. Mechanisms of extinction in birds: phylogeny, ecology and threats. In Phylogeny and Conservation, Purvis A, Gittleman JL, Brooks TM (eds). Cambridge: Cambridge University Press; 317–336.

Bibby CJ. 1995. Recent, past and future extinctions in birds. In *Extinction Rates*, Lawton JH, May RM (eds). Oxford: Oxford University Press; 98–110.

BirdLife International. 2008. *State of the World's Birds: Indicators for our Changing World*. Cambridge: BirdLife International.

Both C, Visser ME. 2001. Adjustment to climate change is constrained by arrival date in a long-distance migrant bird. *Nature* **411**: 296–298.

Both C, Visser ME. 2005. The effect of climate change on the correlation between avian life history traits. *Global Change Biology* **11**: 1606–1613.

Both C, Bouwhuis S, Lessells CM, Visser ME. 2006. Climate change and population declines in a long-distance migratory bird. *Nature* **441**: 81–83.

Both C, van Asch M, Bijlsma RG, van den Burg AB, Visser ME. 2009. Climate change and unequal phenological changes across four trophic levels: constraints or adaptations? *Journal of Animal Ecology* **78**: 73–83.

Brook BW, Sodhi NS, Bradshaw CJA. 2008. Synergies among extinction drivers under global change. *Trends in Ecology and Evolution* **23**: 453–460.

Brooks TM, Pilgrim JD, Rodrigues ASL, Da Fonseca GAB (eds). 2005. *Conservation Status and Geographic Distribution of Avian Evolutionary History*. Cambridge: Cambridge University Press.

Butchart SH, Stattersfield MAJ, Brooks TM. 2006a. Going or gone: defining "Possibly Extinct" species to give a truer picture of recent extinctions. *Bulletin of the British Ornithologists Club* **126A**: 7–24.

Butchart SH, Stattersfield MAJ, Collar NJ. 2006b. How many bird extinctions have we prevented? *Oryx* **40**: 266–278.

Carpenter SR, Pingali PL, Bennett EM, Zurek MB. 2005. *Ecosystems and Human Well-being: Scenarios, Volume 2*. Washington, DC: Island Press.

Casas-Criville A, Valera F. 2005. The European bee-eater (*Merops apiaster*) as an ecosystem engineer in arid environments. *Journal of Arid Environments* **60**: 227–238.

Chatterjee S. 1997. *The Rise of Birds: 225 Million Years of Evolution*. Baltimore: The John Hopkins University Press.

Collar NJ, Crosby MJ, Stattersfield AJ. 1994. Birds to Watch 2: The World List of Threatened Birds. Cambridge: Birdlife International.

Conway D. 1998. Recent climate variability and future climate change scenarios for Great Britain. *Progress in Physical Geography* **22**: 350–374.

Cordeiro NJ, Howe HF. 2003. Forest fragmentation severs mutualism between seed dispersers and an endemic African tree. *Proceedings of the National Academy of Sciences USA* **100**: 14052–14056.

Cotton P. 2003. Avian migration phenology and global climate change. *Proceedings of the National Academy of Sciences USA* **100**: 12219–12222.

Cresswell W, McCleery R. 2003. How great tits maintain synchronization of their hatch date with food supply in response to to long-term variability in temperature. *Journal of Animal Ecology* **72**: 356–366.

Crick HQP. 2004. The impact of climate change on birds. *The Ibis* **146**: 48–56.

Crick HQP, Sparks TH. 1999. Climate change related to egg-laying trends. *Nature* **399**: 423–424.

Crick HQP, Dudley C, Glue DE, Thomson DL. 1997. UK birds are laying eggs earlier. *Nature* **388**: 526.

Crick HQP, Baillie SR, Leech DI. 2003. The UK Nest Record Scheme: its value for science and conservation. *Bird study* **50**: 254–270.

Croxall JP, Trathan PN, Murphy EJ. 2002. Environmental change and Antarctic seabird populations. *Science* **297**: 1510–1514.

Croziat L, Nelson G, Rosen DE. 1974. Centers of origin and related concepts. *Systematic Zoology* **23**: 265–287.

Crozier R. 1997. Preserving the information cotent of species: genetic diversity, phylogeny, and conservation. *Annual Review of Ecology and Systematics* **28**: 243–268.

Daily GC, Ehrlich PR, Haddad NM. 1993. Double keystone bird in a keystone species complex. *Proceedings of the National Academy of Sciences USA* **90**: 592–594.

Davies RG, Orme CDL, Olson VA, Thomas GH, Ross SG, Ding TS, Rasmussen PC, Stattersfield AJ, Bennett PM, Blackburn TM, Owens IPF, Gaston KJ. 2006. Human impacts and the global distribution of extinction risk. *Proceedings of the Royal Society of London Series B (Biological Sciences)* **273**: 2127–2133.

Davies RG, Orme CDL, Storch D, Olson VA, Thomas GH, Ross SG, Ding TS, Rasmussen PC, Bennett PM, Owens IPF, Blackburn TM, Gaston KJ. 2007. Topography, energy and the global distribution of bird species richness. *Proceedings of the Royal Society of London Series B (Biological Sciences)* **274**: 1189–1197.

Davis MB, Shaw RG, Etterson JR., 2005. Evolutionary responses to changing climate. *Ecology* **86**: 1704–1714.

DeVault TL. 2003. Scavenging by vertebrates: behavioral, ecological, and evolutionary perspectives on an important energy transfer pathway in terrestrial ecosystems. *Oikos* **102**: 225–234.

Devictor V, Julliard R, Couvet D, Jiguet F. 2008. Birds are tracking climate warming, but not fast enough. *Proceedings of the Royal Society of London Series B (Biological Sciences)* **275**: 2743–2748.

Diniz-Filho JAF. 2004. Phylogenetic diversity and conservation prioriies under distinct models of phenotypic evolution. *Conservation Biology* **18**: 698–704.

Dunn PO, Winkler DW. 1999. Climate change has affected the breeding date of Tree Swallows throughout North America. *Proceedings of the Royal Society of London Series B (Biological Sciences)* **266**: 2487–2490.

Ehrlich PR, Pringle RM. 2008. Where does biodiversity go from here? A grim business-as-usual forecast and a hopeful portfolio of partial solutions. *Proceedings of the National Academy of Sciences, USA* **105**: 11579–11586.

Erwin TL. 1991. An evolutionary basis for conservation strategies. *Science* **253**: 750–752.

Faith DP. 1992. Conservation evaluation and phylogenetic diversity. *Biological Conservation* **61**: 1–10.

Faith DP. 2002. Quantifying biodiversity: a phylogenetic perspective. *Conservation Biology* **16**: 248–252.

Faith DP. 2008. Threatened species and the potential loss of phylogenetic diversity: conservation scenarios based on estimated extinction probabilities and phylogenetic risk analysis. *Conservation Biology* **22**: 1461–1470.

Feinsinger P, Wolfe JS, Swarm LA. 1982. Island ecology: reduced hummingbird diversity and the pollination biology of plants, Trinidad and Tobago, West Indies. Ecology **63**: 494–506.

Felsenstein J. 1985. Phylogenies and the comparative method. *The American Naturalist* **125**: 16–24.

Fisher DO, Owens IPF. 2004. The comparative method in conservation biology. *Trends in Ecology and Evolution* **19**: 391–398.

Ford HA. 1985. Nectar-feeding birds and bird pollination: why are they so prevalent in Australia yet absent from Europe? *Proceedings of the Ecological Society of Australia* **14**: 153–158.

Forest F, Grenyer R, Rouget M, Davies TJ, Cowling RM, Faith DP, Hedderson TAJ, Savolainen V. 2007. Preserving the evolutionary potential of floras in biodiversity hotspots. *Nature* **445**: 757–760.

Frankel OH. 1974. Genetic conservation: our evolutionary responsibility. *Genetics* **78**: 53–65.

Frankham R. 2005. Genetics and extinction. *Biological Conservation* **126**: 131–140.

Gage GS, Brooke M de L, Symonds MRE, Wege D. 2004. Ecological correlates of the threat of extinction in Neotropical bird species. *Animal Conservation* **7**: 161–168.

Garant D, Hadfield JD, Kruuk LEB, Sheldon BC. 2008. Stability of genetic variance and covariance for reproductive characters in the face of climate change in a wild bird population. *Molecular Ecology* **17**: 179–188.

Gaston KJ. 1994a. Geographic range sizes and trajectories to extinction. *Biodiversity Letters* **2**: 163–170.

Gaston KJ. 1994b. Rarity. London: Chapman and Hall.

Gaston KJ, Blackburn TM. 1995. Birds, body size and the threat of extinction. *Philosophical Transactions of the Royal Society of London* **B347**: 205–212.

Gaston KJ, Blackburn TM. 1997. Evolutionary age and extinction risk in the global avifauna. *Evolutionary Ecology* **11**: 557–565.

Gudmundsson F. 1951. The effects of the recent climatic changes on the bird life of Iceland. *Proceedings of the 10th Ornithological Congress*, Uppsala; 502–514.

Hadley AS, Betts MG. 2009. Tropical deforestation alters hummingbird movement patterns. *Biology Letters* **5**: 207–210.

Hannah L, Midgley GF, Lovejoy T, Bond WJ, Bush M, Lovett JC, Scott D, Woodward FI. 2002. Conservation of biodiversity in a changing climate. *Conservation Biology* **16**: 264–268.

Harrington R, Woiwod I, Sparks TH. 1999. Climate change and trophic interactions. *Trends in Ecology and Evolution* **14**: 146–150.

Harris G. 1964. Climatic changes since 1860 affecting European birds. *Weather* **19**: 70–79.

Harvey PH, Pagel MD. 1991. *The Comparative Method in Evolutionary Biology*. Oxford: Oxford University Press.

Hawkins BA, Diniz-Filho JAF, Jaramillo CA, Soeller SA. 2007. Climate, niche conservatism, and the global bird diversity gradient. *The American Naturalist* **170**: S16–S27.

Hill GE, Sargent RR, Sargent MB. 1998. Recent changes in the winter distribution of Rufous Hummingbirds. *The Auk* **115**: 240–245.

Holt RD. 1990. The microevolutionary consequences of climate change. *Trends in Ecology and Evolution* **14**: 311–315.

Hughes JB, Daily GC, Ehrlich PR. 1997. Population diversity: its extent and extinction. *Science* **278**: 689–692.

Hüppop O, Hüppop K. 2003. North Atlantic Oscillation and timing of spring migration in birds. *Proceedings of the Royal Society of London Series B (Biological Sciences)* **270**: 233–240.

International Union for Conservation of Nature and Natural Resources. 2003. *IUCN Red List of Threatened Species.*www.redlist.org.

Isaac NJB, Turvey ST, Collen B, Waterman C, Baillie JEM. 2007. Mammals on the EDGE: conservation priorities based on threat and phylogeny. *PLoS ONE* **2**: e296.

Jablonski D. 2001. Lessons from the past: evolutionary impacts of mass extinctions. *Proceedings of the National Academy of Sciences USA* **98**: 5393–5398.

Jetz W, Rahbek C. 2001. Geometric constraints explain much of the species richness pattern in African birds. *Proceedings of the National Academy of Sciences USA* **98**: 5661–5666.

Jetz W, Rahbek C, Colwell RK. 2004. The coincidence of rarity and richness and the potential signature of history in centres of endemism. *Ecology Letters* **7**: 1180–1191.

Jetz W, Wilcove DS, Dobson AP. 2007. Projected impacts of climate and land-use change on the global diversity of birds. *PLoS Biology* **5**: 1211–1219.

Jetz W, Şekercioğlu CH, Böhning-Gaese K. 2008. The worldwide variation in avian clutch size across species and space. *PLoS Biology* **6**: e303.

Kalela O. 1949. Changes in geographic ranges in the avifauna of northern and central Europe in response to recent changes in climate. *Bird Banding* **20**: 77–103.

Kalela O. 1952. Changes in the geographic distribution of Finnish birds and mammals in relation to recent changes in climate. In *The Recent Climatic Fluctuations In Finland And Its Consequences: A Symposium*, Hustichi I (ed.), Helsinki; 38–51.

Keane A, Brook M de L, Mcgowan PJK. 2005. Correlates if extinction risk and hunting pressure in gamebirds (Galliformes). *Biological Conservation* **126**: 216–233.

Kissling WD, Böhning-Gaese K, Jetz W. 2009. The global distribution of frugivory. *Global Ecology and Biogeography* **18**: 150–162.

Krajewski C. 1991. Phylogeny and diversity. *Science* **254**: 918–919.

Krüger O, Radford AN. 2008. Doomed to die? Predicting extinction risk in the true hawks Accipitridae. *Animal Conservation* **11**: 83–91.

Kruuk LEB. 2004. Estimating genetic parameters in natural populations using the "animal model". *Philosophical Transactions of the Royal Society of London* **B359**: 873–890.

Lande R. 1979. Quantitative genetic analysis of multivariate evolution, applied to brain: body size allometry. *Evolution* **33**: 402–416.

Lee TM, Jetz W. 2008. Future battlegrounds for conservation under global change. *Proceedings of the Royal Society of London Series B (Biological Sciences)* **275**: 1261–1270.

Lemoine N, Böhning-Gaese K. 2003. Potential impact of global climate change on species richness of long-distance migrants. *Conservation Biology* **17**: 577–586.

Lemoine N, Schaefer H-C, Böhning-Gaese K. 2006. Species richness of migratory birds is influenced by global climate change. *Global Ecology and Biogeography* **16**: 55–64.

Long PR, Székely T, Kershaw M, O'Connell M. 2007. Ecological factors and human threats both drive wildfowl population declines. *Animal Conservation* **10**: 183–191.

McKinney ML. 1998. Branching models predict loss of many bird and mammal orders within centuries. *Animal Conservation* **1**: 159–164.

McLain DK, Moulton MP, Redfearn TP. 1995. Sexual selection and the risk of extinction of introduced birds on oceanic islands. *Oikos* **74**: 27–34.

McLain DK, Moulton MP, Sanderson JG. 1999. Sexual selection and extinction: the fate of plumage-dimorphic and plumage-monomorphic birds introduced onto islands. *Evolutionary Ecology Research* **1**: 549–565.

Mitchell TD, Carter TR, Jones PD, Hulme M, New M. 2004. *A Comprehensive Set of High-Resolution Grids of Monthly Climate for Europe and the Globe: The Observed Record (1901–2000) and 16 Scenarios (2001–2100)*. Norwich: University of East Anglia, Tyndall Centre for Climate Change Research.

Mols CMM, Viser ME. 2002. Great tits can reduce caterpillar damage in apple orchards. *Journal of Applied Ecology* **39**: 888–899.

Mooers AØ, Atkins RA. 2003. Indonesia's threatened birds: over 500 million years of evolutionary heritage at risk. *Animal Conservation* **6**: 183–188.

Mooers AØ, Heard SB, Chrostowski E. 2005. Evolutionary heritage as a metric for conservation. In *Phylogeny and Conservation*, Purvis A, Gittleman JL, Brooks TM (eds). Cambridge: Cambridge University Press; 120–138.

Morrow EH, Pitcher TE. 2003. Sexual selection and the risk of extinction in birds. *Proceedings of the Royal Society of London Series B (Biological Sciences)* **270**: 1793–1799.

Moyle RG, Filardi CE, Smith CE, Diamond J. 2009. Explosive Pleistocene diversification and hemispheric expansion of a "great speciator". *Proceedings of the National Academy of Sciences USA* **106**: 1863–1868.

Møller AP. 2000. Sexual selection and conservation. In *Behavior and Conservation*, Birkhead TR, Møller AP (eds). London: Academic Press. 161–171.

Myers N. 1985. The end of lines. *Natural History* **94**: 2–12.

Myers N. 1996. The biodiversity crisis and the future of evolution. *The Environmentalist* **16**: 37–47.

Myers N, Knoll AH. 2001. The biotic crisis and the future of evolution. *Proceedings of the National Academy of Sciences USA* **98**: 5389–5392.

Nee S, May RM. 1997. Extinction and the loss of evolutionary history. *Science* **278**: 692–694.

Olson DM, Dinerstein E, Wikramanayake ED, Burgess ND, Powell GVN, Underwood EC, D'amico IItoua JA, Strand HE, Morrison JC, Loucks CJ, Allnutt TF, Ricketts TH, Kura Y, Lamoreux JF, Wettengel WW, Hedao P, Kassem KR. 2001. Terrestrial ecoregions of the world: A new map of life on Earth. *BioScience* **51**: 933–938.

Olson VA, Davies RG, Orme CDL, Thomas GH, Meiri S, Blackburn TM, Gaston KJ, Owens IPF, Bennett PM. 2009. Global biogeography and ecology of body size in birds. *Ecology Letters* **12**: 249–259.

Orme CDL, Davies RG, Burgess M, Eigenbrod F, Pickup N, Olson V, Webster AJ, Ding T-S, Rasmussen PC, Ridgely RS, Stattersfield AJ, Bennett PM, Blackburn TM, Gaston KJ, Owens IPF. 2005. Global hotspots of species richness are not congruent with endemism or threat. *Nature* **436**: 1016–1019.

Owens IPF, Bennett PM. 2000a. Ecological basis of extinction risk in birds: habitat loss versus human predators. *Proceedings of the National Academy of Sciences USA* **97**: 12144–12148.

Owens IPF, Bennett PM. 2000b. Quantifying biodiversity: a phenotypic perspective. *Conservation Biology* **14**: 1014–1022.

Parmesan C. 2006. Ecological and evolutionary reponses to climate change. *Annual Review of Ecology Evolution and Systematics* **37**: 637–669.

Parmesan C, Yohe G. 2003. Globally coherent fingerprint of climate change impacts across natural systems. *Nature* **421**: 37–42.

Pearce-Higgins JW, Yalden DW, Whittingham MJ. 2005. Warmer spring advances the breeding phenology of golden plovers *Pluvialis apricaria* and their prey (*Tipulidae*). *Oecologia* **143**: 470–476.

Perfecto I, Vandermeer JH, Lopez Bautista G, Ibarra Nunez G, Greenberg R, Bichier P, Langridge S. 2004. Greater predation in shaded coffee farms: the role of resident Neotropical birds. *Ecology* **85**: 2677–2681.

Petchey OL, Evans KL, Fishburn IS, Gaston KJ. 2007. Low functional diversity and no redundancy in British avian assemblages. *Journal of Animal Ecology* **76**: 977–985.

Petchey OL, Eklof A, Borrvall C, Ebenman B. 2008. Trophically unique species are vulnerable to cascading extinction. *The American Naturalist* **171**: 568–579.

Pimm SL, Jones HL, Diamond J. 1988. On the risk of extinction. *The American Naturalist* **132**: 757–785.

Pimm SL, Russell GJ, Gittleman, JL, Brooks TM. 1995. The future of biodiversity. *Science* **269**: 347–350.

Pimm SL, Raven P, Peterson A, Şekercioğlu CH, Ehrlich PR. 2006. Human impacts on the rates of recent, present, and future bird extinctions. *Proceedings of the National Academy of Sciences USA* **103**: 10941–10946.

Post E, Forchhammer MC, Stenseth NC, Callaghan TV. 2001. The timing of life-history events in a changing climate. *Proceedings of the Royal Society of London Series B (Biological Sciences)* **268**: 15–23.

Prinzing A, Brändle M, Pfeifer R, Brandl R. 2002. Does sexual selection influence population trends in European birds? *Evolutionary Ecology Research* **4**: 49–60.

Pulido F. 2007. Phenotypic changes in spring arrival: evolution, phenotypic plasticity, effects of weather and condition. *Climate Research* **35**: 5–23.

Purvis A. 2008. Phylogenetic approaches to the study of extinction. *Annual Review of Ecology Evolution and Systematics* **39**: 301–319.

Purvis A, Agapow P-M, Gittleman JL, Mace GM. 2000. Nonrandom extinction and the loss of evolutionary history. *Science* **288**: 328–330.

Raup DM. 1992. *Extinction: Bad Genes or Bad Luck?* New York: W. W. Norton.

Redding DW, Mooers AO. 2006. Incorporating evolutionary measures into conservation prioritization. *Conservation Biology* **20**: 1670–1678.

Ricklefs RE, Schluter D. 1993. Species diversity: regional and historical influences. In Species Diversity in Ecological Communities, Ricklefs, R. E. Schluter D (eds). Chicago: Chicago University Press; 350–363.

Rohde K. 1992. Latitudinal gradients in species diversity: the search for the primary cause. *Oikos* **65**: 514–527.

Root TL. 1988a. Energy expenditures on avian distributions and abundances. *Ecology* **69**: 330–339.

Root TL. 1988b. Environmental factors associated with avian distributional boundaries. *Journal of Biogeography* **15**: 489–505.

Root TL, Price JT, Hall KR, Schneider SH, Rosenzweig C, Pounds JA. 2003. Fingerprints of global warming on wild animals and plants. *Nature* **421**: 57–60.

Rosenzweig ML. 2001. Loss of speciation rate will impoverish future diversity. *Proceedings of the National Academy of Sciences USA* **98**: 5404–5410.

Russell GJ, Brooks TM, McKinney ML, Anderson CG. 1998. Present and future taxonomic selectivity in bird and mammal extinction. *Conservation Biology* **12**: 1365–1376.

Salomonsen F. 1948. The distribution of birds and the recent climatic change in the North Atlantic area. *Dansk Ornitologisk Forening Tidsskrift* **42**: 85–99.

Sanchez-Pinero F, Polis GA. 2000. Bottom-up dynamics of allochthonous input: direct and indirect effects of seabirds on islands. *Ecology* **81**: 3117–3132.

Schaefer H-C, Jetz W, Böhning-Gaese K. 2008. Impact of climate change on migratory birds: community reassembly versus adaptation. *Global Ecology and Biogeography* **17**: 38–49.

Şekercioğlu CH. 2006a. Ecological significance of bird populations. In *Handbook of the Birds of the World*, del Hoyo J, Elliott A, Christie D (eds). Barcelona: Lynx Edicions; 15–51.

Şekercioğlu CH. 2006b. Increasing awareness of avian ecological function. *Trends in Ecology and Evolution* **21**: 464–471.

Şekercioğlu CH, Daily GC, Ehrlich PR. 2004. Ecosystem consequences of bird declines. *Proceedings of the National Academy of Sciences USA* **101**: 18042–18047.

Sibley CG, Ahlquist JE. 1990. *Phylogeny and Classification of Birds: A Study in Molecular Evolution.* New Haven: Yale University Press.

Sibley CG, Monroe BL. 1990. Distribution and Taxonomy of Birds of the World. Yale University Press, New Haven U.S.A.

Sibley CG, Monroe BL. 1993. Distribution and Taxonomy of Birds of the World: Supplement. New Haven: Yale University Press.

Sorci G, Møller AP, Clobert J. 1988. Plumage dichromatism of birds predicts introduction success in New Zealand. *Journal of Animal Ecology* **67**: 263–269.

Sparks TH, Mason CF. 2004. Can we detect change in the phenology of winter migrants in the UK. *The Ibis* **146** (Supplment 1): 48–56.

Stiles FG. 1981. Geographical aspects of bird-flower coevolution, with particular reference to central America. *Annals of the Missouri Botanical Gardens* **68**: 323–351.

Stocker GC, Irvine AK. 1983. Seed dispersal by cassowaries (*Casuarius casuarius*) in north Queensland's rainforests. *Biotropica* **15**: 170–176.

Templeton AR, Robertson RJ, Brisson J, Strasburg J. 2001. Disrupting evolutionary processes: the effect of habitat fragmentation on collared lizards in the Missouri Ozarks. *Proceedings of the National Academy of Sciences USA* **98**: 5426–5432.

Teplitsky C, Mills JA, Alho JS, Yarrall JW, Merilä J. 2008. Bergmann's rule and climate change revisited: Disentangling environmental and genetic responses in a wild bird population. *Proceedings of the National Academy of Sciences USA* **105**: 13492–13496.

Terborgh J. 1974. Preservation of natural diversity: the problem of extinction prone species. *BioScience* **24**: 715–722.

Thomas CD, Lennon JJ. 1999. Birds extend their ranges northwards. *Nature* **399**: 213.

Thomas CD, Cameron A, Green RE, Bakkenes M, Beaumont LJ, Collingham YC, Erasmus BFN, de Siqueira MF, Grainger A, Hannah L, Hughes L, Huntley B, van Jaarsveld AS, Midgley GF, Miles L, Ortega-Huerta MA, Peterson AT, Phillips OL, Williams SE. 2004. Extinction risk from climate change. *Nature* **427**: 145–148.

Thomas GH. 2008. Phylogenetic distributions of British birds of conservation concern. *Proceedings of the Royal Society of London Series B (Biological Sciences)* **275**: 2077–2083.

Thomas GH, Lanctot RB, Szekely T. 2006. Can intrinsic factors explain population declines in North American breeding shorebirds? A comparative analysis. *Animal Conservation* **9**: 252–258.

Valdivia-Hoeflich T, Vega Rivera JH, Stoner KE. 2005. The citreoline trogon as an ecosystem engineer. *Biotropica* **37**: 465–467.

Van Bael SA, Brawn JD, Robinson SK. 2003. Birds defend tree from herbivores in a Neotropical forest canopy. *Proceedings of the National Academy of Sciences USA* **100**: 8304–8307.

Van Vuuren D, Sala OE, Pereira HM. 2006. The future of vascular plant diversity under four global scenarios. *Ecology and Society* **11**: 25.

Visser ME. 2008. Keeping up with a warming world; assessing the rate of adaptation to climate change. *Proceedings of the Royal Society of London Series B (Biological Sciences)* **275**: 649–659.

Visser MA, van Noordwijk AJ, Tinbergen JM, Lessells CM. 1998. Warmer springs lead to mistimed springs in great tits (*Parus major*). *Proceedings of the Royal Society of London Series B (Biological Sciences)* **268**: 1867–1870.

Visser MA, Perdeck AC, van Balen JH, Both C. 2009. Climate change leads to decreasing bird migration distance. *Global Change Biology* **15**: 1859–1865.

Von Euler F. 2001. Selective extinction and rapid loss of evolutionary history in the bird fauna. *Proceedings of the Royal Society of London Series B (Biological Sciences)* **268**: 127–130.

Walker BH. 2003. Biodiversity and ecological redundancy. *Conservation Biology* **6**: 18–23.

Walther G-RE, Post P, Convey MA, Parmesan C, Beebee TJC, Fromentin J-M, Hoegh-Guldberg O, Bairlein F. 2002. Ecological responses to recent climate change. *Nature* **416**: 389–395.

Wenny DG, Levey DJ. 1998. Directed seed dispersal by bellbirds in a tropical cloud forest. *Proceedings of the National Academy of Sciences USA* **95**: 6204–6207.

Woodruff DS. 2001. Declines of biomes and biotas and the future of evolution. *Proceedings of the National Academy of Sciences USA* **98**: 5471–5476.

Yom-Tov Y. 2001. Global warming and body mass decline in Israeli passerine birds. *Proceedings of the Royal Society of London Series B (Biological Sciences)* **268**: 947–952.

Yom-Tov Y, Yom-Tov S, Wright J, Thorne CJR, Du Feu R. 2006. Recent changes in obody weight and wing length among some British passerine birds. *Oikos* **112**: 91–101.

Glossary

altricial Nestlings that are helpless and often blind at hatching and require a period of intense parental care.

alula Feathers associated with the fused remnants of digit I in birds.

Alvarezsauridae A basal group in the Maniraptora, possibly a sister group to the Ornithomimiosauria.

amniote A tetrapod animal whose egg is enclosed by a specialized layer of tissue, the amnion.

angiosperm Plant that produces flowers and enclosed seeds.

antorbital fenestra A large opening in the skull, in front of the eye.

antorbital fossa A shallow trough on the dorsal surface of the skull, in front of the eye.

archosaur Member of a lineage sharing a common ancestor with crocodylians and birds. Shared characteristics include an antorbital fenestra, mandibular fenestra, and a fourth trochanter.

astragalus A small bone primitively situated at the distal end of the tibia. Analog of the talus bone in mammals.

avian theropod Member of the theropod dinosaur lineage, conventionally the clade emcompassed by *Archaeopteryx* and all its descendants.

Bayesian inference A statistical method in which current information is used to update previous parameters or hypotheses.

bipedality Locomotion that uses only the two hindlimbs.

brachial index The ratio of the humerus to the radius.

brood patch or pouch A featherless patch (or pouch in penguins) found on the belly of many birds that is richly supplied with blood vessels, permitting the efficient transfer of body heat to the egg during incubation.

calcaneum A small bone primitively situated at the distal end of the fibula. Analog of the largest tarsal bone in humans.

carpometacarpus Fused single bony element in birds; remnant of the hand (metacarpals).

Carrier's constraint The apparent inability of squamosal animals that use sinusoidal locomotion to breath and travel at the same time.

centromere A centromere is a region of DNA typically found near the middle of a chromosome where two identical sister chromatids come in contact.

choanal position Located posterior and beneath the brain case.

chondrified digit Cartilaginous stage in the formation of a digit.

choristodere An order of Mesozoic semi-aquatic diapsid reptiles that survived into the Miocene.

clade In phylogenetics, a branch that descends from a single ancestor.

cladistics A method of determining the degree of similarity among groups of organisms based on the presence of shared derived features.

cnemial crest A vertical process on the anterior surface of the knee derived from either the patella or the tibia.

coelophysoid radiation Diversification of one lineage of theropods dinosaurs in the Late Triassic and Early Jurassic

Living Dinosaurs: The Evolutionary History of Modern Birds, First Edition. Edited by Gareth Dyke and Gary Kaiser.
© 2011 John Wiley & Sons, Ltd. Published 2011 by John Wiley & Sons, Ltd.

compsognathid A group of small theropods contemporary with *Archaeopteryx*.

coracoid In birds, an important strut-like bone between the sternum and the shoulder.

Crocodylomorpha The lineage that includes crocodilians and their extinct relatives.

crurotarsal ankle The ankle joint as found in crocodylians in which the line of flexion lies between the astragalus and the calcaneum (see mesotarsal).

cynodont Having teeth like a therapsid or their descendants among modern mammals.

diachronous laying Production of a clutch of eggs over a series of days.

dichromatism The existence of two distinct and concurrent colorations in a species, usually according to sex.

dimorphism The occurrence of different sizes within a species, usually according to sex.

Dinosauromorpha The Dinosauria and their earlier close relatives.

dispersal–vicariance analysis Relating the distribution and phylogenetic relationships of birds to continental drift and other geographic factors.

DNA–DNA hybridization A biomolecular technique based on the tendency of closely related samples of DNA to bond more thoroughly when mixed, than distantly related samples of DNA. A mixture containing closely related samples will melt at a relatively high temperature compared to a mixture of distantly related samples.

Dollo's law of irreversability (1893) Once a lineage has lost a complex structure, it cannot be regained.

Dromaeosauridae A family of theropods with a variety of bird-like features.

ectothermy The dependence on external sources for body heat.

Enantiornithes A sister group of the Ornithurae that was abundant in the Cretaceous but has left no living relatives.

encephalization quotient A numerical estimate of the ratio of the relative brain mass to the body mass.

endocast A model of an internal structure achieved by injecting latex or a similar material into the hollow space left by that structure.

endothermy The production and control of body heat by physiological processes.

fenestra A relatively large opening in a bone.

foramen A relatively small hole in a bone, usually for the passage of a nerve or blood vessel.

fossa A broad groove or trough on the surface of a bone.

frame shift An embryonic process during which the primordia of a reduced series of digits develop in adjacent positions. If digit V is lost, digit I develops in position II, digit II in position III, etc. The theory being that birds have retained manual digits I, II, and III but in positions II, III, and IV.

furcula A structure composed of the fused clavicles and the interclavicle bone in theropod dinosaurs and birds. Commonly called the wishbone.

gastralia Long, thin bones that lie across the abdomen of crocodiles and dinosaurs. They function as free-floating ribs.

gastrolith A stone ingested to assist with the mechanical reduction of food.

glenoid The mobile articulation between the bones of the shoulder and the humerus.

Gondwana A continent in the Mesozoic that included the land masses that are now Antarctica, Australia, New Zealand, South America, Africa, Madagascar, India, and parts of the Middle East.

Graculavidae A Cretaceous group of shorebirds of uncertain affinity that share some characteristics with modern Charadriiformes.

hallux Digit I of the hindlimb, often facing to the rear in birds.

hepatic pump Respiration achieved by pulling the liver headwards to compress the lungs (e.g. crocodylians).

hexamer A series of six nucleotides.

homeothermy The maintenance of a stable body temperature.

homoplastic Possessing similarity in form and structure but arising from discrete evolutionary origins.

***Hox* genes** Genes that control the embryonic development of segmented structures such as the manual digits.

hypantrum–hyposphen See hyposphene–hypantrum

hypocleideum A flat piece of bone that grows across the joint between the two clavicles in the furcula or wishbone

hyposphene–hypantrum An interlocking articulation between vertebrae in some nonavian dinosaurs that improved the rigidity of the spinal column.

interclavicle A bone lying between the ventral tips of the clavicles. Fusion to the clavicles is thought to produce the furcula in birds and a ventral extension of this bone may appear as the hypocleideum.

interclavicular angle In birds and nonavian theropods, the angle formed by the two arms of the furcula.

interdental plate A small, boney plate between adjacent teeth.

intron Part of a gene that is not translated into a protein.

K–Pg boundary The Cretaceous–Paleogene boundary, formerly the Cretaceous–Tertiary boundary, referring to the mass extinction that marked the end of the Mesozoic.

LAG (line of arrested growth) A signal in the bone that records a period of slower or arrested growth, possibly in response to a seasonal change in diet or food supply.

lift based swimming Swimming achieved by outward thrusts of the feet, similar the action of a kayaker.

limb bud condensation The formation of precursors to cartilage in the early stages of developing an embryonic limb.

LINE Long insertion of nucleotide elements into the chromosome. Common in mammals but rare or absent in birds.

lineage A group of organisms descended from a single common ancestor.

"lumbar" vertebrae Birds lack distinctively shaped vertebrae in the lower back but the mammalian term "lumbar" is used as a convenient topographic reference.

mandibular fenestra Relatively large opening in the bones of the lower jaw.

mandibular symphysis Fusion of components of the lower jaw.

maniraptora A clade of Coelurosauria more closely related to birds than *Ornithomimus velox*.

manual formula see phalageal formula

medullary bone A calcium-rich formation in the long bones of female birds that can be mobilized for the production of eggshells.

melanosome An organelle within a cell, containing dark (light absorbing) pigments.

mesotarsal ankle joint The ankle joint found in birds in which the line of flexion lies between the metatarsals and other bones' outer limb (see crurotarsal).

metatarsal Long bone of the foot between the digits and the ankle.

Metatheria A group within mammals that includes marsupials and their fossil ancestors.

morphospace Essentially the envelope bounding the possible shape or size of a particular feature or organism.

Morse's Rule When pentadactyl amniotes undergo a reduction in the number of digits, the outermost (V) and innermost (I) are the first to be lost.

mosaic, evolutionary A tendency in some Mesozoic groups of birds to seemingly show characters from several different living lineages.

neontological Relating to the sudy of living animals.

Neornithes A clade that includes all the descendants of the most recent common ancestor of the living birds.

obturator notch A space between the posterior edge of the ischium and the anterior edge of the pubis.

Ornithischia Dinosaurs with a hip structure similar to that of birds but which are not related to birds.

ornithodira A clade that includes the last common ancestor of the pterosaurs and the dinosaurs.

Ornithurine birds Ernst Haekel's group of birds with modern-type tails; as opposed to the Sauriae, *Archaeopteryx*-like birds with long tails.

osteocyte lacunae Spaces in the bone that house living cells (osteocytes).

outgroup In cladistics, organisms compared to the organisms of interest (the ingroup) in order to determine synapomorphies.

parapatric speciation Speciation achieved by exploitation of a novel but contiguous niche.

paraphyletic A group that arose from a single ancestor, but which does not include all the descendants of that ancestor.

parasaggital stance A posture in which the limbs are mounted at right angles to the sagittal plane, the plane that divides humans into a front and a back but other animals into a top and a bottom.

paravian The first common ancestor of birds, dromaeosaurids, and troodontids.

parsimony analysis A statistical analysis that tests the hypothesis that the smallest number of character changes is most likely to be the correct model for phylogenetic relationships.

pennaceous feather A feather supported by a strong central rachis.

peripatric speciation Speciation based on the successful exploitation of a novel and isolated niche.

phalangeal formula A numerical method of describing the phylogenetic significance of an animal's digits. The anterior digit is number one.

plantar surface Essentially the sole of the foot but anatomically, the ventral or posterior suface.

plesiomorphic Primitive characters shared by several different animals but derived from a common ancestor.

pneumatization The investment by air sacs of hollow spaces in portions of the skeleton.

postaxial (as in postaxial cervical vertebrae) behind the axis of the body

precocial Nestlings that can fend for themselves upon hatching and require little or no parental care.

primordia In the embryo, areas of cartilaginous condensation that will eventually become a bony structure.

pseudo-toothed birds Odontopterigiformes: large, extinct seabirds with an array of sharp keratinous projections along the tomia that somewhat resemble teeth.

pubic fusion Fusion of the tips of the pubic bones, creating a ring through which the egg must pass.

pygostyle A series of fused terminal vertebrae. In living birds it is plow-shaped and holds the rectigium, a bulbous structure that carries the tail feathers (rectrices). Comparable structures appear in some short-bodied, nonavian dinosaurs (e.g. Nomingia) and in several lineages of Mesozoic birds including the Confuciusornithidae and the Enantiornithes.

Pygostylia An avian lineage whose synapomorphies include a pygostyle, the absence of a hyposphene–hypantrum articulation, possession of a retroverted pubis, and a bulbous medial condyle on the tibiotarsus.

retroverted pubis pubis entirely tipped backwards

Saurischia see Ornithischia, lizard-like pelvis.

semilunate carpal A carpal in the wrist of *Deinonychus*, recognized by John Ostrom in 1969 as homologous with a fused structure found in the wrist of birds.

septate lung The sac-like or sponge-like lungs of reptiles and birds as opposed to the alveolar lungs of mammals.

SINE Short insertion of nucleotide elements into the chromosome. Fairly common in birds.

sister taxon A taxon derived from the same common ancestor.

speciose Having a large number of species.

splitter plate In birds, an array of short tail feathers that help to maintain the laminar flow of air passing above and below the body.

stapedial morphology Characteristics of the bones of the inner ear (stapes) used to support the phylogeny of passerine birds.

streptostylic quadrate A quadrate with a moveable hinge with the squamosal bone.

sympatric speciation Speciation achieved within the current geographic range of the parent species.

synapomorphy A derived feature that is shared by all members of the group in question.

syringial morphology Characteristics of the voice box (syrinx) used to support the phylogeny of passerine birds.

telomere A region of repetitive DNA at the end of a chromosome, which protects the end from deterioration.

Tetanurae An inclusive lineage of theropod dinosaurs characterized by an enlarged manus loss of the fourth and fifth digits of the hand, and a less flexible tail. This clade includes all theropods (including all modern birds) that are more closely related to modern birds than they are to *Ceratosaurus*.

thecodont 1. A group of primitive reptiles that has been considered ancestral to birds. However, it has no unifying features and is now considered to be an artificial assemblage. 2. Teeth set in sockets.

therizinosaur A clade of theropod dinosaurs characterized by various features of the forelimbs, skull and pelvis that place them as maniraptorans, close relatives to birds.

theropod One clade of bipedal saurischian dinosaurs that encompasses all the meat-eating dinosaurs up to, and including modern birds).

tomia Keratinous edges of the upper and lower beak.

trochlea A deeply grooved or saddle-shaped articular surface.

troodontid A clade of small- to medium-sized maniratoran theropod dinosaurs characterized by long legs (compared to other theropods) and a large, curved claw on a retractable second toe. These dinosaurs are considered likely candidates for the divergence of the avian lineage.

ultrametric tree A phylogenetic tree with equal root-to-tip path lengths for all lineages, for example one built with the assumption of a molecular clock.

uncinate process A bony process that extends caudally from the middle region of a dorsal rib.

wing loading The ratio of body weight to the lifting surface created by the wings and the body surface between them.

X-ray tomography An X-ray technique that uses digital geometry to generate a three-dimensional image of the inside of an object.

Index

Living Dinosaurs: The Evolutionary History of Modern Birds, First Edition. Edited by Gareth Dyke and Gary Kaiser.
© 2011 John Wiley & Sons, Ltd. Published 2011 by John Wiley & Sons, Ltd.